MW00837780

APPLIED NONLINEAR DYNAMICS

WILEY SERIES IN NONLINEAR SCIENCE

Series Editors: **ALI H. NAYFEH, Virginia Tech**
ARUN V. HOLDEN, University of Leeds

APPLIED NONLINEAR DYNAMICS

Analytical, Computational, and Experimental Methods

Ali H. Nayfeh
Virginia Polytechnic Institute and State University

Balakumar Balachandran
University of Maryland

A Wiley-Interscience Publication
John Wiley & Sons, Inc.
New York / Chichester / Brisbane / Toronto / Singapore

This text is printed on acid-free paper.

Copyright © 1995 by John Wiley & Sons, Inc.

All rights reserved. Published simultaneously in Canada.

Reproduction or translation of any part of this work beyond
that permitted by Section 107 or 108 of the 1976 United
States Copyright Act without the permission of the copyright
owner is unlawful. Requests for permission or further
information should be addressed to the Permissions Department,
John Wiley & Sons, Inc., 605 Third Avenue, New York, NY
10158-0012.

Library of Congress Cataloging-in-Publication Data:

Nayfeh, Ali Hasan 1933–
    Applied nonlinear dynamics: analytical, computational, and
experimental methods. / Ali H. Nayfeh and Balakumar Balachandran.
      p. cm. — (Wiley series in nonlinear science)
    Includes bibliographical references and index.
    ISBN 0-471-59348-6
    1. Dynamics. 2. Nonlinear theories. I. Balachandran, Balakumar.
II. Title. III. Series.
QA845.N39   1994
515'.352—dc20   94-3659
                                                   CIP

Printed in the United States of America

10 9 8 7 6 5 4 3

To our wives

Samirah and Sundari

CONTENTS

PREFACE

Systems that can be modeled by nonlinear algebraic and/or nonlinear differential equations are called **nonlinear systems**. Examples of such systems occur in many disciplines of engineering and science. In this book, we deal with the dynamics of nonlinear systems. Poincaré (1899) studied nonlinear dynamics in the context of the n–body problem in celestial mechanics. Besides developing and illustrating the use of perturbation methods, Poincaré presented a geometrically inspired qualitative point of view.

In the nineteenth and twentieth centuries, many pioneering contributions were made to nonlinear dynamics. A partial list includes those due to Rayleigh, Duffing, van der Pol, Lyapunov, Birkhoff, Krylov, Bogoliubov, Mitropolski, Levinson, Kolomogorov, Andronov, Arnold, Pontryagin, Cartwright, Littlewood, Smale, Bowen, Piexoto, Ruelle, Takens, Hale, Moser, and Lorenz. While studying forced oscillations of the van der Pol oscillator, Cartwright and Littlewood (1945) observed a constrained random–like behavior, which is now called **chaos**. Subsequently, Lorenz (1963) studied a deterministic, third–order system in the context of weather dynamics and showed through numerical simulations that this deterministic system displayed random–like behavior too. Unaware of Lorenz's work, Smale (1967) introduced the horseshoe map as an abstract prototype to explain chaos–like behavior. No doubt Poincaré knew about chaos too, but it is only through numerical simulations on modern computers and experiments with physical systems that the presence of chaos has been discovered to be pervasive in many dynamical systems of physical interest. The observation of Poincaré that small differences in the initial conditions may produce great changes in the final phenomena is now known to be a characteristic of systems that exhibit chaotic behavior. The phenomenon of chaos, which has become very popular now, rejuvenated interest in nonlinear dynamics. The growing numbers of books and research papers published in the last two decades reflect a strong interest in nonlinear dynamics at the present time. The many important contributions that have been made through analytical, experimental, and numerical studies have been documented through many books, including those by Collet and Eckmann (1980), Mees (1981), Sparrow (1982), Guckenheimer and Holmes (1983), Lichtenberg and Lieberman (1983, 1992), Bergé, Pomeau, and Vidal (1984), Holden (1986), Kaneko (1986), Thompson and Stewart (1986), Moon (1987, 1992), Arnold (1988), Barnsley (1988), Schuster (1988), Seydel (1988), Wiggins (1988, 1990), Devaney (1989), Jackson (1989, 1990), Nicolis and Prigogine (1989), Parker and Chua (1989), Ruelle (1989a, 1989b), Tabor (1989), Arrowsmith and Place (1990), Baker and Gollub (1990), El Naschie (1990), Rasband (1990), Hale and Kocak (1991), Schroeder (1991), Troger and Steindl (1991), Drazin (1992), Kim and Stringer (1992), Medvéd (1992), Tufillaro, Abbott, and Reilly (1992), Ueda (1992), Mullin (1993), Ott (1993), Palis and Takens (1993), and Ott, Sauer, and Yorke (1994).

We are of the opinion that the books on nonlinear dynamics published thus far have a strong bias toward analytical methods, or experimental methods, or numerical methods. As these methods are complementary to each other, a person being taught nonlinear dynamics should be provided with a flavor of all the different methods. This is one of the intentions in writing this book. Another intention was to include some of the recent developments in the area of control of nonlinear dynamics of systems. In Chapter 1, we introduce dynamical systems. In Chapters 2–5, we address equilibrium solutions, periodic and quasiperiodic solutions, and chaos. We present some relevant theorems and their implications in Chapters 2 and 3. Proofs are not provided in this book, but references that provide them are included. Further, these chapters are not written within a mathematically rigorous framework. Continuation methods for equilibrium and periodic solutions are also presented in some detail in Chapter 6. We examine the different tools that can be used to characterize nonlinear motions in Chapter 7. In Chapter 8, we discuss methods for bifurcation control, chaos control, and synchronization to chaos.

The authors are deeply indebted to several colleagues for helpful comments and criticisms, including, in particular, Professor Sherif Noah and his students, Dr. Marwan Bikdash, Mr. Haider Arafat, Mr. Samir A. Nayfeh, Mr. Ghaleb Abdallah, Professors Jose Baltezar, Anil Bajaj, Eyad Abed, Dean Mook, and Muhammad Hajj. One of us (BB) would like to thank Professors Davinder Anand and Patrick Cunniff of the University of Maryland for their encouragement and support during the final stages of preparation of this book. We wish to thank Dr. Char–Ming Chin for generating many of the figures dealing with crises, intermittency, and Shilnikov chaos. Thanks are due also to fifteen year old Nader Nayfeh for scanning, editing, and preparing the eps files for all the illustrations in this book. Last but not least, we wish to thank Mrs. Sally G. Shrader for her patient typing of the drafts of the manuscript and fine preparation of the final camera–ready copy of this book.

Ali H. Nayfeh
Blacksburg, Virginia
Balakumar Balachandran
College Park, Maryland
October 1994

Chapter 1

INTRODUCTION

A **dynamical system** is one whose state evolves (changes) with time t. The evolution is governed by a set of rules (not necessarily equations) that specifies the state of the system for either discrete or continuous values of t. A **discrete–time evolution** is usually described by a system of algebraic equations (map), while a **continuous–time evolution** is usually described by a system of differential equations.

The asymptotic behavior of a dynamical system as $t \to \infty$ is called the **steady state** of the system. Often, this steady state may correspond to a bounded set, which may be either a static solution or a dynamic solution. The behavior of the dynamical system prior to reaching the steady state is called the **transient state**, and the corresponding solution of the dynamical system is called the **transient solution**.

A solution of a dynamical system can be either constant or time varying. **Fixed points, equilibrium solutions**, and **stationary solutions** are other names for constant solutions, while **dynamic solutions** is another name for time–varying solutions. We explore equilibrium solutions in Chapter 2 and dynamic solutions in Chapters 3–5. In Sections 1.1 and 1.2, we explain the notion of a dynamical system. In Section 1.3, we discuss attracting sets, and in Sections 1.4 and 1.5, we examine the concepts of stability and attractors.

1

1.1   DISCRETE–TIME SYSTEMS

A discrete–time evolution is governed by

$$\mathbf{x}_{k+1} = \mathbf{F}(\mathbf{x}_k) \tag{1.1.1}$$

where $\mathbf{x}$ is a finite–dimensional vector. At the discrete times t_k and t_{k+1}, $\mathbf{x}_k$ and $\mathbf{x}_{k+1}$ represent the states of the system, respectively. Let the dimension of the finite–dimensional state vector be n. Then, we need n real numbers to specify the state of the system. Formally, the state vector $\mathbf{x} \in \mathcal{R}^n$ and the time $t \in \mathcal{R}$, where the symbol $\in$ means **belongs to** and the symbol $\mathcal{R}^n$ refers to an n–dimensional **Euclidean space**; that is, a real–number space equipped with the **Euclidean norm**

$$\| \, \mathbf{x} \, \| = \sqrt{(x_1^2 + x_2^2 + \cdots + x_n^2)} \tag{1.1.2}$$

where the x_i are the scalar components of $\mathbf{x}$. If the discrete values of time correspond to integers rather than real numbers, we say that $t \in \mathcal{Z}$, where $\mathcal{Z}$ is the set of all integers. We note that the evolution of a dynamical system may also be studied in other spaces, such as cylindrical, toroidal, and spherical spaces. In these cases, one or more state variables are angular coordinates. However, according to topological concepts, local regions of these spaces have the structure of a Euclidean space.

Equation (1.1.1) is a **transformation** or a **map** that transforms the current state of the system to the subsequent state. In the literature, the words **map**, **mapping**, and **function** are often used interchangeably. To a certain extent, the words **set** and **space** are also used interchangeably. Formally, a map $\mathbf{F}$ from points in a region M to points in a region N is represented by $\mathbf{F} : M \to N$. We note that M and N are contained in $\mathcal{R}^n$. Formally, $M \subset \mathcal{R}^n$ and $N \subset \mathcal{R}^n$, where the symbol $\subset$ is called the **subset operator** and means **inclusion**. The map $\mathbf{F}$ is said to map M **onto** N if for every point $\mathbf{y} \in N$ there exists at least one point $\mathbf{x} \in M$ that is mapped to $\mathbf{y}$ by $\mathbf{F}$. Furthermore, $\mathbf{F}$ is said to be **one–to–one** if no two points in M map to the same point in N. A map that is one–to–one and onto is **invertible** (e.g., Dugundji,

1966, Chapter I); that is, given $\mathbf{x}_{k+1}$, we can solve (1.1.1) to determine $\mathbf{x}_k$ uniquely. Denoting the inverse of $\mathbf{F}$ in (1.1.1) by $\mathbf{F}^{-1}$, we have

$$\mathbf{x}_k = \mathbf{F}^{-1}(\mathbf{x}_{k+1})$$

The map $\mathbf{F}^{-1}$ is also onto and one–to–one. A map $\mathbf{F}$ that is not invertible is called a **noninvertible map**.

When each of the scalar components of $\mathbf{F}$ is r times continuously differentiable with respect to the scalar components of $\mathbf{x}$, $\mathbf{F}$ is said to be a $\mathcal{C}^r$ function. When each of the scalar components of $\mathbf{F}$ is continuous with respect to the scalar components of $\mathbf{x}$, $\mathbf{F}$ is said to be a $\mathcal{C}^0$ function. For $r \geq 1$, the map $\mathbf{F}$ is called a **differentiable map**. The map $\mathbf{F}$ is called a **homeomorphism** if it is invertible and both $\mathbf{F}$ and $\mathbf{F}^{-1}$ are continuous; that is, $\mathbf{F}$ is $\mathcal{C}^0$. If both $\mathbf{F}$ and $\mathbf{F}^{-1}$ are $\mathcal{C}^r$ functions where $r \geq 1$, then we call the map a $\mathcal{C}^r$ **diffeomorphism**. In subsequent chapters, we discuss what are called **Poincaré maps**. These maps, which are discretized versions of associated systems of ordinary–differential equations, are diffeomorphisms. In one discretized version, a Poincaré map describes the evolution of a system for discrete values of time. The other cases are discussed in detail in Chapters 3, 4, 5, and 7.

An **orbit of an invertible map** initiated at $\mathbf{x} = \mathbf{x}_0$ is made up of the discrete points

$$\Big\{ \cdots, \ \mathbf{F}^{-m}(\mathbf{x}_0), \ \cdots, \ \mathbf{F}^{-2}(\mathbf{x}_0), \ \mathbf{F}^{-1}(\mathbf{x}_0),$$
$$\mathbf{x}_0, \ \mathbf{F}(\mathbf{x}_0), \ \mathbf{F}^2(\mathbf{x}_0), \ \cdots, \ \mathbf{F}^m(\mathbf{x}_0), \ \cdots \Big\}$$

where $m \in \mathcal{Z}^+$ and $\mathcal{Z}^+$ is the set of all positive integers. When $k > 0$, $\mathbf{F}^k$ means the kth successive application of the map $\mathbf{F}$. Similarly, when $k < 0$, $\mathbf{F}^k$ means the kth successive application of the map $\mathbf{F}^{-1}$. An **orbit of a noninvertible map** initiated at $\mathbf{x} = \mathbf{x}_0$ is made up of the discrete points

$$\Big\{ \mathbf{x}_0, \ \mathbf{F}(\mathbf{x}_0), \ \mathbf{F}^2(\mathbf{x}_0), \ \cdots, \ \mathbf{F}^m(\mathbf{x}_0), \ \cdots \Big\}$$

Successive applications of $\mathbf{F}$ are also referred to as the **forward iterates** of the corresponding map.

With reference to (1.1.1), we note that $\mathbf{F}$ is also called an **evolution operator**. Sometimes, we wish to study the evolution as we change or control a certain set of parameters $\mathbf{M}$. To make this explicit, we write the map as

$$\mathbf{x}_{k+1} = \mathbf{F}(\mathbf{x}_k; \mathbf{M}) \tag{1.1.3}$$

where $\mathbf{M}$ is the vector of control parameters.

Example 1.1. For illustration, we consider the one–dimensional map

$$x_{k+1} = 4\alpha x_k(1 - x_k) \tag{1.1.4}$$

where $0 \leq x_k \leq 1$ and $0 < \alpha \leq 1$. For $\alpha = 0.50$, the orbit of the map initiated at $x_0 = 0.25$ is

$$\{0.25, \ 0.375, \ 0.46875, \ \cdots\}$$

Equation (1.1.4) is the famous **logistic map**, which has been the subject of many studies (e.g., May, 1976). This map is a noninvertible map because it is not a one–to–one map. In fact, this map is a **two–to–one map** because it maps the two points x and $(1 - x)$ to the same point $4\alpha x(1 - x)$. Further, (1.1.4) is an example of a differentiable map.

Example 1.2. We consider the **Hénon map** (Hénon, 1976)

$$x_{k+1} = 1 + y_k - \alpha x_k^2 \tag{1.1.5}$$

$$y_{k+1} = \beta x_k \tag{1.1.6}$$

where α and β are scalar parameters. When $\beta = 0$, (1.1.5) and (1.1.6) reduce to the one–dimensional map

$$x_{k+1} = 1 - \alpha x_k^2$$

which is noninvertible. It is called the **quadratic map**. However, when $\beta \neq 0$, the map (1.1.5) and (1.1.6) is invertible. The inverse is

$$x_k = \frac{1}{\beta} y_{k+1}$$

$$y_k = x_{k+1} - 1 + \frac{\alpha}{\beta^2} y_{k+1}^2$$

We note that $\{x_k \ \ y_k\}^T$ uniquely determines $\{x_{k+1} \ \ y_{k+1}\}^T$ and vice versa. Further, because both $\mathbf{F}$ and $\mathbf{F}^{-1}$ are differentiable, the Hénon map is a diffeomorphism when $\beta \neq 0$. For $\alpha = 0.2$ and $\beta = 0.3$, the orbit of the map initiated at

$$\left\{ \begin{array}{c} x_0 \\ y_0 \end{array} \right\} = \left\{ \begin{array}{c} 1.0 \\ 0.0 \end{array} \right\}$$

is

$$\left\{ \cdots, \left\{ \begin{array}{c} -3.33 \\ 1.22 \end{array} \right\}, \left\{ \begin{array}{c} 0.0 \\ -1.0 \end{array} \right\}, \left\{ \begin{array}{c} 0.0 \\ 0.0 \end{array} \right\}, \right.$$
$$\left. \left\{ \begin{array}{c} 1.0 \\ 0.0 \end{array} \right\}, \left\{ \begin{array}{c} 0.8 \\ 0.3 \end{array} \right\}, \left\{ \begin{array}{c} 1.17 \\ 0.24 \end{array} \right\}, \left\{ \begin{array}{c} 0.97 \\ 0.35 \end{array} \right\}, \cdots \right\}$$

In Figure 1.1.1, we show some of the discrete points that make up the orbit of (x_0, y_0).

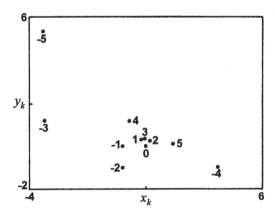

Figure 1.1.1: Some of the discrete points that make up the orbit of $(1, 0)$ of the Hénon map for $\alpha = 0.2$ and $\beta = 0.3$. The index k associated with each point is also shown.

We note that the dynamics of many Poincaré maps show qualitative similarities to the dynamics of the logistic and Hénon maps.

1.2   CONTINUOUS–TIME SYSTEMS

For continuous values of time, the evolution of a system is governed by either an autonomous or a nonautonomous system of differential equations.

1.2.1   Nonautonomous Systems

In the nonautonomous case, the equations are of the form

$$\dot{\mathbf{x}} = \mathbf{F}(\mathbf{x}, t) \qquad\qquad (1.2.1)$$

where $\mathbf{x}$ is finite dimensional, $\mathbf{x} \in \mathcal{R}^n$, $t \in \mathcal{R}$, and $\mathbf{F}$ explicitly depends on t. The vector $\mathbf{F}$ is often referred to as **vector field**, the vector $\mathbf{x}$ is called a **state vector** because it describes the state of the system, and the space $\mathcal{R}^n$ in which $\mathbf{x}$ evolves is called a **state space**. A state space is called a **phase space** when one–half of the states are displacements and the other one–half are velocities. The $(n + 1)$-dimensional space $\mathcal{R}^n \times \mathcal{R}^1$, where the additional dimension corresponds to t, is often referred to as an **extended state space**. In (1.2.1), if $\mathbf{F}$ is a linear function of $\mathbf{x}$ it is called a **linear vector field**, and if $\mathbf{F}$ is a nonlinear function of $\mathbf{x}$ it is called a **nonlinear vector field**.

Let the initial state of the system at time t_0 be $\mathbf{x}_0$, and let I represent a time interval that includes t_0. Then one can think of a solution of (1.2.1) as a map from different points in I into different points in the n–dimensional state space $\mathcal{R}^n$. A graph of a solution of (1.2.1) in the extended state space is known as an **integral curve**. On an integral curve, the vector function $\mathbf{F}$ specifies the tangent vector (velocity vector) at every point $(\mathbf{x}, t)$. A geometric interpretation of a **vector field** is that it is a collection of tangent vectors on different integral curves.

In general, a projection of a solution $\mathbf{x}(t, t_0, \mathbf{x}_0)$ of (1.2.1) onto the n–dimensional state space is referred to as a **trajectory** or an **orbit** of the system through the point $\mathbf{x} = \mathbf{x}_0$. In other words, the solution could be thought of as a point that moves along a trajectory, occupying different positions at different times similar to the way a planet moves through space. We use the symbol $\gamma(\mathbf{x}_0)$ or Γ to denote an orbit. The orbit obtained for times $t \geq 0$ passing through the point $\mathbf{x}_0$ at $t = 0$ is called a **positive orbit** and is denoted by $\gamma^+(\mathbf{x}_0)$; the orbit obtained for times $t \leq 0$ is called a **negative orbit** and is denoted by $\gamma^-(\mathbf{x}_0)$. Also, $\Gamma = \gamma(\mathbf{x}_0) = \gamma^+(\mathbf{x}_0) \cup \gamma^-(\mathbf{x}_0)$, where the symbol $\cup$ stands for the union operator.

Example 1.3. For illustration, we consider the following periodically forced linear oscillator:

$$\ddot{x} + 2\mu\dot{x} + \omega^2 x = F\cos(\Omega t)$$

Letting $x = x_1$ and $\dot{x} = x_2$, we express this second–order equation as a system of two first–order equations in terms of the state variables x_1 and x_2. The result is

$$\dot{x}_1 = x_2 \tag{1.2.2}$$
$$\dot{x}_2 = -\omega^2 x_1 - 2\mu x_2 + F\cos(\Omega t) \tag{1.2.3}$$

For $\omega^2 = 8$, $\mu = 2$, $F = 10$, and $\Omega = 2$, the solution of (1.2.2) and (1.2.3) is

$$x_1 = e^{-2t}\left[a\cos(2t) + b\sin(2t)\right] + 0.5\cos(2t) + \sin(2t)$$

$$x_2 = -2e^{-2t}\left[(a - b)\cos(2t) + (a + b)\sin(2t)\right] - \sin(2t) + 2\cos(2t)$$

where the constants a and b are determined by the initial condition (x_{10}, x_{20}). We note that as $t \to \infty$, the exponential term decays to zero. Therefore, the steady state does not depend on the initial condition. In Figure 1.2.1a, we show an integral curve initiated at $(x_{10}, x_{20}, t_0) = (1, 0, 0)$ in the (x_1, x_2, t) space for $0 \leq t \leq 10$. The arrows on the curve indicate the direction of evolution for positive times. The tangent vector is also shown at two different locations on the integral curve. It should be noted that the apparent intersections in Figure 1.2.1a are a consequence of the chosen viewing angle. In

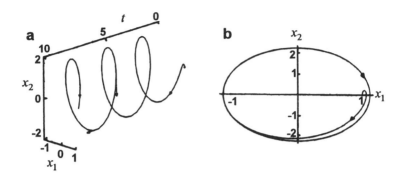

Figure 1.2.1: Solution of (1.2.2) and (1.2.3) initiated from $(1,0)$ at $t = 0$ for $\omega^2 = 8$, $\mu = 2$, $F = 10$, and $\Omega = 2$: (a) integral curve and (b) positive orbit.

Figure 1.2.1b, we show a projection of the integral curve onto the two–dimensional (x_1, x_2) space. This projection is a positive orbit of $(x_{10}, x_{20}) = (1, 0)$.

Again, we remind the reader that besides Euclidean state spaces there are other state spaces, such as cylindrical, toroidal, and spherical spaces. In Figure 1.2.2a, we show a **cylindrical space**. A motion evolving in this space is described by two Cartesian coordinates and an angular coordinate θ. One of the Cartesian coordinates is defined along the cylinder's axis, while the other one is defined along the radius of its cross–section. This cylindrical space is represented by $\mathcal{R}^2 \times S^1$. The variable θ belongs to the space S and is such that $0 \leq \theta < 2\pi$; formally, $\theta \in [0, 2\pi)$. A **toroidal space** is shown in Figure 1.2.2b. Specifically, we call this object a **two–torus**, and a dynamical system evolving in this space is described by two angular coordinates θ_1 and θ_2. We represent this space by $S^1 \times S^1$. One would require n angular coordinates to describe the motion evolving on an n–**torus**. A **spherical space** is shown in Figure 1.2.2c. We need two angular coordinates to describe a motion evolving on the spherical surface.

A local region of the cylindrical, toroidal, or spherical surface of Figure 1.2.2 has the appearance of a flat surface and can be treated as a two–dimensional Euclidean space. Smooth and continuous surfaces, such as those shown in Figure 1.2.2, are called **manifolds**. Manifolds can be thought of as generalized surfaces. (The reader is referred to

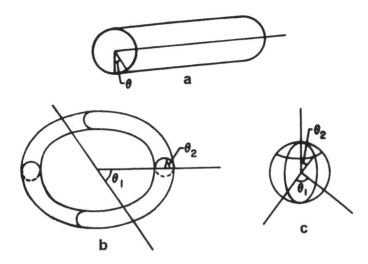

Figure 1.2.2: Different spaces: (a) cylindrical space, (b) toroidal space, and (c) spherical space.

Guillemin and Pollack (1974) for a precise description of a manifold.) In a two–dimensional space, a smooth object, such as a circle, is an example of a manifold, but an object with sharp corners, such as a rectangle, is not an example of a manifold. Locally, the circle may be approximated by a tangent line. Similarly, local regions of toroidal and spherical surfaces can be approximated by tangent planes. We note that an open flat surface is also a manifold.

Returning to (1.2.1), we note that this equation is also referred to as an **evolution equation.** Let the evolution of the system described by this equation be controlled by a set of parameters $\mathbf{M}$. To make this parameter dependence explicit, we describe the evolution by

$$\dot{\mathbf{x}} = \mathbf{F}(\mathbf{x}, t; \mathbf{M}) \qquad (1.2.4)$$

where $\mathbf{M}$ is a vector of control parameters. Formally, $\mathbf{M} \in \mathcal{R}^m$, and the vector function $\mathbf{F}$ can be represented as $\mathbf{F} : \mathcal{R}^n \times \mathcal{R}^1 \times \mathcal{R}^m \to \mathcal{R}^n$.

Next, we state some facts from the theory of ordinary–differential equations. If the scalar components of $\mathbf{F}$ are $\mathcal{C}^0$ (i.e., continuous) in a domain D of $(\mathbf{x}, t)$ space, then a solution $\mathbf{x}(t, \mathbf{x}_0, t_0)$ satisfying the condition $\mathbf{x} = \mathbf{x}_0$ at $t = t_0$ exists in a small time interval around t_0 in D. Moreover, if the scalar components of $\mathbf{F}$ are $\mathcal{C}^1$ in D, then the solution $\mathbf{x}(t, \mathbf{x}_0, t_0)$ is unique in a small time interval around t_0. The uniqueness of solutions is also assured in certain cases where $\mathbf{F}$ is $\mathcal{C}^0$ (Coddington and Levinson, 1955, Chapter 1; Arnold, 1973, 1992, Chapters 2 and 4). If the existence and uniqueness of solutions of a system of the form (1.2.4) are ensured, then this system is **deterministic**. This means that two integral curves starting from two different initial conditions cannot intersect each other in the extended state space. However, the corresponding orbits may intersect each other in the corresponding state space.

If the scalar components of $\mathbf{F}$ are $\mathcal{C}^r$ functions of t and the scalar components of $\mathbf{x}$ and $\mathbf{M}$, then a solution of (1.2.4) satisfying the initial condition $\mathbf{x} = \mathbf{x}_0$ at $t = t_0$ is also a $\mathcal{C}^r$ function of t, t_0, $\mathbf{x}_0$, and $\mathbf{M}$ in a small interval around t_0. Moreover, if a solution of (1.2.4) originating at a certain initial condition exists for all times, then this solution can be extended indefinitely. If a solution exists and is defined only over a finite interval of time, then this solution starting from a location in this interval can be extended up to the boundaries of this interval (Arnold, 1973, 1992, Chapters 2 and 4).

Example 1.4. This system is an example of a deterministic dynamical system. The parameter values used to generate Figure 1.2.3 are the same as those used to generate Figure 1.2.1. In Figure 1.2.3, we graphically show the solutions of (1.2.2) and (1.2.3) initiated at $t = 0$ from (1.0, 0.0) and (1.5, 0.0). From Figure 1.2.3a, we note that the corresponding integral curves do not intersect each other anywhere in the (x_1, x_2, t) space. As in Figure 1.2.1, the apparent intersections in Figure 1.2.3a are a consequence of the chosen viewing angle. From the previous discussion of Example 1.3, it is clear that as $t \to \infty$, both integral curves converge to the steady state

$$x_1 = 0.5 \cos(2t) + \sin(2t)$$
$$x_2 = -\sin(2t) + 2\cos(2t)$$

Although the two integral curves coincide only at $t = \infty$, on the scales of Figure 1.2.3a they are not distinguishable after about $t = 2.5$ units. In

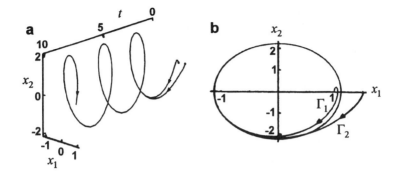

Figure 1.2.3: Solutions of (1.2.2) and (1.2.3) initiated from $(1.0, 0.0)$ and $(1.5, 0.0)$ at $t = 0$ for $\omega^2 = 8$, $\mu = 2$, $F = 10$, and $\Omega = 2$: (a) integral curves and (b) positive orbits. Γ_1 and Γ_2 are the positive orbits of $(1.0, 0.0)$ and $(1.5, 0.0)$, respectively.

Figure 1.2.3b, the positive orbits initiated from $(1.0, 0.0)$ and $(1.5, 0.0)$ are shown. We note the presence of a transverse intersection close to $(0.7, -2.0)$ in Figure 1.2.3b.

1.2.2 Autonomous Systems

In the case of an autonomous system, the equations are of the form

$$\dot{\mathbf{x}} = \mathbf{F}(\mathbf{x}; \mathbf{M}) \tag{1.2.5}$$

where $\mathbf{x}$, $\mathbf{F}$, and $\mathbf{M}$ are as defined before. Here, $\mathbf{F}$ does not explicitly depend on the independent variable t and can be represented by the map $\mathbf{F} : \mathcal{R}^n \times \mathcal{R}^m \to \mathcal{R}^n$. Hence, the system (1.2.5) is **time invariant**, **time independent**, or **stationary**. This means that if $\mathbf{X}(t)$ is a solution of (1.2.5), then $\mathbf{X}(t + \tau)$ is also a solution of (1.2.5) for any arbitrary τ. If the scalar components of $\mathbf{F}$ have continuous and bounded first partial derivatives with respect to the scalar components of $\mathbf{x}$, then the system (1.2.5) has a unique solution for a given initial condition $\mathbf{x}_0$. As a consequence, no two trajectories or orbits of an autonomous system can intersect each other in the n–dimensional state space of the system. Moreover, if the vector field $\mathbf{F}$ is a C^r function of $\mathbf{x}$ and $\mathbf{M}$,

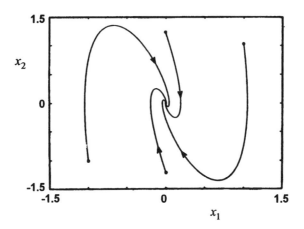

Figure 1.2.4: Positive orbits of (1.2.6) and (1.2.7) initiated at $t = 0$ from $(1.0, 1.0)$, $(0.0, -1.2)$, $(-1.0, -1.0)$, and $(0.0, 1.2)$ for $\omega^2 = 8$ and $\mu = 2$. All four orbits approach the origin as $t \to \infty$.

then the associated solution of (1.2.5) is also a C^r function of t, $\mathbf{x}$, and $\mathbf{M}$ (Arnold, 1973, 1992, Chapters 2 and 4).

Example 1.5. We consider the following autonomous system:

$$\dot{x}_1 = x_2 \qquad\qquad (1.2.6)$$
$$\dot{x}_2 = -\omega^2 x_1 - 2\mu x_2 \qquad\qquad (1.2.7)$$

In Figure 1.2.4, we show positive orbits of (1.2.6) and (1.2.7) initiated from four different initial conditions when $\omega^2 = 8$ and $\mu = 2$. These orbits do not intersect each other anywhere in the plane as they approach the origin, where they all meet. The direction of the orbits in the (x_1, x_2) space is given by

$$\frac{dx_2}{dx_1} = \frac{-(\omega^2 x_1 + 2\mu x_2)}{x_2}$$

which is well defined everywhere except at the origin. Hence, we call $(0, 0)$ a **singular point** of (1.2.6) and (1.2.7). Such solutions are discussed at length in Chapter 2.

It should be noted that nonautonomous systems with time–periodic coefficients can be converted into higher–dimensional autonomous systems. As an example, let us consider

$$\ddot{v}_1 + \omega_1^2 v_1 = f_1(v_1, \dot{v}_1, v_2, \dot{v}_2) + g_1 \cos(\Omega t) \qquad (1.2.8)$$

$$\ddot{v}_2 + \omega_2^2 v_2 = f_2(v_1, \dot{v}_1, v_2, \dot{v}_2) \qquad (1.2.9)$$

By using the four variables v_1, $\dot{v}_1$, v_2, and $\dot{v}_2$, we can rewrite (1.2.8) and (1.2.9) as a system of four first–order equations. This fourth–order system is nonautonomous due to the presence of a time–periodic term. Now, we consider an additional variable θ which is defined such that $\theta = \Omega t \bmod 2\pi$ (A mod B yields the remainder after A is divided by B; for example, 7 mod 3 is equal to 1). Introducing this variable into the fourth–order nonautonomous system and supplementing this system with the equation $\dot{\theta} = \Omega$, we obtain a fifth–order autonomous system. Formally, $\theta \in S^1$, and the space to which the five state variables v_1, v_2, $\dot{v}_1$, $\dot{v}_2$, and θ belong is written as $\mathcal{R}^4 \times S^1$. This space is a cylindrical space.

Equations (1.2.8) and (1.2.9) can also be converted into a five-dimensional autonomous system by considering the five state variables v_1, v_2, $\dot{v}_1$, $\dot{v}_2$, and v_3, where $v_3 = t$ and $\dot{v}_3 = 1$. In this case, the five state variables belong to a five-dimensional Euclidean space; that is, $\mathcal{R}^5$.

1.2.3 Phase Portraits and Flows

One often examines the evolution of a set of trajectories emanating from various initial conditions in the state space. As $t \to \infty$, the evolutions may approach different (asymptotic) solutions of the given system of nonlinear equations. A **phase portrait** is a collection of trajectories that represent the solutions of these equations in the phase space. For instance, Figure 1.2.4 is an example of a phase portrait of (1.2.6) and (1.2.7). In general, a phase portrait contains information about both the transient and the asymptotic behaviors of the solutions of a system. By drawing an analogy to pathlines in fluid mechanics (i.e., lines that trace the movement of different fluid particles in a fluid flow), the orbits starting from different initial conditions are said to describe the **flow** under the given system of equations. In this book, unless otherwise stated, the flow is defined for both positive and negative times. The reader may come across situations in other books or literature where the flow is defined only for either positive or negative times starting from $t = 0$. Such a flow is called a **semiflow**.

Example 1.6. To illustrate the meaning of flow under a given system of equations, we consider (1.2.6) and (1.2.7) when $\omega^2 = 8$ and $\mu = 2$. In Figure 1.2.5, we show positive orbits of (1.2.6) and (1.2.7) initiated at $t = 0.0$ from $(-1.00, 1.00)$, $(-1.25, 1.00)$, $(-1.25, 1.25)$, and $(-1.00, 1.25)$. These four points have been marked as A, B, C, and D, respectively. After 0.5 units of t, these four points are transported to the points A', B', C', and D' by their respective orbits. Any point in the region $ABCD$ is transported to a particular location in the region $A'B'C'D'$ under the flow of (1.2.6) and (1.2.7).

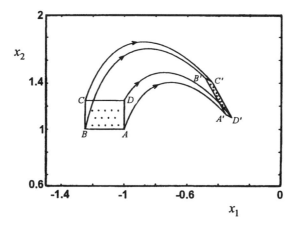

Figure 1.2.5: Four positive orbits of (1.2.6) and (1.2.7) initiated at $t = 0$ when $\mu > 0$. The orbits are shown for $0 \leq t \leq 0.5$. The area $A'B'C'D'$ occupied by the final states is less than the area $ABCD$ occupied by the initial states. Furthermore, the orientation of the corners of the initial area is preserved.

1.3   ATTRACTING SETS

Before introducing the notion of an attracting set, we first explain what is meant by **dissipation** in a flow. Revisiting Figure 1.2.5, we note that the area of the region $A'B'C'D'$ in phase space at a later time is smaller than the area of the initial region $ABCD$ in phase space. In other words, the flow is such that a set of final states occupies a region smaller in size than that occupied by the corresponding initial states. Hence, areas in the phase space of (1.2.6) and (1.2.7) are not **conserved** but **contracted** when $\mu = 2$. This phenomenon is called **dissipation**.

In Figure 1.3.1, we illustrate the flow under (1.2.6) and (1.2.7) when $\omega^2 = 8$ and $\mu = 0$. Four positive orbits initiated at $t = 0$ from $(-1.00, 1.00)$, $(-1.25, 1.00)$, $(-1.25, 1.25)$, and $(-1.00, 1.25)$ are considered, and these four initial points have been marked as A, B, C, and D, respectively. After 0.5 units of t, these four points are transported to the points A', B', C', and D' by their respective orbits. Again, any point in the region $ABCD$ is transported to a particular location in the region $A'B'C'D'$ under the flow of (1.2.6) and (1.2.7). However, unlike the situation in Figure 1.2.5, the set of final states occupies a region equal in size to that occupied by the corresponding initial states. Hence, the flow is said to **conserve** areas in the phase space when $\mu = 0$ in (1.2.6) and (1.2.7).

For dissipation in a general setting, we consider the flow under (1.2.5). At an initial time $t = t_0$, let the volume occupied by a given set of initial conditions (states) be V_0. After a time t_f, the orbits initiated at these conditions reach certain locations (states) in the state space. At $t = t_f$, let the volume occupied by the set of these states be V_f. Then, flows are classified into conservative or dissipative, depending upon whether V_f is equal to or less than V_0; that is, depending upon whether the (local) volumes in the state space stay constant or contract with time.

For dynamical systems governed by (1.2.4) or (1.2.5), we appeal to concepts of fluid mechanics (e.g., Karamcheti, 1976) to determine whether a flow is conservative or dissipative. At an instant in time t, we consider a set of points occupying a small region with volume V and

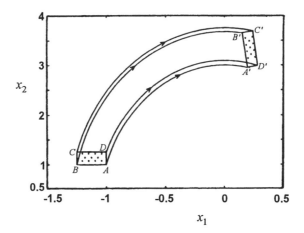

Figure 1.3.1: Four positive orbits of (1.2.6) and (1.2.7) initiated at $t = 0$ when $\mu = 0$. The orbits are shown for $0 \le t \le 0.5$. The area $A'B'C'D'$ occupied by the final states is equal to the area $ABCD$ occupied by the initial states. Furthermore, the orientation of the corners of the initial area is preserved.

surface S in the associated n–dimensional state space. (As the set of points are transported in the state space under the considered flow, V and S change with time.) Considering a small elemental area ΔS, the associated change in the volume V over a time interval Δt is given by

$$\mathbf{v}\Delta t \cdot \mathbf{n}\Delta S \qquad (1.3.1)$$

where $\mathbf{v} = \dot{\mathbf{x}} = \mathbf{F}$ is the velocity vector and $\mathbf{n}$ is the outward unit normal on S. To determine the total change in V, we let $\Delta S \to 0$ and integrate (1.3.1) over the surface S. The result is

$$\Delta V = \Delta t \int \int_S \mathbf{F} \cdot \mathbf{n} dS \qquad (1.3.2)$$

From the divergence theorem (e.g., Karamcheti, 1976), we have

$$\int \int_S \mathbf{F} \cdot \mathbf{n} dS = \int \int \int_V (\nabla \cdot \mathbf{F}) dV \qquad (1.3.3)$$

But, for an infinitesimally small V, the right-hand side of (1.3.3) can be approximated by $(\nabla \cdot \mathbf{F})V$. Therefore, (1.3.2) and (1.3.3) lead to

$$\frac{1}{V}\frac{\Delta V}{\Delta t} = \nabla \cdot \mathbf{F} \qquad (1.3.4)$$

Consequently, a flow is conservative or dissipative, depending on whether the divergence of its vector field is zero or negative. In other words, in conservative systems

$$\sum_{i=1}^{n} \frac{\partial F_i(\mathbf{x})}{\partial x_i} = 0 \qquad (1.3.5)$$

and in dissipative systems (e.g., systems with damping)

$$\sum_{i=1}^{n} \frac{\partial F_i(\mathbf{x})}{\partial x_i} < 0 \qquad (1.3.6)$$

where the F_i and x_i are the scalar components of $\mathbf{F}$ and $\mathbf{x}$ in (1.2.5), respectively. In this book, we shall mainly be interested in dissipative systems.

In mechanics, there are systems called **Hamiltonian systems**, which are governed by

$$\dot{q}_i = \frac{\partial H}{\partial p_i} \qquad (1.3.7)$$

$$\dot{p}_i = -\frac{\partial H}{\partial q_i} \qquad (1.3.8)$$

for $i = 1, 2, \cdots, n$ and $H = H(q_1, q_2, \cdots, q_n, p_1, p_2, \cdots, p_n, t)$. (The function H is called the **Hamiltonian**.) The divergence of the vector field of the system governed by (1.3.7) and (1.3.8) is

$$\nabla \cdot \mathbf{F} = \sum_{i=1}^{n} \left[\frac{\partial}{\partial q_i}\left(\frac{\partial H}{\partial p_i}\right) + \frac{\partial}{\partial p_i}\left(-\frac{\partial H}{\partial q_i}\right) \right] = 0 \qquad (1.3.9)$$

if H is twice continuously differentiable. Therefore, volumes in state space are conserved in Hamiltonian systems and hence they form a subset of the set of conservative systems. The statement about the preservation of volumes in state space of Hamiltonian systems is called the **Liouville theorem** (e.g., Arnold, 1973, Chapter 3; Lichtenberg and Lieberman, 1992, Chapter 1).

In summary, the **the flow in conservative systems is said to preserve volume (locally) in the state space.** Furthermore, as $t \to \infty$, the motion takes place in the full n–dimensional space. For dissipative systems, $V_f < V_0$ and $V_f \to 0$ as $t = t_f \to \infty$. This means that trajectories initiated from different conditions are attracted to a subspace of the state space. This phenomenon is called **attraction**, and the set to which the trajectories are attracted as $t \to \infty$ is called an **attracting set**.

Before we consider the notion of an attracting set in detail, we first explain what is meant by an **invariant set**. A set $\mathbf{P} \subset \mathcal{R}^n$ is called an **invariant set** if for any initial condition $\mathbf{x}(t = t_0) \in \mathbf{P}$ we have $\mathbf{x}(t) \in \mathbf{P}$ for $-\infty < t < \infty$. If this condition is satisfied only for $t \geq 0$ or $t \leq 0$, $\mathbf{P}$ is called a **positive** or **negative invariant set**, respectively. An **attracting set** is an invariant set. Further, it has an open neighborhood such that positive orbits initiated in this neighborhood are attracted to this set. As explained in Section 1.5, a special type of an attracting set is called an **attractor**.

Example 1.7. For the system (1.2.6) and (1.2.7), the divergence of the vector field is -2μ. Hence, if $\mu > 0$ this system is dissipative, and if $\mu = 0$ this system is conservative. Figure 1.3.1 is illustrative of the conservative nature of this system when $\mu = 0$, while Figures 1.2.4 and 1.2.5 are illustrative of the dissipation in this system when $\mu > 0$. In the case of Figure 1.2.4, all four positive orbits are attracted to the origin of the state space.

Example 1.8. We convert the system (1.2.2) and (1.2.3) into a three–dimensional autonomous system by defining an additional state θ such that $\dot{\theta} = \Omega$. Thus, we have

$$\dot{x}_1 = x_2$$
$$\dot{x}_2 = -\omega^2 x_1 - 2\mu x_2 + F\cos(\theta)$$
$$\dot{\theta} = \Omega$$

and the divergence of the vector field is

$$\frac{\partial \dot{x}_1}{\partial x_1} + \frac{\partial \dot{x}_2}{\partial x_2} + \frac{\partial \dot{\theta}}{\partial \theta} = -2\mu$$

So, local volumes in the (x_1, x_2, θ) space are conserved when $\mu = 0$ and contracted when $\mu > 0$. To construct Figure 1.3.2, we used the following parameter values: $\omega^2 = 8$, $\mu = 2$, $F = 10$, and $\Omega = 2$. We show three positive orbits initiated from $(0,1)$, $(0,4)$, and $(0,5)$. All three orbits are **attracted** to the closed orbit Γ encircling the origin as $t \to \infty$ because of the dissipation in the system. In Figure 1.3.2, the area of initial conditions marked A contracts to the area marked B after one unit of time.

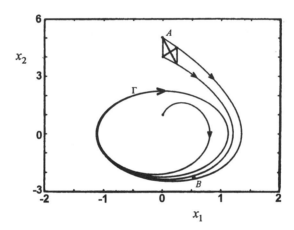

Figure 1.3.2: Three positive orbits of (1.2.2) and (1.2.3) initiated at $t = 0$.

In Figures 1.2.4, 1.2.5, and 1.3.2, we observe the presence of attraction and attracting sets. The origin of the $x_2 - x_1$ space is the attracting set in Figure 1.2.4, and the closed orbit Γ is the attracting set in Figure 1.3.2. Such sets occur only in dissipative systems.

We note that the concepts of dissipation, invariant sets, attracting sets, and attractors also apply to the maps discussed in Section 1.1. The map (1.1.3) is said to be dissipative at $\mathbf{x}_k = \mathbf{x}_0$ if

$$| \det D_{\mathbf{x}_k}\mathbf{F} |< 1 \quad \text{at} \quad \mathbf{x}_k = \mathbf{x}_0 \tag{1.3.10}$$

where det $D_{\mathbf{x}_k}\mathbf{F}$ is the determinant of the $n \times n$ matrix of first partial derivatives of the scalar components of $\mathbf{F}$ with respect to the scalar components of $\mathbf{x}_k$.

Example 1.9. In the case of the Hénon map (1.1.5) and (1.1.6), we have

$$\det D_{\mathbf{x}_k}\mathbf{F} = \det \begin{bmatrix} -2\alpha x_k & 1 \\ \beta & 0 \end{bmatrix} = -\beta$$

Hence, when $\mid \beta \mid < 1$, the Hénon map is dissipative at all $\mathbf{x}_k$. Consequently, any area is contracted by the factor $\mid \beta \mid$ after each iterate.

1.4   CONCEPTS OF STABILITY

In this section, we discuss various types of stability and the concept of attractors. We note that all of the notions of stability discussed below are made in the context of finite–dimensional systems. Structural stability, which deals with the stability of the orbit structure of a dynamical system to small perturbations, is discussed in Section 2.3.7.

1.4.1   Lyapunov Stability

Maps

A solution $\{\mathbf{u}_k\}$ of a map is said to be **Lyapunov stable** if, given a small number $\epsilon > 0$, there exists a number $\delta = \delta(\epsilon) > 0$ such that any other solution $\{\mathbf{v}_k\}$ for which $\parallel \mathbf{u}_k - \mathbf{v}_k \parallel < \delta$ at $k = m$ satisfies $\parallel \mathbf{u}_k - \mathbf{v}_k \parallel < \epsilon$ for all $k > m$, where k and $m \in \mathcal{Z}^+$. For Lyapunov stability, two orbits of a map initiated from two neighboring points at a certain time have to remain "close" to each other for all future times (i.e., $k > m$).

Example 1.10. For illustration, we consider the Lyapunov stability of a solution of the map (1.1.4). When $\alpha = 0.5$, $x_k = 0.5$ is a solution of this map for all $k \in \mathcal{Z}^+$. Such a solution is an example of a **fixed point** of a map. We consider fixed points of a map in more detail in Sections 2.2 and 2.4. An orbit of (1.1.4) initiated at $u_0 = 0.5$ is

$$\{0.5, \; 0.5, \; 0.5, \; 0.5, \; 0.5, \; \cdots, \; 0.5\}$$

An orbit of (1.1.4) initiated at $v_0 = 0.4$ is

$$\{0.4, \; 0.48, \; 0.4992, \; 0.49999872, \; 0.5, \; \cdots, \; 0.5\}$$

Thus, given an $\epsilon > 0$, one can find a $\delta(\epsilon) > 0$ satisfying the conditions of the Lyapunov stability. Hence, $x_k = 0.5$ is a Lyapunov stable solution of (1.1.4).

Continuous–Time Systems

A solution $\mathbf{u}(t)$ of either an autonomous or a nonautonomous system of differential equations is said to be **Lyapunov stable** if, given a small number $\epsilon > 0$, there exists a number $\delta = \delta(\epsilon) > 0$ such that any other solution $\mathbf{v}(t)$ for which $\| \mathbf{u} - \mathbf{v} \| < \delta$ at time $t = t_0$ satisfies $\| \mathbf{u} - \mathbf{v} \| < \epsilon$ for all $t > t_0$. In nonautonomous systems, δ will also be a function of the initial time t_0. If $\mathbf{u}$ is Lyapunov stable, then any other solution that is "close" to it initially remains so and is confined to a tube formed by the union of spheres of radius ϵ centered on points along the trajectory $\mathbf{u}(t)$, the so–called ϵ tube. For a nonautonomous system, we will have an ϵ tube in the extended state space. In Figure 1.4.1, we illustrate the concept of Lyapunov stability for two solutions $\mathbf{u}_1$ and $\mathbf{u}_2$ of a nonautonomous system. In autonomous systems, **Lyapunov**

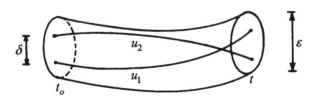

Figure 1.4.1: Illustration of Lyapunov stability for a solution $\mathbf{u}_i$ of a nonautonomous continuous–time system.

stability is also known as **uniform stability** because δ is independent of the initial time t_0.

Example 1.11. Here, we consider the Lyapunov stability of some of the solutions of the Duffing oscillator given by

$$\dot{x}_1 = x_2 \tag{1.4.1}$$
$$\dot{x}_2 = -x_1 + x_1^3 - 2\mu x_2 \tag{1.4.2}$$

We first examine the solutions of (1.4.1) and (1.4.2) when μ, the damping coefficient, is zero. The points $(0,0), (-1,0)$, and $(1,0)$ in the $x_2 - x_1$ plane satisfy (1.4.1) and (1.4.2) for all times and are three solutions of the system. These solutions are called **equilibrium solutions**. The states corresponding to an equilibrium solution are constant in time. More details on them are provided in Section 2.1. In the $x_2 - x_1$ plane, the orbits of solutions in the neighborhood of $(0,0)$ are closed curves surrounding it. These solutions, called **periodic solutions**, are extensively treated in Chapter 3.

We consider the Lyapunov stability of the solution $(0,0)$ and use Figure 1.4.2a in this regard. The solid curve was obtained by numerically integrating the equations from the initial condition $(0.0, 0.3)$. Let (x_{10}, x_{20}) represent the initial condition at time $t = t_0$ for one of the

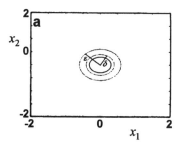
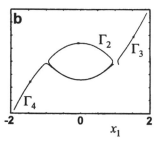

Figure 1.4.2: Illustrations for determining the Lyapunov stability of the following solutions of (1.4.1) and (1.4.2) when $\mu = 0$: (a) $(0,0)$ and (b) $(1,0)$.

periodic solutions surrounding $(0,0)$. If

$$\sqrt{x_{10}^2 + x_{20}^2} < \delta$$

then the corresponding periodic solution is confined to the ϵ tube (whose cross–section has a radius ϵ as shown in Fig. 1.4.2a) for all times. In other words, for the solution $(0,0)$, given a number $\epsilon > 0$, one can always find a number $\delta > 0$ satisfying the conditions for Lyapunov stability. (If we want the motion to remain in an ϵ–neighborhood of $(0,0)$, one can find a δ–neighborhood for the initial condition.) So, the solution $(0,0)$ is uniformly stable.

Next, we consider the Lyapunov stability of the solution $(1,0)$. The trajectories Γ_2, Γ_3, and Γ_4, shown in Figure 1.4.2b, are positive orbits associated with the initial conditions $(0.9,-0.1), (1.1,-0.1)$, and $(0.95,-0.1)$, respectively. All three initial conditions are chosen in a neighborhood of $(1,0)$. The positive orbit Γ_2 is periodic and bounded. The positive orbits Γ_3 and Γ_4 are not bounded and grow indefinitely. All the orbits of Figure 1.4.2b were obtained through numerical integrations. It is not possible to find an ϵ tube around $(1,0)$ within which Γ_3 and Γ_4 will remain confined for all times. Hence, numerical simulations indicate that $(1,0)$ is not uniformly stable or stable in the Lyapunov sense.

1.4.2   Asymptotic Stability

Maps

A solution $\{u_k\}$ of a map is said to be **asymptotically stable** if it is Lyapunov stable and

$$\lim_{k \to \infty} \| \, u_k - v_k \, \| \to 0$$

Example 1.12. When $\alpha = 0.5$, the solution $x_k = 0.5$ of (1.1.4) is asymptotically stable. This is so because $x_k = 0.5$ is Lyapunov stable and the separation between the orbits initiated at $u_0 = 0.5$ and $v_0 = 0.4$ or any other starting point in a small neighborhood of u_0 is very small for $k \geq 4$.

Continuous–Time Systems

A solution $\mathbf{u}(t)$ of an autonomous or nonautonomous system of differential equations is said to be **asymptotically stable** if it is Lyapunov stable and

$$\lim_{t\to\infty} \| \mathbf{u} - \mathbf{v} \| \to 0$$

Example 1.13. We consider (1.4.1) and (1.4.2) when $\mu > 0$, so that we have a dissipative system. In Figure 1.4.3, we show two positive orbits Γ_1 and Γ_2 of this system initiated from $(0.0, 0.6)$ and $(0.0, -0.6)$, respectively. Let $\mathbf{u}$ represent the solution $(0,0)$. As shown in Figure 1.4.3, any solution $\mathbf{v}$ started from an initial condition in a neighborhood of $(0,0)$ tends toward $(0,0)$ or is attracted to it as $t \to \infty$. The separation between the solutions $\mathbf{u}$ and $\mathbf{v}$ goes to zero as $t \to \infty$. Hence, by definition, $(0,0)$ is asymptotically stable. The presence of damping in (1.4.1) and (1.4.2) makes $(0,0)$ an asymptotically stable solution.

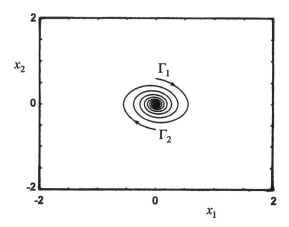

Figure 1.4.3: Illustration of the asymptotic stability of the solution $(0,0)$ of (1.4.1) and (1.4.2) when $\mu = 0.1$. The positive orbits Γ_1 and Γ_2 are initiated from $(0.0, 0.6)$ and $(0.0, -0.6)$, respectively.

1.4.3 Poincaré Stability

This notion of stability is commonly applied to solutions of autonomous or nonautonomous systems of differential equations. The notions described in Sections 1.4.1 and 1.4.2 are typically used to study the stability of equilibrium solutions of (1.2.5). These solutions satisfy the condition $\mathbf{F}(\mathbf{x}; \mathbf{M}) = \mathbf{0}$. In Sections 1.4.1 and 1.4.2, we compared how "close" two integral curves started from two different initial conditions are at the same instant t. This approach is restrictive, and according to it even a periodic solution of a nonlinear autonomous system is unstable. To understand why this is so, let us consider the stability of periodic solutions of the two–dimensional undamped nonlinear autonomous system (1.4.1) and (1.4.2). When $\mu = 0$, using the method of multiple scales, one finds the following approximate solution of (1.4.1) and (1.4.2) for small but finite amplitudes (e.g., Nayfeh, 1981):

$$u \approx a \cos \left[\left(1 - \frac{3}{8} a^2 \right) t + \beta \right]$$

It is clear that the frequency of oscillation $\omega = 1 - \frac{3}{8} a^2$ depends on the amplitude a of oscillation, which in turn is determined by the initial condition. Hence, solutions started from two slightly different initial conditions evolve with two different periods and may not be "close" to each other at a specific large value of t.

For a graphical illustration, we consider two periodic solutions of the undamped system obtained from (1.4.1) and (1.4.2) by setting $\mu = 0$. We are interested in the stability of the periodic solution $\mathbf{u}$ whose orbit is Γ_1 of Figure 1.4.4. This orbit was obtained by numerically integrating (1.4.1) and (1.4.2) from the initial condition $(0.0000, 0.5000)$. Another solution $\mathbf{v}$ of the system corresponds to the orbit Γ_2 in Figure 1.4.4. This orbit was obtained by numerically integrating (1.4.1) and (1.4.2) from the initial condition $(0.0000, 0.6000)$. Due to the different initial conditions, the periods of oscillation of these orbits are different. At the initial time $t = 0$, the separation between the two solutions is $\| \mathbf{u} - \mathbf{v} \| = 0.1$.

After $t = 30$ units, we move to point y on Γ_1 and point z on Γ_2. The coordinates of y are $(0.5263, 0.1068)$ and of z are $(-0.6369, 0.1914)$. The separation between these two solutions is $\| \mathbf{u} - \mathbf{v} \| = 1.1663$ and

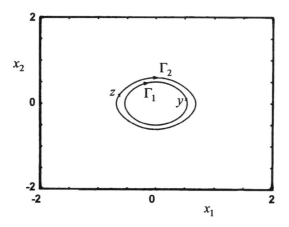

Figure 1.4.4: Periodic solutions of (1.4.1) and (1.4.2) when $\mu = 0$.

is much larger than the initial separation. We may choose two closer initial conditions, but large separations between the respective integral curves in the (x_1, x_2, t) space will eventually occur after a certain number of cycles due to the differences in the associated periods. One cannot make such separations between the integral curves arbitrarily small by choosing initial conditions "close" to each other. Although the two orbits are "close" to each other, according to the definition of Lyapunov stability, the solution $\mathbf{u}$ is unstable because the integral curves are not "close" to each other. To remedy this situation, Poincaré introduced the notion of **orbital stability**. For dynamic solutions, such as periodic solutions of (1.2.5), one uses the notion of **orbital** or **Poincaré stability**.

Let Γ_1 represent the orbit of $\mathbf{u}$ and Γ_2 represent the orbit of $\mathbf{v}$ for all times. The periodic solutions $\mathbf{u}$ and $\mathbf{v}$ have different periods T_1 and T_2 and, hence, the corresponding motions evolve on different time scales. The orbit Γ_1 is said to be **orbitally stable** if, given a small number $\epsilon > 0$, there exists a $\delta = \delta(\epsilon) > 0$ such that if $\| \mathbf{u}(t = 0) - \mathbf{v}(t = \tau) \| < \delta$ for some τ, then there exist t_1 and t_2 for which $\| \mathbf{u}(t_1) - \mathbf{v}(t_2) \| < \epsilon$. Further, if Γ_2 tends to Γ_1 as $t \to \infty$, then we say Γ_1 is asymptotically stable. For Poincaré stability, we examine how "close" orbits are in the state space.

1.4.4 Lagrange Stability (Bounded Stability)

Maps

A solution $\{u_k\}$ of a map is said to be **boundedly stable** if $\| u_k \| \leq L$ for all $k \in \mathcal{Z}$, where L is a finite positive quantity.

Continuous–Time Systems

A solution $u(t)$ of a continuous–time system is said to be **boundedly stable** if $\| u \| \leq L$ for all t, where L is a finite positive quantity.

1.4.5 Stability Through Lyapunov Function

The concept of Lyapunov stability discussed in Section 1.4.1 and the asymptotic stability discussed in Section 1.4.2 are typically used to study the stability of equilibrium solutions of $\dot{\mathbf{x}} = \mathbf{F}(\mathbf{x}; \mathbf{M})$. Let us designate the equilibrium solutions for some given value $\mathbf{M}_0$ of $\mathbf{M}$ as $\mathbf{x} = \mathbf{x}_0$. Further, we assume that there exists a $\mathcal{C}^1$ scalar function $V(\mathbf{x}; \mathbf{M})$ defined in a neighborhood of $\mathbf{x}_0$ such that $V(\mathbf{x}_0; \mathbf{M}_0) = 0$ and $V(\mathbf{x}; \mathbf{M}) > 0$ if $\mathbf{x} \neq \mathbf{x}_0$. The function V is called a **Lyapunov function**. The derivative of V along the solution curves of (1.2.5) is $\dot{V} = \nabla V \cdot \mathbf{F} = \nabla V^T \mathbf{F}$. From the associated stability theorems (Lyapunov, 1947), we have the following:

(a) If $\dot{V} \leq 0$ in the chosen neighborhood of $\mathbf{x}_0$, then $\mathbf{x}_0$ is **stable**.

(b) If $\dot{V} < 0$ in the chosen neighborhood of $\mathbf{x}_0$, then $\mathbf{x}_0$ is **asymptotically stable**.

There are theorems that address the existence of Lyapunov functions (Krasovskii, 1963). For mechanical and structural systems, one can often use the energy as the Lyapunov function. In the context of electrical power systems, some methods have been used to determine Lyapunov functions (e.g., Pai, 1981; Michel, Miller, and Nam, 1982; Michel, Nam, and Vittal, 1984). For Hamiltonian systems, the Hamiltonian can be taken to be the Lyapunov function. However, for other systems, there are no general methods for determining this function.

Example 1.14. For illustration, we consider the stability of the solution $(0,0)$ of (1.4.1) and (1.4.2). We choose the expression for energy as the Lyapunov function and obtain

$$V(x_1, x_2) = \frac{1}{2}x_2^2 + \frac{1}{2}x_1^2 - \frac{1}{4}x_1^4$$

We note that $V(0,0) = 0$ and $V(x_1, x_2) > 0$ in a region around $(0,0)$. In this case, the gradient of V, given by ∇V, is

$$\nabla V^T = \left(\frac{\partial V}{\partial x_1} \ \frac{\partial V}{\partial x_2} \right)$$

and thus

$$\dot{V} = \nabla V^T \mathbf{F} = \frac{\partial V}{\partial x_1} F_1 + \frac{\partial V}{\partial x_2} F_2$$

where the F_i are the scalar components of $\mathbf{F}$. Hence, the derivative of V along the solutions of (1.4.1) and (1.4.2) is given by

$$\dot{V} = -2\mu x_2^2$$

Because μ is positive in a dissipative system, $\dot{V} \le 0$ in any chosen neighborhood of the equilibrium solution $(x_1, x_2) = (0,0)$. Hence, this solution is stable according to Lyapunov's first stability theorem.

Example 1.14 also serves to illustrate a shortcoming of Lyapunov's second stability theorem. Using a number of techniques, such as numerical and perturbation methods, one can show that the origin is an asymptotically stable solution of (1.4.1) and (1.4.2). However, because

$$\dot{V} = -2\mu x_2^2 \le 0$$

in every neighborhood of the origin ($\dot{V} = 0$ all along the x_1 axis; that is, $x_2 = 0$), Lyapunov's second stability theorem cannot be used to conclude that the origin is asymptotically stable. In fact, the condition provided by Lyapunov's second theorem is sufficient but not necessary. This shortcoming can be overcome by using Krasovskii's theorem (Krasovskii, 1963). Let $V(\mathbf{x})$ be a $\mathcal{C}^1$ scalar function, $V(\mathbf{x}_0) = 0$, and

$$D_\ell = \{\mathbf{x} \mid V(\mathbf{x}) \le \ell\}$$

$$V(\mathbf{x}) > 0 \text{ for } \mathbf{x} \epsilon D_\ell \text{ and } \mathbf{x} \neq \mathbf{x}_0$$

$$\dot{V}(\mathbf{x}) \leq 0 \text{ for } \mathbf{x} \epsilon D_\ell$$

If there is no solution $\mathbf{x}^*(t)$ for $\dot{\mathbf{x}} = \mathbf{F}(\mathbf{x}; \mathbf{M}_0)$, other than $\mathbf{x}_0$, which lies completely in D_ℓ and for which $\dot{V}(\mathbf{x}^*(t)) \equiv 0$, then every solution that starts in D_ℓ tends to $\mathbf{x}_0$ and $\mathbf{x}_0$ is asymptotically stable. By finding the largest possible region for which the conditions of Krasovskii's theorem are satisfied, one can determine the **domain of attraction** of $\mathbf{x}_0$; that is, the region in state space such that a trajectory started at any point inside it will tend to $\mathbf{x}_0$ as $t \rightarrow \infty$. For example, using the Lyapunov function

$$V(x_1, x_2) = \frac{1}{2}x_2^2 + \frac{1}{2}x_1^2 - \frac{1}{4}x_1^4$$

and taking D_ℓ to be the region

$$\frac{1}{2}x_2^2 + \frac{1}{2}x_1^2 - \frac{1}{4}x_1^4 < \frac{1}{4}$$

one can show that all the conditions of Krasovskii's theorem are satisfied and hence the origin is an asymptotically stable equilibrium of (1.4.1) and (1.4.2) and that D_ℓ is the domain of attraction of the origin.

1.5 ATTRACTORS

Earlier, in the context of Figure 1.2.4, we noted that the positive orbits of (1.2.6) and (1.2.7) are attracted to the origin of the state space as $t \rightarrow \infty$. By using a Lyapunov function, one can show that the origin is an asymptotically stable solution of (1.2.6) and (1.2.7) when $\mu > 0$. Such asymptotically stable solutions are attractive and examples of **attractors**.

In a general setting, let T^t represent an evolution operator that acts on initial conditions $\mathbf{x}_0$ in $\mathcal{R}^n$ such that $T^t\mathbf{x}_0 = \mathbf{x}(\mathbf{x}_0, t)$, where $\mathbf{x} \in \mathcal{R}^n$. Repeated applications of T^t may take one to a subspace of $\mathcal{R}^n$ called an **attractor**, which is defined by the following properties (Eckmann, 1981):

1. **Invariance**: An attractor X is an invariant set of the flow of the system. Formally, $T^t X \in X$.

2. **Attractivity**: There exists a neighborhood U of the attractor (i.e., $X \subset U$) such that evolutions initiated in U remain in U and approach X as $t \to \infty$. Formally, $T^t U \subset U$ for $t \geq 0$ and $T^t U \to X$ as $t \to \infty$. The symbol $\subset$ stands for the subset operator and $X \subset U$ means that X is included in U.

3. **Recurrence**: Trajectories initiated from a state in an open subset of X repeatedly come arbitrarily close to this initial state for arbitrarily large values of time.

4. **Indecomposability**: An attractor cannot be split up into two nontrivial pieces.

Property 3 rules out unstable solutions and transient solutions from being attractors. Property 4, also referred to as **irreducibility**, implies that an attractor cannot be decomposed into distinct smaller attractors. We note that all attracting sets possess Properties 1 and 2, but only some attracting sets, namely, attractors, possess Properties 3 and 4. Because some unstable solutions can be part of an attracting set, Property 3 can be violated. Further, because more than one attractor can be part of an attracting set, Property 4 can be violated. We note that to verify Property 2 one needs to use a suitable distance measure.

Going back to the origin of Figure 1.2.4, we note that it satisfies Property 1 because it is a solution of (1.2.6) and (1.2.7) for all t. Further, it satisfies Properties 2 and 3 because it is asymptotically stable. In addition, $(0,0)$ satisfies Property 4 because it cannot be split into any smaller sets that satisfy (1.2.6) and (1.2.7). The asymptotically stable equilibrium solution $(0,0)$ of (1.2.6) and (1.2.7) is an example of a **point attractor**. The orbit Γ of Figure 1.3.2 is an attractor of (1.2.2) and (1.2.3). To be specific, it is a **periodic attractor**. Unlike a point attractor, this attractor is a **dynamic solution** because the corresponding state variables are functions of time. Two other attractors, which are characterized by time–varying states, are **quasiperiodic** and **chaotic** attractors. In Chapter 2, we address point attractors, and in Chapters 3, 4, and 5, we address periodic, quasiperiodic, and chaotic attractors, respectively.

The domain $D \subset \mathcal{R}^n$ that includes all the initial conditions $\mathbf{x}_0$ such that $T^t \mathbf{x}_0 \to X$ as $t \to \infty$ is called the **basin** or **domain of attraction** or **stability region** of X. Thus, all evolutions in a basin of attraction of X are attracted to it.

In the literature (e.g., Seydel, 1988), a counter notion of an attractor called a **repellor** is also used. An attractor attracts positive orbits but repels negative orbits, and a repellor repels positive orbits but attracts negative orbits. We noted earlier that the solution $(0,0)$ of (1.2.6) and (1.2.7) is an attractor when $\mu > 0$. This solution is a repellor when $\mu < 0$.

1.6   COMMENTS

In the literature, the method employing Lyapunov functions to determine the stability of a fixed point is also called **Lyapunov's second method**. The existence of Lyapunov functions and applications of Lyapunov's second method are discussed at length by Krasovskii (1963). A detailed exposition of Lyapunov stability theory is also provided by Hagedorn (1988). We note that all the notions of stability discussed above do not provide any explicit schemes to determine the stability of a solution. Explicit schemes for determining the stability of fixed points of a map or an autonomous system of differential equations are discussed in the next chapter, and explicit schemes for determining the stability of periodic and quasiperiodic solutions of a system of differential equations are discussed in Chapters 3 and 4.

1.7   EXERCISES

1.1. Lorenz (1963) used the following equations to study thermally induced fluid convection in the atmosphere:

$$\dot{x} = \sigma(y - x)$$
$$\dot{y} = \rho x - y - xz$$
$$\dot{z} = -\beta z + xy$$

Determine when this system is dissipative and when it is conservative.

1.2. Consider the Rössler system (Rössler, 1976a):

$$\dot{x} = -(y + z)$$
$$\dot{y} = x + ay$$
$$\dot{z} = b + (x - c)z$$

Determine when this system is dissipative and when it is conservative.

1.3. Consider the following two–dimensional map:

$$x_{n+1} = a_{11}x_n + a_{12}y_n$$
$$y_{n+1} = a_{21}x_n + a_{22}y_n$$

Determine the conditions for which this map is dissipative.

1.4. Consider the solutions of (1.4.1) and (1.4.2) for small but finite values of x_1 and x_2. Use a perturbation method to show that

$$x_1 \approx a(t)\cos(t + \beta(t)) \quad \text{and} \quad x_2 \approx -a(t)\sin(t + \beta(t))$$

where

$$\dot{a} = -\mu a$$
$$a\dot{\beta} = -\frac{3}{8}a^3$$

Hence, show that the origin is an asymptotically stable equilibrium solution of (1.4.1) and (1.4.2).

1.5. Determine the exact solution of

$$\dot{x}_1 = x_2$$
$$\dot{x}_2 = -x_1 - 2\mu x_2$$

and show that the origin is asymptotically stable. Let

$$V = \frac{1}{2}(x_1^2 + x_2^2)$$

Show that $\dot{V} = -2\mu x_2^2$ along the solutions of this system and hence one cannot use Lyapunov's second theorem to show that the origin is asymptotically stable. Use Krasovskii's theorem to show that the origin is asymptotically stable and that the whole state space is its domain of attraction.

1.6. Consider the system

$$\dot{x}_1 = x_2$$
$$\dot{x}_2 = -x_1 - \mu x_2^3$$

Use a perturbation method to show that, near the origin,

$$x_1 \approx a(t)\cos(t + \beta(t)) \quad \text{and} \quad x_2 \approx -a(t)\sin(t + \beta(t))$$

where

$$\dot{a} = -\frac{3}{8}\mu a^3$$
$$a\dot{\beta} = 0$$

Hence, show that the origin is asymptotically stable. Let

$$V(x_1, x_2) = \frac{1}{2}(x_1^2 + x_2^2)$$

and use Krasovskii's theorem to show that the origin is asymptotically stable and determine its domain of attraction.

1.7. Consider the system

$$\dot{x}_1 = x_2$$
$$\dot{x}_2 = -x_1 - x_1^3 - \mu x_2 \mid x_2 \mid$$

Using a perturbation method show that, near the origin,

$$x_1 \approx a(t)\cos(t + \beta(t)) \quad \text{and} \quad x_2 \approx -a(t)\sin(t + \beta(t))$$

where

$$\dot{a} = -\frac{4\mu}{3\pi}a^2$$

$$a\dot{\beta} = \frac{3}{8}a^3$$

Hence, show that the origin is asymptotically stable. Let

$$V(x_1, x_2) = \frac{1}{2}x_2^2 + \frac{1}{2}x_1^2 + \frac{1}{4}x_1^4$$

Use Krasovskii's theorem to show that the origin is asymptotically stable and determine its domain of attraction.

Chapter 2

EQUILIBRIUM SOLUTIONS

An important class of solutions of a map, such as $x_{k+1} = \mathbf{F}(x_k; \mathbf{M})$, or a system of differential equations, such as $\dot{x} = \mathbf{F}(x; \mathbf{M})$, are **fixed–point solutions** or **equilibrium solutions**. By and large, in the literature, only fixed–point solutions of a system of differential equations are called **equilibrium solutions**. In Section 2.1, we consider fixed points of continuous–time systems and their stability in detail. The fixed points of maps and their stability are considered in Section 2.2. In Section 2.3, we present the notions of local and global bifurcations, bifurcation diagram, bifurcation set, and structural stability and discuss local bifurcations of equilibrium solutions in the context of differential equations. Many examples are used to explain the different concepts. Furthermore, in Section 2.3, we illustrate how the methods of center-manifold reduction and multiple scales can be used as simplification methods for bifurcation analysis. Local bifurcations of fixed points of maps and their consequences are examined in Section 2.4.

2.1 CONTINUOUS–TIME SYSTEMS

In the case of the autonomous system

$$\dot{x} = \mathbf{F}(x; \mathbf{M}) \tag{2.1.1}$$

the fixed points are defined by the vanishing of the vector field; that is,

$$\mathbf{F}(x; \mathbf{M}) = \mathbf{0} \tag{2.1.2}$$

35

A location in the state space where this condition is satisfied is called a **singular point**. At such a point, the integral curve of the vector field $\mathbf{F}$ corresponds to the point itself. Also, an orbit of a fixed point is the fixed point itself. Fixed points are also called **stationary solutions, critical points, constant solutions**, and sometimes **steady–state solutions**. Physically, a **fixed point corresponds to an equilibrium position of a system**. Further, fixed points are examples of invariant sets of (2.1.1).

2.1.1 Linearization Near an Equilibrium Solution

Let the solution of (2.1.2) for $\mathbf{M} = \mathbf{M}_0$ be $\mathbf{x}_0$, where $\mathbf{x}_0 \in \mathcal{R}^n$ and $\mathbf{M}_0 \in \mathcal{R}^m$. To determine the stability of this equilibrium solution, we superimpose on it a small disturbance $\mathbf{y}$ and obtain

$$\mathbf{x}(t) = \mathbf{x}_0 + \mathbf{y}(t) \tag{2.1.3}$$

Substituting (2.1.3) into (2.1.1) yields

$$\dot{\mathbf{y}} = \mathbf{F}(\mathbf{x}_0 + \mathbf{y}; \mathbf{M}_0) \tag{2.1.4}$$

We note that the fixed point $\mathbf{x} = \mathbf{x}_0$ of (2.1.1) has been transformed into the fixed point $\mathbf{y} = \mathbf{0}$ of (2.1.4). Assuming that $\mathbf{F}$ is at least $\mathcal{C}^2$, expanding (2.1.4) in a Taylor series about $\mathbf{x}_0$, and retaining only linear terms in the disturbance leads to

$$\dot{\mathbf{y}} = \mathbf{F}(\mathbf{x}_0; \mathbf{M}_0) + D_{\mathbf{x}}\mathbf{F}(\mathbf{x}_0; \mathbf{M}_0)\mathbf{y} + O(\| \mathbf{y} \|^2)$$

or

$$\dot{\mathbf{y}} \approx D_{\mathbf{x}}\mathbf{F}(\mathbf{x}_0; \mathbf{M}_0) = {}^{\backprime}\mathbf{y} \tag{1.5}$$

where A, the matrix of first partial derivatives, is called the **Jacobian matrix**. If the components of F are

$$F_1(x_1, x_2, \cdots, x_n), F_2(x_1, x_2, \cdots, x_n), \cdots, F_n(x_1, x_2, \cdots, x_n),$$

then

$$A = \begin{bmatrix} \frac{\partial F_1}{\partial x_1} & \frac{\partial F_1}{\partial x_2} & \cdot & \cdot & \cdot & \frac{\partial F_1}{\partial x_n} \\[2mm] \frac{\partial F_2}{\partial x_1} & \frac{\partial F_2}{\partial x_2} & \cdot & \cdot & \cdot & \frac{\partial F_2}{\partial x_n} \\[2mm] \cdot & \cdot & \cdot & \cdot & \cdot & \cdot \\ \cdot & \cdot & \cdot & \cdot & \cdot & \cdot \\ \cdot & \cdot & \cdot & \cdot & \cdot & \cdot \\ \frac{\partial F_n}{\partial x_1} & \frac{\partial F_n}{\partial x_2} & \cdot & \cdot & \cdot & \frac{\partial F_n}{\partial x_n} \end{bmatrix}$$

Next, we show that the eigenvalues of the constant matrix A provide information about the **local stability** of the fixed point $\mathbf{x}_0$. We say **local** because we have considered a small disturbance and linearized the vector field.

The solution of (2.1.5) that passes through the initial condition $\mathbf{y}_0 \in \mathcal{R}^n$ at time $t_0 \in \mathcal{R}$ can be expressed as

$$\mathbf{y}(t) = e^{(t-t_0)A}\mathbf{y}_0 \tag{2.1.6}$$

where

$$e^{(t-t_0)A} = \sum_{j=0}^{\infty} \frac{(t-t_0)^j}{j!} A^j$$

If the eigenvalues λ_i of the matrix A are distinct, then there exists a matrix P such that $P^{-1}AP = D$, where D is a diagonal matrix with entries $\lambda_1, \lambda_2, \cdots, \lambda_n$; that is,

$$D = \begin{bmatrix} \lambda_1 & 0 & \cdot & \cdot & \cdot & 0 \\ 0 & \lambda_2 & \cdot & \cdot & \cdot & 0 \\ \cdot & \cdot & \cdot & \cdot & \cdot & \cdot \\ \cdot & \cdot & \cdot & \cdot & \cdot & \cdot \\ 0 & 0 & \cdot & \cdot & \cdot & \lambda_n \end{bmatrix}$$

If the eigenvalues are complex, then the matrix P will also be complex. The columns of the matrix P are the right eigenvectors $\mathbf{p}_1, \mathbf{p}_2, \cdots, \mathbf{p}_n$ of the matrix A corresponding to the eigenvalues $\lambda_1, \lambda_2, \cdots, \lambda_n$; that is, $P = [\mathbf{p}_1\, \mathbf{p}_2\, \cdots\, \mathbf{p}_n]$. Hence,

$$AP = [A\mathbf{p}_1\, A\mathbf{p}_2\, \cdots\, A\mathbf{p}_n] = [\lambda_1\mathbf{p}_1\, \lambda_2\mathbf{p}_2\, \cdots\, \lambda_n\mathbf{p}_n] = PD$$

Consequently,

$$D = P^{-1}AP$$

Introducing the transformation $\mathbf{y} = P\mathbf{v}$ into (2.1.5), we obtain

$$P\dot{\mathbf{v}} = AP\mathbf{v} \quad \text{or} \quad \dot{\mathbf{v}} = D\mathbf{v}$$

Hence,

$$\mathbf{v} = e^{(t-t_0)D}\mathbf{v}_0$$

where $\mathbf{v}_0 = \mathbf{v}(t_0) = P^{-1}\mathbf{y}_0$. In terms of $\mathbf{y}$, this solution becomes

$$\mathbf{y}(t) = Pe^{(t-t_0)D}P^{-1}\mathbf{y}_0 \tag{2.1.7}$$

The matrix $e^{(t-t_0)D}$ is a diagonal matrix with entries $e^{(t-t_0)\lambda_i}$. Hence, the eigenvalues of A are also known as the **characteristic exponents** associated with $\mathbf{F}$ at $(\mathbf{x}_0, \mathbf{M}_0)$.

If the eigenvalues of A are not distinct, then there exists a matrix P such that $P^{-1}AP = J$ is a **Jordan canonical form** with off–diagonal entries; that is,

$$J = \begin{bmatrix} J_1 & \phi & \cdot & \cdot & \cdot & \phi \\ \phi & J_2 & \cdot & \cdot & \cdot & \phi \\ \cdot & & \cdot & \cdot & \cdot & \cdot \\ \cdot & & & \cdot & \cdot & \cdot \\ \cdot & & & & \cdot & \cdot \\ \phi & \phi & \cdot & \cdot & \cdot & J_k \end{bmatrix}$$

where ϕ represents a matrix with zero entries and

$$J_m = \begin{bmatrix} \lambda_m & 1 & 0 & \cdot & \cdot & \cdot & 0 \\ 0 & \lambda_m & 1 & \cdot & \cdot & \cdot & 0 \\ 0 & 0 & \lambda_m & \cdot & \cdot & \cdot & \cdot \\ \cdot & \cdot & \cdot & \cdot & & & \cdot \\ 0 & 0 & 0 & \cdot & \cdot & \cdot & \lambda_m \end{bmatrix}$$

In writing the matrix J, we have assumed that A has k distinct eigenvalues. Further, let the (algebraic) multiplicity of the mth eigenvalue λ_m be n_m. Then, the matrix J_m corresponding to the eigenvalue λ_m differs from the diagonal matrix D due to the presence of the elements 1 above the diagonal elements. In this case, the columns $\mathbf{p}_i$ of the matrix

P are the **generalized eigenvectors** corresponding to the eigenvalues λ_i of the matrix A. There are n_m generalized eigenvectors corresponding to the eigenvalue λ_m. These vectors are the nonzero solutions of

$$(A - \lambda_m \mathbf{I})\,\mathbf{p} = \mathbf{0}, \; (A - \lambda_m \mathbf{I})^2\,\mathbf{p} = \mathbf{0}, \; \cdots, \; (A - \lambda_m \mathbf{I})^{n_m}\,\mathbf{p} = \mathbf{0}$$

For an $n \times n$ matrix with n distinct eigenvalues, the generalized eigenvectors are also the eigenvectors of the matrix. The components of $\mathbf{v}$ have terms of the form $t^k e^{(t-t_0)\lambda_i}$, where the integer k depends on the multiplicity n_i of the eigenvalue λ_i.

2.1.2   Classification and Stability of Equilibrium Solutions

When all of the eigenvalues of A have nonzero real parts, the corresponding fixed point is called a **hyperbolic fixed point**, irrespective of the values of the imaginary parts; otherwise, it is called a **nonhyperbolic fixed point**.

There are three types of hyperbolic fixed points: **sinks**, **sources**, and **saddle points**. If all of the eigenvalues of A have negative real parts, then all of the components of the disturbance $\mathbf{y}$ decay in time, and hence $\mathbf{x}$ approaches the fixed point $\mathbf{x}_0$ of (2.1.1) as $t \to \infty$. Therefore, the fixed point $\mathbf{x}_0$ of (2.1.1) is asymptotically stable according to Section 1.4.2. An asymptotically stable fixed point is called a **sink**. If the matrix A associated with a sink has complex eigenvalues, the sink is also called a **stable focus**. On the other hand, if all of the eigenvalues of the matrix A associated with a sink are real, the sink is also called a **stable node**. A sink is stable in forward time (i.e., $t \to \infty$) but unstable in reverse time (i.e., $t \to -\infty$). Further, all sinks qualify as attractors.

If one or more of the eigenvalues of A have positive real parts, some of the components of $\mathbf{y}$ grow in time, and $\mathbf{x}$ moves away from the fixed point $\mathbf{x}_0$ of (2.1.1) as t increases. In this case, the fixed point $\mathbf{x}_0$ is said to be unstable. When all of the eigenvalues of A have positive real parts, $\mathbf{x}_0$ is said to be a **source**. If the matrix A associated with a source has complex eigenvalues, the source is also called an **unstable focus**. On the other hand, if all of the eigenvalues of the matrix A associated with a source are real, the source is also called an **unstable node.** A source is unstable in forward time but stable in reverse time. Because trajectories move away from a source in forward time, the source is an example of a repellor.

When some, but not all, of the eigenvalues have positive real parts while the rest of the eigenvalues have negative real parts, the associated fixed point is called a **saddle point.** Because a saddle point is unstable in both forward and reverse times, some authors call it a **nonstable fixed point** (e.g., Parker and Chua, 1989).

Next, we address nonhyperbolic fixed points. A nonhyperbolic fixed point is unstable if one or more of the eigenvalues of A have positive real parts. If some of the eigenvalues of A have negative real parts while the rest of the eigenvalues have zero real parts, the fixed point $\mathbf{x} = \mathbf{x}_0$ of (2.1.1) is said to be **neutrally** or **marginally stable.** If all of the eigenvalues of A are purely imaginary and nonzero, the corresponding fixed point is called a **center.**

Example 2.1.   For illustration, we consider the classification of the fixed points $(0,0), (-1,0)$, and $(1,0)$ of (1.4.1) and (1.4.2). In the vicinity of a fixed point, we obtain the following system after linearization:

$$\dot{\mathbf{y}} = \begin{bmatrix} 0 & 1 \\ -1 + 3x_1^2 & -2\mu \end{bmatrix} \mathbf{y} \qquad (2.1.8)$$

Hence, the eigenvalues of the Jacobian matrix are

$$\lambda_1 = -\mu - \sqrt{\mu^2 - 1 + 3x_1^2} \quad \text{and} \quad \lambda_2 = -\mu + \sqrt{\mu^2 - 1 + 3x_1^2} \qquad (2.1.9)$$

For all three fixed points, both of the eigenvalues have nonzero real parts when $\mu \neq 0$. Hence, all three fixed points are hyperbolic fixed points.

In the vicinity of the fixed point $(0,0)$, (2.1.8) and (2.1.9) become

$$\dot{\mathbf{y}} = \begin{bmatrix} 0 & 1 \\ -1 & -2\mu \end{bmatrix} \mathbf{y} \qquad (2.1.10)$$

and

$$\lambda_1 = -\mu - \sqrt{\mu^2 - 1} \quad \text{and} \quad \lambda_2 = -\mu + \sqrt{\mu^2 - 1} \qquad (2.1.11)$$

respectively. We conclude from (2.1.11) that the fixed point $(0,0)$ is a center when $\mu = 0$, an unstable node when $\mu \leq -1$, an unstable focus when $-1 < \mu < 0$, a stable focus when $0 < \mu < 1$, and a stable node when $\mu \geq 1$. In Figures 2.1.1a–c, we show phase portraits in the vicinity of the origin of the $x_2 - x_1$ space when the origin is an unstable focus, a center, and a stable focus, respectively. A positive orbit spirals away from a neighborhood of the unstable focus in Figure 2.1.1a, and a positive orbit spirals into the stable focus in Figure 2.1.1c. The orbit of Figure 2.1.1b, which corresponds to a periodic solution, closes on itself.

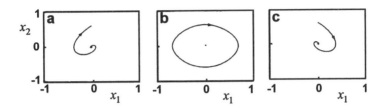

Figure 2.1.1: Phase portraits in the vicinity of the origin of (1.4.1) and (1.4.2): (a) $\mu = -0.4$, (b) $\mu = 0$, and (c) $\mu = 0.4$.

In the vicinity of either the fixed point $(-1,0)$ or the fixed point $(1,0)$, (2.1.8) and (2.1.9) become

$$\dot{\mathbf{y}} = \begin{bmatrix} 0 & 1 \\ 2 & -2\mu \end{bmatrix} \mathbf{y} \qquad (2.1.12)$$

and

$$\lambda_1 = -\mu - \sqrt{\mu^2 + 2} \text{ and } \lambda_2 = -\mu + \sqrt{\mu^2 + 2} \qquad (2.1.13)$$

respectively. We conclude from (2.1.13) that the fixed points $(-1,0)$ and $(1,0)$ are saddles for any value of μ because there is always one eigenvalue that is a positive real number and another eigenvalue that is a negative real number. In Figures 2.1.2a and 2.1.2b, we show the flow in the vicinity of $(0,0)$ of (2.1.12) and the flow in the vicinity of $(1,0)$ of (1.4.1) and (1.4.2), respectively. In both of these figures, we note that the saddle point attracts two positive orbits and repels two positive orbits.

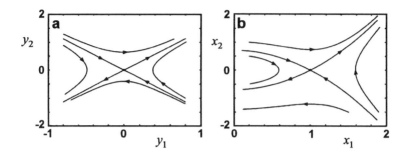

Figure 2.1.2: (a) Flow in the vicinity of the saddle point $(0,0)$ of the linear system (2.1.12) and (b) flow in the vicinity of the saddle point $(1,0)$ of the nonlinear system (1.4.1) and (1.4.2). Both the flows are qualitatively similar.

Many theorems provide precise statements on what the stability of fixed–point solutions of the linearized system (2.1.5) imply for the stability of fixed–point solutions of the full nonlinear system (2.1.1). The **Hartman–Grobman theorem** (e.g., Arnold, 1988, Chapter 3; Wiggins, 1990, Chapter 2) is applicable to hyperbolic fixed points, whereas the **Shoshitaishvili theorem** (e.g., Arnold, 1988, Chapter 6) is applicable to nonhyperbolic fixed points. From these theorems, it follows that (a) the fixed point $\mathbf{x} = \mathbf{x}_0$ of the nonlinear system (2.1.1) is stable when the fixed point $\mathbf{y} = \mathbf{0}$ of the linear system (2.1.5) is asymptotically stable; (b) the fixed point $\mathbf{x} = \mathbf{x}_0$ of the nonlinear system (2.1.1) is unstable when the fixed point $\mathbf{y} = \mathbf{0}$ of the linear system (2.1.5) is unstable; and (c) linearization cannot determine the stability of neutrally stable fixed points (including centers) of (2.1.1). In the case of neutrally stable fixed points, a nonlinear analysis is necessary to determine the stability of $\mathbf{x}_0$. It will be necessary to retain quadratic and, sometimes, higher–order terms in the disturbance $\mathbf{y}$ in the Taylor–series expansion of (2.1.4).

In a topological setting, the **Hartman–Grobman theorem** implies that the trajectories in the vicinity of a hyperbolic fixed point $\mathbf{x} = \mathbf{x}_0$ of (2.1.1) are qualitatively similar to those in the vicinity of the hyperbolic fixed point $\mathbf{y} = \mathbf{0}$ of (2.1.5). In other words, the local nonlinear dynamics near $\mathbf{x} = \mathbf{x}_0$ is qualitatively similar to the linear dynamics near $\mathbf{y} = \mathbf{0}$, and a qualitative change in the local nonlinear dynamics can be detected by examining the associated linear dynamics. In Figure 2.1.2, we observe qualitative similarities between the flow near the hyperbolic fixed point $(0,0)$ of the linear system $(2.1.12)$ and the flow near the hyperbolic fixed point $(1,0)$ of the corresponding nonlinear system $(1.4.1)$ and $(1.4.2)$.

According to the Hartman–Grobman theorem, there exists a continuous coordinate transformation (i.e., a homeomorphism) that transforms the nonlinear flow into the linear flow in the vicinity of a hyperbolic fixed point. In some cases, the method of normal forms (Nayfeh, 1993) may be used to generate a coordinate transformation to transform the nonlinear flow into the linear flow (e.g., Arnold, 1988; Guckenheimer and Holmes, 1983; Nayfeh, 1993). Further, such a coordinate transformation would be a differentiable one because the method of normal forms yields transformations in the form of power–series expansions. Next, we consider an example to illustrate the situations in which the method of normal forms cannot be used to produce a coordinate transformation.

Example 2.2. We consider the planar system

$$\left\{ \begin{array}{c} \dot{x}_1 \\ \dot{x}_2 \end{array} \right\} = \left[\begin{array}{cc} \lambda_1 & 0 \\ 0 & \lambda_2 \end{array} \right] \left\{ \begin{array}{c} x_1 \\ x_2 \end{array} \right\} + \epsilon \left\{ \begin{array}{c} \alpha_1 x_1^2 + \alpha_2 x_1 x_2 + \alpha_3 x_2^2 \\ \alpha_4 x_1^2 + \alpha_5 x_1 x_2 + \alpha_6 x_2^2 \end{array} \right\} \quad (2.1.14)$$

where λ_1 and λ_2 are different from zero, ϵ is a small positive parameter, and the α_i are independent of ϵ. The system $(2.1.14)$ has a hyperbolic fixed point at $(x_1, x_2) = (0,0)$. In the vicinity of this fixed point, the method of normal forms can be used to transform $(2.1.14)$ into (Nayfeh, 1993, Chapter 2)

$$\left\{ \begin{array}{c} \dot{y}_1 \\ \dot{y}_2 \end{array} \right\} = \left[\begin{array}{cc} \lambda_1 & 0 \\ 0 & \lambda_2 \end{array} \right] \left\{ \begin{array}{c} y_1 \\ y_2 \end{array} \right\} \quad (2.1.15)$$

by using the coordinate transformation

$$\left\{ \begin{array}{c} x_1 \\ x_2 \end{array} \right\} = \left\{ \begin{array}{c} y_1 \\ y_2 \end{array} \right\} + \epsilon \left\{ \begin{array}{c} h_{11}(y) \\ h_{12}(y) \end{array} \right\} + \cdots \quad (2.1.16)$$

where

$$\left\{ \begin{array}{c} h_{11} \\ h_{12} \end{array} \right\} = \left[\begin{array}{ccc} \Gamma_1 & \Gamma_2 & \Gamma_3 \\ \Gamma_4 & \Gamma_5 & \Gamma_6 \end{array} \right] \left\{ \begin{array}{c} y_1^2 \\ y_1 y_2 \\ y_2^2 \end{array} \right\} \qquad (2.1.17)$$

and the Γ_i are given by

$$\Gamma_1 = \frac{\alpha_1}{\lambda_1}, \ \Gamma_2 = \frac{\alpha_2}{\lambda_2}, \ \Gamma_3 = \frac{\alpha_3}{2\lambda_2 - \lambda_1}$$
$$\Gamma_4 = \frac{\alpha_4}{2\lambda_1 - \lambda_2}, \ \Gamma_5 = \frac{\alpha_5}{\lambda_1}, \ \Gamma_6 = \frac{\alpha_6}{\lambda_2} \qquad (2.1.18)$$

We note from (2.1.17) and (2.1.18) that the transformation (2.1.16) breaks down when either $\lambda_2 \approx 2\lambda_1$ or $\lambda_1 \approx 2\lambda_2$. These two relationships are examples of **resonances**. Hence, in the presence of either of these resonances, the method of normal forms cannot be used to transform (2.1.14) into (2.1.15).

Coddington and Levinson (1955, Chapter 13) present theorems for the stability of fixed points of the system

$$\dot{\mathbf{x}} = A\mathbf{x} + \mathbf{F}(\mathbf{x}, t) \qquad (2.1.19)$$

where A is an $n \times n$ constant matrix and the vector function $\mathbf{F}$ is continuous in $\mathbf{x}$ and t and **Lipschitz continuous** in $\mathbf{x}$ and is such that $\mathbf{F}$ is $o(\| \mathbf{x} \|)$ as $\| \mathbf{x} \| \to 0$. [The vector function $\mathbf{F}(\mathbf{x}, t)$ is said to be **Lipschitz continuous** in $\mathbf{x}$ and to satisfy a **Lipschitz condition** in the $(n+1)$–dimensional space $\mathcal{R}^n \times \mathcal{R}^1$ of $(\mathbf{x}, t)$ if there exists a positive constant K such that

$$\| \mathbf{F}(\mathbf{y}, t) - \mathbf{F}(\mathbf{z}, t) \| \leq K \| \mathbf{y} - \mathbf{z} \| \qquad (2.1.20)$$

for all $(\mathbf{y}, t)$ and $(\mathbf{z}, t)$ in the $(n+1)$–dimensional space. The constant K is called a **Lipschitz constant** for $\mathbf{F}$. If the scalar components of $\mathbf{F}$ have continuous and bounded first partial derivatives with respect to the scalar components of $\mathbf{x}$ in $D \subset \mathcal{R}^n \times \mathcal{R}^1$, it can be shown that the function $\mathbf{F}$ satisfies a Lipschitz condition in D (Coddington and Levinson, 1955).] The autonomous system (2.1.1) is a special case of (2.1.19).

The **Poincaré–Lyapunov theorem** (e.g., Sanders and Verhulst, 1985, Chapter 1) is concerned with the stability of the fixed point $\mathbf{x} = \mathbf{0}$

of (2.1.19) when

$$\mathbf{F} = \mathbf{B}(t)\mathbf{x} + \mathbf{g}(\mathbf{x}, t)$$

where

$$\lim_{t \to \infty} \| \mathbf{B} \| = 0 \quad \text{and} \quad \| \mathbf{g} \| = o(\| \mathbf{x} \|) \quad \text{as} \quad \| \mathbf{x} \| \to 0$$

Although (2.1.19) is nonautonomous, the above assumption about $\mathbf{g}$ ensures that $\mathbf{x} = \mathbf{0}$ is a fixed point of (2.1.19). For a general $\mathbf{F}$, $\mathbf{x} = \mathbf{0}$ is not a fixed point of (2.1.19). According to the Poincaré–Lyapunov theorem, if all of the eigenvalues of the matrix A have negative real parts, then the fixed point $\mathbf{x} = \mathbf{0}$ of (2.1.19) is asymptotically stable.

Example 2.3. We consider the stability of the fixed point $\mathbf{x} = \mathbf{0}$ of the system

$$\dot{x}_1 = x_2$$

$$\dot{x}_2 = -2\mu x_2 - \sin x_1$$

where $\mu > 0$. To verify if the conditions for applying the Poincaré–Lyapunov theorem are satisfied, we rewrite this planar system in the following form:

$$\left\{ \begin{array}{c} \dot{x}_1 \\ \dot{x}_2 \end{array} \right\} = \left[\begin{array}{cc} 0 & 1 \\ -1 & -2\mu \end{array} \right] \left\{ \begin{array}{c} x_1 \\ x_2 \end{array} \right\} + \left\{ \begin{array}{c} 0 \\ x_1 - \sin x_1 \end{array} \right\}$$

In this case, the eigenvalues of the 2×2 matrix A are

$$\lambda_1 = -\mu - \sqrt{\mu^2 - 1} \quad \text{and} \quad \lambda_2 = -\mu + \sqrt{\mu^2 - 1}$$

Because both of the eigenvalues have negative real parts, according to the Poincaré–Lyapunov theorem, the fixed point $\mathbf{x} = \mathbf{0}$ of the planar system is asymptotically stable.

We note that **although linearization helps in determining whether a hyperbolic fixed point of (2.1.1) is stable, it does not provide any information regarding the size of the domain around the fixed point of (2.1.1) where the conclusion of stability holds.** As noted earlier, this domain is called the **stability region** of the fixed point. If it is possible to construct a Lyapunov function for a given problem, then this function can be used to determine the whole stability region or a subset of this region (e.g., Michel, Miller, and Nam, 1982; Michel, Nam, and Vittal, 1984). Chiang, Hirsch, and Wu (1988) provide analytical results for determining the stability regions of fixed points of autonomous systems satisfying certain generic conditions. The stability region of a fixed point can also be numerically determined by conducting simulations of (2.1.1) for different initial conditions.

Next, we present two examples to illustrate why a nonlinear analysis is necessary to determine the stability of neutrally stable fixed points and centers.

Example 2.4. We consider the stability of the fixed point $\mathbf{x} = \mathbf{0}$ of the system

$$\left\{ \begin{array}{c} \dot{x}_1 \\ \dot{x}_2 \end{array} \right\} = \left[\begin{array}{cc} 0 & 1 \\ -1 & 0 \end{array} \right] \left\{ \begin{array}{c} x_1 \\ x_2 \end{array} \right\} - \left\{ \begin{array}{c} 0 \\ \mu x_2^3 \end{array} \right\} \qquad (2.1.21)$$

The fixed point $\mathbf{x} = \mathbf{0}$ is a center in the linearization of (2.1.21) because the associated eigenvalues are $+i$ and $-i$. The influence of the nonlinear term on the stability of this fixed point is depicted in Figures 2.1.3a and 2.1.3b for $\mu > 0$ and $\mu < 0$, respectively. The origin in Figure 2.1.3a attracts positive orbits whereas the origin in Figure 2.1.3b repels positive orbits. Hence, the fixed point $\mathbf{x} = \mathbf{0}$ of (2.1.21) is stable when $\mu > 0$ and unstable when $\mu < 0$.

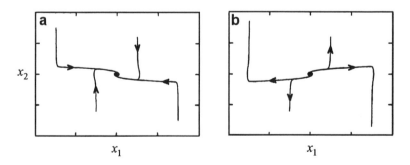

Figure 2.1.3: Phase portraits of (2.1.21): (a) $\mu > 0$ and (b) $\mu < 0$.

Example 2.5. We consider the system

$$\left\{\begin{array}{c} \dot{x} \\ \dot{y} \end{array}\right\} = \left[\begin{array}{cc} 0 & 0 \\ 0 & -1 \end{array}\right]\left\{\begin{array}{c} x \\ y \end{array}\right\} + \left\{\begin{array}{c} kxy \\ x^2 \end{array}\right\} \qquad (2.1.22)$$

The fixed point $(0,0)$ is a neutrally stable fixed point in the linearization of (2.1.22) because the associated eigenvalues are 0 and -1. In Figures 2.1.4a and 2.1.4b, we show the flow in the vicinity of this fixed point for $k = 1$ and $k = -1$, respectively. In both figures, positive orbits initiated on the line $x = 0$ are attracted to the origin. However, in Figure 2.1.4a, some positive orbits initiated in the neighborhood of the origin are repelled, indicating that the origin is an unstable fixed point of (2.1.22) when $k > 0$. All the positive orbits shown in Figure 2.1.4b are attracted to the origin, indicating that the origin is a stable fixed point of (2.1.22) when $k < 0$.

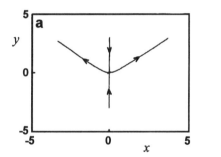 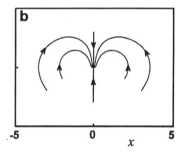

Figure 2.1.4: Phase portraits of (2.1.22): (a) $k = 1$ and (b) $k = -1$.

We determined the orbits shown in Figures 2.1.3 and 2.1.4 through numerical integrations of (2.1.21) and (2.1.22), respectively.

2.1.3   Eigenspaces and Invariant Manifolds

Let the Jacobian matrix corresponding to a fixed point of an n–dimensional autonomous system have s eigenvalues with negative real parts, u eigenvalues with positive real parts, and c eigenvalues with zero parts. Considering the eigenvectors associated with these eigenvalues, we can represent the space $\mathcal{R}^n$ as the direct sum of the three subspaces E^s, E^u, and E^c defined by

$$E^s = \text{span } \{\mathbf{p}_1, \mathbf{p}_2, \cdots, \mathbf{p}_s\}$$
$$E^u = \text{span } \{\mathbf{p}_{s+1}, \mathbf{p}_{s+2}, \cdots, \mathbf{p}_{s+u}\}$$
$$E^c = \text{span } \{\mathbf{p}_{s+u+1}, \mathbf{p}_{s+u+2}, \cdots, \mathbf{p}_{s+u+c}\}$$

where $\mathbf{p}_1, \mathbf{p}_2, \cdots, \mathbf{p}_s$ are the s (generalized) eigenvectors whose corresponding eigenvalues have negative real parts; $\mathbf{p}_{s+1}, \mathbf{p}_{s+2}, \cdots, \mathbf{p}_{s+u}$ are the u (generalized) eigenvectors whose corresponding eigenvalues have positive real parts; and $\mathbf{p}_{s+u+1}, \mathbf{p}_{s+u+2}, \cdots, \mathbf{p}_{s+u+c}$ are the c (generalized) eigenvectors whose corresponding eigenvalues have zero real parts.

Of each s eigenvalues with negative real parts, let 2ℓ eigenvalues be complex conjugates of each other. The eigenvectors corresponding to the 2ℓ complex eigenvalues will be complex and will lie in a complex subspace. However, the real and imaginary parts of each complex eigenvector lie in a real subspace and can be used in forming the basis of this space (e.g., Arnold, 1973). A pair of complex conjugate eigenvectors would correspond to two real vectors, one formed from the real part and the other from the imaginary part. Hence, for the 2ℓ eigenvalues, we have 2ℓ real vectors. These vectors and the generalized eigenvectors corresponding to the real eigenvalues form the basis of E^s.

The spaces E^s, E^u, and E^c are **invariant subspaces** of the corresponding linear system. A solution of the linear system initiated in an invariant subspace remains in this subspace for all times. Thus, solutions initiated in E^s approach the fixed point as $t \to \infty$, solutions initiated in E^u approach the fixed point as $t \to -\infty$, and solutions initiated in E^c neither grow nor decay in time. The subspaces E^s, E^u, and E^c are called **stable**, **unstable**, and **center subspaces** or **manifolds**, respectively, of the considered fixed point of the linear system.

Example 2.6. For illustration, we consider the linear system (2.1.10) obtained through a linearization of (1.4.1) and (1.4.2) in the vicinity of its fixed point $(0,0)$. From (2.1.10) and (2.1.11), we find that the eigenvectors of the Jacobian matrix are

$$\mathbf{p}_1 = \left\{ \begin{array}{c} 1 \\ \lambda_1 \end{array} \right\} \text{ and } \mathbf{p}_2 = \left\{ \begin{array}{c} 1 \\ \lambda_2 \end{array} \right\} \qquad (2.1.23)$$

and the solution of (2.1.10) can be expressed as

$$\mathbf{y} = c_1 e^{\lambda_1 t} \mathbf{p}_1 + c_2 e^{\lambda_2 t} \mathbf{p}_2 \qquad (2.1.24)$$

where the c_i are determined by the initial condition.

When $\mu = 0$, the eigenspace E^c of $(0,0)$ is the space spanned by $\mathbf{p}_1$ and $\mathbf{p}_2$, and the eigenspaces E^s and E^u of $(0,0)$ are empty. We conclude from (2.1.24) that a solution of (2.1.10) initiated in the center eigenspace of $(0,0)$ remains in this subspace for all times. The $y_2 - y_1$ plane is the center eigenspace E^c of $(0,0)$. Similarly, we can also determine the different invariant subspaces of the fixed point $(0,0)$ of (2.1.10) when $\mu \neq 0$. When $\mu < 0$, the $y_2 - y_1$ plane is the unstable eigenspace of $(0,0)$ and the other eigenspaces of $(0,0)$ are empty. When $\mu > 0$, the $y_2 - y_1$ plane is the stable eigenspace of $(0,0)$ and the other eigenspaces of $(0,0)$ are empty.

Next, we consider (2.1.12) obtained through a linearization of (1.4.1) and (1.4.2) in the vicinity of its fixed point $(1,0)$. When $\mu = 0$, we find from (2.1.12) and (2.1.13) that the eigenvalues and eigenvectors of the Jacobian matrix are

$$\lambda_1 = -\sqrt{2}, \ \mathbf{p}_1 = \left\{ \begin{array}{c} 1 \\ -\sqrt{2} \end{array} \right\} \text{ and } \lambda_2 = \sqrt{2}, \ \mathbf{p}_2 = \left\{ \begin{array}{c} 1 \\ \sqrt{2} \end{array} \right\} \qquad (2.1.25)$$

The solution of (2.1.12) is given by (2.1.24), where the λ_i and $\mathbf{p}_i$ are specified by (2.1.25).

Here, the eigenspace E^s of $(0,0)$ is spanned by $\mathbf{p}_1$ and is a one–dimensional manifold, the eigenspace E^u of $(0,0)$ is spanned by $\mathbf{p}_2$ and is a one–dimensional manifold, and the eigenspace E^c of $(0,0)$ is empty. If we start from an initial condition in E^s, then $c_2 = 0$ in (2.1.24) and we remain in E^s for all times. Similarly, if we start from an initial condition in E^u, then $c_1 = 0$ in (2.1.24) and we remain in E^u for all times. In Figure 2.1.5a, the subspaces E^s and E^u associated with the fixed point $(0,0)$ of (2.1.12) are depicted as broken lines in the $y_2 - y_1$ space. The

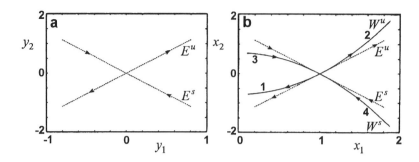

Figure 2.1.5: (a) Stable and unstable eigenspaces of the fixed point $(0,0)$ of (2.1.12) and (b) stable and unstable manifolds of the fixed point $(1,0)$ of (1.4.1) and (1.4.2).

arrows on E^s and E^u indicate the direction of evolution in forward time. In Figure 2.1.5b, the solid lines are numerically determined solutions of (1.4.1) and (1.4.2) when $\mu = 0$. These solutions approach the fixed point $(1,0)$ in either forward or reverse time. At $(1,0)$, we note that the curves W^s and W^u intersect each other and are tangent to the subspaces E^s and E^u, respectively. We note that the union of the trajectories 1 and 2 of Figure 2.1.5b constitutes the curve W^u. Similarly, the union of the trajectories 3 and 4 of Figure 2.1.5b constitutes the curve W^s. The curves W^s and W^u are called the **stable** and **unstable manifolds** of the fixed point $(1,0)$ of (1.4.1) and (1.4.2), respectively.

The **stable manifold** of a fixed point of (2.1.1) is the set of all initial conditions such that the flow initiated at these points asymptotically approaches the fixed point as $t \to \infty$, whereas the **unstable manifold** of a fixed point of (2.1.1) is the set of all initial conditions such that the flow initiated at these points asymptotically approaches the fixed point as $t \to -\infty$. In a nonlinear system, a stable manifold is denoted by W^s, an unstable manifold is denoted by W^u, and a center manifold is denoted by W^c.

Let the vector field of a nonlinear system described by (2.1.1) be $\mathcal{C}^r$, where $r \geq 2$. Then, there are existence theorems that state that if the fixed point $\mathbf{y} = \mathbf{0}$ of the linear system (2.1.5) has s–dimensional stable, u–dimensional unstable, and c–dimensional center manifolds, then the fixed point $\mathbf{x} = \mathbf{x}_0$ of (2.1.1) also has s–dimensional stable, u–dimensional unstable, and c–dimensional center manifolds in a neighborhood of the fixed point. At $x = x_0$, the manifolds associated with the nonlinear system intersect each other and are tangent to their respective invariant subspaces of $y = 0$ of (2.1.5) (Kelley, 1967; Carr, 1981; Guckenheimer and Holmes, 1983). Because these theorems guarantee the existence of manifolds only in a neighborhood of x_0, these manifolds are **local invariant manifolds**. Further, the manifolds W^s and W^u of $x = x_0$ of (2.1.1) have the asymptotic properties of E^s and E^u of $y = 0$, respectively, in the neighborhood of x_0.

In Figure 2.1.5b, as mentioned earlier, we show numerically determined stable and unstable manifolds of the saddle point $(1, 0)$ of the nonlinear system (1.4.1) and (1.4.2). The stable manifold W^s of $(1, 0)$ is one–dimensional, and the unstable manifold W^u of $(1, 0)$ is also one–dimensional. At the saddle, these manifolds are tangent to their corresponding eigenvectors of the linearized system. As we move away from the saddle, the manifolds are no longer tangent to these eigenvectors. Here, to determine the unstable manifold, we chose initial conditions on E^u "close" to the saddle point and computed the positive orbits labeled 1 and 2 in Figure 2.1.5b through numerical integrations. Similarly, to determine the stable manifold of the saddle point, we chose initial conditions on E^s "close" to the considered saddle point and computed the negative orbits labeled 3 and 4 in Figure 2.1.5b through numerical integrations.

Example 2.7. We consider the manifolds of the fixed point $(0, 0)$ of (2.1.22). The center and stable eigenspaces of this fixed point are the one–dimensional manifolds $y = 0$ and $x = 0$, respectively. The numerical results displayed in Figure 2.1.4 indicate that the line $x = 0$ is also the global stable manifold W^s. The global center manifold W^c, which is one–dimensional and tangent to the x axis at the origin, is addressed in Section 2.3.4.

In the next two examples, we introduce some new notions and further discuss the notions of global stable and unstable manifolds of a fixed point.

Example 2.8. We consider the two–dimensional system

$$\dot{x}_1 = x_2 \tag{2.1.26}$$
$$\dot{x}_2 = -(\omega_0^2 x_1 + \alpha_3 x_1^3 + \alpha_5 x_1^5) - (2\mu_1 x_2 + \mu_3 x_2^3) \tag{2.1.27}$$

Equations (2.1.26) and (2.1.27) have been used to study the roll motions of a ship by Wright and Marshfield (1979), Nayfeh and Khdeir (1986a,b), Nayfeh and Sanchez (1990), Falzarano (1990), Kreider (1992), Bikdash, Balachandran, and Nayfeh (1994), and in some of the references provided therein. In this system, the α_i are nonlinear coefficients, μ_1 is the linear damping coefficient, and μ_3 is the nonlinear damping coefficient.

Each fixed point (x_{10}, x_{20}) of this system satisfies $x_{20} = 0$ and

$$\omega_0^2 x_{10} + \alpha_3 x_{10}^3 + \alpha_5 x_{10}^5 = 0 \tag{2.1.28}$$

The solutions of (2.1.28) are

$$x_{10} = 0, \quad x_{10}^2 = \frac{-\alpha_3 \pm \sqrt{\alpha_3^2 - 4\omega_0^2 \alpha_5}}{2\alpha_5} \tag{2.1.29}$$

If $(\alpha_3^2 - 4\omega_0^2 \alpha_5) < 0$, (2.1.29) admits only the real root $x_{10} = 0$. Otherwise, there are either three or five real roots depending on the signs of the α_i in (2.1.29). The stability of a fixed point of (2.1.26) and (2.1.27) depends on the eigenvalues of the matrix

$$D_{\mathbf{x}}\mathbf{F} = \begin{bmatrix} 0 & 1 \\ -(\omega_0^2 + 3\alpha_3 x_{10}^2 + 5\alpha_5 x_{10}^4) & -(2\mu_1 + 3\mu_3 x_{20}^2) \end{bmatrix} \tag{2.1.30}$$

The eigenvalues λ_i are given by

$$\begin{aligned} \lambda_{1,2} = &-\frac{1}{2}(2\mu_1 + 3\mu_3 x_{20}^2) \\ &\pm \frac{1}{2}\sqrt{(2\mu_1 + 3\mu_3 x_{20}^2)^2 - 4(\omega_0^2 + 3\alpha_3 x_{10}^2 + 5\alpha_5 x_{10}^4)} \end{aligned} \tag{2.1.31}$$

Using the fact that $x_{20} = 0$, we simplify (2.1.31) to

$$\lambda_{1,2} = -\mu_1 \pm \sqrt{\mu_1^2 - (\omega_0^2 + 3\alpha_3 x_{10}^2 + 5\alpha_5 x_{10}^4)} \tag{2.1.32}$$

We infer from (2.1.32) that the eigenvalues do not depend on μ_3.

For the parameter values

$$\omega_0 = 5.278, \quad \alpha_3 = -1.402\omega_0^2, \quad \alpha_5 = 0.271\omega_0^2$$

(2.1.29) admits five real solutions. Consequently, we obtain the five fixed points $(-2.0782, 0.0), (-0.9243, 0.0), (0.0, 0.0), (0.9243, 0.0)$, and $(2.0872, 0.0)$. For convenience, we refer to these fixed points as A, B, C, D, and E, respectively.

A few numerically determined trajectories of the undamped system (i.e., $\mu_i = 0$) are plotted in Figure 2.1.6a, and a few numerically determined trajectories of the damped system when $\mu_1 > 0$ and $\mu_3 > 0$ are plotted in Figure 2.1.6b. Using (2.1.31), we find that the fixed points A, C, and E are centers in the undamped case and stable foci in the damped case. Further, the fixed points B and D are saddle points in both the undamped and damped cases. The trajectories do not approach the centers in the undamped case, but they spiral into the stable foci in the damped case. Each saddle repels two positive orbits and attracts two positive orbits, and the global manifolds of each saddle are determined by these orbits, as shown in Figure 2.1.6.

In Figures 2.1.5b, 2.1.6a, and 2.1.6b, the stable and unstable manifolds of each saddle point intersect **transversely** at the saddle point. At a **transversal intersection of manifolds**, the union of the tangent spaces of the intersecting manifolds span the whole space (Guillemin and Pollack, 1974). Here, at each saddle point, the tangent space to each manifold is a one–dimensional space, and the tangent spaces to the stable and unstable manifolds taken together span the two–dimensional space. In Figure 2.1.6a, we note that the global unstable manifold of the saddle point B intersects the global stable manifold of the saddle point D on orbit Γ_1. Such an intersection is an example of a **nontransversal intersection**.

In Figure 2.1.6a, we note that one of the ends of the global stable manifold of the saddle B merges with one of the ends of the global unstable manifold of B to form the closed orbit Γ_3. Similarly, one of the ends of the global stable manifold of the saddle D merges with one of the ends of the global unstable manifold of D to form the closed orbit Γ_4. Closed orbits, such as Γ_3 and Γ_4, which lead to the same saddle

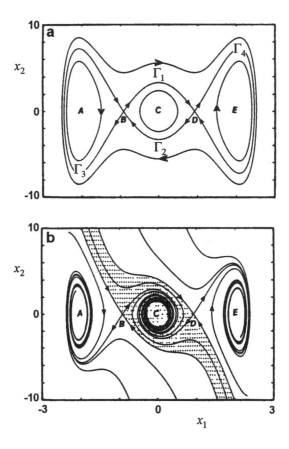

Figure 2.1.6: Phase portraits for the system (2.1.26) and (2.1.27) when $\omega_0 = 5.278, \alpha_3 = -1.402\omega_0^2$, and $\alpha_5 = 0.271\omega_0^2$: (a) $\mu_1 = \mu_3 = 0$ and (b) $\mu_1 = 0.086$ and $\mu_3 = 0.108$.

point in forward and reverse times, are called **homoclinic orbits**. We observe that the trajectories Γ_1 and Γ_2 run between the saddle points B and D. Such trajectories that run toward different saddle points in forward and reverse times are called **heteroclinic half–orbits**. A closed orbit, such as that formed by the union of Γ_1 and Γ_2, is called a **heteroclinic orbit**.

The heteroclinic and homoclinic orbits of Figure 2.1.6a separate the regions of qualitatively different oscillations. The heteroclinic orbit separates the region where oscillations occur only about C from the other regions. The homoclinic orbit Γ_3 separates the region where oscillations occur only about A from the other regions. Similarly, the homoclinic orbit Γ_4 separates the region where oscillations occur only about E from the other regions. Hence, the heteroclinic and homoclinic orbits of Figure 2.1.6a are called **separatrices**. In general, we use the word **separatrices** for curves and surfaces that separate regions of qualitatively different motions.

In the phase portrait of the damped system shown in Figure 2.1.6b, we do not have any homoclinic or heteroclinic orbits. The inclusion of damping breaks up the homoclinic and heteroclinic orbits of Figure 2.1.6a. In Figure 2.1.6b, the global stable manifolds of the saddles B and D separate the basins of attraction of the point attractors A, C, and E. In other words, the stable manifolds of B and D define the **boundaries of basins of attraction** or **stability boundaries** of A, C, and E. The region to the right of the stable manifold of the saddle D is the basin of attraction of E, the region to the left of the stable manifold of the saddle B is the basin of attraction of A, and the region bounded by the stable manifolds of the saddles B and D is the basin of attraction of C. The global stable manifolds of the saddle points in Figure 2.1.6b are also examples of separatrices. According to the analytical results of Chiang, Hirsch, and Wu (1988), in many autonomous systems one can determine the global stable manifolds of the unstable fixed points to define the stability boundaries of the stable fixed points of the system, as seen in Figure 2.1.6.

Before we proceed to the next example, we make some general remarks. We note that there is attraction along a stable manifold of a fixed point. Hence, it follows that saddle points and stable nodes can be part of an attracting set and an unstable node cannot be part of an attracting set. The stable and unstable manifolds of a fixed point and homoclinic and heteroclinic orbits are all examples of invariant sets. However, these invariant sets are formed by the union of many orbits unlike a fixed point or a periodic orbit each of which corresponds to a unique orbit.

In the next example, we illustrate the influence of damping on the stability boundary of a fixed point of a planar system.

Example 2.9. We consider the following second–order system called the **swing equation** in power systems (Anderson and Fouad, 1977):

$$\frac{2H}{\omega_R}\ddot{\theta} + D\dot{\theta} = P_m - \frac{V_G V_B}{X_G}\sin(\theta - \theta_B) \qquad (2.1.33)$$

In (2.1.33), θ represents the position of the generator rotor, ω_R represents the angular velocity, H represents the inertia constant of the rotor, D represents the damping, P_m represents the input mechanical power, and the second term on the right–hand side represents the generated electrical power. The swing equation has been studied by Tamura and Yorino (1987), Nayfeh, Hamdan, and Nayfeh (1990, 1991), and several others. We mention that the equation governing a driven planar pendulum is similar to (2.1.33).

Defining $x_1 = \theta$ and $x_2 = \dot{\theta}$, we rewrite (2.1.33) as the following system of first–order equations:

$$\left\{\begin{array}{c} \dot{x}_1 \\ \dot{x}_2 \end{array}\right\} = \left[\begin{array}{cc} 0 & 1 \\ 0 & -Dc_1 \end{array}\right]\left\{\begin{array}{c} x_1 \\ x_2 \end{array}\right\} + \left\{\begin{array}{c} 0 \\ P_m c_1 - c_3 \sin(x_1 - \theta_B) \end{array}\right\}$$
$$(2.1.34)$$

where

$$c_1 = \frac{\omega_R}{2H} \quad \text{and} \quad c_3 = \frac{V_G V_B}{X_G}c_1 \qquad (2.1.35)$$

For free oscillations of the rotor, we have

$$V_B = V_{B0} \quad \text{and} \quad \theta_B = \theta_{B0} \qquad (2.1.36)$$

where V_{B0} and θ_{B0} are constants. We substitute (2.1.36) into (2.1.34) and find that each fixed point (x_{10}, x_{20}) of (2.1.34) satisfies $x_{20} = 0$ and

$$\sin(x_{10} - \theta_{B0}) = \frac{P_m c_1}{c_3} \qquad (2.1.37)$$

If one assumes that the different parameters are such that $|\frac{P_m c_1}{c_3}| \leq 1$, then there are an infinite number of solutions for x_{10}. Here, we consider only the following three solutions for x_{10}:

$$x_{10}^{(1)} = \arcsin(\frac{P_m c_1}{c_3}) + \theta_{B0}, \quad x_{10}^{(2)} = \pi - x_{10}^{(1)}, \quad x_{10}^{(3)} = -(\pi + x_{10}^{(1)}) \quad (2.1.38)$$

The stability of a fixed point of (2.1.34) is determined by the eigenvalues of the matrix

$$D_{\mathbf{x}}\mathbf{F} = \begin{bmatrix} 0 & 1 \\ -c_3 \cos(x_{10} - \theta_{B0}) & -Dc_1 \end{bmatrix} \quad (2.1.39)$$

The eigenvalues λ_i of this matrix are given by

$$\lambda_{1,2} = \frac{1}{2}\left\{ -Dc_1 \pm \sqrt{D^2 c_1^2 - 4c_3 \cos(x_{10} - \theta_{B0})} \right\} \quad (2.1.40)$$

For the parameter values (Nayfeh, Hamdan, and Nayfeh, 1990)

$$H = 2.37, \ V_{B0} = 1, \ V_G = 1.27, \ \theta_{B0} = 0,$$
$$P_m = 1, \ X_G = 0.645, \ \omega_R = 120\pi \quad (2.1.41)$$

we find that the three fixed points are $(-3.6743, 0.0)$, $(0.5327, 0.0)$, and $(2.6089, 0.0)$ and denote them as B, A, and C, respectively. For $D = 0$, B and C are saddles while A is a center. For $D > 0$, B and C are saddles while A is a stable focus.

In Figures 2.1.7a–e, we have plotted the numerically determined global manifolds of the saddles B and C for different values of D. As seen in Figure 2.1.7a, there is a homoclinic orbit in the undamped case. However, this orbit is destroyed when damping is included. In Figure 2.1.7b, the stability boundary of A is determined only by the stable manifold of C. As damping is increased, the stable manifold of C and the unstable manifold of B approach each other, as seen in Figure 2.1.7c. However, the stability boundary of A is still determined only by the stable manifold of C. In Figure 2.1.7d, one of the ends of the unstable manifold of B merges with one of the ends of the stable manifold of C, resulting in the creation of a heteroclinic half–orbit. The increase in damping has led to the **creation** of a heteroclinic half–orbit. In this case, the stability boundary of A is determined by the

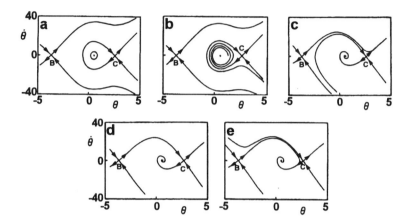

Figure 2.1.7: Phase portraits for the system (2.1.33) when $H = 2.37$, $V_{B0} = 1, V_G = 1.27, \theta_{B0} = 0$, $P_m = 1, X_G = 0.645$, and $\omega_R = 120\pi$: (a) $D = 0$, (b) $D = 0.008$, (c) $D = 0.060$, (d) $D = 0.065573$, and (e) $D = 0.07$.

stable manifolds of B and C and the unstable manifold of B. There is a nontransversal intersection of the manifolds of B and C on the stability boundary of A. As damping is further increased, the heteroclinic half–orbit of Figure 2.1.7d is destroyed, resulting in the scenario shown in Figure 2.1.7e. In this case, the stability boundary of A is determined by the stable manifolds of B and C. From Figures 2.1.7a–e, we note that damping can destroy as well as create nontransversal intersections. Further, as the damping is increased, the stability region of A also increases in size.

2.1.4   Analytical Construction of Stable and Unstable Manifolds

Here, we illustrate analytical construction of stable and unstable manifolds of the saddle point $(1,0)$ of (1.4.1) and (1.4.2). The corresponding linearization and eigenvalues are given by (2.1.12) and (2.1.13). The eigenvalues λ_1 and λ_2 have negative and positive real parts, respectively. We first use the transformation

$$x_1 = 1 + v_1 \text{ and } x_2 = v_2$$

and shift the fixed point $(1,0)$ to the origin. The result is

$$
\left\{ \begin{array}{c} \dot{v}_1 \\ \dot{v}_2 \end{array} \right\} = \left[\begin{array}{cc} 0 & 1 \\ 2 & -2\mu \end{array} \right] \left\{ \begin{array}{c} v_1 \\ v_2 \end{array} \right\} + \left\{ \begin{array}{c} 0 \\ 3v_1^2 + v_1^3 \end{array} \right\} \qquad (2.1.42)
$$

In the second step, we construct the matrix $P = [\mathbf{p}_1 \ \mathbf{p}_2]$, where $\mathbf{p}_1$ and $\mathbf{p}_2$ are the eigenvectors associated with the eigenvalues λ_1 and λ_2 of (2.1.13), respectively. The matrix P has the form

$$
P = \left[\begin{array}{cc} 1 & 1 \\ \lambda_1 & \lambda_2 \end{array} \right] \qquad (2.1.43)
$$

In the third step, we introduce the transformation $\mathbf{v} = P\mathbf{z}$ into (2.1.42) and obtain

$$
\dot{\mathbf{z}} = \left[\begin{array}{cc} \lambda_1 & 0 \\ 0 & \lambda_2 \end{array} \right] \mathbf{z} + P^{-1} \left\{ \begin{array}{c} 0 \\ 3(z_1 + z_2)^2 + (z_1 + z_2)^3 \end{array} \right\}
$$

or

$$
\dot{\mathbf{z}} = \left[\begin{array}{cc} \lambda_1 & 0 \\ 0 & \lambda_2 \end{array} \right] \mathbf{z} + \frac{1}{\lambda_2 - \lambda_1} \left\{ \begin{array}{c} -3(z_1 + z_2)^2 - (z_1 + z_2)^3 \\ 3(z_1 + z_2)^2 + (z_1 + z_2)^3 \end{array} \right\} \qquad (2.1.44)
$$

In (2.1.44), the states z_1 and z_2 are linearly uncoupled but nonlinearly coupled. In the linearization of (2.1.44), there is a fixed point at $(0,0)$ whose stable and unstable subspaces are given by $z_2 = 0$ and $z_1 = 0$, respectively. It follows from the existence theorem on stable and unstable manifolds that there exist corresponding local stable and unstable manifolds of the fixed point $(0,0)$ of the nonlinear system (2.1.44). Hence, there is a local stable manifold $z_2 = h(z_1)$ such that

$$
h(z_1 = 0) = 0 \quad \text{and} \quad \frac{dh}{dz_1}(z_1 = 0) = 0 \qquad (2.1.45)
$$

Similarly, there is a local unstable manifold $z_1 = g(z_2)$ such that

$$
g(z_2 = 0) = 0 \quad \text{and} \quad \frac{dg}{dz_2}(z_2 = 0) = 0 \qquad (2.1.46)
$$

Substituting $z_2 = h(z_1)$ into the second equation in (2.1.44), we obtain

$$
\frac{dh}{dz_1}\dot{z}_1 = \lambda_2 h + \frac{1}{\lambda_2 - \lambda_1} \left\{ 3(z_1 + h)^2 + (z_1 + h)^3 \right\}
$$

which, upon substituting for $\dot{z}_1$ from the first equation in (2.1.44), becomes

$$\frac{dh}{dz_1}\left\{\lambda_1 z_1 - \frac{1}{\lambda_2 - \lambda_1}\left[3(z_1 + h)^2 + (z_1 + h)^3\right]\right\}$$
$$= \lambda_2 h + \frac{1}{\lambda_2 - \lambda_1}\left\{3(z_1 + h)^2 + (z_1 + h)^3\right\} \qquad (2.1.47)$$

To solve (2.1.47), we assume that $h(z_1)$ has the form

$$h(z_1) = b_0 + b_1 z_1 + b_2 z_1^2 + b_3 z_1^3 + \cdots \qquad (2.1.48)$$

Because of (2.1.45), the coefficients b_0 and b_1 turn out to be zero. Thus, substituting (2.1.48) into (2.1.47) and equating the coefficients of z_1^2 and z_1^3 on both sides, we obtain the following equations for b_2 and b_3:

$$(\lambda_2 - 2\lambda_1)b_2 = -\frac{3}{\lambda_2 - \lambda_1} \qquad (2.1.49)$$

$$(\lambda_2 - 3\lambda_1)b_3 + \frac{12}{\lambda_2 - \lambda_1}b_2 = -\frac{1}{\lambda_2 - \lambda_1} \qquad (2.1.50)$$

Solving (2.1.49) and (2.1.50), we obtain

$$b_2 = -\frac{3}{(\lambda_2 - \lambda_1)(\lambda_2 - 2\lambda_1)} , \quad b_3 = -\frac{1 + 12b_2}{(\lambda_2 - \lambda_1)(\lambda_2 - 3\lambda_1)} \qquad (2.1.51)$$

Hence, the local stable manifold of the fixed point $(0,0)$ of (2.1.44) can be represented by

$$z_2 = b_2 z_1^2 + b_3 z_1^3 + \cdots$$

where b_2 and b_3 are given by (2.1.51) provided that λ_2 is away from $\lambda_1, 2\lambda_1$, and $3\lambda_1$.

Similarly, the local unstable manifold of the fixed point $(0,0)$ of (2.1.44) can be determined and represented as

$$z_1 = c_2 z_2^2 + c_3 z_2^3 + \cdots$$

where c_2 and c_3 are given by

$$c_2 = -\frac{3}{(\lambda_1 - \lambda_2)(\lambda_1 - 2\lambda_2)}$$

$$c_3 = -\frac{1}{(\lambda_1 - \lambda_2)(\lambda_1 - 3\lambda_2)}(1 + 12c_2)$$

provided that λ_2 is away from λ_1, $\frac{1}{2}\lambda_1$, and $\frac{1}{3}\lambda_1$.

The above–described procedure is commonly used to construct local invariant manifolds of a fixed point (e.g., Hassard, 1980). A similar procedure is used in Section 2.3 to construct local center manifolds of fixed points.

In general, analytical techniques can be used to determine local invariant manifolds of a fixed point, and numerical techniques are needed to determine global manifolds of a fixed point. However, at the present time, there are no established numerical techniques for determining two– or higher–dimensional manifolds of a fixed point. We provide a brief discussion for numerically determining manifolds of fixed points in Section 5.7. Parker and Chua (1989, Chapter 6) describe numerical algorithms for determining one–dimensional manifolds of fixed points. Guckenheimer and Worfolk (1993) discuss algorithms based on geodesics for computing stable manifolds of fixed points.

2.2 FIXED POINTS OF MAPS

Here, we consider fixed points of the map

$$\mathbf{x}_{k+1} = \mathbf{F}(\mathbf{x}_k; \mathbf{M}) \tag{2.2.1}$$

A fixed point $\mathbf{x}_0$ of this map satisfies the condition

$$\mathbf{x}_0 = \mathbf{F}^m(\mathbf{x}_0; \mathbf{M}_0) \text{ for all } m \in \mathcal{Z} \tag{2.2.2}$$

where $\mathbf{M} = \mathbf{M}_0$ is the value of the vector of control parameters. We note that an orbit of a map initiated at a fixed point of the map is the fixed point itself. Moreover, the fixed points of a map are examples of invariant sets.

To determine the stability of the fixed point $\mathbf{x}_0$, we superimpose on it a disturbance $\mathbf{y}$ and find from (2.2.1) that

$$\mathbf{x}_0 + \mathbf{y}_{k+1} = \mathbf{F}(\mathbf{x}_0 + \mathbf{y}_k; \mathbf{M}_0) \tag{2.2.3}$$

where $k \in \mathcal{Z}$. Expanding $\mathbf{F}$ in a Taylor series around $\mathbf{x}_0$, using (2.2.2), and linearizing in $\mathbf{y}_k$, we obtain

$$\mathbf{y}_{k+1} = D_{\mathbf{x}}\mathbf{F}(\mathbf{x}_0; \mathbf{M}_0)\mathbf{y}_k = A\mathbf{y}_k \tag{2.2.4}$$

where $D_{\mathbf{x}}\mathbf{F}$ is the matrix of the first partial derivatives of $\mathbf{F}$ evaluated at $(\mathbf{x}_0; \mathbf{M}_0)$. Next, we introduce the linear transformation

$$\mathbf{y} = P\mathbf{z} \tag{2.2.5}$$

into (2.2.4) and obtain

$$P\mathbf{z}_{k+1} = AP\mathbf{z}_k \tag{2.2.6}$$

Assuming that P is nonsingular, we multiply (2.2.6) from the left by P^{-1} and arrive at

$$\mathbf{z}_{k+1} = J\mathbf{z}_k, \quad J = P^{-1}AP \tag{2.2.7}$$

We choose P as in the preceding section so that J has a Jordan canonical form. If the eigenvalues ρ_i of A are distinct, J is a diagonal matrix with entries $\rho_1, \rho_2, \cdots, \rho_n$. Then, (2.2.7) can be rewritten as

$$z_{k+1}^{(m)} = \rho_m z_k^{(m)}, \quad m = 1, 2, \cdots, n \tag{2.2.8}$$

where $z^{(m)}$ is the mth component of $\mathbf{z}$. It follows from (2.2.8) that as $k \to \infty$,

$$z_k^{(m)} \to 0 \qquad \text{if } |\rho_m| < 1$$

$$z_k^{(m)} \to \infty \qquad \text{if } |\rho_m| > 1$$

$$z_k^{(m)} = z_0^{(m)} \qquad \text{if } \rho_m = 1$$

$$z_{2k+1}^{(m)} = -z_0^{(m)} \quad \text{if } \rho_m = -1$$

$$z_{2k}^{(m)} = z_0^{(m)} \qquad \text{if } \rho_m = -1$$

Therefore, to ascertain the stability of the fixed point $\mathbf{x}_0$, we examine the location of the eigenvalues of A in the complex plane with respect to the unit circle shown in Figure 2.2.1. If all of the eigenvalues of A are such that they are either inside the unit circle or outside the unit circle, the corresponding fixed point is called a **hyperbolic fixed point**. A hyperbolic fixed point is called a **saddle point** if some eigenvalues are within the unit circle and the rest of them are outside the unit circle. A hyperbolic fixed point is called a

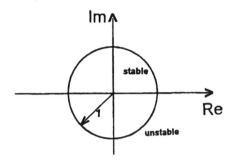

Figure 2.2.1: Unit circle in the complex plane.

sink if all of the eigenvalues are within the unit circle. Similarly, a **source** corresponds to the case where all of the eigenvalues are outside the unit circle. If one or more eigenvalues of A lie on the unit circle, the corresponding fixed point is called a **nonhyperbolic fixed point**. The **Hartman–Grobman theorem** is also applicable to fixed points of maps. From this theorem, it follows that linearization of a map is sufficient to determine the stability of a hyperbolic fixed point. If all of the eigenvalues of A lie within the unit circle, the fixed point x_0 is said to be asymptotically stable. If at least one eigenvalue of A lies outside the unit circle, the fixed point x_0 is unstable, as depicted in Figure 2.2.1. If none of the eigenvalues of A lie outside the unit circle, a linear analysis is not sufficient to determine the stability of a nonhyperbolic fixed point and nonlinear terms have to be included on the right–hand side of (2.2.4).

There are also invariant linear subspaces E^s, E^u, and E^c associated with a fixed point of a map. The subspace E^s is spanned by the eigenvectors corresponding to the eigenvalues ρ_s, where $| \rho_s | < 1$. The subspace E^u is spanned by the eigenvectors corresponding to the eigenvalues ρ_u, where $| \rho_u | > 1$. The subspace E^c is spanned by the eigenvectors corresponding to the eigenvalues ρ_c, where $| \rho_c | = 1$. For hyperbolic fixed points, the subspace E^c is empty. There are also invariant manifolds associated with a fixed point of a map. A stable manifold of a fixed point of a map is the set of all initial conditions such that the orbits initiated at these points asymptotically approach the fixed point as the iterate number $k \to \infty$. On the other hand,

an unstable manifold of a fixed point of a map is the set of all initial
conditions such that the orbits initiated at these points asymptotically
approach the fixed point as the iterate number $k \to -\infty$. Ushiki (1980)
obtained analytical expressions for local manifolds of hyperbolic fixed
points of maps. For certain special cases, Parker and Chua (1989,
Chapter 6) provide numerical algorithms to determine manifolds of
fixed points of a map.

A solution $\mathbf{x}_0$ that satisfies the condition

$$\mathbf{x}_0 = \mathbf{F}^k(\mathbf{x}_0; \mathbf{M}_0) \tag{2.2.9}$$

where $k \geq 1$ is called a **period–k point** or **periodic point of order
k** of the map $\mathbf{F}$. This point is a fixed point of the map $\mathbf{G}$, which is
formed by k successive iterations of $\mathbf{F}$; that is,

$$\mathbf{G}(\mathbf{x}; \mathbf{M}) = \mathbf{F}^k(\mathbf{x}; \mathbf{M})$$

Thus, the stability of period-k points of $\mathbf{F}$ can be studied by investigating the stability of the fixed points of $\mathbf{F}^k$.

Example 2.10. A fixed point of the logistic map (1.1.4) is $x = 0$. It
is stable for $0.0 < \alpha < 0.25$ because $\mid F'(x = 0) \mid < 1$. For $\alpha > 0.25$,
there exists a nontrivial fixed point at $x_0 = 1 - 0.25/\alpha$. Because
$F'(x_0) = 2 - 4\alpha$, this fixed point is stable when $0.25 < \alpha < 0.75$.
At $\alpha = 0.75$, this nontrivial fixed point is nonhyperbolic. In Figure
2.2.2a, we have plotted $F(x)$ versus x when $\alpha = 0.8$. The intersections
of the line $y = x$ with the curve $F(x)$ give the fixed points of $F(x)$.
They occur at $x = 0$ and $x = 0.6875$. The fixed point $x = 0.6875$ is
unstable because $\mid F'(x = 0.6875) \mid > 1$.

Next, we investigate the fixed points of $F^2(x)$, which are given by

$$F^2(x) = F[F(x)] = F[4\alpha x(1 - x)]$$

or

$$F^2(x) = 16\alpha^2 x(1 - x)\left[1 - 4\alpha x(1 - x)\right]$$

Hence, the fixed points of $F^2(x)$ are given by the solutions of

$$16\alpha^2 x(1 - x)\left[1 - 4\alpha x(1 - x)\right] = x$$

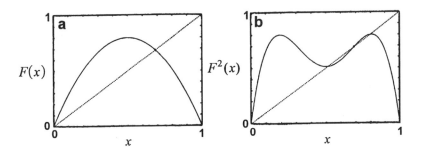

Figure 2.2.2: Graphs to determine solutions of (1.1.4) at $\alpha = 0.8$: (a) fixed points and (b) period–two points.

which are

$$x = 0, \ 1 - \frac{1}{4\alpha}, \ \text{ and } \ x = \frac{1}{2} + \frac{1}{4\alpha}\left[\frac{1}{2} \pm \sqrt{\left(2\alpha - \frac{1}{2}\right)^2 - 1}\right]$$

The first two fixed points are also fixed points of $F(x)$. In Figure 2.2.2b, we have plotted $F^2(x)$ versus x when $\alpha = 0.8$. The intersections of the curve $F^2(x)$ with the curve $y = x$ give the fixed points of $F^2(x)$. We note that there are four intersections. The dot corresponds to the fixed point $x = 1 - 0.25/\alpha$ of $F^2(x)$, which is also a fixed point of $F(x)$. The other three fixed points are $x = 0$, $x = 0.5130$, and $x = 0.7795$. We note that

$$F(0.5130) = 0.7795 \ \text{ and } \ F^2(0.5130) = 0.5130$$

and that

$$F(0.7795) = 0.5130 \ \text{ and } \ F^2(0.7795) = 0.7795$$

Hence, $x = 0.5130$ and $x = 0.7795$ are period–two points of $F(x)$.

To determine the stability of the fixed points of $F^2(x)$, we calculate its Jacobian; that is,

$$\rho = \frac{d}{dx}\left[F^2(x)\right] = 16\alpha^2(1 - 2x)\left[1 - 8\alpha x(1 - x)\right]$$

For the fixed points $x = 0$ and $x = 0.6875$, $\rho = 10.24$ and $\rho = 1.44$, respectively. Hence, these fixed points are unstable. This is expected because these fixed points are unstable fixed points of $F(x)$; $F'(x = 0) = 3.2$ and $F'(x = 0.6875) = 1.2$. For the fixed points $x = 0.5130$ and $x = 0.7795$ of $F^2(x)$, $\rho = 0.159$ and $\rho = 0.573$, respectively. Hence, both of the period–two points of $F(x)$ are stable.

Example 2.11. Each fixed point (x_0, y_0) of the Hénon map $(1.1.5)$ and $(1.1.6)$ is a solution of

$$x_0 = 1 + y_0 - \alpha x_0^2 \qquad (2.2.10)$$
$$y_0 = \beta x_0 \qquad (2.2.11)$$

From $(2.2.10)$ and $(2.2.11)$, we obtain

$$\alpha x_0^2 + (1 - \beta)x_0 - 1 = 0 \qquad (2.2.12)$$

whose solutions are

$$x_0 = \frac{1}{2\alpha}\left\{(\beta - 1) \pm \sqrt{(1 - \beta)^2 + 4\alpha}\right\} \qquad (2.2.13)$$

When $\alpha > -(1 - \beta)^2/4$, $(2.2.13)$ has two real solutions. The corresponding fixed points of $(1.1.5)$ and $(1.1.6)$ are

$$(x_{10}, y_{10}) = \left(\frac{1}{2\alpha}\left\{(\beta - 1) - \sqrt{(1 - \beta)^2 + 4\alpha}\right\}, \beta x_{10}\right)$$
$$\qquad\qquad\qquad\qquad\qquad\qquad\qquad\qquad\qquad (2.2.14)$$
$$(x_{20}, y_{20}) = \left(\frac{1}{2\alpha}\left\{(\beta - 1) + \sqrt{(1 - \beta)^2 + 4\alpha}\right\}, \beta x_{20}\right)$$

The stability of (x_{j0}, y_{j0}) depends on the eigenvalues of the matrix

$$D_{\mathbf{x}}\mathbf{F} = \begin{bmatrix} -2\alpha x_{j0} & 1 \\ \beta & 0 \end{bmatrix} \qquad (2.2.15)$$

The two eigenvalues are

$$\rho_1 = -\alpha x_{j0} - \sqrt{\alpha^2 x_{j0}^2 + \beta}$$
$$\qquad\qquad\qquad\qquad\qquad (2.2.16)$$
$$\rho_2 = -\alpha x_{j0} + \sqrt{\alpha^2 x_{j0}^2 + \beta}$$

When $\alpha = 0.08$ and $\beta = 0.3$, we obtain the fixed points

$$(x_{10}, y_{10}) = (-10.0, -3.0) \quad \text{and} \quad (x_{20}, y_{20}) = (1.25, 0.375)$$

The eigenvalues corresponding to (x_{10}, y_{10}) are -0.1695 and 1.7695, and the eigenvalues corresponding to (x_{20}, y_{20}) are -0.6568 and 0.4568. Therefore, (x_{10}, y_{10}) is a saddle point while (x_{20}, y_{20}) is a stable fixed point.

In Figure 2.2.3, we show some of the discrete points that make up the stable and unstable manifolds of the saddle point $(-10.0, -3.0)$. One of the ends of the unstable manifold W^u of the saddle is attracted to the stable fixed point $(1.25, 0.375)$. To determine the stable and unstable manifolds, we first determined the eigenvectors $\mathbf{p}_1$ and $\mathbf{p}_2$ of $(2.2.15)$ corresponding to the eigenvalues $\rho_1 = -0.1695$ and $\rho_2 = 1.7695$, respectively. Then, we used the eigenvector $\mathbf{p}_1$ to choose initial conditions "close" to $(-10.0, -3.0)$. Successive applications of the map $\mathbf{F}^{-1}$ to these initial conditions yielded the discrete points on W^s. Similarly, we used the eigenvector $\mathbf{p}_2$ to choose initial conditions "close" to $(-10.0, -3.0)$. Successive applications of the map $\mathbf{F}$ to these initial conditions yielded the discrete points on W^u.

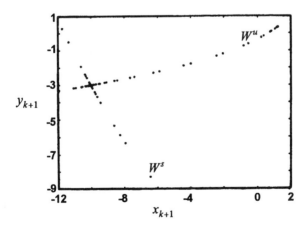

Figure 2.2.3: Stable manifold W^s and unstable manifold W^u of the saddle point $(-10, -3)$ of $(1.1.5)$ and $(1.1.6)$ when $\alpha = 0.08$ and $\beta = 0.3$.

2.3 BIFURCATIONS OF CONTINUOUS SYSTEMS

In this section, we present the concepts of local and global bifurcations, continuous and discontinuous or catastrophic bifurcations, bifurcation diagrams, bifurcation sets, codimension of a bifurcation, structural stability, and simplification methods. **Bifurcation**, a French word introduced into nonlinear dynamics by Poincaré, is used to indicate a qualitative change in the features of a system, such as the number and type of solutions, under the variation of one or more parameters on which the considered system depends. By the terminology **local bifurcation**, we mean a qualitative change occurring in the neighborhood of a fixed point or a periodic solution of the system. We consider any other qualitative change to be a **global bifurcation**.

In bifurcation problems, it is useful to consider a space formed by using the state variables and the control parameters, called the **state–control space**. In this space, locations at which bifurcations occur are called **bifurcation points**. Many branches of similar and/or different solutions merge or emerge from local bifurcation points. A bifurcation that requires at least m control parameters to occur is called a **codimension–m bifurcation**. Here, unless otherwise stated, it should be understood that the control parameters are varied in a stationary sense; that is, the control parameters are varied very slowly so that their instantaneous values can be considered constants.

Shilnikov (1976), Zeeman (1982), Abraham (1985), Thompson and Stewart (1986), and Thompson, Stewart, and Ueda (1994) classify bifurcations into **continuous** and **discontinuous** or **catastrophic bifurcations**, depending on whether the states of the system vary continuously or discontinuously as the control parameter is varied gradually through its critical value. Discontinuous or catastrophic bifurcations can be further subdivided into **dangerous** and **explosive** bifurcations, depending on whether the system response jumps to a remote disconnected attractor or explodes into a larger attractor, with the new attractor including the old (ghost or phantom) attractor as a proper subset.

In an explosive bifurcation, the system response does not jump to a remote attractor. The outcome of such a bifurcation is determinant, independent of the rate of control sweep, and insensitive to the presence of noise. Consequently, upon reversal of the control sweep in a quasistationary manner, the new large attractor implodes to the old small attractor at the same critical bifurcation value, with no hysteresis. The new large attractor may or may not be chaotic. When the new attractor is chaotic, an orbit on it spends long stretches of time near the fixed point with chaotic or turbulent outbreaks and excursions away from the fixed point. The consequence of the explosive bifurcation is an on–off intermittent transition to chaos, as discussed in Section 5.4.

In a dangerous bifurcation, the current attractor suddenly disappears from the state space of the considered system. This event is also known as **blue sky catastrophe** (Abraham, 1985; Thompson and Stewart, 1986; Abraham and Shaw, 1992). The postbifurcation response jumps to a remote attractor, which may be bounded or unbounded. Bounded responses include point, periodic, quasiperiodic, and chaotic attractors. (The unbounded responses have disastrous consequences; examples include capsizing of ships and voltage collapse in power systems.) Typically, upon reversal of the control sweep, a bounded response remains on the path of the new attractor, resulting in hysteresis. The outcome of a dangerous bifurcation may be determinant or indeterminant, depending on whether the system has only one attractor past the critical control value or not. When the system has multiple attractors, the postbifurcation response depends on the rate of control sweep and the presence of noise.

Here, we consider local bifurcations of fixed points of an autonomous system of differential equations as a scalar control parameter is varied. Bifurcations of fixed points under the variation of two control parameters are also briefly addressed. Further, we also discuss methods to simplify a dynamical system in the vicinity of a bifurcation point. Local bifurcations of periodic solutions under the variation of a scalar control parameter are addressed in Chapter 3. Local bifurcations of fixed points of maps are discussed in Section 2.4.

2.3.1   Local Bifurcations of Fixed Points

From Section 2.1, we know that the matrix A in (2.1.5) and the associated eigenvalues are functions of the control parameter vector **M**. Let us suppose that as one or more control parameters are slowly varied, a fixed point becomes nonhyperbolic at a certain location in the state–control space. Then, if the state–space portraits before and after this location are qualitatively different, this location is called a **bifurcation point**, and the accompanying qualitative change is called a **bifurcation**. Furthermore, a bifurcation that requires at least m independent control parameters to occur is called a **codimension–m bifurcation**.

If we start with control parameters corresponding to a stable fixed point of a continuous–time system, such as (2.1.1), and then slowly vary one of the control parameters, this fixed point can lose stability through one of the following bifurcations (e.g., Arnold, 1988): (a) saddle–node bifurcation, (b) pitchfork or symmetry–breaking bifurcation, (c) transcritical bifurcation, or (d) Hopf bifurcation. At bifurcation points associated with saddle–node, pitchfork, and transcritical bifurcations only branches of fixed points or static solutions meet. Hence, these three bifurcations are classified as **static bifurcations**. In contrast, branches of fixed points and periodic solutions meet at a Hopf bifurcation point. Hence, a Hopf bifurcation is classified as a **dynamic bifurcation**.

Static Bifurcations

We consider the static bifurcations of the fixed points of (2.1.1) under the influence of a scalar control parameter α. In the $\mathbf{x} - \alpha$ state–control space, a simple static bifurcation of a fixed point of (2.1.1) is said to occur at $(\mathbf{x}_0; \alpha_c)$ if the following conditions are satisfied:

1. $\mathbf{F}(\mathbf{x}_0; \alpha_c) = \mathbf{0}$,

2. $D_\mathbf{x}\mathbf{F}$ has a zero eigenvalue while all of its other eigenvalues have nonzero real parts at $(\mathbf{x}_0; \alpha_c)$.

The first condition ensures that the considered solution is a fixed point of (2.1.1), and the second condition implies that this fixed point is a nonhyperbolic fixed point. One should note that these conditions are **necessary but not sufficient.** However, in the event a static bifurcation does occur, one can distinguish saddle–node bifurcation points from other static bifurcation points. We let $\mathbf{F}_\alpha$ represent the $n \times 1$ vector of first partial derivatives of the components of $\mathbf{F}$ with respect to the control parameter α and construct the $n \times (n + 1)$ matrix $[D_\mathbf{x}\mathbf{F} \mid \mathbf{F}_\alpha]$. At a **saddle–node bifurcation point,** $\mathbf{F}_\alpha$ does not belong to the range of the matrix $D_\mathbf{x}\mathbf{F}$. On the other hand, at **pitchfork and transcritical bifurcation points,** $\mathbf{F}_\alpha$ belongs to the range of the matrix $D_\mathbf{x}\mathbf{F}$. (The range of an $n \times n$ matrix A consists of all vectors $A\mathbf{z}$, where $\mathbf{z} \in \mathcal{R}^n$.) Hence, the matrix $[D_\mathbf{x}\mathbf{F} \mid \mathbf{F}_\alpha]$ has a rank of n at saddle–node bifurcation points and a rank of $(n - 1)$ at other static bifurcation points. In the state–control space, all of the branches of fixed points that meet at a saddle–node bifurcation point have the same tangent. This is not so at pitchfork and transcritical bifurcation points. Next, we consider many one–dimensional systems to illustrate the different bifurcations.

Example 2.12. We consider the system

$$\dot{x} = F(x; \mu) = \mu - x^2 \qquad (2.3.1)$$

where μ is a scalar control parameter. Here, we have a two–dimensional state–control space. For $\mu < 0$, (2.3.1) does not have any fixed points. However, for $\mu > 0$, (2.3.1) has the two nontrivial fixed points

$$x = \sqrt{\mu} \ \text{ and } \ x = -\sqrt{\mu}$$

In this case, the Jacobian matrix has a single eigenvalue given by

$$\lambda = -2x$$

The fixed point $x = \sqrt{\mu}$ is a stable node because $\lambda < 0$, and the fixed point $x = -\sqrt{\mu}$ is an unstable node because $\lambda > 0$. In Figure 2.3.1, we display the different fixed–point solutions of (2.3.1) and their stability in the $x - \mu$ space. We use broken and solid lines to depict branches of unstable and stable fixed points, respectively. We note the following at $(0, 0)$: (a) $F(x; \mu) = 0$ and (b) $D_x F$ has a zero eigenvalue. Hence, there

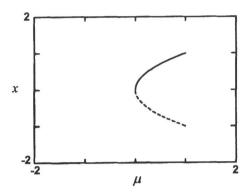

Figure 2.3.1: Scenario in the vicinity of a saddle–node bifurcation.

is a nonhyperbolic fixed point at $\mu = 0$. Further, we note that there is a change in the number of fixed–point solutions as we pass through $\mu = 0$. Hence, the origin of the $x - \mu$ space is a static–bifurcation point. At this bifurcation point, $F_\mu = 1$ and $D_x F$ is a scalar equal to zero. So, F_μ does not belong to the range of $D_x F$, and the rank of $[D_x F \mid F_\mu]$ is one. Therefore, the origin of the $x - \mu$ space is a **saddle–node bifurcation point**. In Figure 2.3.1, both of the branches that meet at the bifurcation point have the same tangent. Moreover, we observe that branches of stable nodes and unstable nodes meet at the saddle–node bifurcation point of the one–dimensional system (2.3.1). Typically, in higher–dimensional systems, branches of saddle points and stable nodes meet at a saddle–node bifurcation point.

Diagrams such as Figure 2.3.1 in which the variation of solutions and their stability are displayed in the state–control space are called **bifurcation diagrams**. In the bifurcation diagram, a branch of stable solutions is called a **stable branch** and a branch of unstable solutions is called an **unstable branch**. In most situations, a branch of solutions either ends or begins at a bifurcation point.

Example 2.13. We consider the system

$$\dot{x} = F(x; \mu) = \mu x + \alpha x^3 \qquad (2.3.2)$$

where μ is again the scalar control parameter. There are three fixed points:

$$x = 0; \qquad \text{trivial fixed point}$$
$$x = \pm\sqrt{-\mu/\alpha}; \quad \text{nontrivial fixed points}$$

In this case, the Jacobian matrix

$$D_x F = \mu + 3\alpha x^2$$

has the single eigenvalue

$$\lambda = \mu \qquad \text{at } x = 0$$
$$\lambda = -2\mu \quad \text{at } x = \pm\sqrt{-\mu/\alpha}$$

Consequently, the trivial fixed point is stable when $\mu < 0$ and unstable when $\mu > 0$. On the other hand, when $\alpha < 0$, nontrivial fixed points exist only when $\mu > 0$ and they are stable. However, when $\alpha > 0$, nontrivial fixed points exist only when $\mu < 0$ and they are unstable. The bifurcation diagrams of Figures 2.3.2a and 2.3.2b correspond to $\alpha = -1$ and $\alpha = 1$, respectively. In both cases, we note the following at $(0,0)$: (a) $F(x,\mu) = 0$, (b) $D_x F$ has a zero eigenvalue, (c) the number of fixed–point solutions for $\mu < 0$ is different from that for $\mu > 0$, and (d) there is a change in the stability of the trivial fixed point as we pass through $\mu = 0$. Hence, the origin of the state–control space is a bifurcation point.

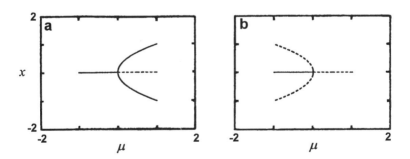

Figure 2.3.2: Local scenarios: (a) supercritical pitchfork bifurcation and (b) subcritical pitchfork bifurcation.

When $\alpha = -1$, two stable branches of fixed points $x = \sqrt{\mu}$ and $x = -\sqrt{\mu}$ bifurcate from the bifurcation point, as shown in Figure 2.3.2a. When $\alpha = 1$, two unstable branches of fixed points $x = \sqrt{\mu}$ and $x = -\sqrt{\mu}$ bifurcate from the bifurcation point, as shown in Figure 2.3.2b. For both $\alpha = 1$ and $\alpha = -1$, the bifurcation point is not a saddle–node bifurcation point because both of the scalars F_μ and $D_x F$ are zero there. The bifurcations observed in Figures 2.3.2a and 2.3.2b are called **pitchfork bifurcations** because the bifurcating nontrivial branches have the geometry of a pitchfork at $(0,0)$. Specifically, the bifurcation in Figure 2.3.2a is called a **supercritical pitchfork bifurcation**, and the bifurcation in Figure 2.3.2b is called a **subcritical** or **reverse pitchfork bifurcation**. In the case of a supercritical pitchfork bifurcation, locally we have a branch of stable fixed points on one side of the bifurcation point and two branches of stable fixed points and a branch of unstable fixed points on the other side of the bifurcation point. In the case of a subcritical pitchfork bifurcation, locally we have two branches of unstable fixed points and a branch of stable fixed points on one side of the bifurcation point and a branch of unstable fixed points on the other side of the bifurcation point. Unlike Figure 2.3.1, all of the branches that meet at the bifurcation points in Figures 2.3.2a and 2.3.2b do not have the same tangent.

Example 2.14. We consider the one–dimensional system

$$\dot{x} = \mu x - x^2 \tag{2.3.3}$$

There are two fixed points:

$$x = 0; \text{ trivial fixed point}$$
$$x = \mu; \text{ nontrivial fixed point}$$

The Jacobian matrix

$$D_x F = \mu - 2x$$

has the single eigenvalue

$$\lambda = \mu \quad \text{at } x = 0$$
$$\lambda = -\mu \quad \text{at } x = \mu$$

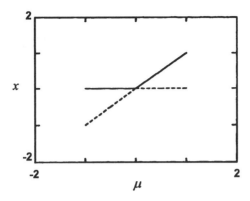

Figure 2.3.3: Scenario in the vicinity of a transcritical bifurcation.

In the corresponding bifurcation diagram shown in Figure 2.3.3, the fixed point $x = 0$ is a nonhyperbolic fixed point at $\mu = 0$. At this point, a static bifurcation occurs because there is an exchange of stability between the trivial and nontrivial branches. We note that the bifurcation point is not a saddle–node bifurcation point because the scalars F_μ and $D_x F$ are both zero at $(0, 0)$. The bifurcation point in Figure 2.3.3 is an example of a **transcritical bifurcation point**. We point out that all of the branches that meet at this bifurcation point do not have the same tangent.

Example 2.15. We consider

$$\dot{x} = F(x, \mu) = \mu - x^5 \qquad (2.3.4)$$

We have only one fixed point, namely,

$$x = \mu^{1/5}$$

This solution is depicted in Figure 2.3.4. At the origin of the $x - \mu$ space, $F(x, \mu) = 0$ and $D_x F$ has a zero eigenvalue, implying that $x = 0$ is a nonhyperbolic fixed point at $\mu = 0$. However, $(0, 0)$ is not a bifurcation point because there is no qualitative change either in the number of fixed–point solutions or in the stability of the fixed–point solutions as we pass through $\mu = 0$ in the state–control space.

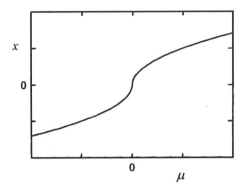

Figure 2.3.4: Fixed–point solutions of (2.3.4).

Hopf Bifurcations

When a scalar control parameter α is varied, a Hopf bifurcation of a fixed point of a system such as (2.1.1) is said to occur at $\alpha = \alpha_c$ if the following conditions (Marsden and McCracken, 1976) are satisfied:

1. $\mathbf{F}(\mathbf{x}_0; \alpha_c) = \mathbf{0}$,

2. The matrix $D_{\mathbf{x}}\mathbf{F}$ has a pair of purely imaginary eigenvalues $\pm i\omega_h$ while all of its other eigenvalues have nonzero real parts at $(\mathbf{x}_0; \alpha_c)$,

3. For $\alpha \simeq \alpha_c$, let the analytic continuation of the pair of imaginary eigenvalues be $\hat{\lambda} \pm i\omega$. Then Real $(d\hat{\lambda}/d\alpha) \neq 0$ at $\alpha = \alpha_c$. This condition implies a transversal or nonzero speed crossing of the imaginary axis and hence is called a **transversality condition**.

Again, the first two conditions imply that the fixed point undergoing the bifurcation is a nonhyperbolic fixed point. When all of the above three conditions are satisfied, a periodic solution of period $2\pi/\omega_h$ is born at $(\mathbf{x}_0; \alpha_c)$; bifurcating periodic solutions can also occur when the transversality condition is not satisfied (e.g., Marsden and McCracken, 1976). It is to be noted that bifurcating periodic solutions can also occur under certain other degenerate conditions (e.g., Golubitsky and Schaeffer, 1985). In such cases, we have **degenerate Hopf bifurcations**.

The Hopf bifurcation is also called the **Poincaré–Andronov-Hopf bifurcation** (e.g., Wiggins, 1990) to give credit to the works of Poincaré and Andronov that preceded the work of Hopf. As pointed out in the literature (e.g., Arnold, 1988, Chapter 6; Abed, 1994), Poincaré (1899) was aware of the conditions for this bifurcation to occur. (Poincaré studied such bifurcations in the context of lunar orbital dynamics.) Andronov and his co–workers studied Hopf bifurcations in planar systems before Hopf studied such bifurcations in general n–dimensional systems (Andronov and Chaikin, 1949; Arnold, 1988). In aeroelasticity, the consequence of a Hopf bifurcation is known as **galloping** or **flutter**.

Example 2.16. We consider the planar system

$$\dot{x} = \mu x - \omega y + (\alpha x - \beta y)(x^2 + y^2) \qquad (2.3.5)$$

$$\dot{y} = \omega x + \mu y + (\beta x + \alpha y)(x^2 + y^2) \qquad (2.3.6)$$

where x and y are the states and μ is the scalar control parameter. The fixed point $(0,0)$ is a solution of $(2.3.5)$ and $(2.3.6)$ for all values of μ. The eigenvalues of the corresponding Jacobian matrix there are

$$\lambda_1 = \mu - i\omega \text{ and } \lambda_2 = \mu + i\omega$$

From these eigenvalues, we note that $(0,0)$ is a nonhyperbolic fixed point of $(2.3.5)$ and $(2.3.6)$ when $\mu = 0$. Further, at $(x, y, \mu) = (0,0,0)$, we note that

$$\frac{d\lambda_1}{d\mu} = 1 \text{ and } \frac{d\lambda_2}{d\mu} = 1$$

Hence, the three conditions required for a Hopf bifurcation are satisfied, and a Hopf bifurcation of the fixed point $(0,0)$ of $(2.3.5)$ and $(2.3.6)$ occurs at $\mu = 0$. The period of the bifurcating periodic solution at $(0,0,0)$ is $2\pi/\omega$.

By using the transformation

$$x = r \cos \theta \text{ and } y = r \sin \theta$$

we transform $(2.3.5)$ and $(2.3.6)$ into

$$\dot{r} = \mu r + \alpha r^3 \qquad (2.3.7)$$

$$\dot{\theta} = \omega + \beta r^2 \qquad\qquad (2.3.8)$$

The trivial fixed point of (2.3.7) corresponds to the fixed point $(0,0)$ of (2.3.5) and (2.3.6), and a nontrivial fixed point (i.e., $r \neq 0$) of (2.3.7) corresponds to a periodic solution of (2.3.5) and (2.3.6). In the latter case, r is the amplitude and $\dot{\theta}$ is the frequency of the periodic solution that is created due to the Hopf bifurcation. A stable nontrivial fixed point of (2.3.7) corresponds to a stable periodic solution of (2.3.5) and (2.3.6). Likewise, an unstable nontrivial fixed point of (2.3.7) corresponds to an unstable periodic solution of (2.3.5) and (2.3.6).

We note that (2.3.7) is identical to (2.3.2), so the Hopf bifurcation at $(0,0,0)$ in the $x - y - \mu$ space is equivalent to a pitchfork bifurcation at $(0,0)$ in the $r - \mu$ space. When $\alpha = -1$, we have a supercritical pitchfork bifurcation in the $r - \mu$ space and, hence, a **supercritical Hopf bifurcation** in the $x - y - \mu$ space. When $\alpha = 1$, we have a subcritical pitchfork bifurcation in the $r - \mu$ space and, hence, a **subcritical Hopf bifurcation** in the $x - y - \mu$ space. The bifurcation diagrams for $\alpha = -1$ and $\alpha = 1$ are shown in Figures 2.3.5a and 2.3.5b, respectively. In the upper half of Figure 2.3.5, the bifurcating periodic solutions in the $x - y - \mu$ space are depicted as parabolic surfaces. In the case of a supercritical Hopf bifurcation, locally we have a branch of stable fixed points on one side of the bifurcation point and a branch of unstable fixed points and a branch of stable periodic solutions on the other side of the bifurcation point. In the case of a subcritical Hopf bifurcation, locally we have a branch of unstable periodic solutions and a branch of stable fixed points on one side of the bifurcation point and a branch of unstable fixed points on the other side of the bifurcation point.

When $\alpha = 0$ in (2.3.5) and (2.3.6), although the conditions for a Hopf bifurcation are satisfied there are no periodic orbits in the vicinity of the bifurcation point. This case is degenerate.

Example 2.17. The following system is a model of a centrifugal governor, which probably is one of the first mechanical systems in which the consequences of a Hopf bifurcation were observed (Pontryagin,

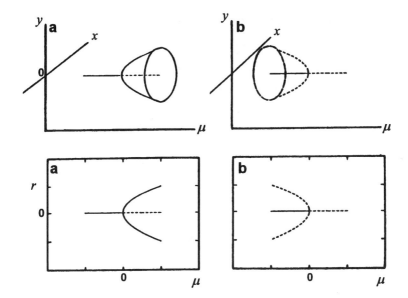

Figure 2.3.5: Local scenarios: (a) supercritical Hopf bifurcation and (b) subcritical Hopf bifurcation.

1962, Chapter 5; Hassard, Kazarinoff, and Wan, 1981, Chapter 3):

$$\dot{x}_1 = x_2 \tag{2.3.9}$$

$$\dot{x}_2 = (\sin x_1 \cos x_1)x_3^2 - \sin x_1 - \mu x_2 \tag{2.3.10}$$

$$\dot{x}_3 = k(\cos x_1 - \rho) \tag{2.3.11}$$

In this system, the control parameter of interest is μ, which is a measure of friction. Further, $\mu > 0$, $k > 0$, and $0 < \rho < 1$. The fixed points of (2.3.9)–(2.3.11) are given by

$$(x_{10}, x_{20}, x_{30}) = (\arccos \rho, \ 0, \ \pm 1/\sqrt{\rho}) \tag{2.3.12}$$

The Jacobian matrix evaluated at the fixed point $(\arccos \rho, 0, 1/\sqrt{\rho})$ is

$$D_x F = \begin{bmatrix} 0 & 1 & 0 \\ \frac{\rho^2 - 1}{\rho} & -\mu & 2\sqrt{\rho}\sqrt{1 - \rho^2} \\ -k\sqrt{1 - \rho^2} & 0 & 0 \end{bmatrix} \tag{2.3.13}$$

Hence, the eigenvalues of this matrix are given by

$$\lambda^3 + \mu\lambda^2 + \frac{1-\rho^2}{\rho}\lambda + 2k\sqrt{\rho}(1-\rho^2) = 0 \qquad (2.3.14)$$

The necessary and sufficient conditions that none of the roots of this equation has a positive real part is provided by the Routh–Hurwitz criterion (e.g., Meirovitch, 1970, Chapter 6), which in this case translates to

$$\mu > 0, \quad \frac{1-\rho^2}{\rho}\left(\mu - 2k\rho^{3/2}\right) > 0, \quad k\sqrt{\rho}(1-\rho^2) > 0 \qquad (2.3.15)$$

The first condition is always satisfied because $\mu > 0$. The last condition is always satisfied because k is positive and $0 < \rho < 1$. However, the second condition is violated when $\mu < 2k\rho^{3/2}$. This violation implies the existence of a pair of complex conjugate roots with a positive real part. Hence, the critical value is

$$\mu = \mu_c = 2k\rho^{3/2} \qquad (2.3.16)$$

Alternatively, this critical value can be determined as follows. At criticality, let the three eigenvalues of (2.3.13) be

$$-i\omega_0, \quad i\omega_0, \quad \text{and} \quad \lambda_3 \qquad (2.3.17)$$

Then, we have

$$\lambda^3 - \lambda_3\lambda^2 + \omega_0^2\lambda - \lambda_3\omega_0^2 = 0 \qquad (2.3.18)$$

Comparing (2.3.18) with (2.3.14), we find that

$$\lambda_3 = -\mu, \quad \omega_0^2 = \frac{1-\rho^2}{\rho}, \quad \lambda_3\omega_0^2 = -2k\sqrt{\rho}(1-\rho^2) \qquad (2.3.19)$$

Eliminating λ_3 and ω_0 from the relations in (2.3.19), we obtain (2.3.16).

Now, we need to check if the transversality condition required for a Hopf bifurcation is satisfied at $\mu = \mu_c$. For $\mu \simeq \mu_c$, we let the analytic continuation of the eigenvalues $-i\omega_0$ and $i\omega_0$ be λ_1 and λ_2, respectively. We differentiate all the terms in (2.3.14) with respect to the control parameter μ, use (2.3.19), and arrive at

$$\frac{d\lambda}{d\mu} = -\frac{1}{[3 + \frac{2\mu}{\lambda} + (\frac{\omega_0}{\lambda})^2]} \qquad (2.3.20)$$

Then, at $\mu = \mu_c$, (2.3.20) leads to

$$\frac{d\lambda_1}{d\mu} = -\frac{\{1 - i\mu_c/\omega_0\}}{2\{1 + \mu_c^2/\omega_0^2\}} \quad \text{and} \quad \frac{d\lambda_2}{d\mu} = -\frac{\{1 + i\mu_c/\omega_0\}}{2\{1 + \mu_c^2/\omega_0^2\}}$$

Because the real parts of $d\lambda_1/d\mu$ and $d\lambda_2/d\mu$ are nonzero, the transversality condition is satisfied. Hence, the fixed point $(\arccos \rho, 0, 1/\sqrt{\rho})$ of (2.3.9)–(2.3.11) experiences a Hopf bifurcation at $\mu = \mu_c$.

Next, we examine if the fixed point $(\arccos \rho, 0, -1/\sqrt{\rho})$ of (2.3.9)–(2.3.11) can experience a Hopf bifurcation. The corresponding Jacobian matrix is

$$D_x F = \begin{bmatrix} 0 & 1 & 0 \\ \frac{\rho^2 - 1}{\rho} & -\mu & -2\sqrt{\rho}\sqrt{1 - \rho^2} \\ -k\sqrt{1 - \rho^2} & 0 & 0 \end{bmatrix}$$

Hence, the eigenvalues of this matrix are given by

$$\lambda^3 + \mu\lambda^2 + \left(\frac{1 - \rho^2}{\rho}\right)\lambda - 2k\sqrt{\rho}(1 - \rho^2) = 0$$

The necessary and sufficient conditions that none of the roots of this equation has a positive real part is provided again by the Routh–Hurwitz criterion, which in this case translates to

$$\mu > 0, \quad \frac{1 - \rho^2}{\rho}\left(\mu + 2k\rho^{3/2}\right) > 0, \quad -k\sqrt{\rho}(1 - \rho^2) > 0 \qquad (2.3.21)$$

Because $\mu > 0$, $k > 0$, and $0 < \rho < 1$, the first and second conditions are always satisfied while the third condition is not satisfied. Hence, the fixed point $(\arccos \rho, 0, -1/\sqrt{\rho})$ is unstable in the parameter range of interest.

2.3.2 Normal Forms for Bifurcations

The **normal form** of a bifurcation is a simplified system of equations that approximates the dynamics of the system in the vicinity of a bifurcation point. The simplification can be achieved by using one of the methods described in Sections 2.3.4–2.3.6. The dimension of the normal form is generally lower than the dimension of the full system of equations that describes the dynamics of the system. Here, we present the normal forms for generic bifurcations of fixed points that occur when a single control parameter is varied. For static bifurcations, the normal form is a one–dimensional autonomous system, and for the Hopf bifurcation, the normal form is a two–dimensional autonomous system.

The **normal form for a generic saddle–node bifurcation of a fixed point** is

$$\dot{x} = \mu + \alpha x^2 \tag{2.3.22}$$

where x is the state variable and μ is the scalar control parameter. The bifurcation diagram of Figure 2.3.1 corresponds to $\alpha = -1$.

The **normal form for a generic transcritical bifurcation of a fixed point** is

$$\dot{x} = \mu x - \alpha x^2 \tag{2.3.23}$$

where μ is again the control parameter. The bifurcation diagram of Figure 2.3.3 corresponds to $\alpha = 1$.

The **normal form for a generic pitchfork bifurcation of a fixed point** is

$$\dot{x} = \mu x + \alpha x^3 \tag{2.3.24}$$

The bifurcation diagrams of Figures 2.3.2a and 2.3.2b correspond to $\alpha = -1$ and $\alpha = 1$, respectively.

The **normal form for a generic Hopf bifurcation of a fixed point** is

$$\dot{x} = \mu x - \omega y + (\alpha x - \beta y)(x^2 + y^2) \tag{2.3.25}$$
$$\dot{y} = \omega x + \mu y + (\beta x + \alpha y)(x^2 + y^2) \tag{2.3.26}$$

and its corresponding polar form is

$$\dot{r} = \mu r + \alpha r^3 \tag{2.3.27}$$
$$\dot{\theta} = \omega + \beta r^2 \tag{2.3.28}$$

The bifurcation diagrams of Figures 2.3.5a and 2.3.5b correspond to $\alpha = -1$ and $\alpha = 1$, respectively.

2.3.3   Bifurcation Diagrams and Sets

In Section 2.3.1, we presented many bifurcation diagrams each display-ing the dynamics near a bifurcation point of the corresponding system. We call such diagrams displaying a local bifurcation as **local bifurca-tion diagrams**. To construct a local bifurcation diagram for a given n–dimensional autonomous system, one first simplifies the considered system to its normal form in the vicinity of a bifurcation point and then uses this normal form to construct the bifurcation diagram.

Here, we construct bifurcation diagrams for two examples. In each case, we display a collection of the different local bifurcations that take place in a given range of the chosen control parameter. Further, we explain what is meant by a bifurcation set.

Example 2.18. We consider the forced Duffing oscillator (Nayfeh and Mook, 1979, Chapter 4)

$$\ddot{u} + u + \epsilon \left(2\mu\dot{u} + \alpha u^3 \right) = 2\epsilon k \cos(\Omega t) \qquad (2.3.29)$$

for the case of primary resonance. In (2.3.29), ϵ is a small positive parameter, μ is the damping coefficient and a positive parameter, α is the coefficient of the cubic nonlinearity, $2\epsilon k$ is the amplitude of forcing, and Ω is the excitation frequency. The parameters μ, α, and k are independent of ϵ. Further, the excitation frequency is such that

$$\Omega = 1 + \epsilon\sigma \qquad (2.3.30)$$

where the parameter σ is called the **external detuning**.

By using the method of multiple scales (Nayfeh, 1973, 1981), one obtains a first–order approximation for the solution of (2.3.29) as

$$u = a\cos(\Omega t - \gamma) + O(\epsilon) \qquad (2.3.31)$$

where the amplitude a and phase γ are governed by

$$a' = -\mu a + k\sin\gamma \qquad (2.3.32)$$

$$a\gamma' = \sigma a - \frac{3}{8}\alpha a^3 + k\cos\gamma \qquad (2.3.33)$$

The primes in (2.3.32) and (2.3.33) indicate derivatives with respect to the time scale τ, where $\tau = \epsilon t$.

The system (2.3.32) and (2.3.33) is a planar autonomous dynamical system. The fixed points (a_0, γ_0) of this system are given by

$$-\mu a_0 + k \sin \gamma_0 = 0 \tag{2.3.34}$$

$$\sigma a_0 - \frac{3}{8}\alpha a_0^3 + k \cos \gamma_0 = 0 \tag{2.3.35}$$

From (2.3.34) and (2.3.35), we obtain the so-called **frequency–response equation**

$$\left[\mu^2 + \left(\sigma - \frac{3}{8}\alpha a_0^2\right)^2\right] a_0^2 = k^2 \tag{2.3.36}$$

Equation (2.3.36) is an implicit equation for the amplitude of the response a_0 as a function of the external detuning σ (i.e., the excitation frequency) and the amplitude of the excitation k.

The stability of (a_0, γ_0) is determined by the eigenvalues of the Jacobian matrix

$$D_{\mathbf{x}}\mathbf{F} = \begin{bmatrix} -\mu & -a_0\left(\sigma - \frac{3}{8}\alpha a_0^2\right) \\ \frac{1}{a_0}\left(\sigma - \frac{9}{8}\alpha a_0^2\right) & -\mu \end{bmatrix} \tag{2.3.37}$$

The corresponding eigenvalues λ_i are the roots of

$$\lambda^2 + 2\mu\lambda + \mu^2 + \left(\sigma - \frac{3}{8}\alpha a_0^2\right)\left(\sigma - \frac{9}{8}\alpha a_0^2\right) = 0 \tag{2.3.38}$$

From (2.3.38), we find that the sum of the eigenvalues is -2μ. This sum is negative because $\mu > 0$. Consequently, at least one of the two eigenvalues will always have a negative real part. This fact eliminates the possibility of a pair of purely imaginary eigenvalues and, hence, a Hopf bifurcation. However, static bifurcations can occur. To this end, we find that one of the eigenvalues is zero when

$$\mu^2 + \left(\sigma - \frac{3}{8}\alpha a_0^2\right)\left(\sigma - \frac{9}{8}\alpha a_0^2\right) = 0 \tag{2.3.39}$$

For fixed $\mu > 0$, $\sigma > 0$, and $\alpha > 0$, we show the variation of a_0 with k in Figure 2.3.6. In this bifurcation diagram, the solid and broken

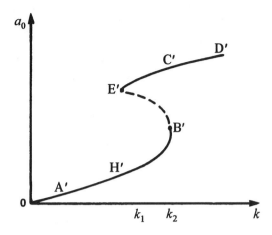

Figure 2.3.6: Response amplitude a_0 versus the amplitude of the forcing k.

lines correspond to the stable and unstable fixed points of (2.3.32) and (2.3.33), respectively. As the control parameter k is gradually increased from a zero value, a_0 follows the curve $A'H'B'$ until the critical value $k = k_2$ is reached. At $k = k_2$, a saddle–node bifurcation occurs, and, locally, there are no other solutions for $k > k_2$.

If we start from the point D' and decrease k gradually, a_0 follows the curve $D'C'E'$. At the critical point $k = k_1$, a saddle–node bifurcation occurs, and, locally, there are no other solutions for $k < k_1$. We note that points B' and E' are points of vertical tangencies. To this end, we find from (2.3.36) that

$$\frac{dk^2}{da_0^2} = \mu^2 + \left(\sigma - \frac{3}{8}\alpha a_0^2\right)\left(\sigma - \frac{9}{8}\alpha a_0^2\right)$$

which is zero by virtue of (2.3.39). Because saddle–node bifurcation points are locations of vertical tangencies, they are called **tangent bifurcations**. Further, because of the geometry at such points, they are also called **turning points** and **folds**. Yet another name for these bifurcation points is **limit points** (e.g., Kubicek and Marek, 1983; Keller, 1987).

Bifurcation diagrams, such as Figure 2.3.6, are known as **force–response curves** because they show the variation of the response amplitude as a function of the forcing amplitude. Let us suppose that an experiment is conducted to construct Figure 2.3.6. Then, as k is gradually increased from a zero value, a_0 follows the curve $A'H'B'$ until the critical value $k = k_2$ is reached. Here, a jump occurs from the stable branch $A'H'B'$ to the stable branch $E'C'D'$. As k is increased beyond k_2, a_0 follows the curve $C'D'$. Consequently, as k is slowly increased, the state of the system (e.g., a_0) evolves continuously except at $k = k_2$, where it experiences a discontinuous (jump) or catastrophic change. Therefore, saddle–node bifurcations are examples of **discontinuous** or **catastrophic bifurcations** (Shilnikov, 1976; Zeeman, 1982; Abraham and Shaw, 1992). If we start from the point D' and decrease k gradually, a_0 follows the curve $D'C'E'$. At $k = k_1$, a jump occurs from point E' to point H'. As k is decreased below k_1, a_0 follows the curve $H'A'$. Again, there is a discontinuous or catastrophic bifurcation at the saddle–node value $k = k_1$ at which the state–control function is discontinuous. We note that in the range $k_1 < k < k_2$, the realized response depends on the direction of sweep of the control parameter. This phenomenon is called the **hysteresis phenomenon**.

In Figure 2.3.6, for all values of $k < k_1$, there is only one stable fixed point of (2.3.32) and (2.3.33) in the $a - \gamma$ space. Hence, all evolutions in this space are attracted to the stable fixed point. Again, for all values of $k > k_2$, there is only one stable fixed point in the $a - \gamma$ space. In the interval $k_1 < k < k_2$, two stable branches of solutions coexist. Therefore, this interval is referred to as an interval of **bistability**. In the bistability interval, there are two stable and one unstable fixed points for each value of k. We let the fixed points on the branches $A'H'B'$, $B'E'$, and $E'C'D'$ be $A, B,$ and C, respectively. Then, evolutions in the $a - \gamma$ space approach either A or C as time $t \to \infty$, depending on the initial condition. In Figure 2.3.7, we show a qualitative sketch of the global stable and unstable manifolds of B. One end of the unstable manifold of B is attracted to A, while the other end is attracted to C. The stable manifold partitions the $a - \gamma$ plane into two regions. The evolutions initiated in the region to the left of the stable manifold of B are attracted to A, while the evolutions initiated in the region to the right of the stable manifold are attracted to C.

We note that the state–space portraits for $k < k_1$ and $k_1 < k < k_2$

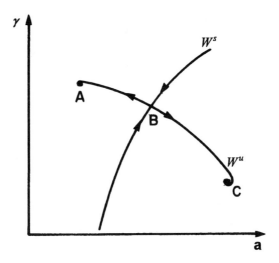

Figure 2.3.7: Basins of attraction of the stable fixed points A and C in the $a - \gamma$ space. The stable manifold of the saddle point B divides this space into two regions.

are **structurally different** because there is only one stable fixed point in one case and two stable and one unstable fixed points in the other case. Similarly, the state–space portraits for $k > k_2$ and $k_1 < k < k_2$ are **structurally different**.

Next, we consider the bifurcation diagrams when σ is used as a control parameter for $k > 0$ and $\mu > 0$. Figures 2.3.8a and 2.3.8b correspond to $\alpha < 0$ and $\alpha > 0$, respectively. Again, the solid and broken lines correspond to the stable and unstable fixed points of (2.3.32) and (2.3.33), respectively. For both $\alpha < 0$ and $\alpha > 0$, saddle–node bifurcations occur at $\sigma^{(1)}$ and $\sigma^{(2)}$.

To verify if these bifurcation points are points of vertical tangencies, we determine from (2.3.36) that

$$\frac{d\sigma}{da_0^2} = \frac{3}{8}\alpha - \frac{\mu^2 + (\sigma - \frac{3}{8}\alpha a_0^2)^2}{2a_0^2(\sigma - \frac{3}{8}\alpha a_0^2)} \tag{2.3.40}$$

Because (2.3.39) is satisfied at the bifurcation points, we substitute for

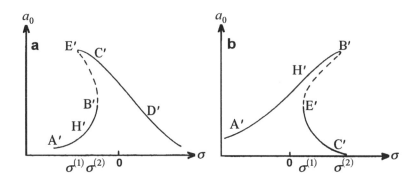

Figure 2.3.8: Bifurcation diagram constructed by using σ as a control parameter: (a) $\alpha < 0$ and (b) $\alpha > 0$. In each case, saddle–node bifurcations occur at B' and E'.

μ^2 from (2.3.39) into (2.3.40). After simplifying, we find that

$$\frac{d\sigma}{da_0^2} = 0$$

Bifurcation diagrams such as Figures 2.3.8a and 2.3.8b are called **frequency–response curves**. Further, the response displayed in Figure 2.3.8a for $\alpha < 0$ is called a **softening–type response**, while the response displayed in Figure 2.3.8b for $\alpha > 0$ is called a **hardening–type response**. If experiments were conducted to construct the frequency–response curves, jumps would be observed at the saddle–node bifurcation points, which are catastrophic bifurcations.

Example 2.19. Here, we consider the parametrically excited Duffing oscillator

$$\ddot{u} + u + \epsilon \left[2\mu \dot{u} + \alpha u^3 + 2ku \cos(\Omega t) \right] = 0 \qquad (2.3.41)$$

The parameters μ, α, and k are all independent of ϵ, while the parameter Ω is such that

$$\Omega = 2 + \epsilon\sigma \qquad (2.3.42)$$

This type of resonance, where the frequency of the parametric excitation is close to twice the natural frequency of the system, is called a **principal parametric resonance** (e.g., Nayfeh and Mook, 1979).

By using the method of multiple scales, one obtains the following first–order approximation for the solution of (2.3.41):

$$u = a \cos\left(\frac{1}{2}\Omega t - \frac{1}{2}\gamma\right) + O(\epsilon) \tag{2.3.43}$$

where the amplitude a and phase γ are governed by

$$a' = -\left\{\mu a + \frac{1}{2}ka \sin\gamma\right\} \tag{2.3.44}$$

$$a\gamma' = -\left\{-\sigma a + \frac{3}{4}\alpha a^3 + ka \cos\gamma\right\} \tag{2.3.45}$$

and the prime indicates the derivative with respect to the time scale $\tau = \epsilon t$. The first–order approximation can alternatively be expressed in the form

$$u = p \cos\left(\frac{1}{2}\Omega t\right) + q \sin\left(\frac{1}{2}\Omega t\right) + O(\epsilon) \tag{2.3.46}$$

where

$$p' = -\left\{\mu p + \frac{1}{2}(\sigma + k)q - \frac{3}{8}\alpha q(p^2 + q^2)\right\} \tag{2.3.47}$$

$$q' = -\left\{\mu q - \frac{1}{2}(\sigma - k)p + \frac{3}{8}\alpha p(p^2 + q^2)\right\} \tag{2.3.48}$$

It follows from (2.3.43) and (2.3.46) that

$$p = a \cos\left(\frac{1}{2}\gamma\right) \quad \text{and} \quad q = a \sin\left(\frac{1}{2}\gamma\right) \tag{2.3.49}$$

Equations (2.3.44) and (2.3.45) represent the so–called polar form of the modulation equations, while (2.3.47) and (2.3.48) represent the so–called Cartesian form of the modulation equations. We also observe that (2.3.44) and (2.3.45) are invariant under the transformation

$$(a, \gamma) \Leftrightarrow (-a, \gamma) \tag{2.3.50}$$

while (2.3.47) and (2.3.48) are invariant under the transformation

$$(p, q) \Leftrightarrow (-p, -q) \tag{2.3.51}$$

Next, we examine the fixed points (a_0, γ_0) of (2.3.44) and (2.3.45). They are solutions of the algebraic system

$$\mu a_0 + \frac{1}{2} k a_0 \sin \gamma_0 = 0 \qquad (2.3.52)$$

$$-\sigma a_0 + \frac{3}{4}\alpha a_0^3 + k a_0 \cos \gamma_0 = 0 \qquad (2.3.53)$$

There are two types of fixed points: (a) trivial fixed points corresponding to $a_0 = 0$ and (b) nontrivial fixed points corresponding to $a_0 \neq 0$. The corresponding fixed points (p_0, q_0) of (2.3.47) and (2.3.48) can be found by using (2.3.49). We find from (2.3.52) and (2.3.53) that the nontrivial fixed points are given by

$$a_0^2 = \frac{4}{3\alpha}\left\{\sigma \pm \sqrt{k^2 - 4\mu^2}\right\} \quad \text{and} \quad \gamma_0 = \arcsin\left(-\frac{2\mu}{k}\right) \qquad (2.3.54)$$

It is also obvious that the trivial fixed points share the symmetry property of the equations while the nontrivial fixed points do not share this symmetry property. Hence, we call the trivial and nontrivial fixed points **symmetric** and **asymmetric solutions**, respectively.

Next, we consider the stability of the trivial and nontrivial fixed points. For a trivial fixed point, we note that (2.3.52) and (2.3.53) become identities. In such cases, where one or more of the state variables of the system assume a zero value and cause one or more of the fixed–point equations to become identities, it is not convenient to determine the stability of the fixed point from the polar form of the modulation equations (e.g., Nayfeh and Mook, 1979; Nayfeh and Asfar, 1988; see also Section 4.5). In these and other cases, the stability of a fixed point can be conveniently determined from the Cartesian form of the modulation equations. From (2.3.47) and (2.3.48), we find that the Jacobian matrix associated with (p_0, q_0) is given by

$$D_\mathbf{x}\mathbf{F} = \begin{bmatrix} -\mu + \frac{3}{4}\alpha p_0 q_0 & -\frac{1}{2}(\sigma + k) + \frac{3}{8}\alpha(p_0^2 + 3q_0^2) \\ \frac{1}{2}(\sigma - k) - \frac{3}{8}\alpha(3p_0^2 + q_0^2) & -\mu - \frac{3}{4}\alpha p_0 q_0 \end{bmatrix}$$

$$(2.3.55)$$

We note from (2.3.55) that the sum of the two eigenvalues is -2μ, which is negative. Hence, at least one of the two eigenvalues will always have a negative real part. This fact eliminates the possibility of a pair of purely imaginary eigenvalues and, hence, a Hopf bifurcation. However, static bifurcations can occur. When k is used as a control parameter, the rank of the augmented matrix $[D_{\mathbf{x}}\mathbf{F} \mid \mathbf{F}_k]$ can be used to decide if a static bifurcation is a saddle–node bifurcation. To this end, we find from (2.3.47) and (2.3.48) that

$$\mathbf{F}_k = -\frac{1}{2}\left\{ \begin{array}{c} q \\ p \end{array} \right\} \tag{2.3.56}$$

For the trivial fixed point $(p_0, q_0) = (0,0)$, (2.3.55) reduces to

$$D_{\mathbf{x}}\mathbf{F} = \left[\begin{array}{cc} -\mu & -\frac{1}{2}(\sigma + k) \\ \frac{1}{2}(\sigma - k) & -\mu \end{array} \right] \tag{2.3.57}$$

and its corresponding eigenvalues are the roots of

$$\lambda^2 + 2\mu\lambda + \mu^2 + \frac{1}{4}(\sigma^2 - k^2) = 0 \tag{2.3.58}$$

One of these eigenvalues is zero when

$$k = k_2 = \sqrt{4\mu^2 + \sigma^2} \tag{2.3.59}$$

At $k = k_2$, (2.3.56) becomes

$$\mathbf{F}_k = \left\{ \begin{array}{c} 0 \\ 0 \end{array} \right\}$$

and the rank of the augmented matrix $[D_{\mathbf{x}}\mathbf{F} \mid \mathbf{F}_k]$ is one. Hence, the associated static bifurcation is not a saddle–node bifurcation.

For $\alpha > 0$ and $\sigma < 0$, it follows from (2.3.54) that nontrivial fixed points are possible only when $k > k_2$. They are given by

$$(a_{10}, \gamma_{10}) = \left(\sqrt{a_0^2}, \gamma_0 \right) \quad \text{and} \quad (a_{20}, \gamma_{20}) = \left(-\sqrt{a_0^2}, \gamma_0 \right) \tag{2.3.60}$$

where a_0^2 and γ_0 are defined by (2.3.54). The corresponding fixed points of (2.3.47) and (2.3.48) are

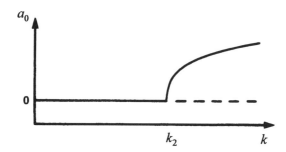

Figure 2.3.9: Bifurcation diagram constructed by using k as a control parameter for $\alpha > 0$ and $\sigma < 0$.

$$(p_{10}, q_{10}) = \left[a_{10} \cos\left(\frac{1}{2}\gamma_{10}\right), a_{10} \sin\left(\frac{1}{2}\gamma_{10}\right)\right]$$
$$(p_{20}, q_{20}) = \left[a_{20} \cos\left(\frac{1}{2}\gamma_{20}\right), a_{20} \sin\left(\frac{1}{2}\gamma_{20}\right)\right] \qquad (2.3.61)$$

In Figure 2.3.9, we show the bifurcation diagram when k is used as a control parameter for $\alpha > 0$ and $\sigma < 0$. In this and following figures, the solid and broken lines correspond to the stable and unstable fixed points of (2.3.47) and (2.3.48), respectively. As k is gradually increased from zero, the trivial fixed point remains stable until the critical value $k = k_2$ is reached. Here, a static bifurcation of the trivial fixed point occurs. For $k > k_2$, the trivial fixed point is a saddle point, and the stable nontrivial fixed points (a_{10}, γ_{10}) and (a_{20}, γ_{20}) are created. The fixed point (a_{20}, γ_{20}) is not shown in Figure 2.3.9. Both of the newly created fixed points are asymmetric solutions. Because the static bifurcation at $k = k_2$ leads to the creation of asymmetric solutions, it is called a **symmetry–breaking bifurcation**. Specifically, this bifurcation is a supercritical pitchfork bifurcation. In Sections 2.3.4 and 2.3.6, we derive the normal form for (2.3.47) and (2.3.48) in the vicinity of $k = k_2$ using two different methods and verify that we have a supercritical pitchfork bifurcation.

If an experiment were to be conducted to construct Figure 2.3.9, a gradual transition from a trivial response amplitude to a nontrivial response amplitude would be observed at $k = k_2$. Consequently, supercritical pitchfork bifurcations are continuous or safe bifurcations (Abraham and Shaw, 1992; Thompson, Stewart, and Ueda, 1994).

For $\alpha > 0$ and $\sigma > 0$, it follows from (2.3.54) that nontrivial fixed points are possible for

$$k > k_1 = 2\mu \qquad (2.3.62)$$

At $k = k_1$, (2.3.60) and (2.3.61) become

$$(a_{10}, \gamma_{10}) = \left(\sqrt{\frac{4\sigma}{3\alpha}}, -\frac{1}{2}\pi \right)$$

$$(a_{20}, \gamma_{20}) = \left(-\sqrt{\frac{4\sigma}{3\alpha}}, -\frac{1}{2}\pi \right) \qquad (2.3.63)$$

and

$$(p_{10}, q_{10}) = \left(\sqrt{\frac{2\sigma}{3\alpha}}, -\sqrt{\frac{2\sigma}{3\alpha}} \right)$$

$$(p_{20}, q_{20}) = \left(-\sqrt{\frac{2\sigma}{3\alpha}}, \sqrt{\frac{2\sigma}{3\alpha}} \right) \qquad (2.3.64)$$

For (p_{10}, q_{10}), (2.3.55) reduces to

$$D_{\mathbf{x}}\mathbf{F} = \begin{bmatrix} -(\mu + \frac{1}{2}\sigma) & -\mu + \frac{1}{2}\sigma \\ -(\mu + \frac{1}{2}\sigma) & -\mu + \frac{1}{2}\sigma \end{bmatrix} \qquad (2.3.65)$$

which obviously has a zero eigenvalue. Further, (2.3.56) becomes

$$\mathbf{F}_k = -\frac{1}{2} \left\{ \begin{array}{c} q_{10} \\ p_{10} \end{array} \right\} \qquad (2.3.66)$$

Because $\mathbf{F}_k$ does not belong to the range of $D_{\mathbf{x}}\mathbf{F}$, the rank of the matrix $[D_{\mathbf{x}}\mathbf{F} \mid \mathbf{F}_k]$ is two. Hence, at $k = k_1$, the fixed point (p_{10}, q_{10}) experiences a saddle–node bifurcation. Due to symmetry, the fixed point (p_{20}, q_{20}) also experiences a saddle–node bifurcation.

In Figure 2.3.10, we show the bifurcation diagram when k is used

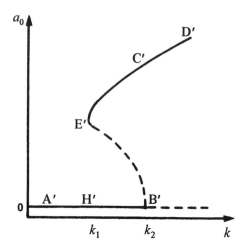

Figure 2.3.10: Bifurcation diagram constructed by using k as a control parameter for $\alpha > 0$ and $\sigma > 0$.

as a control parameter for $\alpha > 0$ and $\sigma > 0$. Again, as k is gradually increased from zero, the trivial fixed point loses stability at $k = k_2$ due to a symmetry–breaking bifurcation. Here, locally, there are no other stable solutions for $k > k_2$, forcing the system to jump in a fast dynamic transient to C'. Therefore, we have a subcritical pitchfork bifurcation, which is another example of a discontinuous or catastrophic bifurcation.

Next, let us suppose that we are on the stable branch of nontrivial fixed points at a point where $k > k_2$. Then, as the control parameter k is decreased gradually, the nontrivial fixed point remains stable until the critical value $k = k_1$ is reached. Here, a saddle–node bifurcation occurs. To verify that this point is a point of vertical tangency, we find from (2.3.54) that

$$\frac{dk^2}{da_0^2} = \frac{3\alpha}{2}\left(\frac{3\alpha}{4}a_0^2 - \sigma\right) \tag{2.3.67}$$

which is zero at $k = k_1$ because

$$a_0^2 = \frac{4\sigma}{3\alpha}$$

In Sections 2.3.4 and 2.3.6, we derive the normal form for (2.3.47) and (2.3.48) in the vicinity of the static–bifurcation points at $k = k_1$ and

$k = k_2$ and verify that they are saddle–node and subcritical pitchfork
bifurcation points, respectively. We point out that in Figures 2.3.9
and 2.3.10, branches with different tangents meet at the pitchfork
bifurcation points. This situation also occurs at transcritical bifurcation
points. Hence, we refer to transcritical and pitchfork bifurcation points
as **branch points**. Our usage of this terminology is consistent with
that of Parker and Chua (1989) and not consistent with that of Seydel
(1979a, 1988) and Kubicek and Marek (1983), who refer to all static
bifurcation points as branch points.

If an experiment were to be conducted for constructing Figure
2.3.10, jumps would be observed at $k = k_2$ during a forward sweep
of the control parameter and at $k = k_1$ during a reverse sweep of the
control parameter. Therefore, the bifurcations at $k = k_1$ and k_2 are
catastrophic bifurcations. Thus, subcritical pitchfork bifurcations are
also catastrophic bifurcations.

In Figure 2.3.11, we show the loci of the possible bifurcation points
of (2.3.47) and (2.3.48) in the space of the control parameters k and
σ for $k > 0$ and $\mu = 0.1$. The broken and solid lines correspond to
pitchfork and saddle–node bifurcation points, respectively. Here, we do
not have any transcritical or Hopf bifurcation points.

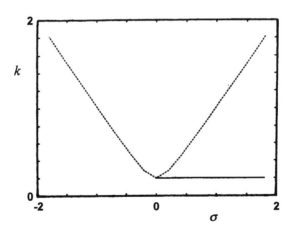

Figure 2.3.11: Bifurcation set in the $k - \sigma$ plane.

The pitchfork bifurcation points fall on the curve

$$k^2 = 4\mu^2 + \sigma^2$$

The saddle–node bifurcation points occur only for $\sigma > 0$, and they fall on the line

$$k = 2\mu$$

When $\sigma = 0$, the pitchfork and saddle–node bifurcation points coalesce. Sets, such as Figure 2.3.11, that consist of the different bifurcation points in the control–parameter space are called **bifurcation sets**. In other systems, one may need to use the numerical schemes described in Chapter 6 to determine the loci of static and dynamic bifurcation points.

In the above example, it was possible to determine the fixed points of the given dynamical system in closed form. However, it may not always be possible to find the fixed points of a given dynamical system in closed form. One must then resort to algorithms such as the Newton–Raphson algorithm (e.g., Stoer and Bulirsch, 1980) and homotopy algorithms (Watson, Billups, and Morgan, 1987; Watson, 1990) to determine the fixed points numerically. Many software packages, such as the IMSL MATH/LIBRARY (1989) and HOMPACK (Watson, Billups, and Morgan, 1987), are available for determining the real solutions of a given system of nonlinear equations. Furthermore, to trace branches of fixed points systematically in a state–control space of a general system, one may need to use the **continuation schemes** described in Chapter 6.

2.3.4   Center Manifold Reduction

In studies of dynamical systems, simplification methods are often used to reduce the order of the system of equations and/or eliminate as many nonlinearities as possible in the system of equations. Perturbation methods, such as the methods of multiple scales and averaging, may be perceived as simplification methods because there is a reduction in the dimension as one goes from the original system to the averaged system. Here, three methods are considered for conducting local bifurcation analysis in the vicinity of a fixed point of a system such as (2.1.1).

We discuss center manifold reduction in this section, the Lyapunov–Schmidt method in Section 2.3.5, and the method of multiple scales in Section 2.3.6.

Recalling the concept of the center manifold of a fixed point, we note that there is a center manifold associated with the fixed point undergoing the bifurcation. This manifold is a curved m–dimensional surface that is tangent at the bifurcation point to the subspace spanned by the m eigenvectors $\mathbf{p}_1$, $\mathbf{p}_2$, $\cdots$, $\mathbf{p}_m$ corresponding to the m eigenvalues λ_1, λ_2, $\cdots$, λ_m with zero real parts. The dimension m is less than the dimension n of the full system. In center manifold analyses (Carr, 1981; Arnold, 1988), based on the Shoshitaishvili theorem, one effectively reduces the dynamics of the considered n–dimensional system to the dynamics on the m–dimensional center manifold. Commonly, in local bifurcation analysis, a center manifold reduction is used to reduce the order of the dynamical system first, and then the method of normal forms is used to simplify the (nonlinear) structure of the reduced system.

To determine the center manifold associated with a fixed point $\mathbf{x} = \mathbf{x}_0$ of (2.1.1) at $\mathbf{M} = \mathbf{M}_0$, we first use the transformation (2.1.3) to shift the fixed point in question to the origin and obtain (2.1.4). Next, assuming that $\mathbf{F}$ in (2.1.4) is sufficiently smooth (i.e., $\mathcal{C}^r$, where r is as large as needed), we expand (2.1.4) in a Taylor series for small $\| \mathbf{y} \|$ and obtain

$$\dot{\mathbf{y}} = A\mathbf{y} + \mathbf{F}_2(\mathbf{y}) + \mathbf{F}_3(\mathbf{y}) + O(\| \mathbf{y} \|^4) \qquad (2.3.68)$$

where $A = D_{\mathbf{x}}\mathbf{F}(\mathbf{x}_0; \mathbf{M}_0)$ is the $n \times n$ matrix of first partial derivatives of $\mathbf{F}$ evaluated at $(\mathbf{x}_0; \mathbf{M}_0)$ and the scalar components of the $n \times 1$ vector $\mathbf{F}_N(\mathbf{y})$ are polynomials of degree N in the components y_1, y_2, $\cdots$, y_n of $\mathbf{y}$.

Second, we arrange the eigenvalues of A so that λ_1, $\lambda_2, \cdots, \lambda_m$ are the m eigenvalues with zero real parts, and we let $\mathbf{p}_1, \mathbf{p}_2, \cdots, \mathbf{p}_m$ be their corresponding generalized eigenvectors. Moreover, we let λ_{m+1}, λ_{m+2}, $\cdots$, λ_n be the $(n-m)$ eigenvalues with nonzero real parts and $\mathbf{p}_{m+1}$, $\mathbf{p}_{m+2}$, $\cdots$, $\mathbf{p}_n$ be their corresponding generalized eigenvectors.

Third, we introduce the linear transformation $\mathbf{y} = P\mathbf{v}$, where $P = [\mathbf{p}_1\, \mathbf{p}_2\, \cdots\, \mathbf{p}_n]$, in (2.3.68) and obtain

$$\dot{\mathbf{v}} = J\mathbf{v} + P^{-1}\mathbf{F}_2(P\mathbf{v}) + P^{-1}\mathbf{F}_3(P\mathbf{v}) + \cdots \qquad (2.3.69)$$

where $J = P^{-1}AP$. We note that J can be rewritten as

$$J = \begin{bmatrix} J_c & 0 \\ 0 & J_s \end{bmatrix} \tag{2.3.70}$$

where J_c is an $m \times m$ matrix whose eigenvalues are λ_1, λ_2, $\cdots$, λ_m and J_s is an $(n-m) \times (n-m)$ matrix whose eigenvalues are λ_{m+1}, λ_{m+2}, $\cdots$, λ_n.

Fourth, we let $\mathbf{v}_c$ be the m–dimensional vector with the components v_1, v_2, $\cdots$, v_m and $\mathbf{v}_s$ be the $(n - m)$–dimensional vector with the components v_{m+1}, v_{m+2}, $\cdots$, v_n. In terms of $\mathbf{v}_c$ and $\mathbf{v}_s$, we rewrite (2.3.69) as

$$\dot{\mathbf{v}}_c = J_c\mathbf{v}_c + \mathbf{G}_2(\mathbf{v}_c, \mathbf{v}_s) + \mathbf{G}_3(\mathbf{v}_c, \mathbf{v}_s) + \cdots \tag{2.3.71}$$

and

$$\dot{\mathbf{v}}_s = J_s\mathbf{v}_s + \mathbf{H}_2(\mathbf{v}_c, \mathbf{v}_s) + \mathbf{H}_3(\mathbf{v}_c, \mathbf{v}_s) + \cdots \tag{2.3.72}$$

We note that $\mathbf{v}_c$ and $\mathbf{v}_s$ are linearly uncoupled but nonlinearly coupled. Further, $\mathbf{H}_N(0,0) = 0$, $\mathbf{G}_N(0,0) = 0$, and the Jacobian matrices $D\mathbf{H}_N(0,0)$ and $D\mathbf{G}_N(0,0)$ are matrices with zero entries. Because $\mathbf{H}_N$ and $\mathbf{G}_N$ are polynomials and hence infinitely differentiable, there exists a **local center manifold** of the form (Carr, 1981)

$$\mathbf{v}_s = \mathbf{h}(\mathbf{v}_c)$$

where $\mathbf{h}$ is a polynomial function of $\mathbf{v}_c$. Because $\mathbf{h}$ satisfies (2.3.71) and (2.3.72) for only small $\| \mathbf{v}_c \|$, it is a **local invariant manifold**. (We recall that a solution of [2.3.71] and [2.3.72] initiated on the local invariant manifold is likely to remain on this manifold only for a finite length of time.) Further, this manifold is a **local center manifold** because

$$\mathbf{h}(0) = 0 \quad \text{and} \quad D_{\mathbf{v}_c}h_i(0) = 0 \tag{2.3.73}$$

where the h_i are the scalar components of $\mathbf{h}$.

Fifth, we determine the $(n - m)$–dimensional function $\mathbf{h}$ by constraining the center manifold to be m–dimensional in the n–dimensional space. Substituting for $\mathbf{v}_s$ in (2.3.72) and using (2.3.71), we arrive at

$$D_{\mathbf{v}_c}\mathbf{h}(\mathbf{v}_c) \{J_c\mathbf{v}_c + \mathbf{G}_2[\mathbf{v}_c, \mathbf{h}(\mathbf{v}_c)] + \mathbf{G}_3[\mathbf{v}_c, \mathbf{h}(\mathbf{v}_c)] + \cdots\}$$
$$= J_s\mathbf{h}(\mathbf{v}_c) + \mathbf{H}_2[\mathbf{v}_c, \mathbf{h}(\mathbf{v}_c)] + \mathbf{H}_3[\mathbf{v}_c, \mathbf{h}(\mathbf{v}_c)] + \cdots \tag{2.3.74}$$

To solve (2.3.74), one approximates the components of $\mathbf{h}(\mathbf{v}_c)$ with polynomials. The polynomial approximations are usually taken to be quadratic to the first approximation and do not contain constant and linear terms so that the conditions (2.3.73) are satisfied. Substituting the assumed quadratic polynomial approximations into (2.3.74) and equating the coefficients of the different terms in the polynomials on both sides, one obtains a system of algebraic equations for the coefficients of the polynomials. Solving these equations, we obtain a first approximation to the center manifold $\mathbf{v}_s = \mathbf{h}(\mathbf{v}_c)$. Finally, substituting this approximation into (2.3.71), we obtain the m–dimensional system of equations

$$\dot{\mathbf{v}}_c = J_c \mathbf{v}_c + \mathbf{G}_2[\mathbf{v}_c, \mathbf{h}(\mathbf{v}_c)] + \mathbf{G}_3[\mathbf{v}_c, \mathbf{h}(\mathbf{v}_c)] + \cdots \qquad (2.3.75)$$

describing the dynamics of the system (2.1.1) on the center manifold. Thus, with this process, one reduces the number of equations from n to m. Later, (2.3.75) is transformed by using the method of normal forms to a "simpler" form, which is not necessarily unique.

There are stability theorems (Carr, 1981) that imply that if the trivial fixed point of (2.3.75) is stable (unstable), then the corresponding fixed point of (2.1.1) is stable (unstable). Carr (1981) has also discussed center manifold analysis in the context of infinite–dimensional systems. There is also some work on construction of center manifolds in a nonlocal scenario (e.g., Knobloch, 1990). Next, we consider three examples.

Example 2.20. Here, we construct a center manifold for the fixed point $(0,0)$ of (2.1.22). The eigenvalues of the associated linearization are 0 and -1. In Figure 2.3.12, the corresponding one–dimensional center and stable eigenspaces are depicted by broken lines. Now, we determine the local center manifold for the nonlinear system by using the steps outlined above. It is not necessary to carry out the first three steps because (2.1.22) is already in the form of (2.3.71) and (2.3.72). The local center manifold of $(0,0)$ is tangent to the x-axis at the origin. Hence, we assume that it is given by

$$y = h(x) = ax^2 + bx^3 + \cdots \qquad (2.3.76)$$

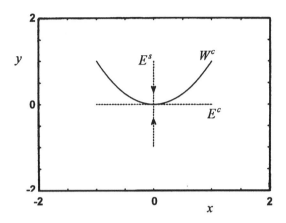

Figure 2.3.12: Center manifold W^c and stable manifold W^s of the fixed point $(0,0)$ of $(2.1.22)$.

Substituting $(2.3.76)$ into the second equation in $(2.1.22)$ yields

$$(2ax + 3bx^2)\dot{x} = x^2 - ax^2 - bx^3$$

But, it follows from the first equation in $(2.1.22)$ that

$$\dot{x} = kxy = kx(ax^2 + bx^3)$$

Consequently,

$$(2ax + 3bx^2 + \cdots)kx(ax^2 + bx^3 + \cdots) = (1 - a)x^2 - bx^3 + \cdots \quad (2.3.77)$$

Equating the coefficients of x^2 and x^3 on both sides of $(2.3.77)$, we obtain

$$a = 1 \quad \text{and} \quad b = 0 \qquad (2.3.78)$$

Therefore, $(2.3.76)$ has the form

$$y = x^2 + \cdots \qquad (2.3.79)$$

This local center manifold is depicted by using a solid line in Figure 2.3.12. We substitute $(2.3.79)$ into the first equation of $(2.1.22)$ and find that the dynamics on this local center manifold is governed by

$$\dot{x} = kx^3 + \cdots \qquad (2.3.80)$$

Thus, (2.1.22) is reduced to (2.3.80) in the vicinity of the fixed point $(0,0)$. It follows from (2.3.80) that the flow on the center manifold is attracted to the fixed point $x = 0$ when $k < 0$ and repelled away from it when $k > 0$, in agreement with the numerical results shown in Figure 2.1.4.

Example 2.21. We simplify the dynamical system (2.3.47) and (2.3.48) to its normal form in the vicinity of the bifurcation point at $k = k_2$ in Figures 2.3.9 and 2.3.10. To this end, we rewrite (2.3.47) and (2.3.48) in the form

$$\left\{ \begin{array}{c} p' \\ q' \end{array} \right\} = - \left\{ \begin{array}{c} \mu p + \frac{1}{2}(\sigma + k_2)q + \frac{1}{2}\hat{k}q - \frac{3}{8}\alpha q(p^2 + q^2) \\ \mu q - \frac{1}{2}(\sigma - k_2)p + \frac{1}{2}\hat{k}p + \frac{3}{8}\alpha p(p^2 + q^2) \end{array} \right\} \tag{2.3.81}$$

where

$$k_2 = \sqrt{4\mu^2 + \sigma^2} \quad \text{and} \quad \hat{k} = k - k_2$$

The fixed point $(p_0, q_0) = (0,0)$ experiences a static bifurcation at $k = k_2$ or $\hat{k} = 0$. The associated Jacobian matrix is given by (2.3.57), whose eigenvalues and eigenvectors are

$$\lambda_1 = 0, \ \mathbf{z}_1 = \left\{ \begin{array}{c} 1 \\ -\eta \end{array} \right\}, \quad \text{and} \quad \lambda_2 = -2\mu, \ \mathbf{z}_2 = \left\{ \begin{array}{c} 1 \\ \eta \end{array} \right\} \tag{2.3.82}$$

where

$$\eta = \frac{2\mu}{\sigma + k_2} = \frac{2\mu}{\sigma + \sqrt{4\mu^2 + \sigma^2}} \tag{2.3.83}$$

By using the eigenvectors, we construct the matrix

$$Z = [\mathbf{z}_1 \ \mathbf{z}_2] = \left[\begin{array}{cc} 1 & 1 \\ -\eta & \eta \end{array} \right] \tag{2.3.84}$$

whose inverse is

$$Z^{-1} = \frac{1}{2} \left[\begin{array}{cc} 1 & -\eta^{-1} \\ 1 & \eta^{-1} \end{array} \right] \tag{2.3.85}$$

Next, we introduce the transformation

$$\left\{ \begin{array}{c} p \\ q \end{array} \right\} = Z \left\{ \begin{array}{c} x \\ y \end{array} \right\} \tag{2.3.86}$$

into (2.3.81) and multiply throughout from the left with the matrix Z^{-1}. This results in the following equations:

$$
\begin{aligned}
x' = &-\frac{1}{4}\hat{k}\left\{\eta(y-x) - \frac{1}{\eta}(x+y)\right\} \\
&+\frac{3}{16}\alpha\left\{\eta(y-x)\left[(x+y)^2 + \eta^2(y-x)^2\right]\right\} \\
&+\frac{3}{16}\alpha\left\{\frac{1}{\eta}(x+y)\left[(x+y)^2 + \eta^2(y-x)^2\right]\right\}
\end{aligned}
\tag{2.3.87}
$$

$$
\begin{aligned}
y' = &-2\mu y - \frac{1}{4}\hat{k}\left\{\eta(y-x) + \frac{1}{\eta}(x+y)\right\} \\
&+\frac{3}{16}\alpha\left\{\eta(y-x)\left[(x+y)^2 + \eta^2(y-x)^2\right]\right\} \\
&-\frac{3}{16}\alpha\left\{\frac{1}{\eta}(x+y)\left[(x+y)^2 + \eta^2(y-x)^2\right]\right\}
\end{aligned}
\tag{2.3.88}
$$

Now, the system (2.3.87) and (2.3.88) has the form of (2.3.71) and (2.3.72). To determine the dependence of the center manifold on the parameter $\hat{k}$, we need to augment (2.3.87) and (2.3.88) with the additional equation

$$
\hat{k}' = 0 \tag{2.3.89}
$$

This augmentation, suggested by Carr (1981), is called the **suspension trick** because the two–dimensional system (2.3.87) and (2.3.88) is suspended in the three–dimensional $(x, y, \hat{k})$ space. In (2.3.87)–(2.3.89), terms such as $\hat{k}x$ and $\hat{k}y$ are treated as nonlinear terms.

The center eigenspace of the fixed point $(x, y, \hat{k}) = (0, 0, 0)$ of (2.3.87)–(2.3.89) is spanned by the generalized eigenvectors

$$
\{1\ 0\ 0\}^T \quad \text{and} \quad \{0\ 0\ 1\}^T
$$

The center manifold of $(0, 0, 0)$ is tangent to this center eigenspace at the fixed point. For small $|\,x\,|$ and $|\,\hat{k}\,|$, the center manifold is described by

$$
y = h(x, \hat{k}) \tag{2.3.90}
$$

where the function h is such that, at the fixed point $(0, 0, 0)$,

$$h = 0, \quad \frac{\partial h}{\partial x} = 0, \quad \text{and} \quad \frac{\partial h}{\partial \hat{k}} = 0 \qquad (2.3.91)$$

Therefore, we represent the local center manifold by the polynomial expansion

$$y = b_1 x^2 + b_2 x^3 + b_3 \hat{k} x + O(x^4) \qquad (2.3.92)$$

where it is assumed that $\hat{k}$ is $O(x^2)$. The reason for this assumption will become clear as we proceed further.

Substituting (2.3.92) into (2.3.88) and using (2.3.87) and (2.3.89), we arrive at

$$-2\mu \left\{ b_1 x^2 + b_2 x^3 + b_3 \hat{k} x + \cdots \right\} + \frac{1}{4} \hat{k} x \left(\eta - \frac{1}{\eta} \right)$$
$$- \frac{3}{16} \alpha x^3 \left(\eta + \frac{1}{\eta} \right) (1 + \eta^2) + \cdots = 0 \qquad (2.3.93)$$

Equating each of the coefficients of x^2, x^3, and $\hat{k}x$ to zero, we find that

$$b_1 = 0$$
$$b_2 = -\frac{3\alpha}{32\mu} \left(\eta + \frac{1}{\eta} \right) (1 + \eta^2) \qquad (2.3.94)$$
$$b_3 = \frac{1}{8\mu} \left(\eta - \frac{1}{\eta} \right)$$

Then, substituting (2.3.94) into (2.3.92), we obtain

$$y = \frac{1}{8\mu} \left(\eta - \frac{1}{\eta} \right) \hat{k} x - \frac{3\alpha}{32\mu} \left(\eta + \frac{1}{\eta} \right) (1 + \eta^2) x^3 + \cdots \qquad (2.3.95)$$

Next, substituting (2.3.95) into (2.3.87) and (2.3.89), we find that the dynamics on the center manifold is governed by

$$x' = \frac{1}{4} \hat{k} \left(\eta + \frac{1}{\eta} \right) x - \frac{3\alpha}{16} (\eta^2 + 1) \left(\eta - \frac{1}{\eta} \right) x^3 + \cdots \qquad (2.3.96)$$

$$\hat{k}' = 0 \qquad (2.3.97)$$

Finally, substituting (2.3.83) into (2.3.96), we obtain

$$x' = \frac{\hat{k}k_2}{4\mu}\, x \,+\, \frac{3\alpha\sigma}{16\mu}\left[1 + \frac{4\mu^2}{(\sigma + k_2)^2}\right]x^3 + \cdots \qquad (2.3.98)$$

We note that the first and second terms on the right–hand side of (2.3.98) are of the same order when $\hat{k}$ is $O(x^2)$, thus justifying the assumption made earlier. If we did not have any prior knowledge of the bifurcation, we would have assumed that $\hat{k}$ is $O(x)$ and carried out the analysis. Furthermore, here it was not necessary to use the method of normal forms to simplify (2.3.98). In general, this may not be the case.

Equation (2.3.98) has the same form as (2.3.2); that is, the normal form for a pitchfork bifurcation. This bifurcation is supercritical when $\alpha\sigma < 0$ and subcritical when $\alpha\sigma > 0$. Hence, we have supercritical and subcritical pitchfork bifurcations at $k = k_2$ in Figures 2.3.9 and 2.3.10, respectively. When $\sigma = 0$, the coefficient of x^3 in (2.3.98) is zero, and there may be a degenerate bifurcation. In this case, it will be necessary to determine higher–order terms for describing the dynamics on the center manifold.

Example 2.21 (continued). We simplify (2.3.47) and (2.3.48) in the vicinity of the saddle–node bifurcation that takes place at $k = k_1$ in Figure 2.3.10. We rewrite (2.3.47) and (2.3.48) as

$$\left\{\begin{array}{c} p' \\ q' \end{array}\right\} = -\left\{\begin{array}{c} \mu p + \frac{1}{2}(\sigma + k_1)q + \frac{1}{2}\hat{k}q - \frac{3}{8}\alpha q(p^2 + q^2) \\[2mm] \mu q - \frac{1}{2}(\sigma - k_1)p + \frac{1}{2}\hat{k}p + \frac{3}{8}\alpha p(p^2 + q^2) \end{array}\right\} \qquad (2.3.99)$$

where

$$k_1 = 2\mu \quad \text{and} \quad \hat{k} = k - k_1$$

The fixed points (p_{10}, q_{10}) and (p_{20}, q_{20}) of (2.3.99) at $k = k_1$ or $\hat{k} = 0$ are given by (2.3.64). Here, we restrict our attention to the bifurcation experienced by (p_{10}, q_{10}). As a first step, we use the linear transformation

$$p = p_{10} + v = \sqrt{\frac{2\sigma}{3\alpha}} + v$$

$$q = q_{10} + w = -\sqrt{\frac{2\sigma}{3\alpha}} + w$$

(2.3.100)

to shift the fixed point (p_{10}, q_{10}) of (2.3.99) to the origin. The result is

$$v' = \frac{1}{2}\hat{k}\sqrt{\frac{2\sigma}{3\alpha}} - \left(\mu + \frac{1}{2}\sigma\right)v + \left(\frac{1}{2}\sigma - \mu - \frac{1}{2}\hat{k}\right)w$$
$$- \frac{3\alpha}{8}\sqrt{\frac{2\sigma}{3\alpha}}\left(v^2 - 2vw + 3w^2\right) + \frac{3\alpha}{8}\left(v^2w + w^3\right)$$

(2.3.101)

$$w' = -\frac{1}{2}\hat{k}\sqrt{\frac{2\sigma}{3\alpha}} - \left(\frac{1}{2}\sigma + \mu + \frac{1}{2}\hat{k}\right)v + \left(\frac{1}{2}\sigma - \mu\right)w$$
$$- \frac{3\alpha}{8}\sqrt{\frac{2\sigma}{3\alpha}}\left(3v^2 - 2vw + w^2\right) - \frac{3\alpha}{8}\left(v^3 + vw^2\right)$$

(2.3.102)

At $\hat{k} = 0$, the associated Jacobian matrix is given by (2.3.65), and its corresponding eigenvalues and eigenvectors are

$$\lambda_1 = 0, \ \mathbf{z}_1 = \begin{Bmatrix} \eta_1 \\ \eta_2 \end{Bmatrix} \ \text{and} \ \lambda_2 = -2\mu, \ \mathbf{z}_2 = \begin{Bmatrix} 1 \\ 1 \end{Bmatrix}$$

(2.3.103)

where

$$\eta_1 = -\mu + \frac{1}{2}\sigma \ \text{and} \ \eta_2 = \mu + \frac{1}{2}\sigma$$

(2.3.104)

By using the eigenvectors, we construct the matrix

$$Z = [\mathbf{z}_1 \ \mathbf{z}_2] = \begin{bmatrix} \eta_1 & 1 \\ \eta_2 & 1 \end{bmatrix}$$

(2.3.105)

whose inverse is

$$Z^{-1} = \frac{1}{2\mu}\begin{bmatrix} -1 & 1 \\ \eta_2 & -\eta_1 \end{bmatrix}$$

(2.3.106)

Next, we introduce the transformation

$$\left\{ \begin{array}{c} v \\ w \end{array} \right\} = Z \left\{ \begin{array}{c} x \\ y \end{array} \right\} \tag{2.3.107}$$

into the system (2.3.101) and (2.3.102) and multiply it throughout from the left with the matrix Z^{-1}. This results in the following equations:

$$x' = -\frac{1}{2\mu}\hat{k}\sqrt{\frac{2\sigma}{3\alpha}} + \frac{1}{4\mu}\hat{k}(\eta_2 - \eta_1)x + \frac{3\alpha}{8\mu}\sqrt{\frac{2\sigma}{3\alpha}}(w^2 - v^2)$$

$$-\frac{3\alpha}{16\mu}(v + w)(v^2 + w^2) \tag{2.3.108}$$

$$y' = -2\mu y + \frac{1}{4\mu}\hat{k}\sqrt{\frac{2\sigma}{3\alpha}}(\eta_2 + \eta_1)$$

$$-\frac{1}{4\mu}\hat{k}\left[\eta_2(\eta_2 x + y) - \eta_1(\eta_1 x + y)\right]$$

$$\tag{2.3.109}$$

$$-\frac{3\alpha}{16\mu}\sqrt{\frac{2\sigma}{3\alpha}}\left[\eta_2(v^2 - 2vw + 3w^2) - \eta_1(3v^2 - 2vw + w^2)\right]$$

$$+\frac{3\alpha}{16\mu}\left[\eta_2(v^2 w + w^3) + \eta_1(v^3 + w^2 v)\right]$$

We augment (2.3.108) and (2.3.109) with the equation

$$\hat{k}' = 0 \tag{2.3.110}$$

to capture the dependence of the center manifold on the parameter $\hat{k}$. In the three–dimensional system (2.3.108)–(2.3.110), terms such as $\hat{k}x$ and $\hat{k}y$ are treated as nonlinear terms.

The center eigenspace of $(0,0,0)$ of (2.3.108)–(2.3.110) is spanned by the generalized eigenvectors

$$\{1\ 0\ 0\}^T \quad \text{and} \quad \left\{ 0 \ \ \sqrt{\frac{2\sigma}{3\alpha}}\frac{\eta_2 + \eta_1}{8\mu^2} \ \ 1 \right\}^T$$

The center manifold of $(0,0,0)$ is tangent to this eigenspace at the fixed point. For small $| x |$ and $| \hat{k} |$, we describe the manifold by the polynomial expansion

$$y = b_1 \hat{k} + b_2 x^2 + \cdots \tag{2.3.111}$$

where $\hat{k}$ is assumed to be $O(x^2)$. Substituting (2.3.111) into (2.3.108)–(2.3.110), we arrive at

$$-2\mu \left\{ b_1 \hat{k} + b_2 x^2 + \cdots \right\} + \frac{1}{4\mu} \hat{k} \sqrt{\frac{2\sigma}{3\alpha}} (\eta_2 + \eta_1)$$
$$-\frac{9\alpha}{16\mu} \sqrt{\frac{2\sigma}{3\alpha}} (\eta_2 - \eta_1)(\eta_1^2 + \eta_2^2) x^2 + \cdots = 0 \tag{2.3.112}$$

Equating the coefficients of x^2 and $\hat{k}$ on both sides of (2.3.112), we find that

$$b_1 = \sqrt{\frac{2\sigma}{3\alpha}} \frac{\eta_2 + \eta_1}{8\mu^2}$$
$$b_2 = -\frac{9\alpha}{32\mu^2} \sqrt{\frac{2\sigma}{3\alpha}} (\eta_2 - \eta_1)(\eta_1^2 + \eta_2^2) \tag{2.3.113}$$

Therefore, the local center manifold is described by (2.3.111), where b_1 and b_2 are given by (2.3.113). Substituting (2.3.111) into (2.3.108) and (2.3.110) and using (2.3.104) and (2.3.113), we find that the dynamics on the center manifold is governed by

$$x' = -\frac{\hat{k}}{2\mu} \sqrt{\frac{2\sigma}{3\alpha}} + \frac{3\alpha\sigma}{4} \sqrt{\frac{2\sigma}{3\alpha}} x^2 + \cdots \tag{2.3.114}$$

$$\hat{k}' = 0 \tag{2.3.115}$$

It is clear that the first and second terms on the right–hand side of (2.3.114) are of the same order when $\hat{k}$ is $O(x^2)$, as assumed earlier. Equation (2.3.114) has the normal form for a saddle–node bifurcation. When $\alpha\sigma > 0$, fixed points of (2.3.114) exist only for $\hat{k} > 0$, in agreement with the results shown in Figure 2.3.10.

2.3.5 The Lyapunov–Schmidt Method

The Lyapunov–Schmidt method can also be used to simplify a dynamical system in the vicinity of a static bifurcation point. We consider the autonomous system (2.1.1) and let $(\mathbf{x}_0, \mathbf{M}_0)$ be a static bifurcation point. In the vicinity of this point, the system $\mathbf{F}(\mathbf{x}; \mathbf{M}) = \mathbf{0}$ is simplified to a "simpler" system of algebraic equations after carrying out projections onto the range and null spaces associated with the matrix $D_{\mathbf{x}}\mathbf{F}(\mathbf{x}_0; \mathbf{M}_0)$. Subsequently, the system of equations in the range space is solved, and then the system of equations in the null space is treated. The "simple" system of algebraic equations obtained by using the Lyapunov–Schmidt method is essentially the one obtained by setting the time derivatives equal to zero in the normal form associated with the considered bifurcation. We note that the Lyapunov–Schmidt method is also applicable to infinite–dimensional systems. The reader is referred to Vainberg and Trenogin (1962, 1974) for more information on the Lyapunov–Schmidt method.

2.3.6 The Method of Multiple Scales

In this section, we use two examples to illustrate how the method of multiple scales can be used as a simplification method in the vicinity of a static bifurcation point.

Example 2.22. We consider the bifurcation that occurs at $k = k_2$ in Figures 2.3.9 and 2.3.10. In the vicinity of these bifurcation points, we expand the states in (2.3.47) and (2.3.48) according to

$$p = p_0 + \sum_{n=1}^{3} \tilde{\epsilon}^n p_n(T_0, T_1, T_2) + O(\tilde{\epsilon}^4) \qquad (2.3.116)$$

$$q = q_0 + \sum_{n=1}^{3} \tilde{\epsilon}^n q_n(T_0, T_1, T_2) + O(\tilde{\epsilon}^4) \qquad (2.3.117)$$

where

$$(p_0, q_0) = (0, 0) \qquad (2.3.118)$$

and the time scales T_n are defined by

$$T_n = \tilde{\epsilon}^n \tau \tag{2.3.119}$$

Further, $\tilde{\epsilon}$ is a small positive dimensionless parameter that is artificially introduced to establish the different orders of magnitude. The results obtained are independent of this parameter, and it is ultimately absorbed back into the solution, which is equivalent to setting it equal to unity at the end. The transformation of the time derivative is given by

$$\frac{d}{d\tau} = D_0 + \tilde{\epsilon}D_1 + \tilde{\epsilon}^2 D_2 + \cdots \tag{2.3.120}$$

where $D_n = \partial/\partial T_n$. To express the nearness of the control parameter to the bifurcation value k_2, we introduce a detuning parameter $\hat{k}$ according to

$$k = k_2 + \tilde{\epsilon}^2 \hat{k} \tag{2.3.121}$$

The deviation of k from k_2 is ordered at $O(\tilde{\epsilon}^2)$ so that its influence and that of the cubic nonlinearities in (2.3.47) and (2.3.48) are realized at the same order in the perturbation analysis.

We substitute (2.3.116)–(2.3.121) into (2.3.47) and (2.3.48), expand the results, equate coefficients of like powers of $\tilde{\epsilon}$, and obtain the following hierarchy of equations:

$O(\tilde{\epsilon})$:

$$\left\{ \begin{array}{c} D_0 p_1 \\ D_0 q_1 \end{array} \right\} = D_{\mathbf{x}}\mathbf{F} \left\{ \begin{array}{c} p_1 \\ q_1 \end{array} \right\} \tag{2.3.122}$$

$O(\tilde{\epsilon}^2)$:

$$\left\{ \begin{array}{c} D_0 p_2 \\ D_0 q_2 \end{array} \right\} = D_{\mathbf{x}}\mathbf{F} \left\{ \begin{array}{c} p_2 \\ q_2 \end{array} \right\} - \left\{ \begin{array}{c} D_1 p_1 \\ D_1 q_1 \end{array} \right\} \tag{2.3.123}$$

$O(\tilde{\epsilon}^3)$:

$$\left\{ \begin{array}{c} D_0 p_3 \\ D_0 q_3 \end{array} \right\} = D_{\mathbf{x}}\mathbf{F} \left\{ \begin{array}{c} p_3 \\ q_3 \end{array} \right\} - \left\{ \begin{array}{c} D_1 p_2 + D_2 p_1 \\ D_1 q_2 + D_2 q_1 \end{array} \right\}$$

$$- \frac{1}{2}\hat{k} \left\{ \begin{array}{c} q_1 \\ p_1 \end{array} \right\} + \frac{3\alpha}{8} \left\{ \begin{array}{c} q_1(p_1^2 + q_1^2) \\ -p_1(p_1^2 + q_1^2) \end{array} \right\} \qquad (2.3.124)$$

In (2.3.122)–(2.3.124), $D_{\mathbf{x}}\mathbf{F}$ is given by (2.3.57), and the corresponding eigenvalues and eigenvectors are given by (2.3.82).

The solution of the linear system (2.3.122) is

$$\left\{ \begin{array}{c} p_1 \\ q_1 \end{array} \right\} = x(T_1, T_2)\mathbf{z}_1 + y(T_1, T_2)\mathbf{z}_2 e^{-2\mu T_0}$$

Because the decaying part of this solution goes to zero as $T_0 \to \infty$, we retain only the nondecaying part and obtain

$$\left\{ \begin{array}{c} p_1 \\ q_1 \end{array} \right\} = x(T_1, T_2)\mathbf{z}_1 = x(T_1, T_2) \left\{ \begin{array}{c} 1 \\ -\eta \end{array} \right\} \qquad (2.3.125)$$

where η is defined in (2.3.83). The function x in (2.3.125) is determined by imposing the solvability conditions at the subsequent levels of approximation.

Because p_1 and q_1 given by (2.3.125) are independent of T_0, the higher–order terms in (2.3.116) and (2.3.117) are also independent of this time scale. Then, the solvability condition for (2.3.123) amounts to finding the solvability condition for

$$D_{\mathbf{x}}\mathbf{F} \left\{ \begin{array}{c} p_2 \\ q_2 \end{array} \right\} = \mathbf{c} \qquad (2.3.126)$$

where

$$\mathbf{c} = \left\{ \begin{array}{c} D_1 p_1 \\ D_1 q_1 \end{array} \right\} \qquad (2.3.127)$$

The system (2.3.126) is solvable if $\mathbf{c}$ is in the range of the matrix $D_{\mathbf{x}}\mathbf{F}$. If we recall from linear algebra that the range of $D_{\mathbf{x}}\mathbf{F}$ forms an orthogonal complement to the null space associated with the adjoint system, the solvability condition is (e.g., Nayfeh, 1981)

$$\mathbf{c}^T \bar{\mathbf{v}} = 0 \qquad (2.3.128)$$

for every nonzero solution $\mathbf{v}$ of the adjoint system

$$\overline{D_{\mathbf{x}}\mathbf{F}}^T\mathbf{v} = \mathbf{0} \tag{2.3.129}$$

where the overbar denotes the complex conjugate. Here, we have only one nonzero solution for (2.3.129). This is given by

$$\mathbf{v} = \left\{ \begin{array}{c} 1 \\ \frac{2\mu}{\sigma - \sqrt{4\mu^2 + \sigma^2}} \end{array} \right\} \tag{2.3.130}$$

Then, the solvability condition (2.3.128) translates to

$$D_1 x = 0 \tag{2.3.131}$$

which implies that $x = x(T_2)$. Therefore, the solution of (2.3.123) is

$$\left\{ \begin{array}{c} p_2 \\ q_2 \end{array} \right\} = \left\{ \begin{array}{c} 0 \\ 0 \end{array} \right\} \tag{2.3.132}$$

Because x is still an undetermined function, we proceed to the equations at $O(\tilde{\epsilon}^3)$. After substituting (2.3.132) into (2.3.124), we find that the solvability condition for (2.3.124) is

$$\mathbf{b}^T\bar{\mathbf{v}} = 0 \tag{2.3.133}$$

where $\mathbf{v}$ is given by (2.3.130) and

$$\mathbf{b} = \left\{ \begin{array}{c} D_2 p_1 + \frac{1}{2}\hat{k}q_1 - \frac{3\alpha}{8}q_1(p_1^2 + q_1^2) \\ D_2 q_1 + \frac{1}{2}\hat{k}p_1 + \frac{3\alpha}{8}p_1(p_1^2 + q_1^2) \end{array} \right\} \tag{2.3.134}$$

We substitute for $\mathbf{v}$ and $\mathbf{b}$ in (2.3.133), use (2.3.83), set $\tilde{\epsilon} = 1$, carry out algebraic manipulations, and obtain

$$x' = \frac{\hat{k}k_2}{4\mu}x + \frac{3\alpha\sigma}{16\mu}\left[1 + \frac{4\mu^2}{(\sigma + k_2)^2}\right]x^3 \tag{2.3.135}$$

We note that (2.3.135) is the same as (2.3.98) obtained by using center manifold reduction. After determining x from (2.3.135), we can determine third–order expansions for the states p and q by using (2.3.116)–(2.3.118), (2.3.125), and (2.3.132).

In this example, a prior knowledge of the normal form for the bifurcation guided us in ordering the deviation of k from $\hat{k}$ at $O(\tilde{\epsilon}^2)$ in (2.3.121). Let us suppose that we did not have this prior knowledge and that the following expansion was used instead of (2.3.121):

$$k = k_2 + \tilde{\epsilon}\tilde{k} + \tilde{\epsilon}^2\hat{k}$$

Then, the solvability condition for the system of equations at $O(\tilde{\epsilon}^2)$ would have dictated that $\tilde{k} = 0$.

Example 2.22 (continued). Here, we simplify (2.3.47) and (2.3.48) in the vicinity of the bifurcation point $k = k_1$ in Figure 2.3.10. The fixed point undergoing the bifurcation is

$$(p_{10}, q_{10}) = \left(\sqrt{\frac{2\sigma}{3\alpha}}, -\sqrt{\frac{2\sigma}{3\alpha}} \right) \tag{2.3.136}$$

We use the linear transformation (2.3.100) and rewrite the system (2.3.47) and (2.3.48) as

$$v' = \left(\frac{1}{2}k - \mu \right)\sqrt{\frac{2\sigma}{3\alpha}} - \left(\mu + \frac{1}{2}\sigma \right)v + \frac{1}{2}(\sigma - k)w$$
$$-\frac{3\alpha}{8}\sqrt{\frac{2\sigma}{3\alpha}}\left\{ v^2 - 2vw + 3w^2 \right\} + \frac{3\alpha}{8}\left\{ v^2w + w^3 \right\} \tag{2.3.137}$$

$$w' = -\left(\frac{1}{2}k - \mu \right)\sqrt{\frac{2\sigma}{3\alpha}} - \frac{1}{2}(\sigma + k)v + \left(\frac{1}{2}\sigma - \mu \right)w$$
$$-\frac{3\alpha}{8}\sqrt{\frac{2\sigma}{3\alpha}}\left\{ 3v^2 - 2vw + w^2 \right\} - \frac{3\alpha}{8}\left\{ v^3 + vw^2 \right\} \tag{2.3.138}$$

At $k = k_1 = 2\mu$, the fixed point $(0,0)$ of (2.3.137) and (2.3.138) experiences a saddle–node bifurcation.

In the neighborhood of the bifurcation point, we seek expansions of the form

$$v = v_0 + \sum_{n=1}^{2} \tilde{\epsilon}^n v_n(T_1, T_2) + O(\tilde{\epsilon}^3) \qquad (2.3.139)$$

$$w = w_0 + \sum_{n=1}^{2} \tilde{\epsilon}^n w_n(T_1, T_2) + O(\tilde{\epsilon}^3) \qquad (2.3.140)$$

where

$$(v_0, w_0) = (0, 0) \qquad (2.3.141)$$

Again, $\tilde{\epsilon}$ is an artificially introduced bookkeeping parameter and the time scales T_n are defined by (2.3.119). The transformation of the time derivative is given by (2.3.120). The nearness of the control parameter k to k_1 is expressed by

$$k = k_1 + \tilde{\epsilon}^2 \hat{k} \qquad (2.3.142)$$

We substitute (2.3.139)–(2.3.142) and (2.3.120) into (2.3.137) and (2.3.138), expand the results, equate coefficients of like powers of $\tilde{\epsilon}$, and obtain the following hierarchy of equations:

$O(\tilde{\epsilon})$:

$$\left\{ \begin{array}{c} D_0 v_1 \\ D_0 w_1 \end{array} \right\} = D_{\mathbf{x}} \mathbf{F} \left\{ \begin{array}{c} v_1 \\ w_1 \end{array} \right\} \qquad (2.3.143)$$

$O(\tilde{\epsilon}^2)$:

$$\left\{ \begin{array}{c} D_0 v_2 \\ D_0 w_2 \end{array} \right\} = D_{\mathbf{x}} \mathbf{F} \left\{ \begin{array}{c} v_2 \\ w_2 \end{array} \right\} - \left\{ \begin{array}{c} D_1 v_1 \\ D_1 w_1 \end{array} \right\} + \frac{1}{2} \hat{k} \sqrt{\frac{2\sigma}{3\alpha}} \left\{ \begin{array}{c} 1 \\ -1 \end{array} \right\}$$

$$- \frac{3\alpha}{8} \sqrt{\frac{2\sigma}{3\alpha}} \left\{ \begin{array}{c} v_1^2 - 2v_1 w_1 + 3w_1^2 \\ 3v_1^2 - 2v_1 w_1 + w_1^2 \end{array} \right\} \qquad (2.3.144)$$

In (2.3.143) and (2.3.144), $D_{\mathbf{x}} \mathbf{F}$ is given by (2.3.65), and the corresponding eigenvalues and eigenvectors are given by (2.3.103).

The nondecaying solution of (2.3.143) is

$$\left\{ \begin{array}{c} v_1 \\ w_1 \end{array} \right\} = x(T_1)\mathbf{z}_1 = x(T_1) \left\{ \begin{array}{c} \eta_1 \\ \eta_2 \end{array} \right\} \tag{2.3.145}$$

where the η_i are specified by (2.3.104). The function x is determined by imposing the solvability condition at the next level of approximation.

Because v_1 and w_1 given by (2.3.145) are independent of T_0, the higher–order terms in (2.3.139) and (2.3.140) are also independent of this time scale. Then, the solvability condition for (2.3.144) amounts to finding the solvability condition for

$$D_{\mathbf{x}}\mathbf{F} \left\{ \begin{array}{c} v_2 \\ w_2 \end{array} \right\} = \mathbf{c} \tag{2.3.146}$$

where

$$\mathbf{c} = \left\{ \begin{array}{c} D_1 v_1 - \frac{1}{2}\hat{k}\sqrt{\frac{2\sigma}{3\alpha}} + \frac{3\alpha}{8}\sqrt{\frac{2\sigma}{3\alpha}}(v_1^2 - 2v_1 w_1 + 3w_1^2) \\[2mm] D_1 w_1 + \frac{1}{2}\hat{k}\sqrt{\frac{2\sigma}{3\alpha}} + \frac{3\alpha}{8}\sqrt{\frac{2\sigma}{3\alpha}}(3v_1^2 - 2v_1 w_1 + w_1^2) \end{array} \right\} \tag{2.3.147}$$

The solvability condition is

$$\mathbf{c}^T \bar{\mathbf{v}} = 0 \tag{2.3.148}$$

for every nonzero solution $\mathbf{v}$ of the adjoint system

$$\overline{D_{\mathbf{x}}\mathbf{F}}^T \mathbf{v} = \mathbf{0} \tag{2.3.149}$$

In this case, we have only one nonzero solution for (2.3.149), namely,

$$\bar{\mathbf{v}} = \left\{ \begin{array}{c} 1 \\ -1 \end{array} \right\} \tag{2.3.150}$$

We substitute for $\mathbf{c}$ and $\mathbf{v}$ in (2.3.148), use (2.3.104) and (2.3.145), set $\tilde{\epsilon} = 1$, carry out algebraic manipulations, and obtain

$$x' = -\frac{\hat{k}}{2\mu}\sqrt{\frac{2\sigma}{3\alpha}} + \frac{3\alpha\sigma}{4}\sqrt{\frac{2\sigma}{3\alpha}}x^2 \tag{2.3.151}$$

We note that (2.3.151) is the same as (2.3.114) obtained by using center manifold reduction. After determining x from (2.3.151), we can determine second–order approximations for the states v and w by using (2.3.139)–(2.3.151), and (2.3.145).

The method of multiple scales has been used as a simplification method in many studies (e.g., Nayfeh, 1970; Moroz, 1986). With this method, it is only necessary to determine the eigenvectors associated with the zero eigenvalues at the bifurcation point. Further, **one effectively determines the dynamics on the center manifold without determining this manifold.** By contrast, in the case of center manifold reduction, it is necessary to determine all the eigenvectors at the bifurcation point and the inverse of the matrix of eigenvectors. Here, we determine the center manifold before determining the dynamics on this manifold. Hence, from a computational point of view, the method of multiple scales may be more appealing for systems with large dimensions.

Many software packages, such as MACSYMA (1988), IMSL MATH/ LIBRARY (1989), and MATLAB (1989), are available for symbolically/numerically determining eigenvalues and eigenvectors.

2.3.7   Structural Stability

In Section 1.4, we discussed various types of stability of solutions of a given system. Here, we study the stability of the **orbit structure** of a dynamical system to perturbations. By **orbit structure**, we mean the different orbits of the considered system and their number and stability for a given set of parameter values. The term **orbits** includes fixed points, periodic orbits, quasiperiodic orbits, and homoclinic and heteroclinic orbits.

The Hartman–Grobman theorem, discussed earlier in this chapter, states that vector fields are **structurally stable** in the vicinity of hyperbolic fixed points. On the other hand, in the vicinity of a bifurcation point, the orbit structure is unstable and the associated vector field is **structurally unstable**. For illustration, we consider the following examples:

Example 2.23. We consider the bifurcation diagram of Figure 2.3.9 for the dynamical system (2.3.44) and (2.3.45). A static bifurcation occurs at $k = k_2$. For $k < k_2$, there is only one stable fixed point in the (a, γ) space. This fixed point corresponds to $a_0 = 0$. For $k > k_2$, we have three fixed points in the (a, γ) space. Two of these fixed points correspond to $a_0 \neq 0$ and are stable, while the other fixed point corresponds to $a_0 = 0$ and is unstable. Hence, the portraits in the (a, γ) space for $k < k_2$ and $k > k_2$ are **structurally different** because there are differences in the number and stability of the orbits of the system. However, the state–space portraits for all $k < k_2$ or $k > k_2$ are similar. Because there is a qualitative change in the solution structure as k is increased or decreased through k_2, the vector field of (2.3.44) and (2.3.45) is said to be structurally unstable for $k = k_2$. This vector field is structurally stable for all $k \neq k_2$.

Example 2.24. Here, we examine the phase portraits shown in Figure 2.1.6. At each saddle point of Figure 2.1.6a, the respective stable and unstable manifolds intersect transversely. Further, on the homoclinic and heteroclinic orbits of Figure 2.1.6a, the corresponding stable and unstable manifolds intersect nontransversely. Comparing Figures 2.1.6a and 2.1.6b, we note that the transversal intersections at the saddle points B and D are stable to the damping perturbation, whereas the nontransversal intersections are not stable to this perturbation. In general, a vector field with nontransversal intersections is not structurally stable to all perturbations.

The discussions in this section clearly show that the subject of structural stability relies on geometric concepts. It is quite complex in n–dimensional spaces, where n is greater than two. An interesting and detailed exposition of structural stability is provided by Arnold (1988, Chapter 3).

2.3.8   Stability of Bifurcations to Perturbations

Although presented as a separate topic, this topic comes under the domain of structural stability. We state that a bifurcation is stable to a perturbation if there are no qualitative differences between the bifurcation diagrams in the perturbed and unperturbed systems. In the following examples, we consider the stability of pitchfork, saddle–node, and transcritical bifurcations to perturbations:

Example 2.25. We consider the stability of the saddle–node bifurcation in (2.3.1). The perturbed system has the form

$$\dot{x} = \mu - x^2 + \epsilon x \qquad (2.3.152)$$

where $\epsilon = 0$ yields the unperturbed system (2.3.1). In Figures 2.3.13a–c, we show the bifurcation diagrams for $\epsilon < 0$, $\epsilon = 0$, and $\epsilon > 0$, respectively. Although the location of the saddle–node bifurcation point for $\epsilon \neq 0$ differs from that for $\epsilon = 0$, the qualitative character of all three bifurcation diagrams is the same. Hence, the saddle–node bifurcation in (2.3.1) is stable to the perturbation ϵx.

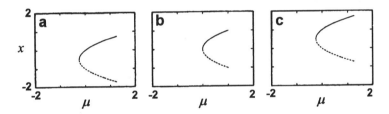

Figure 2.3.13: Stability of saddle–node bifurcation: (a) $\epsilon = -1$, (b) $\epsilon = 0$, and (c) $\epsilon = 1$.

Example 2.26. We consider the stability of the transcritical bifurcation in (2.3.3). The perturbed system has the form

$$\dot{x} = \mu x - x^2 + \epsilon \qquad (2.3.153)$$

where ϵ is the perturbation added to the normal form (2.3.3). In Figures 2.3.14a–c, we show the bifurcation diagrams for $\epsilon < 0$, $\epsilon = 0$, and $\epsilon > 0$, respectively. When $\epsilon < 0$, the transcritical bifurcation in Figure 2.3.14b is replaced by a pair of saddle–node bifurcations, as seen in Figure 2.3.14a. When $\epsilon > 0$, there are isolated branches of stable

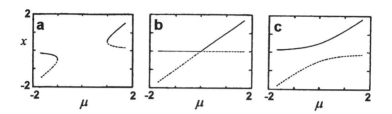

Figure 2.3.14: Stability of transcritical bifurcation: (a) $\epsilon = -0.25$, (b) $\epsilon = 0$, and (c) $\epsilon = 0.25$.

and unstable solutions and no bifurcations, as seen in Figure 2.3.14c. Because there are qualitative differences between Figures 2.3.14a and 2.3.14b and Figures 2.3.14c and 2.3.14b, the transcritical bifurcation in (2.3.3) is not stable to the considered perturbation.

Example 2.27. We consider the stability of the supercritical pitchfork bifurcation in (2.3.2). The perturbed system is

$$\dot{x} = \mu x - x^3 + \epsilon \qquad\qquad (2.3.154)$$

In Figures 2.3.15a–c, we show the bifurcation diagrams for $\epsilon < 0$, $\epsilon = 0$, and $\epsilon > 0$, respectively. In each of the Figures 2.3.15a and 2.3.15c, there is a saddle–node bifurcation point at

$$\mu = (27\epsilon^2/4)^{1/3}$$

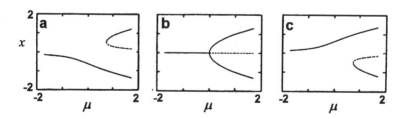

Figure 2.3.15: Stability of supercritical pitchfork bifurcation: (a) $\epsilon = -0.25$, (b) $\epsilon = 0$, and (c) $\epsilon = 0.25$.

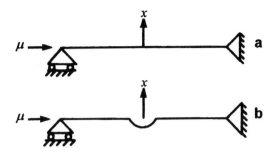

Figure 2.3.16: Buckling under a static loading: (a) perfect beam and (b) imperfect beam.

and two isolated branches of solutions. From Figure 2.3.15, we conclude that the supercritical pitchfork bifurcation in (2.3.2) is not stable to the considered perturbation. In Figures 2.3.16a and 2.3.16b, we show perfect and slightly bent beams subjected to a static loading μ, respectively. The variable x provides a measure of the displacement of the center of the beam. In Figure 2.3.16b, the beam has an imperfection in the direction $x < 0$. As μ is varied, the static buckling of a straight and perfect beam from the upright position to the buckled positions can be explained in terms of the pitchfork bifurcation shown in Figure 2.3.15b. In this diagram, the trivial and nontrivial fixed points correspond to the upright and buckled positions, respectively. If the imperfection in Figure 2.3.16b is small, the corresponding bifurcation diagram would resemble that shown in Figure 2.3.15b. However, when the imperfection is large, as the beam is loaded, the beam will deform in the direction $x < 0$ of the initial bending and there would not be any buckling. In this case, we can use (2.3.154) and consider the parameter ϵ as a measure of the imperfection. Then, Figure 2.3.15a would correspond to the imperfect beam of Figure 2.3.16b.

2.3.9 Codimension of a Bifurcation

Let us assume that we have a p–dimensional surface contained in an n–dimensional space. The **codimension** of this surface is $(n - p)$, and it represents the number of independent equations required to constrain the surface to be p–dimensional in the n–dimensional space. The bifurcations described thus far occur as a single control parameter is varied in the state–control space. In other words, these bifurcations depend on a single parameter. Because the associated bifurcation points lie on a surface whose codimension is one, the corresponding bifurcations are called **codimension–one bifurcations**. Extending this concept, a **codimension–k bifurcation** depends on k (independent) control parameters, and the associated bifurcation points lie on a surface whose codimension is k.

The structure of the Jacobian matrix associated with the fixed point depends on the codimension of a bifurcation. At a **codimension–one bifurcation point**, all the eigenvalues of the Jacobian matrix have nonzero real parts except that either one eigenvalue is zero or the real part of a pair of complex conjugate eigenvalues is zero (i.e., there are two purely imaginary eigenvalues). At a **codimension–m bifurcation point**, the Jacobian matrix has k zero eigenvalues and $(m - k)$ pairs of purely imaginary eigenvalues, where $0 \le k \le m$. To determine all the possible codimension–m bifurcations a fixed point can experience, one can examine the associated bifurcation set. For example, the locations in the bifurcation set, where the loci of two codimension–one bifurcation points intersect, are usually **codimension–two bifurcation points**.

As was noted earlier in Section 2.3.2, the normal form for a bifurcation depends on the codimension of a bifurcation. For a codimension–one bifurcation, one of the terms in each of the equations of the associated normal form vanishes at the bifurcation point. For example, in the normal form for a transcritical bifurcation; that is, (2.3.23), the linear term vanishes at the bifurcation point. If the quadratic term also vanishes, there is a codimension–two bifurcation. However, in this case, the associated normal form will have higher–order terms.

Example 2.28. In the context of (2.3.44) and (2.3.45) or (2.3.47) and (2.3.48), the sum of the eigenvalues λ_i of the 2×2 Jacobian matrix is -2μ, which is always negative. Hence, at least one of the two eigenvalues always has a negative real part. Therefore, it is not possible to have codimension–two bifurcations in this system.

To understand the consequences of some codimension–two bifurcations of fixed points, one needs to carry out a nonlocal analysis. For a thorough consideration of different possible codimension–two bifurcations of fixed points and their consequences, the reader is referred to Guckenheimer and Holmes (1983, Chapter 7) and Arnold (1988, Chapter 6).

2.3.10   Global Bifurcations

Global bifurcations are **nonlocal bifurcations**. They are associated with global changes in the state space and can occur as a control parameter is gradually varied. To go from the state of the system preceding the bifurcation to that subsequent to the bifurcation, a global change in the state variables is necessary.

To give an idea of what is meant by a global bifurcation, we return to Figure 2.1.6. When damping is included, the homoclinic and heteroclinic orbits of Figure 2.1.6a are destroyed. A global change occurs in the state space when the damping coefficients assume positive values, as illustrated in Figure 2.1.6b. In Figure 2.1.7, we examined how the stability region of a fixed point varied with damping. Here, a global bifurcation takes place as we go from Figure 2.1.7c to Figure 2.1.7e. In Figure 2.1.7d, which represents the state–space scenario at the bifurcation point, there is a heteroclinic half–orbit.

Typically, at a global bifurcation point, there is either a homoclinic or a heteroclinic orbit. We call a bifurcation that leads to the destruction of a homoclinic (heteroclinic) orbit as a **homoclinic (heteroclinic) bifurcation**. Global bifurcations are of considerable interest because they can result in the creation of chaotic orbits, as discussed in Chapter 5.

2.4   BIFURCATIONS OF MAPS

As a single control parameter is varied, a fixed point of a map can experience a bifurcation if it is nonhyperbolic. From Section 2.2, we recall that there are three conditions in which a fixed point $x = x_0$ of the map (2.2.1) can be nonhyperbolic at $M = M_0$. These conditions are

1. $D_{\mathbf{x}}\mathbf{F}(\mathbf{x}_0; \mathbf{M}_0)$ has one eigenvalue equal to 1, with the remaining $(n-1)$ eigenvalues being within the unit circle,

2. $D_{\mathbf{x}}\mathbf{F}(\mathbf{x}_0; \mathbf{M}_0)$ has one eigenvalue equal to -1, with the remaining $(n-1)$ eigenvalues being within the unit circle,

3. $D_{\mathbf{x}}\mathbf{F}(\mathbf{x}_0; \mathbf{M}_0)$ has a pair of complex conjugate eigenvalues on the unit circle, with the remaining $(n-2)$ eigenvalues being within the unit circle.

As illustrated in the next three examples, under the first of the above three conditions, the bifurcations of fixed points of maps are analogous to codimension–one static bifurcations of fixed points of continuous–time systems.

Example 2.29. We consider the one–dimensional map

$$x_{k+1} = x_k + \mu - x_k^2 \tag{2.4.1}$$

where μ is a scalar control parameter. When $\mu > 0$, it is clear from Figure 2.4.1a that this map intersects the identity map $x_{k+1} = x_k$ in two points. In other words, for $\mu > 0$, (2.4.1) has the nontrivial fixed points

$$x_{10} = \sqrt{\mu} \text{ and } x_{20} = -\sqrt{\mu}$$

The Jacobian matrix associated with the fixed point x_{j0} has the single eigenvalue

$$\rho = 1 - 2x_{j0}$$

The fixed point x_{20} is an unstable node for all $\mu > 0$ because $|\rho| > 1$. On the other hand, the fixed point x_{10} is a stable node for $0 < \mu < 1$ because $|\rho| < 1$. When μ is decreased to zero, the two fixed points approach each other and coalesce at the single fixed point $x = 0$, as shown in Figure 2.4.1b. The map (2.4.1) is tangent to the identity map and, hence, the associated bifurcation is called **tangent bifurcation**. When $\mu < 0$, the map (2.4.1) does not intersect the identity map, as shown in Figure 2.4.1c, and hence for $\mu < 0$, (2.4.1)

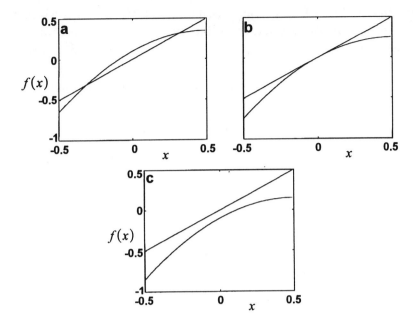

Figure 2.4.1: The functions $f(x) = \mu + x - x^2$ and x for (a) $\mu = 0.1$, (b) $\mu = 0.0$, and (c) $\mu = -0.1$.

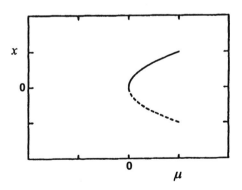

Figure 2.4.2: Scenario in the vicinity of a saddle–node bifurcation.

does not have any fixed points. In Figure 2.4.2, we show the different fixed points of (2.4.1) and their stability in the vicinity of the origin of the $x - \mu$ space. Broken and solid lines are used to represent branches of unstable and stable fixed points, respectively. A **saddle–node** or **tangent bifurcation** occurs at $(x, \mu) = (0, 0)$. The scenario is identical to that shown in Figure 2.3.1 for a continuous–time system.

Example 2.30. We consider the one–dimensional map

$$x_{k+1} = x_k + \mu x_k - x_k^2 \qquad (2.4.2)$$

where μ is again a scalar control parameter. This map has the two fixed points

$$x_{10} = 0 : \quad \text{trivial fixed point}$$
$$x_{20} = \mu : \quad \text{nontrivial fixed point}$$

For the fixed point x_{j0}, the Jacobian matrix has the single eigenvalue

$$\rho = 1 + \mu - 2x_{j0}$$

Hence, it follows that the trivial fixed point is stable for $-2 < \mu < 0$ and unstable for all $\mu > 0$. On the other hand, the nontrivial fixed point is unstable for all $\mu < 0$ and stable for $0 < \mu < 2$. The scenario in the vicinity of $(x, \mu) = (0, 0)$ is illustrated in Figure 2.4.3. A **transcritical bifurcation** occurs at the origin. The scenario is identical to that depicted in Figure 2.3.3 for a continuous–time system.

Example 2.31. As the third example, we consider the one–dimensional map

$$x_{k+1} = x_k + \mu x_k + \alpha x_k^3 \qquad (2.4.3)$$

where μ is a scalar control parameter. This map has the fixed points

$$x_{10} = 0 : \qquad\qquad \text{trivial fixed point}$$
$$x_{20,30} = \pm\sqrt{-\mu/\alpha} : \quad \text{nontrivial fixed points}$$

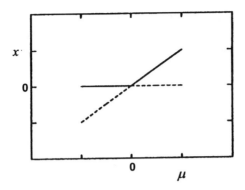

Figure 2.4.3: Scenario in the vicinity of a transcritical bifurcation.

The Jacobian matrix associated with the fixed point x_{j0} has the single eigenvalue

$$\rho = 1 + \mu + 3\alpha x_{j0}^2$$

Therefore, the trivial fixed point is stable for $-2 < \mu < 0$ and unstable for all $\mu > 0$. For $\alpha < 0$, nontrivial fixed points exist only for $\mu > 0$, and they are stable for $0 < \mu < 1$. For $\alpha > 0$, nontrivial fixed points exist only for $\mu < 0$, and they are unstable. The scenarios for $\alpha = -1$ and $\alpha = 1$ near the origin $(x, \mu) = (0, 0)$ are shown in Figures 2.4.4a and 2.4.4b, respectively. There is a **supercritical pitchfork bifurcation** at the origin in Figure 2.4.4a and a **subcritical pitchfork bifurcation**

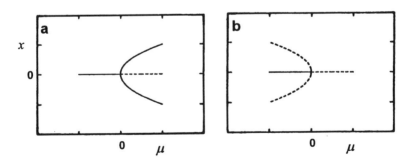

Figure 2.4.4: Local scenarios: (a) supercritical pitchfork bifurcation and (b) subcritical pitchfork bifurcation.

at the origin in Figure 2.4.4b. The scenarios are identical to those shown in Figures 2.3.2a and 2.3.2b for a continuous–time system.

When a fixed point of (2.2.1) is nonhyperbolic because an eigenvalue is at -1, the associated bifurcation does not have any analog with a bifurcation of a fixed point of a continuous–time system. To illustrate this bifurcation, we consider the following example.

Example 2.32. We consider the logistic map

$$x_{k+1} = F(x_k) = 4\alpha x_k(1 - x_k)$$

From Section 2.2, we know that for $\alpha > 0.25$ this map has the fixed point

$$x_{10} = 1 - \frac{1}{4\alpha}$$

The associated eigenvalue is

$$\rho_1 = 2 - 4\alpha$$

Hence, the fixed point x_{10} of F is nonhyperbolic at $\alpha = 0.75$ because $\rho_1 = -1$. For $\alpha > 0.75$, the fixed point x_{10} is unstable, and we have the following two new period–two points of F:

$$x_{20,30} = \frac{1}{2} + \frac{1}{4\alpha}\left[\frac{1}{2} \pm \sqrt{\left(2\alpha - \frac{1}{2}\right)^2 - 1}\right]$$

These period–two points of F are fixed points of

$$F^2(x_k) = F[F(x_k)] = F[4\alpha x_k(1 - x_k)]$$

or

$$F^2(x_k) = 16\alpha^2 x_k(1 - x_k)[1 - 4\alpha x_k(1 - x_k)]$$

The eigenvalue associated with a fixed point x_{j0} of this map is

$$\rho_2 = 16\alpha^2(1 - 2x_{j0})[1 - 8\alpha x_{j0}(1 - x_{j0})]$$

We numerically determined that the fixed points x_{20} and x_{30} of F^2 or equivalently the period–two points of F are stable for $\alpha < 0.85$. In

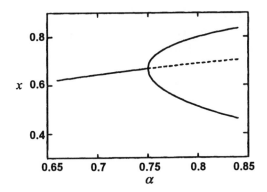

Figure 2.4.5: Scenario in the vicinity of a period–doubling bifurcation of the fixed point of (1.1.4).

Figure 2.4.5, we show the different solutions of F and their stability for $0.65 < \alpha < 0.85$. The branches of stable and unstable solutions are depicted by solid and broken lines, respectively. The fixed point x_{10} of F experiences a **period–doubling bifurcation** at $\alpha = 0.75$. As a consequence, two branches of period–two points emerge from this bifurcation point. We note that an iterate of F initiated at either of the two period–two points flips back and forth between them because

$$F(x_{20}) = x_{30} \text{ and } F(x_{30}) = x_{20}$$

For this reason, the period–doubling bifurcation of a fixed point of a map is also called a **flip bifurcation**. The fixed point x_{10} and the period–two points x_{20} and x_{30} of the map F are all fixed points of the map F^2. At $\alpha = 0.75$, the fixed point x_{10} of F^2 is nonhyperbolic because $\rho_2 = 1$. Hence, from Figure 2.4.5, we infer that this fixed point of F^2 experiences a pitchfork bifurcation at $\alpha = 0.75$. In this case, the pitchfork bifurcation is supercritical. In other cases, it is possible that a period–doubling bifurcation of F can correspond to a subcritical pitchfork bifurcation of the map F^2. The associated period–two points that arise due to this bifurcation will be unstable.

When a fixed point of (2.2.1) is nonhyperbolic with a pair of complex conjugate eigenvalues on the unit circle, the fixed point of the map can experience what is called the **Neimark–Sacker bifurcation** or

Hopf bifurcation (Iooss, 1979, Chapter III; Wiggins, 1990, Chapter 3). This bifurcation can occur in two– and higher–dimensional maps. We note that center manifold analysis can be used to simplify maps in the vicinity of bifurcation points (Carr, 1981). For more information on bifurcation analyses for maps, the reader is referred to the books of Iooss (1979), Guckenheimer and Holmes (1983), Arnold (1988), and Wiggins (1990).

2.5   EXERCISES

2.1. Consider the following map over the interval $[0, 1]$:

$$x_{n+1} = \begin{cases} 2ax_n; & 0 \le x_n \le \frac{1}{2} \\ 2a(1 - x_n); & \frac{1}{2} \le x_n \le 1 \end{cases}$$

This map, which is called the **tent map**, is a piece–wise linear map.

(a) Verify that this map is a noninvertible map.

(b) For $a = \frac{3}{8}$ and $a = \frac{5}{8}$, determine the period–one and period–two points and discuss their stability.

(c) For $a = \frac{5}{8}$, check if this map has any period–k points, where k is odd.

(d) Show that the fixed points for the tent map are $x_0^* = 0$ and, for $a > \frac{1}{2}, x_1^* = 2a/(1 + 2a)$.

(e) Show that x_1^* is always unstable and x_0^* is asymptotically stable when $0 < a < \frac{1}{2}$.

2.2. Consider the logistic map

$$x_{n+1} = 4\alpha x_n(1 - x_n)$$

Use the transformation

$$x_n = c_1 y_n + c_0$$

to show that this map can be written as

(a) $y_{n+1} = 1 - a y_n^2$

if $c_0 = \frac{1}{2}, c_1 = \frac{1}{2}(2\alpha - 1)$, and $a = 2\alpha(2\alpha - 1)$

and as

(b) $y_{n+1} = a - y_n^2$

if $c_0 = \frac{1}{2}, c_1 = \frac{1}{4\alpha}$, and $a = 2\alpha(2\alpha - 1)$.

2.3. Show that the logistic map

$$x_{n+1} = 4\alpha x_n(1 - x_n)$$

can be transformed into

$$\theta_{n+1} = 2\theta_n \bmod (1)$$

when $\alpha = -\frac{1}{2}$ by using the transformation

$$x_n = \frac{1}{2} + \cos 2\pi\theta_n$$

2.4. For each of the following maps, examine the stability of $x = 0$ and show that it is a nonhyperbolic fixed point:

(a) $x_{n+1} = x_n - x_n^3$

(b) $x_{n+1} = x_n + x_n^3$

(c) $x_{n+1} = -x_n - x_n^3$

(d) $x_{n+1} = -x_n + x_n^3$

2.5. Show that a fixed point of the map $f(x; \lambda) = (1 + \lambda)x + x^2$ undergoes a transcritical bifurcation at $\lambda = 0$.

2.6. Show that the map $f(x; \lambda) = e^x - \lambda$ undergoes a saddle–node bifurcation at $\lambda = 1$.

2.7. Determine the fixed points of the following cubic map and discuss their stability:
$$x_{n+1} = (1 - \lambda)x_n + \lambda x_n^3$$
For what value of λ does the first period–doubling bifurcation occur?

2.8. Consider the one–dimensional map
$$x_{n+1} = x_n - ax_n^3$$
Show that $x = 0$ is the only fixed point and that this fixed point is stable when $a > 0$ and unstable when $a < 0$.

2.9. Determine the fixed points of the following maps and their stability:

(a) $x_{n+1} = ax_n - x_n^3$

(b) $x_{n+1} = ax_n + x_n^3$

2.10. Consider the one–dimensional map
$$x_{n+1} = x_n \left[\frac{\mu}{x_n^2 + (\mu - x_n^2)e^{-4\pi\mu}} \right]^{1/2}$$
Show that if $\mu < 0$, this map has a stable fixed point $x = 0$, and if $\mu > 0$, this map has an unstable fixed point at $x = 0$ and two stable fixed points at $x = \pm\sqrt{\mu}$.

2.11. Consider the one–dimensional map
$$z_{n+1} = az_n^2 + 2bz_n + c$$
in the complex plane when $a \neq 0$. Let $w_n = az_n + b$ and reduce this map to
$$w_{n+1} = w_n^2 - d$$
where $d = b^2 - ac - b$.

2.12. Consider the following one–dimensional systems:

(a) $\dot{x} = \mu x + x^2$

(b) $\dot{x} = -\mu + x^2$

(c) $\dot{x} = -\mu x + x^3$

(d) $\dot{x} = -\mu x - x^3$

(e) $\dot{x} = \mu - x^3$

In each case, x is the state variable and μ is the control parameter. Construct the bifurcation diagrams for all cases and discuss them.

2.13. Consider the following system (Drazin, 1992, p. 64):

$$\dot{x} = x^3 + \delta x^2 - \mu x$$

Determine the fixed points of this system and study the bifurcations in the (x, μ) plane for zero and nonzero values of δ. Show that the pitchfork bifurcation at $(0,0)$ for $\delta = 0$ becomes a transcritical bifurcation for small δ and that there is a turning point at $(-\frac{1}{2}\delta, -\frac{1}{4}\delta^2)$. Sketch the bifurcation diagram in the (x, μ) plane for $\delta > 0$.

2.14. Consider the one–dimensional system (Drazin, 1992, p. 64)

$$\dot{x} = x^3 - 2ax^2 - (b - 3)x + c$$

for real a, b, and c.

(a) Show that if $c = 0$, then there is a transcritical bifurcation, but if $c \neq 0$, there are two (nonbifurcating) branches of equilibria.

(b) Show that the loci of the bifurcation points is given by the curve

$$(27c - 18b + 38)^2 = 4(3b - 5)^3$$

which has a cusp at $b = \frac{3}{5}$ and $c = -\frac{8}{27}$.

2.15. Consider the system

$$\dot{x} = x^4 + ax^2 + bx + c$$

Sketch the intersection of the bifurcation set and three planes a = constant for $a < 0$, $a = 0$, and $a > 0$ (Thom, 1975).

2.16. Consider the one–dimensional system

$$\dot{x} = ax + bx^3 + cx^5$$

Determine the fixed points and their stability.

2.17. Sketch the bifurcation diagram of

$$\dot{x} = -x(x^2 - 2bx - a)$$

in the (a, x) plane for a given positive b, indicating which solutions are stable.

2.18. Sketch the bifurcation diagram of

$$\dot{x} = 2(a^2 - x^2) - (a^2 + x^2)^2$$

in the (a, x) plane, indicating which equilibrium solutions are stable, and identifying two turning points and a transcritical bifurcation.

2.19. Sketch the bifurcation diagram of

$$\dot{x} = x^3 + a^3 - 3ax$$

in the (a, x) plane, indicating which equilibrium solutions are stable and identifying a turning point and a pitchfork bifurcation.

2.20. Consider the following map analyzed by Holmes (1979):

$$x_{k+1} = y_k$$
$$y_{k+1} = -bx_k + dy_k - y_k^3$$

This map was proposed by Holmes as an approximation to the Poincaré map of a periodically forced Duffing oscillator.

(a) Verify that this map is a diffeomorphism when $b \neq 0$.

(b) Assume that $d > 0$ and $b > 0$. Determine the fixed points of this map and their stability in each of the following cases: (i) $d < (1 + b)$ and (ii) $d > (1 + b)$.

(c) Is $d = 2(1 + b)$ a bifurcation point?

(d) Set $b = \frac{1}{2}$ and examine the bifurcations that take place as d is varied in the range $[2, 4]$.

2.21. Consider the Hénon map described by (1.1.5) and (1.1.6). Assume that $\beta \neq 0$ and $\alpha > 0$.

(a) Verify that this map has a stable fixed point and an unstable fixed point when

$$\alpha < \frac{3}{4}(1 - \beta)^2$$

(b) Is the map dissipative for $\beta = 0.3$? For this case, determine the period–one and period–two points of this map for $\alpha = 0.1$, $\alpha = 0.5$, and $\alpha = 1.3$. Discuss their stability.

(c) For the above values of α, plot the iterates of this map.

(d) Examine the bifurcation that takes place at

$$(x_k, y_k, \alpha) = \left[\frac{(1 - \beta)}{2\alpha}, \frac{\beta(1 - \beta)}{2\alpha}, \frac{3}{4}(1 - \beta)^2 \right]$$

2.22. Consider the two one–dimensional maps

$$x_{n+1} = e^{x_n} - \lambda$$
$$y_{n+1} = -\frac{1}{2}\lambda \tan^{-1} y_n$$

Find the fixed points and their stability. Show that x undergoes a saddle–node bifurcation at $\lambda = 1$ whereas y undergoes a period–doubling bifurcation at $\lambda = 2$.

2.23. Consider the following three–dimensional map of Klein, Baier, and Rössler (1991):

$$x_{n+1} = \alpha - \alpha y_n^2 + dz_n$$
$$y_{n+1} = x_n + \beta + \gamma z_n$$
$$z_{n+1} = y_n$$

(a) Examine if this map is dissipative in each of the following cases:
(i) $d < 1$, (ii) $d = 1$, and (iii) $d > 1$.

(b) Determine the fixed points of this map and discuss their stability.

2.24. Determine the fixed points and their types for the following systems, and for each case sketch the trajectories and the separatrices in the phase plane:

(a) $\ddot{x} + 2\dot{x} + x + x^3 = 0$

(b) $\ddot{x} + 2\dot{x} + x - x^3 = 0$

(c) $\ddot{x} + 2\dot{x} - x + x^3 = 0$

(d) $\ddot{x} + 2\dot{x} - x - x^3 = 0$

(e) $\ddot{x} - a + x^2 = 0$ for $a > 0, a = 0$, and $a < 0$

2.25. Consider the following speed–control system investigated by Fallside and Patel (1965):

$$\dot{x}_1 = x_2$$
$$\dot{x}_2 = K_d x_2 - x_1 - G x_1^2 \left(-\frac{x_2}{K_d} + x_1 + 1 \right)$$

(a) For $K_d = -1$ and $G = 6$, determine the fixed points and their stability.

(b) Plot the stable manifolds of the unstable fixed points and a few other trajectories in the (x_1, x_2) space.

(c) Discuss the phase portrait.

2.26. Consider the following single–degree–of–freedom system with quadratic and cubic nonlinearities:

$$\ddot{x} + x + \delta x^2 + \alpha x^3 = 0$$

Sketch the potential energy $V(x)$ for the system and the associated phase portrait for each of the following cases: (i) $\delta = 3$ and $\alpha = 4$, (ii) $\delta = \alpha = 4$, and (iii) $\delta = 5$ and $\alpha = 4$.

It is common to refer to the first case as a **single–well potential system** because there is a **well** in the graph of $V(x)$ versus x. The third case is referred to as a **two–well potential system**. From the phase portraits, one can discern a qualitative change as one goes from the first case to the third case.

2.27. Consider the system

$$\dot{x}_1 = x_2$$
$$\dot{x}_2 = x_1 + \lambda x_2 - x_1^2$$

Discuss the bifurcations of the fixed points of this system as a function of the control parameter λ. Sketch the state space for $\lambda = 0, \lambda > 0$, and $\lambda < 0$.

2.28. Consider the system

$$\dot{x}_1 = x_2 - \frac{1}{3}\lambda(x_1^3 - 3x_1)$$
$$\dot{x}_2 = -x_1$$

Show that the origin is the only equilibrium point. Determine its stability as a function of λ.

2.29. Show that the origin is an unstable equilibrium point for each of the two systems

(a) $\dot{x}_1 = x_2, \quad \dot{x}_2 = x_1 + 2x_2^3$

(b) $\dot{x}_1 = x_1 + 5x_2 + x_1^2 x_2, \quad \dot{x}_2 = 5x_1 + x_2 - x_2^3$

2.30. Show that the origin is a saddle point for each of the two systems

(a) $\dot{x}_1 = -x_2, \quad \dot{x}_2 = x_2 + x_1^3$

(b) $\dot{x}_1 = x_2, \quad \dot{x}_2 = x_1 + x_1^3$

Determine the stable and unstable manifolds of the origin for the linearized as well as the nonlinear systems.

2.31. Determine an approximation to the stable and unstable manifolds of the saddles of the system

$$\dot{x}_1 = 1 - x_1 x_2$$
$$\dot{x}_2 = x_1 - x_2^3$$

2.32. Consider the system

$$\dot{x}_1 = \lambda + 2x_1 x_2$$
$$\dot{x}_2 = 1 + x_1^2 - x_2^2$$

Show that there are two saddles when $\lambda = 0$ and that the x_2-axis is invariant. Hence, show that there is a heteroclinic connection. Sketch the phase plane. Show that, when $|\lambda| \neq 0$ and is small, there are still two saddle points but the saddle connection is no longer present. Sketch the phase plane for $1 \gg \lambda > 0$ and $-1 \ll \lambda < 0$.

2.33. Show that the trivial solution is the only equilibrium solution of

$$\dot{x} = xy^2 + x^2y + x^3 \text{ and } \dot{y} = y^3 - x^3$$

and that it is unstable.

2.34. Show that the trivial solution of the system

$$\dot{x} = 2xy^2 - x^3 \text{ and } \dot{y} = \frac{2}{5}x^2y - y^3$$

is asymptotically stable.

2.35. Show that the origin is a stable equilibrium point of the system

$$\dot{x} = y - x^3 \text{ and } \dot{y} = -x^2$$

2.36. Show that the origin is an asymptotically stable equilibrium point of the system

$$\dot{x} = y - x(x^4 + y^4) \text{ and } \dot{y} = -x - y(x^4 + y^4)$$

2.37. The origin of each of the following systems is a **degenerate saddle point**.

(a) $\dot{x} = x^2$, $\dot{y} = -y$

(b) $\dot{x} = x^2 - y^2$, $\dot{y} = 2xy$

Sketch the phase portrait for each case.

2.38. Show that the origin is an unstable equilibrium point of the system

$$\dot{x} = 2x^2y \text{ and } \dot{y} = -2xy^2$$

Hint: Show that xy is constant on each orbit.

2.39. Consider the system

$$\dot{x} = xy \text{ and } \dot{y} = 2 - x - y$$

Find the fixed points and determine their stability.

2.40. Consider the planar system

$$\dot{x} = \frac{1}{2}(-x + x^3)$$

$$\dot{y} = \frac{2y}{1 - 2x^2} + \epsilon$$

where $\epsilon \ll 1$. Show that there are three saddle points. For $\epsilon = 0$ and $\epsilon \neq 0$, sketch the phase portrait and indicate any heteroclinic connections.

2.41. The free oscillation about the upright position of an inverted pendulum constrained to oscillate between two closely spaced rigid barriers is described by (e.g., Shaw and Rand, 1989)

$$\begin{matrix} \ddot{x} + 2\mu\dot{x} - x = 0 & \quad |\,x\,| < 1 \\ \dot{x} \to -r\dot{x} & \quad |\,x\,| = 1 \end{matrix}$$

where x describes the position of the pendulum; the locations of the rigid barriers are $x = -1$ and $x = 1$; μ is a measure of the friction; and $r \leq 1$ is a reflection coefficient representing energy loss during impact with either of the rigid barriers.

Assuming elastic impact (i.e., $r = 1$), construct and discuss the phase portraits for the following two cases: (i) $\mu = 0$ and (ii) $\mu > 0$.

2.42. In studying the forced response of a van der Pol oscillator with delayed amplitude limiting, Nayfeh (1968) encountered the following system of equations:

$$\dot{x}_1 = x_1(1 - x_1^2) + F \cos x_2$$

$$\dot{x}_2 = \sigma + \nu x_1^2 - \frac{F}{x_1} \sin x_2$$

(a) Show that the fixed points (x_{10}, x_{20}) of this system satisfy

$$\rho \left[(1 - \rho)^2 + (\sigma + \nu\rho)^2 \right] = F^2$$

where $\rho = x_{10}^2$.

(b) For $\nu = -0.15$, plot the loci of fixed points in the $\rho - \sigma$ plane for $F^2 = 1, \frac{1}{3}, \frac{4}{27}$, and $\frac{1}{10}$. What is the significance of the value $\frac{4}{27}$?

(c) Show that the interior points of the ellipse defined by

$$(1 - \rho)(1 - 3\rho) + (\sigma + \nu\rho)(\sigma + 3\nu\rho) = 0$$

are saddle points and hence unstable. Also, show that the exterior points are nodes if $D \geq 0$ and foci if $D < 0$, where

$$D = 4 \left[(1 - 3\nu^2)\rho^2 - 4\nu\rho\sigma - \sigma^2 \right]$$

(d) Finally, show that the exterior points are stable if $\rho > \frac{1}{2}$ and unstable if $\rho < \frac{1}{2}$.

2.43. Elzebda, Nayfeh, and Mook (1989) used the following planar system to model the subsonic wingrock phenomenon of slender delta wings:

$$\dot{x}_1 = x_2$$
$$\dot{x}_2 = -\omega^2 x_1 + \mu x_2 + b_1 \mid x_1 \mid x_2 + b_2 \mid x_2 \mid x_2 + b_3 x_1^3$$

The state x_1 represents the roll angle. The coefficients ω, μ, b_1, b_2, and b_3 are functions of the angle of attack α of the wing. For one particular wing, at $\alpha = \alpha_1$,

$$\omega^2 = 0.00362949; \quad \mu = -0.00858295;$$
$$b_1 = 0.02020694; \quad b_2 = -0.0219083; \quad b_3 = -0.051880962$$

For the same wing, at an angle of attack $\alpha_2 > \alpha_1$, we have

$$\omega^2 = 0.01477963; \quad \mu = 0.004170843;$$
$$b_1 = -0.02381943; \quad b_2 = 0.02977157; \quad b_3 = 0.016297021$$

(a) Determine the equilibrium positions and their stability for $\alpha = \alpha_1$ and $\alpha = \alpha_2$.

(b) Construct the phase portraits numerically for both angles of attack and discuss them.

At a critical angle of attack α_c, the equilibrium position at the origin loses stability due to a Hopf bifurcation. The ensuing oscillatory motion about the origin is called **wingrock**. Here, $\alpha_1 < \alpha_c < \alpha_2$.

2.44. Mingori and Harrison (1974) studied the following system for analyzing the motion of a particle constrained to move on a circular path that is spinning and coning:

$$\dot{u} = v$$
$$\dot{v} = -\mu_1(v - 1) + \mu_2\mu_3 \sin u + \frac{1}{2}\mu_3^2 \sin 2u$$

Let $\mu_1 = 0.1$ and $\mu_2 = 2.0$. Then, as μ_3 is varied from zero, bifurcations take place at 0.0502, 0.3, and 2.265. Examine the qualitative changes that take place in the (u, v) space due to these bifurcations.

2.45. A bead of mass m sliding on a rotating circular hoop of radius R is described by

$$\ddot{\theta} + 2\mu\dot{\theta} + \frac{g}{R} \sin \theta - \omega^2 \sin \theta \cos \theta = 0$$

Here, θ describes the angular position of the bead on the hoop, g is the acceleration due to gravity, μ is a measure of the friction experienced by the bead, and ω is the angular velocity of the hoop.

(a) For $\mu = 0$, determine the fixed points (equilibrium positions) of the system and sketch the phase portrait in each of the following cases: (i) $\omega^2 < g/R$, (ii) $\omega^2 = g/R$, and (iii) $\omega^2 > g/R$.

(b) For $\mu > 0$, choose ω as a control parameter and examine the different local bifurcations of fixed points that occur as ω is increased from zero. Construct appropriate bifurcation diagrams.

2.46. Consider the Rössler (1976a) equations:

$$\dot{x} = -(y + z)$$
$$\dot{y} = x + ay$$
$$\dot{z} = b + (x - c)z$$

Assume that the parameters a, b, and c are positive.

(a) When a is used as a control parameter, verify that a fixed point of this system experiences a saddle–node bifurcation at

$$(x, y, z, a) = \left(\frac{c}{2}, -\frac{c}{2a}, \frac{c}{2a}, \frac{c^2}{4b} \right)$$

(b) Simplify the three–dimensional system to the normal form for a saddle–node bifurcation in the vicinity of the bifurcation point.

2.47. Consider the Lorenz equations (Lorenz, 1963):

$$\dot{x} = \sigma(y - x)$$
$$\dot{y} = \rho x - y - xz$$
$$\dot{z} = -\beta z + xy$$

Assume that the parameters σ, β, and ρ are positive.

(a) Choose ρ as the control parameter and examine the different local bifurcations experienced by the different fixed points. Verify that a Hopf bifurcation of a fixed point occurs at

$$\rho_c = \frac{\sigma(\sigma + \beta + 3)}{\sigma - \beta - 1}$$

(b) Construct the bifurcation diagram for $0 < \rho \le \rho_c$.

(c) Simplify the three–dimensional system to the normal form for a pitchfork bifurcation in the vicinity of $(x, y, z, \rho) = (0, 0, 0, 1)$ and obtain

$$\dot{u} = \frac{\sigma(1 - \rho)}{\sigma + 1} u - \frac{\sigma}{\beta(\sigma + 1)} u^3$$

2.48. Consider the system

$$\dot{x}_1 = -x_1 + h(x_3)$$
$$\dot{x}_2 = -h(x_3)$$
$$\dot{x}_3 = -ax_1 + bx_2 - ch(x_3)$$

where a, b, and c are positive constants and $h(0) = 0$ and

$$yh(y) > 0 \text{ for } 0 < |y| < k \text{ for some } k > 0$$

(a) Show that the origin is an isolated equilibrium point.

(b) Is the origin an asymptotically stable equilibrium point?

(c) Suppose that $yh(y) > 0$. Is the origin globally asymptotically stable?

2.49. Consider the nonlinear oscillator

$$\ddot{u} + u + \epsilon \left[2\mu_1 \dot{u} + \mu_2 \dot{u} \, | \, \dot{u} \, | + \alpha u^3 + 2K u \cos(\Omega t) \right] = 0$$

where ϵ is a small, positive parameter. Further, the parameters μ_1, μ_2, and K are all independent of ϵ while the parameter Ω is such that

$$\Omega = 2 + \epsilon\sigma$$

A first approximation obtained for this system has the form

$$u = p\cos\left(\frac{1}{2}\Omega t\right) + q\sin\left(\frac{1}{2}\Omega t\right) + O(\epsilon)$$

where

$$p' = -\mu_1 p - \frac{1}{2}(\sigma + K)q + \frac{3\alpha}{8}q(p^2 + q^2) - \frac{4\mu_2}{3\pi}p\sqrt{p^2 + q^2}$$

$$q' = -\mu_1 q + \frac{1}{2}(\sigma - K)p - \frac{3\alpha}{8}p(p^2 + q^2) - \frac{4\mu_2}{3\pi}q\sqrt{p^2 + q^2}$$

In the above equations, the prime denotes the derivative with respect to the time scale $\tau = \epsilon t$.

(a) Simplify the dynamical system governing p and q to the normal form for a transcritical bifurcation in the vicinity of

$$(p, q, K_c) = \left(0, 0, \sqrt{4\mu_1^2 + \sigma^2}\right)$$

(b) What happens to the bifurcation at the above–mentioned bifurcation point when $\mu_2 = 0$?

(c) Construct the frequency–response curves when $\mu_2 = 0$ and discuss them.

2.50. Consider the map (2.4.1) and examine the bifurcation that occurs at $(x_k, \mu) = (1, 1)$.

2.51. Consider the map (2.4.2) and study the bifurcation that takes place at $(x, \mu) = (0, -2)$.

2.52. Consider the following map:

$$x_{k+1} = \mu + x_k^2$$

Examine the bifurcation that takes place at $(x, \mu) = \left(\frac{1}{2}, \frac{1}{4}\right)$?

2.53. Consider the following map:

$$x_{k+1} = \mu x_k + x_k^2$$

Examine the bifurcation that takes place at $(x, \mu) = (0, 1)$?

2.54. Consider the following map:

$$x_{k+1} = \mu - x_k + x_k^2$$

Verify that the period–doubled solutions bifurcating from the point $(x, \mu) = (0, 0)$ are stable. (The associated period–doubling bifurcation is **supercritical**.)

2.55. Consider the following system:

$$x_{n+1} = x_n(1 - x_n) \text{ for } n = 0, 1, \cdots$$

Show that if $0 < x_0 < 1$ then $x_n \sim 1/n$ as $n \to \infty$.

2.56. Consider the following system (Bender and Orszag, 1978):

$$x_{n+1} = 2x_n(1 - x_n) \text{ for } n = 0, 1, \cdots$$

Show that if $y_n = 1 - 2x_n$, then $y_{n+1} = y_n^2$. Hence show that

$$x_n = \frac{1}{2}\left[1 - (1 - 2x_0)^{2n}\right]$$

Deduce that $\lim_{n \to \infty} x_n = \frac{1}{2}$ if and only if $0 < x_0 < 1$.

2.57. Consider the following system:

$$x_{n+1} = \frac{1}{2 - x_n} \text{ for } n = 0, 1, \cdots$$

If $x_n \neq 2$, show that

$$x_n = \frac{n - (n - 1)x_0}{n + 1 - nx_0}$$

and therefore $\lim_{n \to \infty} x_n = 1$.

2.58. The free oscillations of a planar pendulum are described by

$$\ddot{\theta} + 2\mu\dot{\theta} + \sin\theta = 0; \quad \mu \in [0, 1)$$

The origin $(0, 0)$ is a center for $\mu = 0$ and an asymptotically stable fixed point for $\mu > 0$. Use central–difference schemes to approximate the time derivatives at $t_n = nh$ as

$$\ddot{\theta} \simeq \frac{\theta_{n+1} - 2\theta_n + \theta_{n-1}}{4h^2} \quad \text{and} \quad \dot{\theta} \simeq \frac{\theta_{n+1} - \theta_{n-1}}{2h}$$

where h is the size of the time step. Set $x_n = \theta_n$ and $y_n = \theta_{n-1}$ and obtain the map

$$x_{n+1} = \frac{4h^2}{1+4\mu h}\left[\frac{1}{2h^2}x_n - \sin x_n + \frac{4\mu h - 1}{4h^2}y_n\right]$$

$$y_{n+1} = x_n$$

Examine the stability of the fixed point $(0,0)$ of this map and discuss if this numerical scheme preserves the stability properties for $\mu = 0$ and $\mu > 0$. In the context of this exercise, the interesting discussion of Hale and Kocak (1991, Section 15.3) on numerical algorithms and maps is worth noting.

2.59. Consider the two–dimensional map

$$x_{n+1} = \lambda + x_n + \lambda y_n + x_n^2$$

$$y_{n+1} = \frac{1}{2}y_n + \lambda x_n + x_n^2$$

Determine the fixed points and then show that the origin undergoes a saddle–node bifurcation at $\lambda = 0$.

2.60. Consider the two–dimensional map

$$x_{n+1} = y_n$$

$$y_{n+1} = -\frac{1}{2}x_n + \lambda y_n - y_n^3$$

Determine the fixed points. Show that the origin undergoes a pitchfork bifurcation at $\lambda = \frac{3}{2}$. Analyze the bifurcation at $\lambda = 3$.

Chapter 3

PERIODIC SOLUTIONS

In this chapter, we consider periodic solutions of dynamical systems, especially continuous–time systems. Unlike equilibrium solutions, periodic solutions are characterized by time–varying states. A periodic solution is a dynamic solution that is characterized by one basic frequency f. The spectrum of a periodic signal consists of a spike at the frequency 0 and spikes at integer multiples of f. The amplitudes of some of the frequency components may be zero. After defining periodic solutions in Section 3.1, we study their stability by using Floquet theory in Section 3.2 and Poincaré maps in Section 3.3. We examine different local bifurcations of periodic solutions and their consequences in Section 3.4. Again, bifurcations of the periodic solutions of dissipative systems can be classified into continuous and discontinuous or catastrophic bifurcations, depending on whether the states of the system vary continuously or discontinuously as the control parameter is varied through its critical value. In Section 3.5, the methods of center manifold reduction and multiple scales are used to construct periodic solutions analytically in the vicinity of a Hopf bifurcation point.

3.1 PERIODIC SOLUTIONS

A solution $\mathbf{x} = \mathbf{X}(t)$ of a continuous–time system is **periodic** with least period T if $\mathbf{X}(t + T) = \mathbf{X}(t)$ and $\mathbf{X}(t + \tau) \neq \mathbf{X}(t)$ for $0 < \tau < T$.

3.1.1   Autonomous Systems

A periodic solution $\mathbf{X}$ of least finite period $T > 0$ of the system

$$\dot{\mathbf{x}} = \mathbf{F}(\mathbf{x}; \mathbf{M}) \qquad (3.1.1)$$

corresponds to a closed orbit Γ in $\mathcal{R}^n$ and is such that $\mathbf{X}(t_0) = \mathbf{X}(t_0+T)$ and $\mathbf{X}(t_0 + \tau) \neq \mathbf{X}(t_0)$ for $0 < \tau < T$. By specifying the initial time t_0, one specifies a location $\mathbf{x} = \mathbf{x}_0$ on the orbit. For a periodic solution initiated at $\mathbf{x} = \mathbf{x}_0$, the positive and negative orbits $\gamma^+(\mathbf{x}_0)$ and $\gamma^-(\mathbf{x}_0)$ are such that $\gamma^+(\mathbf{x}_0) = \gamma^-(\mathbf{x}_0) = \Gamma$. As explained later, a periodic solution of (3.1.1) can be treated as a fixed point of an appropriately defined map called the **Poincaré map**. We note that periodic solutions of an autonomous system are examples of invariant sets.

A periodic solution of (3.1.1) is called a **limit cycle** if there are no other periodic solutions sufficiently close to it. In other words, a limit cycle is an **isolated periodic solution** and corresponds to an isolated closed orbit in the state space. Every trajectory initiated near a limit cycle approaches it either as $t \to \infty$ or as $t \to -\infty$.

Example 3.1. We consider the system

$$\dot{x} = \mu x - \omega y + (\alpha x - \beta y)(x^2 + y^2) \qquad (3.1.2)$$
$$\dot{y} = \omega x + \mu y + (\beta x + \alpha y)(x^2 + y^2) \qquad (3.1.3)$$

where x and y are the states and μ, ω, α, and β are constants. As illustrated in Section 2.3.1, this system takes the simple form

$$\dot{r} = \mu r + \alpha r^3 \qquad (3.1.4)$$
$$\dot{\theta} = \omega + \beta r^2 \qquad (3.1.5)$$

under the transformation

$$x = r \cos \theta \quad \text{and} \quad y = r \sin \theta \qquad (3.1.6)$$

Multiplying (3.1.4) with $2r$ yields

$$\frac{d}{dt}(r^2) = 2\mu r^2 + 2\alpha r^4 \qquad (3.1.7)$$

Assuming that $\mu \neq 0$ and using separation of variables, we integrate (3.1.7) and obtain

$$r = \left[\left(\frac{\alpha}{\mu} + \frac{1}{r_0^2}\right) e^{-2\mu t} - \frac{\alpha}{\mu}\right]^{-1/2} \tag{3.1.8}$$

where $r_0 \neq 0$ is the value of r at $t = 0$. Letting $\theta = \omega t + \phi$ in (3.1.5) gives

$$\dot{\phi} = \beta r^2 \tag{3.1.9}$$

Then, it follows from (3.1.7) and (3.1.9) that

$$\frac{d\phi}{dr^2} = \frac{\beta}{2\mu + 2\alpha r^2}$$

when $r^2 \neq -\mu/\alpha$. Hence, for $\alpha \neq 0$,

$$\phi = \frac{\beta}{2\alpha}\ln(2\mu + 2\alpha r^2) + c \tag{3.1.10}$$

where c is a constant. Substituting for r and θ in (3.1.6), we obtain a closed–form solution of (3.1.2) and (3.1.3).

When $\mu > 0$ and $\alpha < 0$, it follows from (3.1.8) that

$$\lim_{t \to \infty} r = (-\mu/\alpha)^{1/2}$$

irrespective of the value of r_0 as long as it is different from zero. Consequently, it follows from (3.1.5) that

$$\lim_{t \to \infty} \dot{\theta} = \omega - \beta\mu/\alpha$$

Therefore, we have

$$\lim_{t \to \infty} x = \left(-\frac{\mu}{\alpha}\right)^{1/2} \cos\left[\left(\omega - \frac{\beta\mu}{\alpha}\right)t + \theta_0\right]$$

$$\lim_{t \to \infty} y = \left(-\frac{\mu}{\alpha}\right)^{1/2} \sin\left[\left(\omega - \frac{\beta\mu}{\alpha}\right)t + \theta_0\right] \tag{3.1.11}$$

where θ_0 is the initial value of θ. Equations (3.1.11) represent a closed

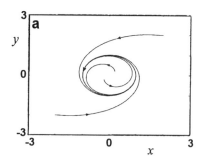
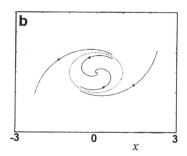

Figure 3.1.1: Periodic solution and adjacent orbits of (3.1.2) and (3.1.3): (a) $\mu = 1$, $\alpha = -1$, $\omega = 1$, and $\beta = 1$; and (b) $\mu = -1$, $\alpha = 1$, $\omega = 1$, and $\beta = 1$.

orbit in the $x - y$ plane. This orbit is a circle whose center is at the origin and radius is $\sqrt{-\mu/\alpha}$; that is,

$$x^2 + y^2 = -\mu/\alpha \qquad (3.1.12)$$

In Figure 3.1.1a, the closed orbit corresponds to the periodic solution of (3.1.2) and (3.1.3) when $\mu > 0$ and $\alpha < 0$. We also display four positive orbits in the figure, with the arrow on each orbit indicating the direction of evolution. Because there are no closed orbits sufficiently near this periodic solution (in fact, there are no other closed orbits in the entire planar space), the closed orbit of Figure 3.1.1a is a limit cycle. It is also an invariant set because an orbit initiated from any point on the closed trajectory remains on this orbit for all times. Further, we observe that positive orbits initiated from different points in the state space $x - y$, other than the origin, are attracted to the limit cycle. Hence, it is a **stable limit cycle** or a **periodic attractor**. Its basin of attraction is the entire $x - y$ space excluding the origin, which is an unstable fixed point of (3.1.2) and (3.1.3).

When $\mu < 0$ and $\alpha > 0$, we infer from (3.1.8) that

$$\lim_{t \to -\infty} r = (-\mu/\alpha)^{1/2}$$

Then, (3.1.12) still describes the corresponding closed orbit in the $x - y$ plane. In Figure 3.1.1b, we show some orbits of (3.1.2) and (3.1.3) when $\mu < 0$ and $\alpha > 0$. Here, the closed orbit is depicted by broken lines. Again, because there are no other closed trajectories sufficiently near this periodic solution, it is a limit cycle. Further, it is also an invariant set. Moreover, this limit cycle is said to be **unstable** because all positive orbits initiated from nearby points spiral away from it as $t \to \infty$ while all negative orbits initiated from nearby points spiral toward it as $t \to -\infty$. Such an unstable limit cycle is an example of a **repellor**. In Figure 3.1.1b, the origin is a **point attractor** of (3.1.2) and (3.1.3), and its basin of attraction is bound by the closed orbit.

Example 3.2. We consider the system

$$\ddot{x} + x + 2x^3 = 0 \tag{3.1.13}$$

Multiplying (3.1.13) with $2\dot{x}$ and integrating, we obtain

$$\dot{x}^2 + x^2 + x^4 = H = \dot{x}_0^2 + x_0^2 + x_0^4 \tag{3.1.14}$$

where H is a constant that represents the total energy of the system and $x_0 = x(0)$ and $\dot{x}_0 = \dot{x}(0)$. Thus, for any initial condition $(x_0, \dot{x}_0)$, (3.1.14) represents a closed trajectory in the $\dot{x} - x$ plane and, hence, a periodic solution. In Figure 3.1.2, we show four closed orbits of (3.1.13) obtained by choosing four different initial conditions. We note that the periodic solutions in Figure 3.1.2 are not isolated but form a continuum. Because there exist an infinite number of closed trajectories in the vicinity of any closed trajectory, a periodic solution of (3.1.13) is not a limit cycle.

There is a large body of theoretical work on periodic solutions of two–dimensional or planar autonomous systems of the form

$$\dot{x}_1 = f_1(x_1, x_2) \quad \text{and} \quad \dot{x}_2 = f_2(x_1, x_2) \tag{3.1.15}$$

where f_1 and f_2 are $\mathcal{C}^1$. Let $D \subset \mathcal{R}^2$ represent a simply connected domain in the $x_2 - x_1$ space. (A simply connected domain does not contain any holes or disjoint regions.) According to **Bendixson's criterion**, if the divergence

$$\frac{\partial f_1}{\partial x_1} + \frac{\partial f_2}{\partial x_2}$$

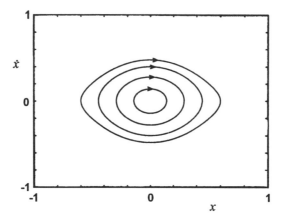

Figure 3.1.2: Periodic solutions of (3.1.13).

of the vector field does not change sign or does not vanish identically in some region of D, then periodic solutions are not possible in D. Thus, Bendixson's criterion can be used as a tool to exclude the existence of periodic solutions of systems like (3.1.15). This criterion can be proved by using **Green's theorem**. Let us assume that there exists a closed orbit Γ in D that encloses a certain region S. Then,

$$\int\int_S \left(\frac{\partial f_1}{\partial x_1} + \frac{\partial f_2}{\partial x_2}\right) dx_1 dx_2 = \int_\Gamma (f_1 dx_2 - f_2 dx_1)$$
$$= \int_\Gamma (f_1 \dot{x}_2 - f_2 \dot{x}_1)\, dt \qquad (3.1.16)$$

Because the last integral in (3.1.16) is zero on account of (3.1.15), the divergence of the vector field should either change sign or be zero in S for the double integral to vanish, thereby proving the criterion. Next, we present four examples to illustrate the use of Bendixson's criterion.

Example 3.3. We again consider the system of equations (3.1.2) and (3.1.3). The divergence of their vector field is given by

$$\frac{\partial \dot{x}}{\partial x} + \frac{\partial \dot{y}}{\partial y} = 2\mu + 4\alpha(x^2 + y^2)$$

Hence, according to Bendixson's criterion, periodic solutions are possible only when $\alpha = \mu = 0$ or $\alpha\mu < 0$. As illustrated in Figure 3.1.1, periodic solutions exist when $\mu > 0$ and $\alpha < 0$ and when $\mu < 0$ and $\alpha > 0$, in agreement with Bendixson's criterion.

Example 3.4. We consider the system

$$\dot{x}_1 = x_2 \tag{3.1.17}$$
$$\dot{x}_2 = -x_1 + x_1^3 - 2\mu x_2 \tag{3.1.18}$$

The divergence of this planar vector field is -2μ, which is a real constant. Hence, according to Bendixson's criterion, periodic solutions are only possible when $\mu = 0$. In Figure 1.4.2b, the orbit Γ_2 corresponds to a periodic solution of (3.1.17) and (3.1.18) when $\mu = 0$.

Example 3.5. Next, we consider the system of equations

$$\dot{x}_1 = x_2 \tag{3.1.19}$$
$$\dot{x}_2 = -\omega_0^2 \left(x_1 - 1.402 x_1^3 + 0.271 x_1^5\right) - \left(2\mu_1 x_2 + \mu_3 x_2^3\right) \tag{3.1.20}$$

where $\omega_0 = 5.278$. The divergence of this vector field is

$$\frac{\partial \dot{x}_1}{\partial x_1} + \frac{\partial \dot{x}_2}{\partial x_2} = -2\mu_1 - 3\mu_3 x_2^2$$

Hence, according to Bendixson's criterion, periodic solutions are possible only when either $\mu_1 = \mu_3 = 0$ or $\mu_1\mu_3 < 0$.

In Figure 2.1.6a, we show some of the numerically computed orbits of (3.1.19) and (3.1.20) when $\mu_1 = \mu_3 = 0$. One can numerically verify that inside the heteroclinic orbit (and, similarly, within the homoclinic orbits) there is a continuum of periodic orbits whose periods increase monotonically and approach infinity as the heteroclinic orbit is approached. Therefore, homoclinic and heteroclinic orbits are called **infinite–period orbits** in the literature.

In Figure 2.1.6b, we show some of the computed trajectories when $\mu_1 > 0$ and $\mu_3 > 0$. The divergence of the corresponding vector field is negative definite and, hence, periodic solutions are not possible according to Bendixson's criterion. The introduction of damping transformed the centers at $(-2.0782, 0.0), (0.0, 0.0)$, and $(2.0782, 0.0)$ in Figure 2.1.6a into stable foci in Figure 2.1.6b and destroyed all closed orbits, including the homoclinic and heteroclinic orbits. The stable foci of Figure 2.1.6b are point attractors, and their basins of attraction are separated by the stable manifolds of the saddles at $(-0.9243, 0.0)$ and $(0.9243, 0.0)$.

In Figure 3.1.3, we show some of the computed trajectories when $\mu_1 < 0$ and $\mu_3 > 0$. Here, $2\mu_1 x_2$ and $\mu_3 x_2^3$ can be thought of as negative and positive damping terms, respectively. Because the divergence of the vector field changes sign, periodic solutions may be possible. The introduction of linear negative damping transformed the centers at $(-2.0782, 0.0), (0.0, 0.0)$, and $(2.0782, 0.0)$ in Figure 2.1.6a into unstable foci in Figure 3.1.3 and destroyed the homoclinic and heteroclinic orbits. However, the positive nonlinear damping limits the growth of disturbances initiated near the foci, resulting in the limit cycles seen in

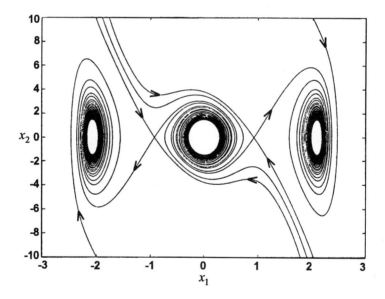

Figure 3.1.3: Phase portraits of (3.1.19) and (3.1.20) for $\mu_1 = -0.086$ and $\mu_3 = 0.108$.

Figure 3.1.3. Again, the stable manifolds of the saddles $(-0.9243, 0.0)$ and $(0.9243, 0.0)$ separate the basins of attraction of the three limit cycles.

We consider the next example to point out that the Bendixson criterion is **a necessary but not sufficient condition** for the existence of periodic solutions in a planar system.

Example 3.6. We consider the two–dimensional system (Hale and Kocak, 1991)

$$\dot{x}_1 = -x_2 + x_1^2 - x_1 x_2 \tag{3.1.21}$$

$$\dot{x}_2 = x_1 + x_1 x_2 \tag{3.1.22}$$

In this case, the divergence of the vector field is $3x_1 - x_2$, which changes sign whenever the line $x_2 = 3x_1$ in the $x_2 - x_1$ space is crossed. However, as shown in Figure 3.1.4, there are no periodic solutions in the $x_2 - x_1$ space.

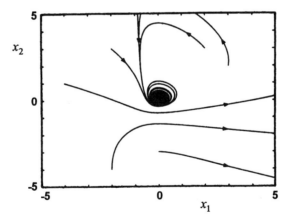

Figure 3.1.4: Phase portrait of (3.1.21) and (3.1.22).

The **Poincaré–Bendixson theorem** (e.g., Coddington and Levinson, 1955, Chapter 16; Hale, 1969, Chapter II), which is formulated in a topological setting, provides precise statements on when an asymptotic state of a two–dimensional autonomous system is a periodic solution.

However, the Poincaré–Bendixson theorem is not applicable to three– and higher–dimensional systems.

Recently, Li and Muldowney (1993) generalized Bendixson's criterion for precluding the existence of periodic orbits in three– and higher–dimensional systems. To ascertain the presence of periodic solutions in planar and higher–dimensional autonomous systems, one can use numerical and perturbation methods. The relevant numerical methods are discussed in Chapter 6.

3.1.2 Nonautonomous Systems

A periodic solution $\mathbf{X}$ of least period T of the n–dimensional nonautonomous system

$$\dot{\mathbf{x}} = \mathbf{F}(\mathbf{x}, t; \mathbf{M}) \qquad (3.1.23)$$

also describes a closed orbit Γ in $\mathcal{R}^n$. If we assume that $\mathbf{x} = \mathbf{x}_0$ on this orbit at $t = t_0$, then the periodicity of the solution implies that $\mathbf{X}(\mathbf{x}_0; t_0) = \mathbf{X}(\mathbf{x}_0; t_0 + T)$. A T–periodic solution can also be expressed in terms of a map $\mathbf{T}^t$, such that $\mathbf{T}^T\mathbf{x}_0 = \mathbf{x}_0$ and $\mathbf{T}^t\mathbf{x}_0 \neq \mathbf{x}_0$ for $t_0 < t < t_0 + T$. Then, the orbit $\Gamma = (\mathbf{T}^t\mathbf{x}_0 : \ t_0 \leq t \leq t_0 + T)$. We note that $\mathbf{T}^t$ maps $\mathcal{R}^n$ into $\mathcal{R}^n$ and that the periodic solution $\mathbf{X}$ corresponds to a fixed point of the map $\mathbf{T}^T$. Any point on the orbit Γ is a fixed point of this map, which depends on the parameters t_0, T, and $\mathbf{x}_0$.

Example 3.7. In the case of (1.2.2) and (1.2.3), the closed orbit of Figure 1.2.1b corresponds to a periodic solution.

We note that a closed orbit in the state space of a nonautonomous system does not always represent a periodic solution, as illustrated in Example 3.8.

Example 3.8. We consider the system

$$\dot{x} = nt^{n-1}y \qquad (3.1.24)$$
$$\dot{y} = -nt^{n-1}x \qquad (3.1.25)$$

where $n > 1$. Its solution is

$$x = a\sin t^n + b\cos t^n \ \text{ and } \ y = a\cos t^n - b\sin t^n \qquad (3.1.26)$$

where a and b are constants. Although this solution is not periodic, it corresponds to the following closed orbit in the $y - x$ plane:

$$x^2 + y^2 = a^2 + b^2$$

There are theorems that state that periodic solutions exist for n–dimensional weakly nonlinear systems of the form (Hale, 1963, Chapter 6)

$$\dot{\mathbf{x}} = A\mathbf{x} + \epsilon\mathbf{G}(\mathbf{x}, t; \epsilon) \tag{3.1.27}$$

where $\mathbf{x} \in \mathcal{R}^n$; $t \in \mathcal{R}^1$; $\epsilon \in \mathcal{R}^1$; A is a constant matrix with some zero eigenvalues; $\mathbf{G}$ is periodic in t with finite period T; $\mathbf{G}$ is continuous in $\mathbf{x}$, t, and ϵ; $\mathbf{G}$ has a continuous first partial derivative with respect to ϵ; and $\mathbf{G}$ has continuous first and second partial derivatives with respect to $\mathbf{x}$ in the domain of interest in the $(n + 2)$–dimensional space of $(\mathbf{x}, t; \epsilon)$.

Often, as illustrated by the following example, the equations obtained in standard form by using the method of variation of parameters form a special case of (3.1.27), with all of the entries of the matrix A being zero.

Example 3.9. We consider a forced Duffing oscillator governed by

$$\dot{x}_1 = x_2 \tag{3.1.28}$$

$$\dot{x}_2 = -x_1 - \epsilon(\mu x_2 + \alpha x_1^3 - F\cos\Omega t) \tag{3.1.29}$$

where the parameters μ, α, and F are all independent of the small parameter ϵ and

$$\Omega^2 = 1 + \epsilon\sigma$$

We are interested in ascertaining the existence of periodic solutions of period $2\pi/\Omega$ of (3.1.28) and (3.1.29). Following the method of variation of parameters, we introduce the transformation

$$x_1 = p\cos\Omega t + q\sin\Omega t \tag{3.1.30}$$

$$x_2 = \Omega(-p\sin\Omega t + q\cos\Omega t) \tag{3.1.31}$$

into (3.1.28) and (3.1.29) and obtain

$$\dot{p} = -\frac{\epsilon}{\Omega}\left(\sigma x_1 - \mu x_2 - \alpha x_1^3 + F\cos\Omega t\right)\sin\Omega t \qquad (3.1.32)$$

$$\dot{q} = \frac{\epsilon}{\Omega}\left[\sigma x_1 - \mu x_2 - \alpha x_1^3 + F\cos\Omega t\right]\cos\Omega t \qquad (3.1.33)$$

The system (3.1.32) and (3.1.33) is in the form (3.1.27) if we identify $\mathbf{x}$ with the two–dimensional vector $\{p\ q\}^T$. Further, the corresponding matrix A is a 2×2 matrix with zero entries. Because the right–hand sides of (3.1.32) and (3.1.33) satisfy all the required smoothness conditions, periodic solutions of this system do exist.

3.1.3   Comments

There are fixed–point theorems, such as the **contraction mapping theorem** and the **Brouwer fixed–point theorem**, that provide the conditions under which a map from $\mathcal{R}^n$ to $\mathcal{R}^n$ has at least one fixed point (e.g., Hale, 1969; Arnold, 1973). When this map is associated with a continuous–time system, such as $\dot{\mathbf{x}} = \mathbf{F}(\mathbf{x}, t; \mathbf{M})$ or $\dot{\mathbf{x}} = \mathbf{F}(\mathbf{x}; \mathbf{M})$, there is a periodic solution of the differential system corresponding to the fixed point of the map. We note that fixed–point theorems are useful in determining the existence of periodic solutions in $\mathcal{R}^n$. However, the use of all of the theorems mentioned thus far is limited because they do not provide much information on the location and number of periodic solutions of an n–dimensional system. In practice, one often has to use numerical and perturbation methods to ascertain the existence of periodic solutions and their number and location in an n–dimensional space.

3.2   FLOQUET THEORY

From the stability concepts discussed in Section 1.4, we recall that periodic solutions of a nonlinear continuous–time system can be stable in the sense of Poincaré but unstable in the sense of Lyapunov. However, these concepts do not provide any explicit schemes for determining the stability of periodic solutions. Next, we discuss explicit schemes based on Floquet theory (Floquet, 1883) in this section and Poincaré maps in Section 3.3.

3.2.1   Autonomous Systems

We consider the stability of periodic solutions of the autonomous system

$$\dot{\mathbf{x}} = \mathbf{F}(\mathbf{x}; \mathbf{M}) \qquad (3.2.1)$$

where $\mathbf{x}$ is an n–dimensional state vector and $\mathbf{M}$ is an m–dimensional parameter vector. We let the periodic solution of (3.2.1) at $\mathbf{M} = \mathbf{M}_0$ be denoted by $\mathbf{X}_0(t)$ and have the minimal period T. Then, a disturbance $\mathbf{y}$ is superimposed on $\mathbf{X}_0$, resulting in

$$\mathbf{x}(t) = \mathbf{X}_0(t) + \mathbf{y}(t) \qquad (3.2.2)$$

Substituting (3.2.2) into (3.2.1), assuming that $\mathbf{F}$ is at least twice continuously differentiable (i.e., $\mathcal{C}^2$), expanding the result in a Taylor series about $\mathbf{X}_0$, and retaining only linear terms in the disturbance, we obtain

$$\dot{\mathbf{y}} = D_{\mathbf{x}}\mathbf{F}(\mathbf{X}_0; \mathbf{M}_0)\mathbf{y} + O(\|\,\mathbf{y}\,\|^2) \text{ or } \dot{\mathbf{y}} \simeq A(t; \mathbf{M}_0)\mathbf{y} \qquad (3.2.3)$$

where A is the matrix of first partial derivatives of $\mathbf{F}$. The stability analysis is local, because we linearized in the disturbance $\mathbf{y}$. The matrix A is periodic in time and has a period T, which is the period of the periodic solution $\mathbf{X}_0(t)$. However, T may not be the minimal period of A. For instance, when $\mathbf{F}$ has only odd nonlinearities, the minimal period of A is $\frac{1}{2}T$. Floquet theory deals with linear systems, such as (3.2.3), with periodic coefficients.

The n–dimensional linear system (3.2.3) has n linearly independent solutions $\mathbf{y}_i$, where $i = 1, 2, \cdots, n$. These solutions are usually called a **fundamental set of solutions**. This fundamental set can be expressed in the form of an $n \times n$ matrix called a **fundamental matrix solution** as

$$Y(t) = [\mathbf{y}_1(t)\ \mathbf{y}_2(t)\ \cdots\ \mathbf{y}_n(t)] \qquad (3.2.4)$$

Clearly, Y satisfies the matrix equation

$$\dot{Y} = A(t; \mathbf{M_0})Y \tag{3.2.5}$$

Changing the dependent variable in (3.2.5) from t to $\tau = t + T$, we arrive at

$$\frac{dY}{d\tau} = A(\tau - T; \mathbf{M_0})Y = A(\tau; \mathbf{M_0})Y \tag{3.2.6}$$

on account of the fact that $A(\tau - T; \mathbf{M_0}) = A(\tau; \mathbf{M_0})$. Hence, if

$$Y(t) = [\mathbf{y_1}(t)\ \mathbf{y_2}(t)\ \cdots\ \mathbf{y_n}(t)]$$

is a fundamental matrix solution, then

$$Y(t + T) = [\mathbf{y_1}(t + T)\ \mathbf{y_2}(t + T)\ \cdots\ \mathbf{y_n}(t + T)]$$

is also a fundamental matrix solution. Because (3.2.3) has at most n linearly independent solutions and because the $\mathbf{y_i}(t)$ are such n linearly independent solutions, the $\mathbf{y_i}(t + T)$ must be linear combinations of $\mathbf{y_1}(t)$, $\mathbf{y_2}(t)$, $\cdots$, $\mathbf{y_n}(t)$; that is,

$$Y(t + T) = Y(t)\Phi \tag{3.2.7}$$

where Φ is an $n \times n$ constant matrix. We note that Φ depends on the chosen fundamental matrix solution and is not unique. This matrix may be thought of as a map or a transformation that maps an initial vector in $\mathcal{R}^n$ at $t = 0$ to another vector in $\mathcal{R}^n$ at $t = T$. Specifying the initial condition

$$Y(0) = \mathbf{I} \tag{3.2.8}$$

where $\mathbf{I}$ is the $n \times n$ identity matrix and setting $t = 0$ in (3.2.7), we obtain

$$\Phi = Y(T) \tag{3.2.9}$$

The matrix Φ, defined by (3.2.7)–(3.2.9), is called the **monodromy matrix**.

Introducing the transformation $Y(t) = V(t)P^{-1}$, where P is a nonsingular $n \times n$ constant matrix and P^{-1} is the inverse of P, we rewrite (3.2.7) as

$$V(t + T) = V(t)J \tag{3.2.10}$$

where

$$J = P^{-1}\Phi P \tag{3.2.11}$$

We choose P such that J has the simplest possible form. This form depends on the eigenvalues and eigenvectors of the matrix Φ.

As discussed in Section 2.1.1, when the eigenvalues ρ_m of the monodromy matrix Φ are distinct, one can choose the matrix P such that its columns $\mathbf{p}_1$, $\mathbf{p}_2$, $\cdots$, $\mathbf{p}_n$ are the right eigenvectors of Φ corresponding to the eigenvalues ρ_1, ρ_2, $\cdots$, ρ_n ; that is,

$$\Phi\mathbf{p}_m = \rho_m\mathbf{p}_m \tag{3.2.12}$$

Thus,

$$P = [\mathbf{p}_1\ \mathbf{p}_2\ \cdots\ \mathbf{p}_n] \tag{3.2.13}$$

With this choice,

$$\begin{aligned}
J &= P^{-1}\Phi P = P^{-1}\Phi\,[\mathbf{p}_1\ \mathbf{p}_2\ \cdots\ \mathbf{p}_n]\\
&= P^{-1}\,[\Phi\mathbf{p}_1\ \Phi\mathbf{p}_2\ \cdots\ \Phi\mathbf{p}_n]\\
&= P^{-1}\,[\rho_1\mathbf{p}_1\ \rho_2\mathbf{p}_2\ \cdots\ \rho_n\mathbf{p}_n] = P^{-1}PD = D
\end{aligned} \tag{3.2.14}$$

where

$$D = \begin{bmatrix} \rho_1 & 0 & \cdot & \cdot & \cdot & 0 \\ 0 & \rho_2 & \cdot & \cdot & \cdot & 0 \\ \cdot & \cdot & \cdot & \cdot & \cdot & \cdot \\ \cdot & \cdot & \cdot & \cdot & \cdot & \cdot \\ \cdot & \cdot & \cdot & \cdot & \cdot & \cdot \\ 0 & 0 & \cdot & \cdot & \cdot & \rho_n \end{bmatrix} \tag{3.2.15}$$

We note that if ρ_m is complex, then $\mathbf{p}_m$ is also complex. When the eigenvalues of Φ are not distinct, the matrix P in (3.2.11) is formed by using the generalized eigenvectors of Φ. The corresponding matrix J may be either diagonal or nondiagonal with off–diagonal entries of one.

The eigenvalues ρ_m of Φ are called **Floquet** or **characteristic multipliers**. There is a unique set of characteristic multipliers associated with the matrix A of (3.2.3). Each ρ_m provides a measure of the **local orbital divergence** or **convergence** along a particular direction over one period of the closed orbit of (3.2.1). To visualize this point, in Figure 3.2.1, we consider a closed orbit Γ of a three–dimensional

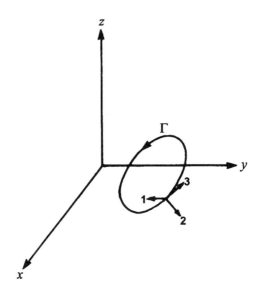

Figure 3.2.1: Closed orbit Γ of a three–dimensional autonomous system. At a point on Γ, the three directions are marked 1, 2, and 3.

autonomous system. A point has been marked on this orbit, and three directions labeled 1, 2, and 3 are shown. The three directions 1, 2, and 3 correspond to the three associated Floquet multipliers ρ_1, ρ_2, and ρ_3, respectively. Along a direction j, the characteristic multiplier ρ_j describes locally the convergence or divergence of nearby orbits with respect to Γ.

When the Floquet multipliers are distinct, (3.2.10) can be written in component form as

$$\mathbf{v}_m(t+T) = \rho_m \mathbf{v}_m(t) \ \text{ for } \ m = 1, 2, \cdots, n \tag{3.2.16}$$

It follows from (3.2.16) that

$$\mathbf{v}_m(t + NT) = \rho_m^N \mathbf{v}_m(t) \tag{3.2.17}$$

where N is an integer. Hence, as $t \to \infty$ (i.e., $N \to \infty$),

$$\mathbf{v}_m(t) \to 0 \ \ \text{if} \ \ |\rho_m| < 1$$
$$\mathbf{v}_m(t) \to \infty \ \text{if} \ |\rho_m| > 1$$

When $\rho_m = 1$, $\mathbf{v}_m(t)$ is periodic with the period T, and when $\rho_m = -1$, $\mathbf{v}_m(t)$ has the period $2T$.

The case where the Floquet multipliers are not distinct is treated by Nayfeh and Mook (1979, Chapter 5).

It is important to note that **one of the Floquet multipliers associated with a periodic solution $\mathbf{X}_0(t)$ of an autonomous system of equations, such as (3.2.1), is always unity.** To show this, we differentiate (3.2.1) once with respect to t and obtain

$$\ddot{\mathbf{x}} = D_\mathbf{x}\mathbf{F}(\mathbf{x}, \mathbf{M})\dot{\mathbf{x}} \qquad (3.2.18)$$

Consequently, if $\mathbf{x}$ is a solution of (3.2.1), then $\dot{\mathbf{x}}$ is a solution of (3.2.18) and hence of (3.2.3). Because $\mathbf{X}_0(t)$ is a solution of (3.2.1), $\dot{\mathbf{X}}_0(t)$ is a solution of (3.2.3). Moreover, because $\mathbf{X}_0(t) = \mathbf{X}_0(t + T)$ then $\dot{\mathbf{X}}_0(t) = \dot{\mathbf{X}}_0(t + T)$ and hence

$$\dot{\mathbf{X}}_0(0) = \dot{\mathbf{X}}_0(T) \qquad (3.2.19)$$

Furthermore, because $\dot{\mathbf{X}}_0(t)$ is a solution of (3.2.3), it must be a linear combination of $\mathbf{y}_1(t)$, $\mathbf{y}_2(t), \cdots, \mathbf{y}_n(t)$; that is,

$$\dot{\mathbf{X}}_0(t) = Y(t)\alpha \qquad (3.2.20)$$

where α is a constant vector. Evaluating (3.2.20) at $t = 0$ and $t = T$ yields

$$\dot{\mathbf{X}}_0(0) = Y(0)\alpha \text{ and } \dot{\mathbf{X}}_0(T) = Y(T)\alpha \qquad (3.2.21)$$

Considering (3.2.19) and (3.2.21), we obtain

$$Y(T)\alpha = Y(0)\alpha \qquad (3.2.22)$$

Using (3.2.8) and (3.2.9), we rewrite (3.2.22) as

$$\Phi\alpha = \alpha \qquad (3.2.23)$$

Therefore, one is an eigenvalue of Φ corresponding to the eigenvector $\alpha = \dot{\mathbf{X}}_0(0) = \mathbf{F}[\mathbf{X}_0(0); M_0]$.

Alternatively, one can show that one of the Floquet multipliers is unity as follows. Because (3.2.1) is autonomous, if $\mathbf{x}(t)$ is a solution, then $\mathbf{x}(t + \tau)$ for any τ is also a solution. Hence, if $\mathbf{X}_0(t)$ is a periodic solution, then $\mathbf{X}_0(t + \tau)$ is also a periodic solution. Letting

$$\mathbf{y}(t) = \mathbf{X}_0(t + \tau) - \mathbf{X}_0(t) \tag{3.2.24}$$

we conclude that

$$\mathbf{y}(0) = \mathbf{X}_0(\tau) - \mathbf{X}_0(0) \tag{3.2.25}$$

is an initial disturbance provided along the orbit of the periodic solution $\mathbf{X}_0(t)$. It follows from (3.2.24) that

$$
\begin{aligned}
\mathbf{y}(t + NT) &= \mathbf{X}_0(t + NT + \tau) - \mathbf{X}_0(t + NT) \\
&= \mathbf{X}_0(t + \tau) - \mathbf{X}_0(t)
\end{aligned} \tag{3.2.26}
$$

Hence,

$$\mathbf{y}(t + NT) = \mathbf{y}(t) \tag{3.2.27}$$

Therefore, for any integer N, the Floquet multiplier associated with this disturbance is unity and, hence, a disturbance provided along a direction tangent to the periodic orbit neither grows nor decays. In Figure 3.2.2, we show a planar projection of the orbit of a periodic solution $\mathbf{X}_0(t)$ of (3.2.1) and two points $\mathbf{c} = \mathbf{X}_0(0)$ and $\mathbf{d} = \mathbf{X}_0(\tau)$ on this orbit. Due to the periodicity of $\mathbf{X}_0(t)$, starting from points $\mathbf{c}$ and $\mathbf{d}$, one returns back to $\mathbf{c}$ and $\mathbf{d}$ after a time NT for any integer N. Consequently, the separation between points $\mathbf{c}$ and $\mathbf{d}$ remains the same after any N periods of oscillation.

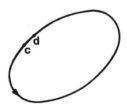

Figure 3.2.2: Planar projection of a closed orbit of an n–dimensional autonomous system. Two nearby points on this orbit are labeled $\mathbf{c}$ and $\mathbf{d}$.

A periodic solution of (3.2.1) is known as a **hyperbolic periodic solution** if only one Floquet multiplier is located on the unit circle in the complex plane. A hyperbolic periodic solution is either **stable** or **unstable**. We infer from (3.2.17) that a hyperbolic periodic solution is **asymptotically stable** if there are no Floquet multipliers outside the unit circle. In all of the directions not tangent to an asymptotically stable periodic orbit, neighboring positive orbits are attracted toward the periodic orbit. Hence, this solution is called a **stable limit cycle** or a **periodic attractor**. A hyperbolic periodic solution is **unstable** if one or more of the Floquet multipliers lie outside the unit circle. In this case, if all of the Floquet multipliers other than the one that is unity lie outside the unit circle, then all neighboring trajectories of the periodic solution are repelled from it in positive times. Hence, this solution is called an **unstable limit cycle** or a **periodic repellor**. When some of the Floquet multipliers associated with an unstable hyperbolic solution lie inside the unit circle, the periodic solution is called an **unstable limit cycle of the saddle type**.

If two or more Floquet multipliers are located on the unit circle, the periodic solution is called a **nonhyperbolic periodic solution**. A nonhyperbolic periodic solution is unstable if one or more of the associated Floquet multipliers lie outside the unit circle. If none of the Floquet multipliers lies outside the unit circle, a nonlinear analysis is necessary to determine the stability of a nonhyperbolic periodic solution. In the nonlinear analysis, one must retain higher–order terms in (3.2.3).

Multiplying (3.2.16) with $e^{-\gamma_m(t+T)}$ yields

$$e^{-\gamma_m(t+T)}\mathbf{v}_m(t+T) = \rho_m e^{-\gamma_m(t+T)}\mathbf{v}_m(t) \qquad (3.2.28)$$

Defining γ_m such that

$$\rho_m = e^{\gamma_m T} \quad \text{or} \quad \gamma_m = \frac{1}{T}\ln\rho_m \qquad (3.2.29)$$

we rewrite (3.2.28) as

$$e^{-\gamma_m(t+T)}\mathbf{v}_m(t+T) = e^{-\gamma_m t}\mathbf{v}_m(t) \qquad (3.2.30)$$

Consequently, $e^{-\gamma_m t}\mathbf{v}_m(t)$ is a periodic vector with the period T. Hence, if $\rho_m \neq 0$, every $\mathbf{v}_m$ can be expressed in the **normal** or **Floquet form**

$$\mathbf{v}_m(t) = e^{\gamma_m t}\boldsymbol{\phi}_m(t) \qquad (3.2.31)$$

where $\phi_m(t + T) = \phi_m(t)$. The γ_m are called **characteristic exponents**; they are unique to within an integer multiple of $2\pi i/T$, where $i = \sqrt{-1}$, according to (3.2.29). It follows from (3.2.31) that

$$\mathbf{v}_m(t) \to 0 \text{ as } t \to \infty$$

if the real part of γ_m is negative and

$$\mathbf{v}_m(t) \to \infty \text{ as } t \to \infty$$

if the real part of γ_m is positive.

The Floquet form (3.2.31) can be used to determine the solutions of (3.2.3). To this end, one seeks a solution of the form

$$\mathbf{y}(t) = e^{\gamma t}\phi(t) \tag{3.2.32}$$

where $\phi(t + T) = \phi(t)$, substitutes (3.2.32) into (3.2.3), and obtains

$$\dot{\phi} = [A(t; \mathbf{M}_0) - \gamma \mathbf{I}]\,\phi \tag{3.2.33}$$

Then, expanding ϕ in a Fourier series according to

$$\phi = \sum_{k=-\infty}^{k=\infty} \mathbf{a}_k \exp\left(\frac{2ik\pi t}{T}\right) \tag{3.2.34}$$

substituting the result into (3.2.33), and equating the coefficients of each harmonic on both sides, we obtain an infinite–dimensional eigenvalue problem for γ and $\mathbf{a}_k$. The determinant of the coefficient matrix is called **Hill's determinant**. In practice, one truncates the series in (3.2.34) to a finite number of terms. When the coefficients of the time–varying terms in $A(t; \mathbf{M}_0)$ are small, one can use a perturbation method, such as the method of multiple scales or averaging (Nayfeh, 1973, 1981), to determine an approximate solution of (3.2.3) and hence determine the behavior of $\mathbf{y}$ as $t \to \infty$.

Example 3.10. We consider the stability of the periodic solution

$$X_0(t) = \left(-\frac{\mu}{\alpha}\right)^{1/2} \cos\left[\left(\omega - \frac{\beta\mu}{\alpha}\right)t + \theta_0\right]$$

$$Y_0(t) = \left(-\frac{\mu}{\alpha}\right)^{1/2} \sin\left[\left(\omega - \frac{\beta\mu}{\alpha}\right)t + \theta_0\right]$$

(3.2.35)

of (3.1.2) and (3.1.3). Substituting

$$x(t) = X_0(t) + \zeta_1(t) \quad \text{and} \quad y(t) = Y_0(t) + \zeta_2(t)$$ (3.2.36)

into (3.1.2) and (3.1.3), expanding the result for small $|\zeta_1|$ and $|\zeta_2|$, and linearizing in the disturbance, we obtain

$$\dot{\zeta} = A(\theta)\zeta$$ (3.2.37)

where ζ is a two–dimensional vector with the components ζ_1 and ζ_2,

$$\theta = (\omega - \frac{\beta\mu}{\alpha})t + \theta_0$$

$$A = -\frac{\mu}{\alpha}\left[\begin{array}{c} \alpha + \alpha c(2\theta) - \beta s(2\theta) \\ -\omega\alpha\mu + 2\beta + \beta c(2\theta) + \alpha s(2\theta) \end{array}\right.$$

$$\left.\begin{array}{c} \frac{\omega\alpha}{\mu} - 2\beta + \beta c(2\theta) + \alpha s(2\theta) \\ \alpha - \alpha c(2\theta) + \beta s(2\theta) \end{array}\right]$$

(3.2.38)

and

$$c(2\theta) \equiv \cos(2\theta) \quad \text{and} \quad s(2\theta) \equiv \sin(2\theta)$$

Clearly, the components of the matrix A are periodic with the period $T = 2\pi/(\omega - \beta\mu/\alpha)$, the period of the limit cycle. However, we note that the minimal period of A is $\frac{1}{2}T$ because there are only odd nonlinearities in (3.1.2) and (3.1.3). Replacing the independent variable t by θ and recalling that $\dot{\theta} = \omega - \beta\mu/\alpha$, we rewrite (3.2.37) as

$$\frac{d\zeta}{d\theta} = \left(\omega - \frac{\beta\mu}{\alpha}\right)^{-1} A(\theta)\zeta$$ (3.2.39)

Consequently, the problem of determining the stability of the periodic solution (3.2.35) of (3.1.2) and (3.1.3) has been replaced with the problem of determining the behavior of the solutions of (3.2.39) as $\theta \to \infty$. To accomplish this, we calculate two linearly independent solutions $\zeta^{(1)}(\theta)$ and $\zeta^{(2)}(\theta)$ of (3.2.39) by integrating this system from $\theta = 0$ with the initial conditions

$$\zeta^{(1)}(0) = \begin{bmatrix} 1 \\ 0 \end{bmatrix} \quad \text{and} \quad \zeta^{(2)}(0) = \begin{bmatrix} 0 \\ 1 \end{bmatrix} \tag{3.2.40}$$

Then, we form the monodromy matrix

$$\Phi = \begin{bmatrix} \zeta^{(1)}(2\pi) & \zeta^{(2)}(2\pi) \end{bmatrix} \tag{3.2.41}$$

The fact that one of the two eigenvalues of Φ must be unity provides a check for the calculations. If the modulus of the other eigenvalue is less than unity, the periodic solution is stable. On the other hand, if the modulus of the other eigenvalue is greater than unity, the periodic solution is unstable.

When $\omega = 1.0$, $\beta = 1.0$, $\mu = 1.0$, and $\alpha = -1.0$,

$$\Phi = \begin{bmatrix} 0.00187 & 0.00085 \\ 0.99813 & 1.00000 \end{bmatrix}$$

whose eigenvalues are 1.00085 and 0.00102. Hence, the corresponding periodic solution is stable. When $\omega = 1.0$, $\beta = 1.0$, $\mu = -1.0$, and $\alpha = 1.0$,

$$\Phi = \begin{bmatrix} 535.49200 & 0.00000 \\ 534.49200 & 1.00000 \end{bmatrix}$$

whose eigenvalues are 1.00000 and 535.49200. Hence, the corresponding periodic solution is unstable. These results are in agreement with those presented in Figure 3.1.1.

Example 3.11. We consider the system of equations

$$\dot{x}_1 = x_2 \quad \text{and} \quad \dot{x}_2 = -x_1 - 2x_1^3 \tag{3.2.42}$$

For the solution initiated from $(0.1, 0)$, we use the method of multiple scales and obtain the analytical approximation

$$x_{10}(t) \simeq 0.1 \cos (1.0075t)$$
$$x_{20}(t) \simeq -0.10075 \sin (1.0075t) \tag{3.2.43}$$

To study the stability of this periodic solution, we substitute

$$x_1(t) = x_{10}(t) + \zeta_1(t) \quad \text{and} \quad x_2(t) = x_{20}(t) + \zeta_2(t) \tag{3.2.44}$$

into (3.2.42), expand the result for small $|\zeta_1|$ and $|\zeta_2|$, linearize in the disturbance, and obtain

$$\left\{ \begin{array}{c} \dot{\zeta}_1 \\ \dot{\zeta}_2 \end{array} \right\} = \left[\begin{array}{cc} 0 & 1 \\ -1 - 6x_{10}^2 & 0 \end{array} \right] \left\{ \begin{array}{c} \zeta_1 \\ \zeta_2 \end{array} \right\} \tag{3.2.45}$$

To determine the stability of the periodic solution (3.2.43) of (3.2.42), we numerically determine two linearly independent solutions $\zeta^{(1)}(t)$ and $\zeta^{(2)}(t)$ of (3.2.45) by using the initial conditions (3.2.40) from $t = 0$ to $t = 2\pi/1.0075$ and form the monodromy matrix. The result is

$$\Phi = \left[\begin{array}{cc} 1.00000 & -0.00009 \\ -0.09380 & 1.00000 \end{array} \right]$$

The eigenvalues of this matrix are both very close to 1.00000. Hence, the corresponding periodic solution is nonhyperbolic, and a nonlinear analysis is necessary to determine its stability.

For a general autonomous system, as described in Chapter 6, we numerically integrate (3.2.1) and the matrix equation (3.2.5) simultaneously to determine periodic solutions and their stability. Alternatively, when one has a weakly nonlinear system, one can determine an approximation to a periodic solution of (3.2.1) by using a perturbation method (Nayfeh, 1973, 1981) and then solve (3.2.5) either analytically by using a perturbation method or numerically.

3.2.2 Nonautonomous Systems

We consider the stability of periodic solutions of the nonautonomous system

$$\dot{\mathbf{x}} = \mathbf{F}(\mathbf{x}, t; \mathbf{M}) \tag{3.2.46}$$

where $\mathbf{x} \in \mathcal{R}^n$, $t \in \mathcal{R}^1$, $\mathbf{M} \in \mathcal{R}^m$, and $\mathbf{F}$ is a periodic function. Again, we let the periodic solution of (3.2.46) at $\mathbf{M} = \mathbf{M}_0$ be denoted by $\mathbf{X}_0(t)$ and have a period T. This period is related to the period of $\mathbf{F}$. We superimpose a disturbance $\mathbf{z}(t)$ on $\mathbf{X}_0(t)$ and obtain

$$\mathbf{x}(t) = \mathbf{X}_0(t) + \mathbf{z}(t) \qquad (3.2.47)$$

Substituting (3.2.47) into (3.2.46), assuming that $\mathbf{F}$ is sufficiently smooth (i.e., at least $\mathcal{C}^2$), expanding $\mathbf{F}$ in a Taylor series about $\mathbf{X}_0$, and retaining only linear terms in the disturbance leads to

$$\dot{\mathbf{z}} = D_{\mathbf{x}}\mathbf{F}(\mathbf{X}_0, t; \mathbf{M}_0)\mathbf{z} + O(\| \mathbf{z} \|^2) \quad \text{or} \quad \dot{\mathbf{z}} = A(t; \mathbf{M}_0)\mathbf{z} \qquad (3.2.48)$$

where A is the matrix of first partial derivatives of $\mathbf{F}$. This matrix is periodic in time with period T. Proceeding along the lines discussed earlier, we use Floquet theory to treat (3.2.48) and determine the monodromy matrix Φ associated with a periodic solution of (3.2.46).

Then, the eigenvalues of the monodromy matrix provide information on the stability of the periodic solution. In contrast with the autonomous case for which one of the Floquet multipliers is always unity, in the nonautonomous case, such a condition is not satisfied. If none of the Floquet multipliers lies on the unit circle, the periodic solution is called **hyperbolic**; otherwise, it is called **nonhyperbolic**. If all of the n Floquet multipliers are within the unit circle, then the corresponding solution is **asymptotically stable** and is called a **stable limit cycle** or a **periodic attractor**. If at least one of the Floquet multipliers is outside the unit circle, the associated solution is **unstable**. If all of the Floquet multipliers are outside the unit circle, the periodic solution is called a **repellor**. If some but not all of the Floquet multipliers are outside the unit circle, the periodic solution is of the **saddle type**. If none of the Floquet multipliers associated with a nonhyperbolic solution of (3.2.46) lies outside the unit circle, a nonlinear analysis is necessary to determine the stability.

Example 3.12. We consider the stability of a periodic solution $\mathbf{X}_0$ of the system (3.1.28) and (3.1.29). In this case, the matrix equation corresponding to (3.2.48) takes the form

$$\dot{Z} = \begin{bmatrix} 0 & 1 \\ -1 - 3\epsilon\alpha X_{10}^2 & -\epsilon\mu \end{bmatrix} Z \qquad (3.2.49)$$

where X_{i0} is the ith scalar component of $\mathbf{X}_0$. Once $\mathbf{X}_0$ is known, one can numerically integrate (3.2.49) from $t = 0$ to $t = T$ with the initial condition $Z(t = 0) = \mathbf{I}$, where $\mathbf{I}$ is the identity matrix, and determine the monodromy matrix $Z(T)$.

For a general nonautonomous system, as described in Chapter 6, we numerically integrate (3.2.46) and the matrix equation corresponding to (3.2.48) simultaneously to determine periodic solutions and their stability. Alternatively, when the nonlinearities are weak, one can determine an approximation to a periodic solution of (3.2.46) by using a perturbation method (Nayfeh, 1973, 1981) and then solve the matrix equation corresponding to (3.2.48) either analytically by using a perturbation method or numerically.

3.2.3 Comments on the Monodromy Matrix

When the minimal period of the matrix A in (3.2.5) is $\frac{1}{2}T$, we change the dependent variable from t to $\hat{\tau} = t + \frac{1}{2}T$ and obtain

$$\frac{dY}{d\hat{\tau}} = A(\hat{\tau}; \mathbf{M}_0)Y \qquad (3.2.50)$$

because $A(\hat{\tau} + \frac{1}{2}T; \mathbf{M}_0) = A(\hat{\tau}; \mathbf{M}_0)$. Thus, if

$$Y(t) = [\mathbf{y}_1(t)\ \mathbf{y}_2(t)\ \cdots\ \mathbf{y}_n(t)]$$

is a fundamental matrix solution, then

$$Y(t + \frac{1}{2}T) = \left[\mathbf{y}_1(t + \frac{1}{2}T)\ \mathbf{y}_2(t + \frac{1}{2}T)\ \cdots\ \mathbf{y}_n(t + \frac{1}{2}T)\right]$$

is also a fundamental matrix solution. Because (3.2.50) has at most n linearly independent solutions and because the $\mathbf{y}_i(t)$ are such n linearly independent solutions, the $\mathbf{y}_i(t + \frac{1}{2}T)$ must be linear combinations of $\mathbf{y}_1(t)$, $\mathbf{y}_2(t)$, $\cdots$, $\mathbf{y}_n(t)$; that is,

$$Y(t + \frac{1}{2}T) = Y(t)\hat{\Phi} \qquad (3.2.51)$$

where $\hat{\Phi}$ is an $n \times n$ constant matrix. If we specify

$$Y(t = 0) = \mathbf{I} \tag{3.2.52}$$

we obtain

$$\hat{\Phi} = Y(\frac{1}{2}T) \tag{3.2.53}$$

It follows from (3.2.51) that

$$Y(t + T) = Y(t + \frac{1}{2}T)\hat{\Phi} \tag{3.2.54}$$

Hence,

$$Y(T) = Y\left(\frac{1}{2}T\right)\hat{\Phi} = \hat{\Phi}^2 = \Phi \tag{3.2.55}$$

because $Y(T) = \Phi$, according to (3.2.9). Therefore, the eigenvalues of Φ are the squares of the corresponding eigenvalues of $\hat{\Phi}$. Proceeding along the same lines, we can also construct a matrix $\hat{\Phi}$ for a periodic solution of the nonautonomous system (3.2.46).

3.2.4 Manifolds of a Periodic Solution

As in the case of a fixed point, there are also sets called **stable**, **unstable**, and **center manifolds** of a periodic solution (e.g., Medvéd, 1992). The **stable manifold of a periodic solution** is the set of all initial conditions such that evolutions initiated at these initial conditions approach the periodic solution as $t \to \infty$. This manifold is associated with the Floquet multipliers that lie inside the unit circle. The **unstable manifold of a periodic solution** is the set of all initial conditions such that evolutions initiated at these initial conditions approach the periodic solution as $t \to -\infty$. This manifold is associated with the Floquet multipliers that lie outside the unit circle. The **center manifold of a periodic solution** is associated with the Floquet multipliers that lie on the unit circle.

3.3 POINCARÉ MAPS

Before addressing how a Poincaré map can be used to determine the stability of a periodic solution, we define and discuss Poincaré sections and maps.

A **Poincaré section** is a hypersurface in the state space that is transverse to the flow of a given system of equations. (In an n–dimensional space, a hypersurface is a surface whose dimension is less than n.) In the context of (3.2.1), by **transverse to the flow**, we mean

$$\mathbf{n}(\mathbf{x}) \cdot \mathbf{F}(\mathbf{x}) \neq 0 \quad \text{or} \quad \mathbf{n}^T(\mathbf{x})\mathbf{F}(\mathbf{x}) \neq 0$$

where $\mathbf{n}(\mathbf{x})$ is a vector normal to the section located at $\mathbf{x}$, $\mathbf{F}(\mathbf{x})$ is the vector field describing the flow, and the dot indicates the dot product. Similarly, in the context of (3.2.46), by **transverse to the flow**, we mean

$$\mathbf{n}^T[\mathbf{x}(t)]\mathbf{F}(\mathbf{x};t) \neq 0$$

Example 3.13. We consider the construction of a Poincaré section for orbits of the following system:

$$\begin{bmatrix} \dot{x} \\ \dot{y} \\ \dot{z} \end{bmatrix} = \begin{bmatrix} F_1(x,y,z) \\ F_2(x,y,z) \\ F_3(x,y,z) \end{bmatrix}$$

A trajectory of this system in the $x-y-z$ space and a two–dimensional section Σ transverse to it are shown in Figure 3.3.1a. This section is defined by

$$\Sigma = \left\{ (x,y,z) \in \mathcal{R}^1 \times \mathcal{R}^1 \times \mathcal{R}^1 \mid y = y_0 \right\}$$

The normal $\mathbf{n}$ to Σ is given by

$$\mathbf{n} = \begin{bmatrix} 0 \\ 1 \\ 0 \end{bmatrix}$$

The section Σ is transverse to the trajectories of the three–dimensional autonomous system whenever

$$[0\ 1\ 0] \begin{bmatrix} F_1(x,y_0,z) \\ F_2(x,y_0,z) \\ F_3(x,y_0,z) \end{bmatrix} = F_2(x,y_0,z) \neq 0$$

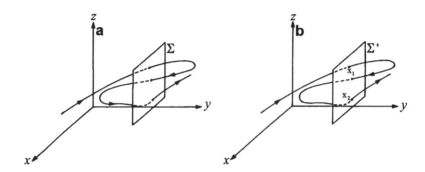

Figure 3.3.1: Poincaré section of a three–dimensional orbit: (a) two-sided section Σ and (b) one-sided section Σ^+.

In Figure 3.3.1a, there are three intersections: Two of them correspond to $F_2 > 0$, and one of them corresponds to $F_2 < 0$. Sections, such as Σ, where $\mathbf{n}^T\,\mathbf{F}$ does not have the same sign for all intersections, are called **two–sided sections**. In Figure 3.3.1b, the section Σ^+ is defined by

$$\Sigma^+ = \Big\{(x,y,z) \in \mathcal{R}^1 \times \mathcal{R}^1 \times \mathcal{R}^1 \mid y = y_0, F_2 > 0\Big\}$$

The first and second intersections of the trajectory with Σ^+ have been marked as $\mathbf{X}_1$ and $\mathbf{X}_2$, respectively, in Figure 3.3.1b. Sections, such as Σ^+, where $\mathbf{n}^T\,\mathbf{F}$ has the same sign for all intersections, are called **one–sided sections**.

It is worth noting that the time interval between two successive intersections of a trajectory with a chosen Poincaré section is not a constant in all situations.

In a general setting, let the successive intersections of a trajectory of an autonomous system with a one–sided Poincaré section be $\mathbf{X}_1$, $\mathbf{X}_2$, $\mathbf{X}_3$, and so on. If the trajectory evolves in an n–dimensional space, it follows that the Poincaré section is an $(n-1)$–dimensional surface and that each point on this section is specified by $(n-1)$ coordinates. The transformation or map that maps the current intersection to the subsequent intersection on a Poincaré section is called a **Poincaré map**. This map, which is $(n-1)$–dimensional, is described by

$$\mathbf{X}_{m+1} = \mathbf{P}(\mathbf{X}_m)$$

where $m \in \mathcal{Z}$. (For an n–dimensional nonautonomous system, the Poincaré section is an n–dimensional surface and the associated map $\mathbf{P}$ is n–dimensional.) The iterates of $\mathbf{P}$ are given by

$$\{\mathbf{X}_1, \mathbf{X}_2, \mathbf{X}_3, \cdots\}$$

Because $\mathbf{X}_{j+1}$ uniquely specifies $\mathbf{X}_j$ and vice versa, a Poincaré map is an invertible map. So, we have

$$\mathbf{X}_m = \mathbf{P}^{-1}(\mathbf{X}_{m+1})$$

Furthermore, a Poincaré map shares the different properties of the flow described by the associated continuous–time system. Consequently, this map is an **orientation preserving diffeomorphism** (e.g., Arnold, 1988). The orientation preservation property implies that the determinant of the Jacobian matrix $D\mathbf{P}$ is always positive. In addition, if the flow is dissipative, the determinant of $D\mathbf{P}$ has a magnitude less than 1.

The Poincaré map $\mathbf{P}$ for Figure 3.3.1b is two–dimensional and is such that

$$\mathbf{P} : \Sigma^+ \rightarrow \Sigma^+$$

and

$$\mathbf{X}_2 = \mathbf{P}(\mathbf{X}_1) \text{ and } \mathbf{X}_1 = \mathbf{P}^{-1}(\mathbf{X}_2)$$

In Figures 3.3.2a and 3.3.2b, we show periodic orbits of an autonomous system in the corresponding state space. The periodic orbit of Figure

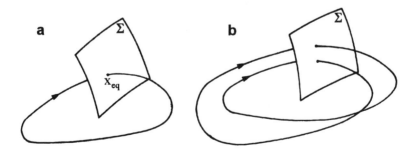

Figure 3.3.2: Poincaré sections of periodic orbits: (a) one intersection with Σ and (b) two intersections with Σ.

3.3.2a intersects the hypersurface Σ transversely once at the point $\mathbf{X} = \mathbf{X}_{eq}$ before closing on itself. On the other hand, the periodic orbit of Figure 3.3.2b intersects the hypersurface Σ twice before closing on itself. Thus, the periodic orbit of Figure 3.3.2a is reduced to a point on the chosen Poincaré section, whereas the periodic orbit of Figure 3.3.2b is reduced to two points on the chosen section. In the case of Figure 3.3.2a, the point $\mathbf{X} = \mathbf{X}_{eq}$ is a fixed point of the associated Poincaré map $\mathbf{P}$; that is,

$$\mathbf{X}_{eq} = \mathbf{P}(\mathbf{X}_{eq})$$

In the case of Figure 3.3.2b, each point on the section Σ is a period–two point of the associated Poincaré map $\mathbf{P}$ and a fixed point of the map $\mathbf{P}^2$. In a general setting, a periodic orbit of a continuous–time system may intersect a Poincaré section k times before closing on itself. Let one of these k intersections be $\mathbf{X}_{eq}$. Then, the corresponding Poincaré map $\mathbf{P}$ is such that

$$\mathbf{X}_{eq} = \mathbf{P}^k(\mathbf{X}_{eq})$$

implying that $\mathbf{X}_{eq}$ is a period–k point of $\mathbf{P}$ or a fixed point of $\mathbf{P}^k$. Hence, the stability of a periodic orbit of a continuous–time system may be determined by examining the stability of the fixed point of an associated map.

From Figure 3.3.2, we note that one can construct Poincaré sections at different locations on the periodic orbit. Consequently, one can obtain different Poincaré maps for the considered orbit. However, in most cases, there exists a differentiable coordinate transformation from one Poincaré map to another, and the maps on the different sections exhibit the same qualitative dynamics; that is, the same number of fixed points, similar stability properties of fixed points, and so forth (e.g., Wiggins, 1988, 1990; Medvéd, 1992).

3.3.1   Nonautonomous Systems

In nonautonomous systems, the period associated with a periodic orbit is usually explicitly known. If we let the vector field $\mathbf{F}$ in (3.2.46) be periodic in time with period T, then a periodic solution of (3.2.46) has a period that is either an integer multiple or integer submultiple of the period T. This period can be used to construct a Poincaré section as illustrated in the next example.

Example 3.14. The system of interest has the form

$$\begin{bmatrix} \dot{x}_1 \\ \dot{x}_2 \end{bmatrix} = \begin{bmatrix} F_1(x_1, x_2) \\ F_2(x_1, x_2, t) \end{bmatrix}$$

where F_2 is a periodic function of time. Starting at an initial time $t = t_0$, we can collect the points on the Poincaré section by stroboscopically monitoring the state variables at intervals of the period T, as illustrated in Figure 3.3.3a. The above system of equations can be posed as the following three–dimensional autonomous system:

$$\begin{bmatrix} \dot{x}_1 \\ \dot{x}_2 \\ \dot{\theta} \end{bmatrix} = \begin{bmatrix} F_1(x_1, x_2) \\ F_2(x_1, x_2, \theta) \\ 2\pi/T \end{bmatrix}$$

where $\theta = 2\pi t/T \pmod{2\pi}$. Next, as shown in Figure 3.3.3b, we

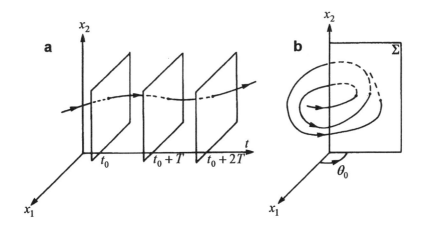

Figure 3.3.3: Poincaré section Σ of an orbit of a two–dimensional nonautonomous system with time–periodic terms: (a) (x_1, x_2, t) space and (b) (x_1, x_2, θ) space.

construct the Poincaré section

$$\Sigma = \left\{ (x_1, x_2, \theta) \in \mathcal{R}^1 \times \mathcal{R}^1 \times S^1 \mid \theta = \theta_0 = \frac{2\pi(t - t_0)}{T} \ (\text{mod } 2\pi) \right\}$$

The normal $\mathbf{n}$ to Σ is given by

$$\mathbf{n} = [0\ 0\ 1]^T$$

and the section Σ is transverse to the flow everywhere in the cylindrical state space because

$$[0\ 0\ 1] \begin{bmatrix} F_1(x_1, x_2) \\ F_2(x_1, x_2, \theta_0) \\ 2\pi/T \end{bmatrix} = \frac{2\pi}{T}$$

A section, such as the one above, that is transverse to the flow everywhere in the state space is called a **global section**. On the other hand, a section that is transverse to the flow only locally in the state space is called a **local section**.

Example 3.15. We consider (1.2.2) and (1.2.3) and set $\omega^2 = 8$, $\mu = 2$, $F = 10$, and $\Omega = 2$. The corresponding solution initiated from $(0, 0)$ is given by

$$x_1 = e^{-2t} \left[a \cos(2t) + b \sin(2t) \right] + 0.5 \cos(2t) + \sin(2t)$$
$$x_2 = -2e^{-2t} \left[(a - b) \cos(2t) + (a + b) \sin(2t) \right] - \sin(2t) + 2\cos(2t)$$

where $a = -0.5$ and $b = -1.5$. To construct a Poincaré section Σ, we start at time $t = 0$ and collect discrete points at time intervals of $2\pi/\Omega = \pi$. Formally, the section is defined as

$$\Sigma = \left\{ (x_1, x_2, \theta) \in \mathcal{R}^1 \times \mathcal{R}^1 \times S^1 \mid \theta = \theta_0 = 0 \right\}$$

On Σ, at $t_n = 2n\pi/\Omega = n\pi$, we have

$$x_{1n} = 0.5 \left(1 - e^{-2n\pi} \right)$$
$$x_{2n} = 2 \left(1 - e^{-2n\pi} \right)$$

Hence, it follows that

$$\lim_{n\to\infty}(x_{1n}, x_{2n}) = (0.5, 2)$$

For the given parameters, the flow described by (1.2.2) and (1.2.3) is attracted to the periodic orbit

$$x_1 = 0.5\cos(2t) + \sin(2t)$$
$$x_2 = -\sin(2t) + 2\cos(2t)$$

while the iterates on the chosen Poincaré section are attracted to the point $(0.5, 2)$ on this orbit.

Returning back to (3.2.46), we consider a periodic solution with period T and let $x_0 = \eta_0$ represent a point on the periodic orbit. Then, a trajectory initiated from η_0 at $t = t_0$ is represented by $x(\eta_0, t, t_0)$. To construct a Poincaré section, we collect discrete points at intervals of the period T. Then, if η represents a point on this section, the Poincaré map $\mathbf{P}$ is defined by

$$\mathbf{P}(\eta) = x(\eta, t_0 + T, t_0) \tag{3.3.1}$$

For the periodic solution, we have

$$x(\eta_0, t_0 + T, t_0) = \eta_0 \tag{3.3.2}$$

and consequently

$$\mathbf{P}(\eta_0) = \eta_0 \tag{3.3.3}$$

It follows from (3.3.3) that the periodic orbit of (3.2.46) corresponds to the fixed point of the map $\mathbf{P}(\eta)$.

In a neighborhood of the fixed point of the map, we have

$$\mathbf{P}(\eta_0 + v) = \mathbf{P}(\eta_0) + D_\eta \mathbf{P}(\eta_0)\, v + O(\|\, v\, \|^2) \tag{3.3.4}$$

where $\|\, v\, \|$ is the norm of the deviation from the fixed point on the Poincaré section. Information on the stability of the fixed point η_0 can be obtained by studying the eigenvalues of the Jacobian matrix $D_\eta \mathbf{P}(\eta_0)$, which is determined by using (3.2.46) and (3.3.2) as follows.

Differentiating the scalar components of $\mathbf{x}$ in (3.2.46) with respect to the scalar components of $\boldsymbol{\eta}$ leads to

$$\frac{d}{dt}(D_{\boldsymbol{\eta}}\mathbf{x}) = (D_{\mathbf{x}}\mathbf{F})\left(D_{\boldsymbol{\eta}}\mathbf{x}\right) \quad \text{or} \quad \dot{Y} = (D_{\mathbf{x}}\mathbf{F})Y \tag{3.3.5}$$

where $Y = D_{\boldsymbol{\eta}}\mathbf{x}$. From the initial condition $\mathbf{x}(t_0) = \boldsymbol{\eta}$, we find after differentiation with respect to $\boldsymbol{\eta}$ that $Y(t_0) = \mathbf{I}$, where $\mathbf{I}$ is the identity matrix. Solving (3.3.5) subject to the initial condition $Y(t_0) = \mathbf{I}$, we obtain $D_{\boldsymbol{\eta}}\mathbf{x}(\boldsymbol{\eta}_0, t_0 + T, t_0)$. Then, we note from (3.3.1) that

$$D_{\boldsymbol{\eta}}\mathbf{P}(\boldsymbol{\eta}_0) = D_{\boldsymbol{\eta}}\mathbf{x}(\boldsymbol{\eta}_0, t_0 + T, t_0) \tag{3.3.6}$$

The Jacobian matrix $D_{\boldsymbol{\eta}}\mathbf{P}$ can only be obtained by generating a fundamental matrix solution of (3.3.5) either analytically or numerically. For a general system, determining an analytical solution of (3.3.5) requires perturbation methods or the use of special functions or the method of harmonic balance and is difficult. Hence, one usually has to integrate (3.3.5) numerically from $t = t_0$ to $t = t_0 + T$ with the initial condition $Y(t_0) = \mathbf{I}$ to obtain the matrix $Y(t_0 + T)$ whose eigenvalues determine the stability of the fixed point of $\mathbf{P}$ and, consequently, the stability of the associated periodic solution.

If all of the n eigenvalues of $D_{\boldsymbol{\eta}}\mathbf{P}$ are within the unit circle, the corresponding fixed point of $\mathbf{P}$ is asymptotically stable. Hence, the associated periodic orbit of the nonautonomous system is asymptotically stable and is an attracting limit cycle. If all of the n eigenvalues of $D_{\boldsymbol{\eta}}\mathbf{P}$ are outside the unit circle, the corresponding fixed point of $\mathbf{P}$ is unstable. Therefore, the associated periodic orbit is an unstable limit cycle and a repellor. If some but not all of the eigenvalues of $D_{\boldsymbol{\eta}}\mathbf{P}$ are outside the unit circle, then the corresponding fixed point of $\mathbf{P}$ is a saddle. Hence, the corresponding periodic solution of (3.2.46) is an unstable limit cycle of the saddle type. In all of the above cases, because none of the eigenvalues of $D_{\boldsymbol{\eta}}\mathbf{P}$ lie on the unit circle, the corresponding fixed points of $\mathbf{P}$ are hyperbolic fixed points. The corresponding solutions of (3.2.46) are also hyperbolic. If one or more of the eigenvalues of $D_{\boldsymbol{\eta}}\mathbf{P}$ lie on the unit circle, the corresponding fixed point of $\mathbf{P}$ is nonhyperbolic. Therefore, the corresponding periodic solution of (3.2.46) is also nonhyperbolic. In this case, a linearization of the Poincaré map

P may not be sufficient for determining the stability and a nonlinear analysis of (3.3.4) with higher–order terms may be necessary.

Example 3.16. We consider a Poincaré map **P** associated with a periodic orbit of (3.1.28) and (3.1.29) initiated at time $t = 0$. In this case, the map **P** is two–dimensional and the vector η in (3.3.1) has the components η_1 and η_2. Then, equation (3.3.5) takes the form

$$\dot{Y} = \begin{bmatrix} 0 & 1 \\ -1 - 3\epsilon\alpha x_1^2 & -\epsilon\mu \end{bmatrix} Y \qquad (3.3.7)$$

where the matrix Y is given by

$$Y = \begin{bmatrix} \partial x_1/\partial\eta_1 & \partial x_1/\partial\eta_2 \\ \partial x_2/\partial\eta_1 & \partial x_2/\partial\eta_2 \end{bmatrix} \qquad (3.3.8)$$

We note that (3.3.7) is identical to (3.2.49), and hence their solutions should be the same.

To determine the stability of a periodic solution of (3.1.28) and (3.1.29), we first compute the periodic solution by using either a numerical scheme or a perturbation method. Then, we either numerically or analytically solve (3.3.7) subject to $Y(0) = \mathbf{I}$ and determine $Y(T)$. Then, the eigenvalues of $Y(T)$ provide information on the stability of the periodic orbit.

3.3.2 Autonomous Systems

In autonomous systems, the period associated with a periodic orbit is not usually explicitly known. Hence, the construction of a Poincaré section differs from that described in the previous section for nonautonomous systems. A Poincaré section for an orbit of the autonomous system (3.2.1) is usually taken to be the hyperplane

$$\mathbf{x}^T \mathbf{n} = c$$

where c is a constant and $\mathbf{n}$ is the normal vector to this hyperplane chosen so that

$$\mathbf{F}^T(\mathbf{x}) \mathbf{n} \neq 0$$

at the location of interest. In some cases, we may be able to choose the
hyperplane $x_k = \eta_k$, where η_k is a constant and x_k is the kth component
of the state vector $\mathbf{x}$.

Next, we consider the stability of a periodic orbit $\mathbf{x} = \mathbf{x}(t, \boldsymbol{\eta})$ with
least period T of the autonomous system (3.2.1) initiated at $\mathbf{x} = \boldsymbol{\eta}$ at
time $t = 0$. This orbit is denoted as Γ in Figure 3.3.4. As illustrated,
we construct an $(n-1)$–dimensional oriented hypersurface Σ in the n–
dimensional state space that is transverse to the vector field $\mathbf{F}(\mathbf{x}; \mathbf{M})$ at
$\boldsymbol{\eta}$. Let $\boldsymbol{\zeta}$ denote a point on this hypersurface. In a small neighborhood
of $\boldsymbol{\eta}$, the trajectory initiated from $\boldsymbol{\zeta}$ at $t = 0$ returns to the location $\tilde{\boldsymbol{\zeta}}$
on this section after a unique time $\tau(\boldsymbol{\zeta})$; that is,

$$\tilde{\boldsymbol{\zeta}} = \mathbf{P}(\boldsymbol{\zeta}) = \mathbf{x}[\tau(\boldsymbol{\zeta}), \boldsymbol{\zeta}], \quad \tilde{\boldsymbol{\zeta}} \in \Sigma \qquad (3.3.9)$$

where $\mathbf{P}$ is the **return** or **Poincaré map**. We note that $\tau(\boldsymbol{\eta}) = T$. The
existence of the unique real–valued function $t = \tau(\boldsymbol{\zeta})$ is a consequence
of the implicit function theorem (Hirsch and Smale, 1974).

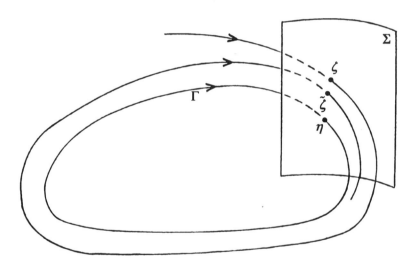

Figure 3.3.4: Poincaré section Σ of the periodic orbit Γ and an adjacent
trajectory.

It is clear that $\boldsymbol{\eta}$ is a fixed point of the map $\mathbf{P}$ and that the stability
of this fixed point reflects the stability of the periodic orbit initiated

at η. The stability of the fixed point of $\mathbf{P}$ in turn depends on the eigenvalues of the matrix $D_\zeta \mathbf{P}(\zeta)\,|_{\zeta = \eta}$.

To compute $D_\zeta \mathbf{P}(\zeta)$, we note that

$$\mathbf{P}(\zeta) = \mathbf{x}[\tau(\zeta), \zeta], \quad \zeta \in \Sigma \tag{3.3.10}$$

Differentiating (3.3.10) with respect to ζ, we obtain

$$D_\zeta \mathbf{P} = D_\zeta \mathbf{x} + \frac{\partial \mathbf{x}}{\partial t} \frac{\partial \tau}{\partial \zeta} \tag{3.3.11}$$

Differentiating the scalar components of $\mathbf{x}$ in (3.2.1) with respect to the scalar components of ζ yields

$$\frac{d}{dt}(D_\zeta \mathbf{x}) = (D_\mathbf{x} \mathbf{F})(D_\zeta \mathbf{x}) \tag{3.3.12}$$

and differentiating $\mathbf{x}(0) = \zeta$ with respect to ζ yields

$$D_\zeta \mathbf{x}(0) = \mathbf{I} \tag{3.3.13}$$

where $\mathbf{I}$ is the $n \times n$ identity matrix. Evaluating the solution of (3.3.12) at $t = T$ subject to the initial condition (3.3.13) yields the first term on the right–hand side of (3.3.11).

In the second term on the right–hand side of (3.3.11), the vector $\partial \mathbf{x}/\partial t$ is determined by evaluating the vector field $\mathbf{F}$ at η. The evaluation of the vector $\partial \tau / \partial \zeta$ is not straightforward and depends on how Σ is constructed. For example, if the Poincaré section is defined by (Curry, 1980)

$$\Sigma = \left\{ (x_1, x_2, \cdots, x_n) \in \mathcal{R}^1 \times \mathcal{R}^1 \cdots \times \mathcal{R}^1 \mid x_n = c \right\}$$

where c is a constant, then the nth component P_n of the map (3.3.9) is given by

$$P_n(\zeta) = \zeta_n = c$$

Consequently, it follows that

$$\frac{\partial P_n}{\partial \zeta_j} = 0 \quad \text{or} \quad \frac{\partial x_n}{\partial \zeta_j} + \frac{\partial x_n}{\partial t} \frac{\partial \tau}{\partial \zeta_j} = 0$$

for $j \neq n$. Thus, we find that

$$\frac{\partial \tau}{\partial \zeta_j} = -\frac{\partial x_n/\partial \zeta_j}{\partial x_n/\partial t}$$

In this scheme, all of the elements of the nth row of $D_\zeta \mathbf{P}$ are zero except the diagonal element, which is unity. Consequently, one of the eigenvalues of $D_\zeta \mathbf{P}$ is unity. Now, let J represent the $(n-1) \times (n-1)$ matrix obtained from $D_\zeta \mathbf{P}$ by deleting the nth row and the nth column. Then, if all of the $(n-1)$ eigenvalues of J are inside the unit circle, the fixed point η of $\mathbf{P}$ is asymptotically stable and, consequently, the corresponding solution of (3.2.1) is an attracting limit cycle. If all of the eigenvalues of J are outside the unit circle, the fixed point of $\mathbf{P}$ is a repellor and, hence, the associated solution of (3.2.1) is a repelling limit cycle. If some but not all of the eigenvalues of J are outside the unit circle, the fixed point of $\mathbf{P}$ is a saddle point and, therefore, the corresponding solution of (3.2.1) is a saddle limit cycle. If one or more of the eigenvalues of J are on the unit circle while the rest of them are inside the unit circle, a nonlinear analysis is necessary to determine the stability of the fixed point of $\mathbf{P}$.

In general, it is not possible to determine explicitly the Poincaré map $\mathbf{P}$ associated with a periodic solution of (3.2.1). Hence, one will have to resort to numerical methods (Hénon, 1982) or analytical approximations to construct $\mathbf{P}$. Even if such a map is constructed, the evaluation of $D\mathbf{P}$ is not straightforward for all Poincaré sections (see Parker and Chua, 1989, Appendix D).

Example 3.17. We consider the stability of a periodic solution of the autonomous system (3.1.2) and (3.1.3) when $\beta = 0$. The corresponding equations are

$$\dot{x} = \mu x - \omega y + \alpha x (x^2 + y^2) \qquad (3.3.14)$$

$$\dot{y} = \omega x + \mu y + \alpha y (x^2 + y^2) \qquad (3.3.15)$$

Using the transformation $(x, y) = (r \cos \theta, r \sin \theta)$ in (3.3.14) and (3.3.15), we obtain

$$\dot{r} = \mu r + \alpha r^3 \qquad (3.3.16)$$

$$\dot{\theta} = \omega \qquad (3.3.17)$$

When $\mu\alpha < 0$, there is a periodic orbit in the $x - y$ plane corresponding to the nontrivial fixed point $r = \sqrt{-\mu/\alpha}$ of (3.3.16). Equations (3.3.16) and (3.3.17) have the closed–form solution

$$r = \left[\left(\frac{\alpha}{\mu} + \frac{1}{r_0^2} \right) e^{-2\mu(t-t_0)} - \frac{\alpha}{\mu} \right]^{-1/2} \qquad (3.3.18)$$

$$\theta = \omega(t - t_0) + \theta_0 \qquad (3.3.19)$$

where $(r, \theta) = (r_0, \theta_0)$ at $t = t_0$.

We define an associated Poincaré section by

$$\Sigma = \left\{ (r, \theta) \in \mathcal{R}^1 \times S^1 \mid \theta = \theta_0 \right\}$$

A trajectory described by (3.3.18) and (3.3.19) initiated at (r_0, θ_0) intersects the Poincaré section Σ at the times $t_k = t_0 + 2k\pi/\omega$, where k is an integer. In other words, the time of flight from one intersection to the subsequent intersection on the section is $2\pi/\omega$. The associated one–dimensional Poincaré map is given by

$$P(r_k) = \left[\left(\frac{\alpha}{\mu} + \frac{1}{r_k^2} \right) e^{-2\mu(t_{k+1}-t_k)} - \frac{\alpha}{\mu} \right]^{-1/2}$$

which is equivalent to

$$P(r) = \left[\left(\frac{\alpha}{\mu} + \frac{1}{r^2} \right) e^{-4\pi\mu/\omega} - \frac{\alpha}{\mu} \right]^{-1/2} \qquad (3.3.20)$$

The fixed point of (3.3.20) is

$$r_0 = \sqrt{-\mu/\alpha} \qquad (3.3.21)$$

which corresponds to the periodic orbit of interest.

To determine the stability of this fixed point, we need to compute the eigenvalue of the Jacobian matrix associated with this fixed point. Differentiating (3.3.20) with respect to r yields

$$\frac{dP}{dr} = r^{-3} e^{-4\pi\mu/\omega} \left[\left(\frac{\alpha}{\mu} + \frac{1}{r^2} \right) e^{-4\pi\mu/\omega} - \frac{\alpha}{\mu} \right]^{-3/2} \qquad (3.3.22)$$

Hence, at the fixed point given by (3.3.21), we have

$$\frac{dP}{dr}\Big|_{r=r_0} = e^{-4\pi\mu/\omega} \tag{3.3.23}$$

When $\mu > 0$, the fixed point of the Poincaré map (3.3.20) is asymptotically stable because $|\, dP/dr \,| < 1$. Hence, the corresponding periodic solution of (3.3.14) and (3.3.15) is asymptotically stable. On the other hand, when $\mu < 0$, the fixed point of the Poincaré map (3.3.20) is unstable because $|\, dP/dr \,| > 1$. Consequently, the corresponding periodic solution is unstable.

In Figure 3.3.5, we show a periodic orbit of (3.3.14) and (3.3.15) in the $x-y$ plane when $\mu > 0$ and $\alpha < 0$. The Poincaré section is specified by $\theta_0 = \frac{1}{2}\pi$. The trajectory initiated at $(0.0, 0.8)$ intersects this section first at location 1, then at location 2, and so on, as it approaches the stable periodic orbit. This closed orbit intersects the section at $(0.0, 0.3162)$, which is marked by a dot. This point is a fixed point of the associated Poincaré map.

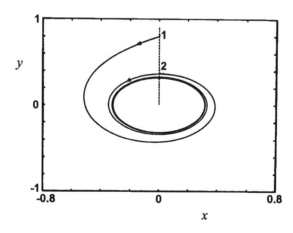

Figure 3.3.5: Periodic orbit of (3.3.14) and (3.3.15) realized for $\mu = 0.1$, $\alpha = -1.0$, and $\omega = 1.0$, along with a trajectory initiated from $(0.0, 0.8)$. The chosen Poincaré section is displayed as a broken line.

3.4 BIFURCATIONS

From Section 3.2, we know that the matrix A in either (3.2.3) or (3.2.48) is a function of the control parameter vector $\mathbf{M}$. Consequently, the monodromy matrix Φ and the associated Floquet multipliers depend on $\mathbf{M}$. Let us suppose that as one or more control parameters are varied, a periodic solution becomes nonhyperbolic at a certain location in the state–control space. Then, if the state–space portraits before and after this location are qualitatively different, this location is called a **bifurcation point** and the accompanying qualitative change is called a **bifurcation**. Furthermore, a bifurcation that requires at least m independent control parameters to occur is called a **codimension–m bifurcation**.

If we start with control parameters corresponding to a stable periodic solution of a continuous–time system and then vary one of the control parameters until this periodic solution loses stability, the resulting solution depends on the manner in which the Floquet multipliers leave the unit circle. (It is useful to remember that one of the Floquet multipliers associated with a periodic solution of an autonomous system is always unity.) There are three possible scenarios, as depicted in Figure 3.4.1 (e.g., Arnold, 1988). First, a Floquet multiplier leaves the unit circle through +1, resulting in one of the

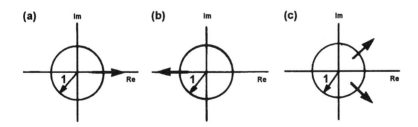

Figure 3.4.1: Scenarios depicting how the Floquet multipliers leave the unit circle for different local bifurcations: (a) transcritical, symmetry-breaking, and cyclic-fold bifurcations; (b) period–doubling bifurcation; and (c) secondary Hopf or Neimark bifurcation.

following three bifurcations: **transcritical, symmetry–breaking,** and **cyclic–fold** bifurcations. Second, a Floquet multiplier leaves the unit circle through -1, resulting in a **period–doubling** bifurcation. Third, two complex conjugate Floquet multipliers leave the unit circle away from the real axis, resulting in a **secondary Hopf** or **Neimark** bifurcation. These bifurcations are discussed in Sections 3.4.1–3.4.5.

Because the Floquet multipliers are the eigenvalues of the monodromy matrix, which is obtained by solving a linearized system of equations around the periodic solution whose stability is being investigated, the associated bifurcations of the periodic solution are all **local bifurcations.** These local considerations may not be sufficient for determining the postbifurcation state of the system, and global considerations may be necessary. In the context of continuous–time systems, the transcritical, symmetry–breaking, cyclic–fold, and Hopf bifurcations of periodic solutions are analogous to the transcritical, symmetry–breaking, saddle–node, and Hopf bifurcations of fixed points, respectively. We note that the three scenarios of Figure 3.4.1 are similar to those encountered in the context of local bifurcations of fixed points of maps. If a Poincaré map associated with a periodic solution can be constructed, the local bifurcation of the considered periodic solution can be studied by examining the local bifurcation of the fixed point of the map.

As in the case of bifurcations of fixed points, bifurcations of limit cycles can be classified into **continuous** and **discontinuous** or **catastrophic bifurcations.** In the case of continuous bifurcations, the motion of the system evolves continuously onto another motion as a control parameter is varied in a quasi–stationary manner. As in the case of fixed points, discontinuous or catastrophic bifurcations may be dangerous or explosive. In a dangerous bifurcation, the system response jumps to a remote attractor which may be infinity, as a control parameter is varied in a quasi–stationary manner. Dangerous bifurcations are typically accompanied by hysteresis. The outcome of these bifurcations may be determinant or indeterminant, depending on whether the system has a single attractor past the bifurcation value or not. In an explosive bifurcation, the old attractor explodes into a larger attractor, with the old attractor being a proper subset of the new attractor. Again, the new attractor may or may not be chaotic.

3.4.1 Symmetry–Breaking Bifurcation

When a Floquet multiplier leaves the unit circle through $+1$, as shown in Figure 3.4.1a, the associated bifurcation depends on the nature of the periodic solution prior to the bifurcation. Let us suppose that the periodic solution prior to the bifurcation is a **symmetric solution;** that is, it possesses a symmetry property. Then, if the bifurcation breaks the symmetry of the periodic solution, it is called a **symmetry–breaking bifurcation.**

The scenarios near **supercritical** and **subcritical symmetry– breaking bifurcation points** are shown in Figures 3.4.2a and 3.4.2b, respectively. In each of these figures, we show the amplitude r of the periodic solution versus the scalar control parameter α. We use solid and broken lines to denote branches of stable and unstable periodic solutions, respectively. At each bifurcation point, branches of symmetric and asymmetric periodic solutions meet. In both Figures 3.4.2a and 3.4.2b, the stable branch of symmetric periodic solutions that exists prior to the bifurcation continues as an unstable branch of symmetric periodic solutions after the bifurcation. Further, in the case of a supercritical bifurcation, locally stable asymmetric periodic solutions coexist with unstable symmetric periodic solutions on one side of the bifurcation point. In the case of a subcritical bifurcation, locally unstable asymmetric periodic solutions coexist with stable

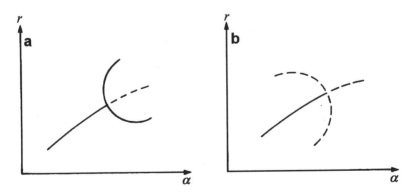

Figure 3.4.2: Local scenarios: (a) supercritical symmetry–breaking bifurcation and (b) subcritical symmetry–breaking bifurcation.

symmetric periodic solutions on one side of the bifurcation point. We note that the local scenarios depicted in Figure 3.4.2 are similar to those depicted for symmetry–breaking bifurcations of fixed points of a continuous–time system in Figure 2.3.2. We note that supercritical and subcritical symmetry–breaking bifurcations are examples of continuous and discontinuous bifurcations, respectively.

Example 3.18. Following Moon and Rand (1985), Rand (1989), and Nayfeh and Balachandran (1990a), we consider the following three-dimensional model of a feedback control system:

$$\dot{x} = \mu x - y - xz \qquad\qquad (3.4.1)$$
$$\dot{y} = \mu y + x \qquad\qquad (3.4.2)$$
$$\dot{z} = -z + y^2 + x^2 z \qquad\qquad (3.4.3)$$

In (3.4.1)–(3.4.3), x, y, and z are the states and μ is the control parameter. We note that this system is invariant under the transformation

$$(x, y, z) \Longleftrightarrow (-x, -y, z)$$

Therefore, if (x, y, z) is a solution of (3.4.1)–(3.4.3), then $(-x, -y, z)$ is also a solution. Hence, all solutions occur in pairs because of the transformation. A solution of (3.4.1)–(3.4.3) that is invariant under this transformation is called a **symmetric solution**. The projections of symmetric solutions onto the $x - y$ plane remain invariant under a 180° rotation about the origin. The fixed point $(0, 0, 0)$ is an obvious symmetric solution. If a solution of (3.4.1)–(3.4.3) is not invariant under the transformation, it is called an **asymmetric solution**.

A two–dimensional projection of the limit–cycle solution realized at $\mu = 0.2$ is shown in Figure 3.4.3a; this solution is symmetric. Furthermore, it is an attractor because one of the associated Floquet multipliers is +1 and the other two lie inside the unit circle. As μ is gradually increased, the symmetric limit cycle deforms smoothly, with two of the Floquet multipliers remaining inside the unit circle until we reach the critical value $\mu^{(1)} = 0.300$. At this critical point, the symmetric periodic solution is nonhyperbolic with two of the associated

Floquet multipliers at $+1$. For values of $\mu > \mu^{(1)}$, we find that the stable periodic solutions are no longer invariant under the symmetry transformation, implying that the symmetry has been broken through the bifurcation. Here, this bifurcation is a supercritical symmetry–breaking bifurcation. In Figures 3.4.3b and 3.4.3c, we show the two asymmetric limit cycles realized at $\mu = 0.350$. The stable symmetric solutions that exist for $\mu < \mu^{(1)}$ continue as unstable symmetric solutions for $\mu > \mu^{(1)}$.

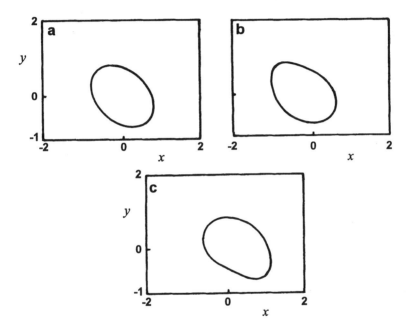

Figure 3.4.3: Two–dimensional projections of the periodic attractors of (3.4.1)–(3.4.3): (a) symmetric solution at $\mu = 0.2$, (b) one of the asymmetric solutions at $\mu = 0.35$, and (c) the other asymmetric solution at $\mu = 0.35$. Reprinted with permission from Nayfeh and Balachandran (1990a).

Example 3.19. We consider the following four–dimensional system treated in the context of surface waves in a closed basin (Nayfeh and

Nayfeh, 1990):

$$p'_1 = -\frac{1}{4}(2\sigma_1 + \sigma_2)q_1 - \mu_1 p_1 + p_2 q_1 - p_1 q_2 \qquad (3.4.4)$$

$$q'_1 = -\frac{1}{4}(2\sigma_1 + \sigma_2)p_1 - \mu_1 q_1 + p_1 p_2 + q_1 q_2 \qquad (3.4.5)$$

$$p'_2 = -\frac{1}{2}\sigma_2 q_2 - \mu_2 p_2 - 2p_1 q_1 - f q_2 \qquad (3.4.6)$$

$$q'_2 = \frac{1}{2}\sigma_2 p_2 - \mu_2 q_2 + p_1^2 - q_1^2 - f p_2 \qquad (3.4.7)$$

In this four–dimensional system, the p_i and q_i are the states while the σ_i, μ_i, and f are the control parameters. This system is invariant under the transformations

$$(p_1, q_1, p_2, q_2) \Longleftrightarrow (q_1, -p_1, -p_2, -q_2) \Longleftrightarrow (-q_1, p_1, -p_2, -q_2)$$
$$\Longleftrightarrow (-p_1, -q_1, p_2, q_2)$$

A solution of (3.4.4)–(3.4.7) that is invariant under the above four transformations is called a **symmetric solution**; otherwise, the solution is called an **asymmetric solution**. Consequently, for a specified set of control parameters, if an asymmetric solution is found, three other solutions can be obtained by applying the above transformations.

In Figure 3.4.4, we show the projection of a limit–cycle solution onto the three–dimensional $p_1 - p_2 - q_2$ space and the two–dimensional $p_2 - q_2$ space. This periodic solution is a symmetric solution. In fact, the two–dimensional projection does reveal the symmetry $(p_2, q_2) \Longleftrightarrow (-p_2, -q_2)$. Further, this solution is an attractor because one of the associated Floquet multipliers is at $+1$ and the other three multipliers are inside the unit circle. As σ_2 is gradually increased while holding the other control parameters fixed, this limit cycle deforms smoothly and remains symmetric and stable until we reach the critical value $\sigma_2 = -1.6897$. Here, the symmetric periodic solution is nonhyperbolic with two of the associated Floquet multipliers at $+1$. This symmetric solution is unstable for $\sigma > \sigma_2$. Furthermore, Nayfeh and Nayfeh (1990) numerically found that there are four stable asymmetric solutions of (3.4.4)–(3.4.7). Hence, a supercritical symmetry–breaking bifurcation occurs at $\sigma = \sigma_2$. In Figure 3.4.5, three–dimensional projections of

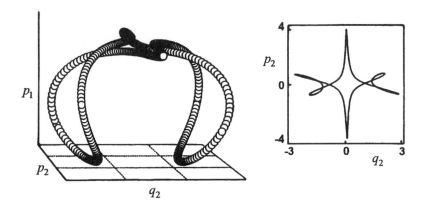

Figure 3.4.4: Symmetric periodic attractor of (3.4.4)–(3.4.7) realized for $f = 1.0$, $\mu_1 = \mu_2 = 0.02$, $\sigma_1 = 0.266$, and $\sigma_2 = -1.80$. Reprinted with permission from Nayfeh and Nayfeh (1990).

the quadruple limit cycles onto the $p_1 - q_1 - q_2$ space are shown when $\sigma_2 = -1.68$ and the other parameters are the same as before. For these parameter values, there also exist unstable fixed points of (3.4.4)–(3.4.7). They are indicated by the plus signs in Figure 3.4.5. The apparent intersections seen in Figures 3.4.4 and 3.4.5 are a consequence of the projection of the four–dimensional trajectories onto a three–dimensional space. All of the periodic solutions shown in Figures 3.4.3–3.4.5 were numerically determined by using the **shooting method** of Section 6.5.

In certain continuous–time systems with odd nonlinearities, an example of which is (3.1.28) and (3.1.29), a solution possesses the symmetry

$$\mathbf{x}(t) = -\mathbf{x}\left(t + \frac{1}{2}T\right)$$

where T is the period of the solution. This symmetry property is called **inversion symmetry**. The spectra of the scalar components of $\mathbf{x}$ contain only odd harmonics of the frequency $2\pi/T$. However, this symmetry is broken when a bifurcation introduces a zero frequency component and/or an even harmonic of the frequency $2\pi/T$. The occurrence of the associated symmetry–breaking bifurcation can be determined by examining the eigenvalues of the matrix $\hat{\Phi}$ in (3.2.53).

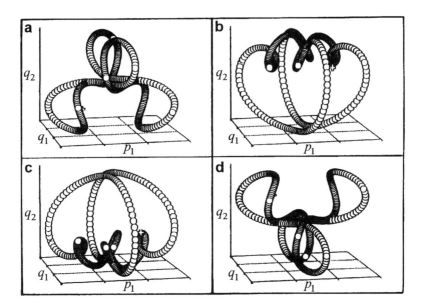

Figure 3.4.5: Four asymmetric periodic attractors of (3.4.4)–(3.4.7) realized for $f = 1.0$, $\mu_1 = \mu_2 = 0.02$, $\sigma_1 = 0.266$, and $\sigma_2 = -1.68$. Reprinted with permission from Nayfeh and Nayfeh (1990).

We note that one of the eigenvalues of $\hat{\Phi}$ associated with a symmetric periodic solution of an autonomous system is always -1. When an eigenvalue of $\hat{\Phi}$ leaves the unit circle through $+1$, a symmetry–breaking bifurcation occurs, and usually this bifurcation precedes a period–doubling bifurcation (Swift and Wiesenfeld, 1984). In the next example (Nayfeh and Sanchez, 1989), we illustrate a supercritical symmetry–breaking bifurcation in a two–dimensional nonautonomous system with an odd nonlinearity.

Example 3.20. The system of interest is

$$\ddot{x} + 0.4\dot{x} + x - x^3 = F\cos(\Omega t) \tag{3.4.8}$$

In Figure 3.4.6a, the periodic attractor realized at $F = 0.350$ and $\Omega = 0.8$ is shown along with the associated power spectrum of the state x. (A power spectrum provides a measure of the energy of a

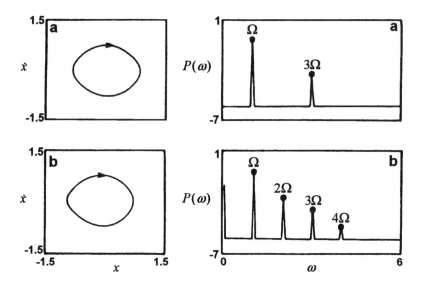

Figure 3.4.6: Phase portraits and power spectra for periodic attractors of (3.4.8): (a) before symmetry–breaking bifurcation and (b) after symmetry–breaking bifurcation.

system at different frequencies.) Both the phase portrait and power spectrum of Figure 3.4.6a are indicative of the inversion symmetry of the periodic solution. As Ω is held constant and F is gradually increased, the symmetric periodic solution experiences a supercritical symmetry–breaking bifurcation, resulting in a pair of stable asymmetric solutions. In Figure 3.4.6b, the phase portrait and power spectrum associated with one of these asymmetric attractors at $F = 0.380$ and $\Omega = 0.8$ are shown. We note the presence of even harmonics of Ω and a zero–frequency component.

3.4.2 Cyclic–Fold Bifurcation

In Figure 3.4.7, we illustrate the scenario near a **cyclic–fold bifurcation point**. Again, r is the amplitude of the periodic solution, and α is the scalar control parameter. A branch of stable periodic solutions and a branch of unstable periodic solutions, which exist for $\alpha < \alpha_c$, coalesce and obliterate each other at the bifurcation point α_c. Typically, the unstable periodic solutions are of the saddle type. In Figure 3.4.7, we note that locally there are no other solutions in the vicinity of the bifurcation point on one side (i.e., $\alpha > \alpha_c$). Therefore, cyclic–fold bifurcations are discontinuous or catastrophic bifurcations. The scenario of Figure 3.4.7 is similar to the scenario of Figure 2.3.1, where a saddle–node bifurcation of a fixed point of a continuous–time system is illustrated. (In an associated one–dimensional map, this bifurcation is called a **tangent bifurcation** because the map is tangent to the identity map at the bifurcation value.) In the continuation literature (e.g., Holodniok and Kubicek, 1984a), cyclic–fold bifurcation points are also called **turning points**.

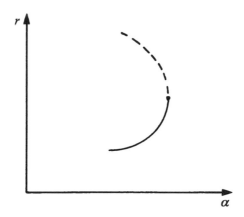

Figure 3.4.7: Scenario in the vicinity of a cyclic–fold bifurcation point.

When a dynamical system undergoes a cyclic–fold bifurcation, the system will be in a state corresponding to an attracting limit cycle for $\alpha < \alpha_c$. For $\alpha > \alpha_c$, the system behavior cannot be determined from local considerations alone; global considerations are necessary. The postbifurcation state is usually determined through numerical simulations. There are two possibilities. First, the system evolution may be attracted to a distant solution, which is either bounded or unbounded. The bounded solution may be a point attractor, or a periodic attractor, or an aperiodic attractor. The bifurcation is dangerous and its outcome may be determinant or indeterminant.

Second, the old attractor may explode into a new larger attractor with the old attractor being a proper subset of the new attractor. Such a bifurcation is an example of an explosive bifurcation. An orbit on this attractor spends long stretches of time near the destroyed limit cycle (**ghost** or **phantom limit cycle**), with interruptions in the form of excursions or outbreaks away from the ghost limit cycle. When the attractor is chaotic or strange, the stretches of time spent near the ghost limit cycle are called **laminar phases**, and the excursions away from the ghost limit cycle are called **turbulent** or **chaotic bursts**. As α is varied to approach α_c from above, the mean time between the irregular bursts approaches infinity and the chaotic attractor implodes to the old limit cycle at α_c. The above described transition from periodicity to chaotic behavior following a cyclic–fold bifurcation has been termed **intermittent transition of type I to chaos** by Manneville and Pomeau (1979, 1980) and Pomeau and Manneville (1980). This transition, which is described in more detail in Section 5.4, is one of the experimentally and analytically established transitions to chaos (Swinney, 1983; Bérge, Pomeau, and Vidal, 1984).

Example 3.21. For the system given by (3.4.4)–(3.4.7), the stable periodic solutions I and II coexist with the unstable limit cycle III when $f = 1.0$, $\mu_1 = 0.02$, $\mu_2 = 0.02$, $\sigma_1 = 0.266$, and $\sigma_2 = -1.9980$. Projections of these three solutions onto the $p_2 - q_2$ plane are shown in Figure 3.4.8. The orbit shown in the inset corresponds to solution I, the orbit depicted by broken lines corresponds to solution III, and the remaining orbit corresponds to solution II. All three solutions continue to exist as σ_2 is gradually increased to -1.9978. At this critical value, periodic solution I becomes nonhyperbolic with two of the associated Floquet multipliers at $+1$. At this critical point, a cyclic–fold bifurcation occurs. As a consequence, for $\sigma > -1.9978$, the periodic solution I does not exist. However, the periodic solution II continues to exist. In the state–control space, if one is on the branch corresponding to periodic solution I, there is a jump from this branch to the branch corresponding to periodic solution II after the bifurcation takes place.

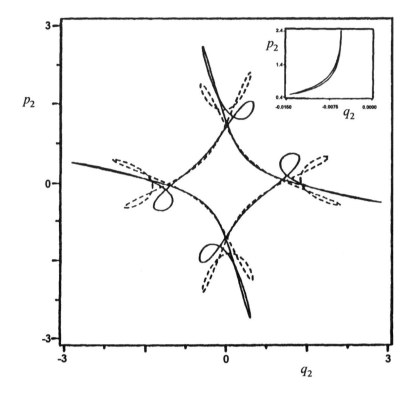

Figure 3.4.8: Three periodic solutions of (3.4.4)–(3.4.7) realized at $\sigma_2 = -1.998$. The two stable solutions are depicted by solid lines, while the unstable solution is depicted by broken lines. Reprinted with permission from Nayfeh and Nayfeh (1990).

As noted earlier in this section, a cyclic–fold bifurcation of a periodic solution can lead to a chaotic solution. Returning to the quadruple solutions of Figure 3.4.5, we find that as σ_2 is gradually increased beyond -1.68, the asymmetric quadruple solutions deform smoothly until the critical value $\sigma_2 = -1.6543$ is reached. At this value, each of the quadruple solutions loses stability through a cyclic–fold bifurcation with a Floquet multiplier leaving the unit circle through $+1$, resulting in a jump to another solution of this system. Nayfeh and Nayfeh (1990) numerically ascertained this solution to be chaotic by using the tools discussed in Chapter 7.

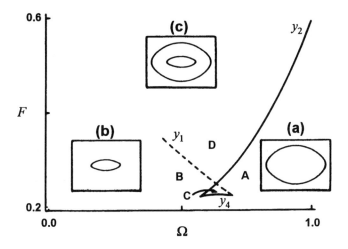

Figure 3.4.9: Bifurcation set for (3.4.8). Each of the curves y_i is the loci of either cyclic–fold or pitchfork bifurcation points of periodic solutions with period $2\pi/\Omega$. The periodic attractors in regions A, B, and C are shown in the insets.

Example 3.22. The bifurcation set numerically generated by Nayfeh and Sanchez (1989) for (3.4.8) in the $F-\Omega$ control space is displayed in Figure 3.4.9. The region below the curve y_2 is marked as A; the region below the curve y_1 is marked as B; the region enclosed by the curves y_1, y_2, and y_4 is marked as C; and the region enclosed by the curves y_1 and y_2 above their intersection point is marked as D. In region D, apart from a narrow strip along y_2 where we have bounded solutions, we do not have any bounded solutions of (3.4.8). We display the different periodic attractors of (3.4.8) found in regions A, B, and C in insets (a), (b), and (c), respectively. The attractor found in region A is larger in size than that found in region B. Both the small and large attractors coexist in region C, indicating bistability.

The curve y_1 is the locus of the cyclic–fold bifurcation points corresponding to the small attractor of region B; the curve y_2 is the locus of the pitchfork bifurcation points corresponding to the large attractor of region A; and the curve y_4 is the locus of the cyclic–fold bifurcation points corresponding to the large attractor of region A.

If one were to start from the small attractor in region C and gradually vary one of the control parameters while holding the other parameter constant so as to cross y_1 into region A, a cyclic–fold bifurcation occurs, leading to a jump to the large attractor of region A. On the other hand, if one were to start from the large attractor in region C and gradually vary one of the control parameters so as to cross y_4, a cyclic–fold bifurcation occurs, leading to a jump to the small attractor of region B. If one were to start from the small attractor in region B and gradually vary one of the control parameters so as to cross y_1, a cyclic–fold bifurcation occurs, leading to a jump to the large attractor of region A if the crossing is into region A and to an unbounded solution if the crossing is into region D. If one were to start from the large attractor of region A and gradually vary one of the control parameters so as to cross the curve y_2 at a location above y_1, a pitchfork bifurcation occurs, leading to a bounded asymmetric solution that exists all along a small strip adjacent to y_2.

The **Y–shaped structure** seen in the bifurcation set of Figure 3.4.9 in the context of the continuous–time system (3.4.8) has also been observed in the bifurcation sets of other dynamical systems (e.g., Abraham, Gollub, and Swinney, 1984). We also note from this section that a cyclic–fold bifurcation of a periodic solution of a dynamical system may lead to a chaotic solution of the considered dynamical system.

3.4.3 Period–Doubling or Flip Bifurcation

When a Floquet multiplier leaves the unit circle through -1 (Fig. 3.4.1b), a period–doubling bifurcation takes place and the branch of stable periodic solutions that exists before the bifurcation (say $\alpha < \alpha_c$) continues as an unstable branch of periodic solutions after the bifurcation (at $\alpha = \alpha_c$). A branch of stable period–doubled solutions is created if the bifurcation is **supercritical,** while a branch of unstable period–doubled solutions is destroyed if the bifurcation is **subcritical.** These scenarios are similar to those discussed in Section 2.4 in the context of period–doubling bifurcations of fixed points of maps.

Local considerations are sufficient to understand the consequences of a supercritical bifurcation, which is a continuous bifurcation, while global considerations are necessary to understand the consequences of a subcritical bifurcation, which is a catastrophic bifurcation. In the subcritical case, an unstable limit cycle (period–two cycle) collides with a stable limit cycle of one–half its period, and the two are replaced with an unstable limit cycle of the lower period. When a system undergoes a subcritical period–doubling bifurcation, the local state of the dynamical system will be an attracting limit cycle for $\alpha < \alpha_c$. However, for $\alpha > \alpha_c$, the postbifurcation state of the system cannot be determined by local considerations alone; global considerations are necessary. There are two possibilities. First, the system evolution may be attracted to a remote solution, which is either bounded (fixed point, periodic solution, quasiperiodic solution, chaos) or unbounded. Such a bifurcation is dangerous and is typically accompanied by a hysteresis. The outcome of the bifurcation may be determinant or indeterminant. In the latter case, the postbifurcation response depends on the rate of control sweep and is sensitive to the presence of noise. Second, the system response may explode into a new attractor when α is slowly varied past α_c, with the old attractor being a proper subset of the new attractor. The outcome of the bifurcation is determinant, independent of the rate of control sweep, and insensitive to the presence of noise. An orbit on this attractor may spend long stretches of time near the destroyed period–two limit cycle (ghost or phantom limit cycle), with interruptions in the form of outbreaks or excursions away from the ghost limit cycle. When the new attractor is chaotic or strange, the stretches of time near the ghost limit cycle are called **laminar phases** and the excursions away from the ghost attractor are called **turbulent** or **chaotic bursts**. As α is varied to approach α_c from above, the bursts become more and more infrequent, with the mean time between bursts approaching infinity, and the large attractor implodes to the small attractor at α_c. The above–described transition from a periodic state to a chaotic state following a subcritical period–doubling bifurcation has been termed **intermittent transition of type III to chaos** by Pomeau and Manneville (1980) and Manneville and Pomeau (1980). This scenario is addressed in more detail in Section 5.4.

In Figure 3.4.10a, we show a periodic orbit of a continuous–time system and its intersection with a one–sided Poincaré section Σ. The scenario after this periodic orbit undergoes a period–doubling bifurcation is depicted in Figure 3.4.10b. The period–doubled orbit

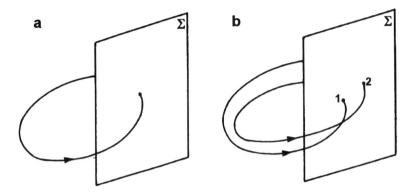

Figure 3.4.10: Periodic orbit and associated Poincaré section: (a) before period–doubling bifurcation and (b) after period–doubling bifurcation.

intersects the Poincaré section two times, once at the point labeled 1 and another time at the point labeled 2. During the course of evolution on the period–doubled orbit, we successively flip between points 1 and 2 on Σ. Therefore, a period–doubling bifurcation is also called a **flip bifurcation**. A second period–doubling bifurcation would mean four points on the corresponding Poincaré section and so on. After k successive period–doubling bifurcations, we would have $2k$ points on the corresponding Poincaré section.

Example 3.23. In the autonomous system (3.4.1)–(3.4.3), as μ is gradually increased beyond $\mu = 0.35$, the two asymmetric limit cycles, shown in Figure 3.4.3, deform smoothly and remain stable until $\mu = 0.4405$. At this critical value, the limit cycles lose stability with one of the Floquet multipliers leaving the unit circle through -1, resulting in a doubling of the periods of these limit cycles. The period–doubling bifurcation was numerically ascertained to be a supercritical bifurcation. One of these period–doubled limit cycles is shown in Figure 3.4.11 when $\mu = 0.445$. The apparent intersection seen in this plot is a consequence of the projection of the three–dimensional orbit onto a two–dimensional plane.

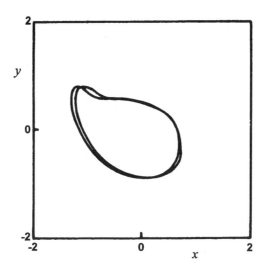

Figure 3.4.11: Stable period–doubled orbit of (3.4.1)–(3.4.3) realized at $\mu = 0.445$. Reprinted with permission from Nayfeh and Balachandran (1990a).

Example 3.24. In the nonautonomous system (3.4.8), a period–doubling bifurcation occurs as F is gradually increased beyond $F = 0.380$, keeping Ω fixed at 0.8. The asymmetric periodic solution at $F = 0.380$ is shown in Figure 3.4.12a. This periodic solution deforms smoothly under the variation of F until a period–doubling bifurcation occurs, leading to the periodic solution shown in Figure 3.4.12b at $F = 0.386$. In the corresponding power spectrum, we note the presence of a peak at $\frac{1}{2}\Omega$, indicating that the basic frequency of the state x is one–half of the excitation frequency. The broken lines in Figures 3.4.12a and 3.4.12b represent Poincaré sections. There is one intersection with the Poincaré section in Figure 3.4.12a, while there are two intersections with the Poincaré section in Figure 3.4.12b.

In many studies (e.g., Feigenbaum, 1978; Novak and Frelich, 1982; Swinney, 1983; Bérge, Pomeau, and Vidal, 1984; Räty, Isomäki, and von Boehm, 1984; Nayfeh and Sanchez, 1989; Nayfeh, Hamdan, and Nayfeh, 1990), **a periodic solution of a dynamical system experiences an infinite sequence of period–doubling bifurcations under the variation of a single control parameter, culminating in**

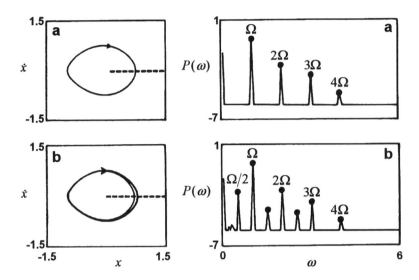

Figure 3.4.12: Phase portraits and power spectra for periodic attractors of (3.4.8): (a) before period–doubling bifurcation and (b) after period–doubling bifurcation. We note the appearance of subharmonics after the period–doubling bifurcation.

a chaotic solution. This route, which is another experimentally and analytically well–established mechanism of transition to chaos, is discussed in Section 5.3.

3.4.4   Transcritical Bifurcation

A transcritical bifurcation of a periodic solution may occur when a Floquet multiplier leaves the unit circle through $+1$, as shown in Figure 3.4.1a. In each of Figures 3.4.13a and 3.4.13b, we have plotted the amplitude r of the periodic solution of a dynamical system versus the scalar control parameter α in the vicinity of a **transcritical bifurcation point**. Again, the solid and broken lines correspond to stable and unstable periodic solutions, respectively. The branches of stable and unstable periodic solutions that exist before the transcritical bifurcation continue as branches of unstable and stable periodic solutions, respectively, after the bifurcation. Hence, a transcritical bi-

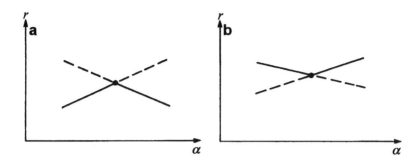

Figure 3.4.13: Two possible local scenarios for a transcritical bifurcation.

furcation leads to an **exchange of stability**. Figure 3.4.13b is similar to Figure 2.3.3 in which a transcritical bifurcation of a fixed point of a continuous–time system is depicted.

3.4.5   Secondary Hopf or Neimark Bifurcation

In Section 2.3.1, we noted that a Hopf bifurcation of a fixed point of a continuous–time system leads to a periodic solution of this system. So, essentially, a Hopf bifurcation introduces a new frequency (possibly incommensurate) with the first one in the bifurcating solution. A Hopf bifurcation of a periodic solution is called a **secondary Hopf** or **Neimark bifurcation**, and it occurs when two complex conjugate eigenvalues exit the unit circle away from the real axis, as shown in Figure 3.4.1c. The bifurcating solution may be periodic or two–period quasiperiodic, depending on the relationship between the newly introduced frequency and the frequency of the periodic solution that exists prior to the bifurcation. Similar to subcritical and supercritical Hopf bifurcations of fixed points, there are subcritical and supercritical Neimark bifurcations of periodic solutions. In both bifurcations, the branch of stable periodic solutions that exists prior to the Neimark bifurcation continues as a branch of unstable periodic solutions after the bifurcation. A branch of stable quasiperiodic solutions is created if the bifurcation is supercritical. This bifurcation is an example of a continuous bifurcation. On the other hand, a branch of unstable quasiperiodic solutions is destroyed if the bifurcation is subcritical; this bifurcation is another example of a catastrophic bifurcation. When the system undergoes a subcritical Hopf bifurcation as a control parameter α exceeds the threshold value α_c, the state of the system is periodic for $\alpha < \alpha_c$. For $\alpha > \alpha_c$, the postbifurcation state of the system cannot be determined by local considerations alone; global considerations are necessary. There are two possibilities. First, the system evolution may be attracted to a distant solution, which is either bounded (point, periodic, quasiperiodic, or chaotic attractor) or unbounded. Again, such a bifurcation is dangerous and is typically accompanied by hysteresis. Second, the state of the system may explode into a larger attractor, with the old attractor being a proper subset of the new attractor. An orbit on this attractor may spend a long time on the destroyed quasiperiodic (ghost or phantom) solution before it is interrupted by outbreaks or excursions on the attractor far away from the ghost quasiperiodic solution. Again, if the new attractor is strange or chaotic, the time stretches that the attractor spends near the ghost quasiperiodic attractor are called **laminar phases**, and the outbreaks away from the ghost attractor are called **turbulent** or **chaotic bursts**. The transition from periodicity to chaos following a subcritical Hopf bifurcation has been termed **intermittent transition of type II to chaos** by Pomeau and Manneville (1980) and Manneville and Pomeau (1980). We discuss this scenario at length in Section 5.4.

Example 3.25. We consider the following sixth–order system (Nayfeh, Asrar, and Nayfeh, 1992):

$$p_1' = -\mu_1 p_1 - \nu_1 q_1 - p_1 q_2 + p_2 q_1 \tag{3.4.9}$$

$$q_1' = -\mu_1 q_1 + \nu_1 p_1 + p_1 p_2 + q_1 q_2 \tag{3.4.10}$$

$$p_2' = -\mu_2 p_2 - \nu_2 q_2 - 2p_1 q_1 - p_2 q_3 + p_3 q_2 \tag{3.4.11}$$

$$q_2' = -\mu_2 q_2 + \nu_2 p_2 + p_1^2 - q_1^2 + p_2 p_3 + q_2 q_3 \tag{3.4.12}$$

$$p_3' = -\mu_3 p_3 - \nu_3 q_3 - 2\Gamma p_2 q_2 \tag{3.4.13}$$

$$q_3' = -\mu_3 q_3 + \nu_3 p_3 + \Gamma(p_2^2 + q_2^2) + F \tag{3.4.14}$$

Here, the states are p_i and q_i, and the parameters are Γ, F, μ_i, and ν_i. For $\Gamma = 0.625$, $\mu_1 = 0.5$, $\mu_2 = 0.5$, $\mu_3 = 0.1$, $\nu_1 = 0.375$, $\nu_2 = 0.25$, $\nu_3 = 0.5$, and $F = 0.533$, the system (3.4.9)–(3.4.14)

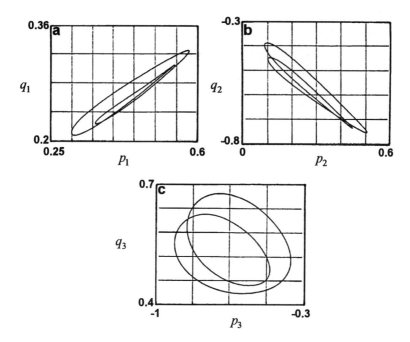

Figure 3.4.14: Two–dimensional projections of the periodic attractor of (3.4.9)–(3.4.14) at $F = 0.533$. Reprinted with permission from Nayfeh, Asrar, and Nayfeh (1992).

has a limit–cycle solution. This solution is depicted in Figure 3.4.14. When F is used as a control parameter and gradually increased beyond 0.5330, this limit cycle deforms smoothly and remains stable until the critical value $F = 0.5366$ is reached. At this value, a complex–conjugate pair of Floquet multipliers crosses the unit circle away from the real axis, resulting in a **Neimark bifurcation**. Because the bifurcating solution is a stable two–period quasiperiodic solution, this Neimark bifurcation is supercritical. Here, the bifurcating solution is a quasiperiodic attractor.

In Figure 3.4.15a, we show a two–dimensional projection of the quasiperiodic solution of (3.4.9)–(3.4.14) realized at $F = 0.537$. (We note that, unlike a periodic orbit, a quasiperiodic orbit of an autonomous system does not close on itself because of the aperiodicity.) Nayfeh, As-

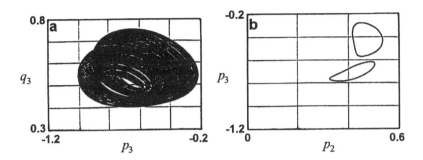

Figure 3.4.15: Quasiperiodic attractor of (3.4.9)–(3.4.14) realized at $F = 0.537$: (a) two–dimensional projection of orbit and (b) two–sided Poincaré section. Reprinted with permission from Nayfeh, Asrar, and Nayfeh (1992).

rar, and Nayfeh (1992) chose the hyperplane

$$q_1 = -0.667p_1 + 0.560$$

to define a two–sided Poincaré section after ascertaining through numerical simulations that this section is locally transverse to the flow; that is, after ascertaining that the quasiperiodic orbit passes through the chosen plane. A two–dimensional projection of the collection of points on the five–dimensional Poincaré section is shown in Figure 3.4.15b. We note the presence of two dense collections of points on two closed loops in Figure 3.4.15b. Such dense collections of points on closed loops of a Poincaré section are characteristic of two–period quasiperiodic solutions.

3.5 ANALYTICAL CONSTRUCTIONS

In this section, we describe how the methods of multiple scales and center manifold reduction can be used to obtain analytical approximations for a periodic solution arising through a Hopf bifurcation of a fixed point of an autonomous system of differential equations. For algebraic ease, we use the system (3.4.1)–(3.4.3) as an example in Sections 3.5.1 and 3.5.2 and state the results for the general case in Section 3.5.3.

3.5.1 Method of Multiple Scales

Here, we follow Nayfeh and Balachandran (1990a) in the treatment of the system (3.4.1)–(3.4.3). The procedure is identical to that carried out in Section 2.3.6.

Example 3.26. Equations (3.4.1)–(3.4.3) possess the fixed–point solution $(x_0, y_0, z_0) = (0, 0, 0)$. The associated Jacobian matrix is

$$\begin{bmatrix} \mu & -1 & 0 \\ 1 & \mu & 0 \\ 0 & 0 & -1 \end{bmatrix}$$

whose eigenvalues are

$$\lambda = \mu \pm i \text{ and } -1$$

In the $x - y - z - \mu$ space, a Hopf bifurcation occurs at $(x, y, z, \mu) = (0, 0, 0, 0)$. Consequently, the system (3.4.1)–(3.4.3) is expected to possess limit–cycle solutions close to the bifurcation point. At the bifurcation point, the complex–conjugate eigenvalues are $\lambda = \pm i$ and hence the period of the limit cycle is 2π. As we move away from the bifurcation point, the period of the limit cycle is given by $2\pi/\omega_m$, where ω_m is one plus a frequency correction. We determine the radius and period of the bifurcating limit cycle through the following analysis.

To determine an approximation to this limit cycle, we seek an expansion of the form

$$x = x_0 + \sum_{n=1}^{3} \epsilon^n x_n(T_0, T_1, T_2) + \cdots \tag{3.5.1}$$

$$y = y_0 + \sum_{n=1}^{3} \epsilon^n y_n(T_0, T_1, T_2) + \cdots \tag{3.5.2}$$

$$z = z_0 + \sum_{n=1}^{3} \epsilon^n z_n(T_0, T_1, T_2) + \cdots \tag{3.5.3}$$

where $T_n = \epsilon^n t$ and ϵ is a small positive nondimensional parameter that is artificially introduced to serve as a bookkeeping device and will be set equal to unity in the final analysis. In terms of the T_n, the time derivative becomes

$$\frac{d}{dt} = D_0 + \epsilon D_1 + \epsilon^2 D_2 + \cdots \qquad (3.5.4)$$

where $D_n = \partial/\partial T_n$. Also, the control parameter is ordered as

$$\mu = \epsilon^2 \mu_2 \qquad (3.5.5)$$

so that the influence of the nonlinear terms and the control parameter μ are realized at the same order.

Substituting (3.5.1)–(3.5.5) into (3.4.1)–(3.4.3), recalling that $x_0 = y_0 = z_0 = 0$, and equating coefficients of like powers of ϵ, we obtain the following hierarchy of equations:

$O(\epsilon)$:

$$D_0 x_1 + y_1 = 0 \qquad (3.5.6)$$
$$D_0 y_1 - x_1 = 0 \qquad (3.5.7)$$
$$D_0 z_1 + z_1 = 0 \qquad (3.5.8)$$

$O(\epsilon^2)$:

$$D_0 x_2 + y_2 = -D_1 x_1 - x_1 z_1 \qquad (3.5.9)$$
$$D_0 y_2 - x_2 = -D_1 y_1 \qquad (3.5.10)$$
$$D_0 z_2 + z_2 = -D_1 z_1 + y_1^2 \qquad (3.5.11)$$

$O(\epsilon^3)$:

$$D_0 x_3 + y_3 = \mu_2 x_1 - x_1 z_2 - x_2 z_1 - D_2 x_1 - D_1 x_2 \qquad (3.5.12)$$
$$D_0 y_3 - x_3 = \mu_2 y_1 - D_2 y_1 - D_1 y_2 \qquad (3.5.13)$$
$$D_0 z_3 + z_3 = x_1^2 z_1 + 2 y_1 y_2 - D_2 z_1 - D_1 z_2 \qquad (3.5.14)$$

The nondecaying solution of (3.5.6)–(3.5.8) is

$$x_1 = i A(T_1, T_2) e^{iT_0} + cc \qquad (3.5.15)$$
$$y_1 = A(T_1, T_2) e^{iT_0} + cc \qquad (3.5.16)$$
$$z_1 = 0 \qquad (3.5.17)$$

where cc is the complex conjugate of the preceding terms and A is determined by imposing the solvability conditions at the next levels of approximation.

Substituting (3.5.15)–(3.5.17) into (3.5.9)–(3.5.11) and eliminating the source of secular terms, we have

$$D_1 A = 0 \quad \text{or} \quad A = A(T_2)$$

Then, the solution of the system (3.5.9)–(3.5.11) is

$$x_2 = y_2 = 0 \tag{3.5.18}$$

$$z_2 = A\bar{A} + \frac{A^2 e^{2iT_0}}{1 + 2i} + cc \tag{3.5.19}$$

where $\bar{A}$ is the complex conjugate of A.

Substituting (3.5.15)–(3.5.19) into (3.5.12) and (3.5.13) and eliminating the terms that produce secular terms, we obtain

$$2A' - 2\mu_2 A + \left(\frac{9}{5} + \frac{2}{5}i\right) A^2 \bar{A} = 0 \tag{3.5.20}$$

where the prime denotes the derivative with respect to T_2. Substituting $A = \frac{1}{2}a \exp(i\beta)$, where a and β are real quantities, into (3.5.20), separating real and imaginary parts, and setting $\epsilon = 1$, we obtain

$$\dot{a} = \mu_2 a - \frac{9}{40}a^3 \tag{3.5.21}$$

$$\dot{\beta} = -\frac{1}{20}a^2 \tag{3.5.22}$$

Referring to Section 2.3.2, we note that the system (3.5.21) and (3.5.22) is in the normal form for a Hopf bifurcation of a fixed point. Further, this system is equivalent to that obtained by Rand (1989) using center manifold reduction and the method of normal forms. In the next section, we illustrate the procedure based on center manifold reduction.

When $\mu_2 < 0$, (3.5.21) has the stable fixed point $a = 0$. On the other hand, when $\mu_2 > 0$, (3.5.21) has the fixed points

$$a = 0, \quad a = \frac{2}{3}\sqrt{10\mu_2}, \quad \text{and} \quad a = -\frac{2}{3}\sqrt{10\mu_2}$$

The zero fixed point is unstable, while the nonzero fixed points are stable. Therefore, the Hopf bifurcation is supercritical because the bifurcating periodic solutions are stable. The approximation for the limit–cycle solution of (3.4.1)–(3.4.3) is

$$x = -a\sin(\omega_m t + \beta_0) + O(a^3) \tag{3.5.23}$$
$$y = a\cos(\omega_m t + \beta_0) + O(a^3) \tag{3.5.24}$$
$$z = \frac{1}{2}a^2\left[1 + \frac{1}{5}\cos(2\omega_m t + 2\beta_0)\right.$$

$$\left. + \frac{2}{5}\sin(2\omega_m t + 2\beta_0)\right] + O(a^3) \tag{3.5.25}$$

where β_0 is a constant,

$$a = \frac{2}{3}\sqrt{10\mu}$$

and

$$\omega_m = (1 + \dot{\beta}) + O(a^3) = \left(1 - \frac{1}{20}a^2\right) + O(a^3) \tag{3.5.26}$$

It should be borne in mind that expansions (3.5.23)–(3.5.26) are determined for small values of μ and, hence, their accuracy is expected to deteriorate as μ becomes large. In Figure 3.5.1, we compare the analytical approximation (broken lines) with the numerical solutions (solid lines) of equations (3.4.1)–(3.4.3) for different values of μ. The numerical solutions are obtained by using the shooting scheme of Section 6.5.2. It is evident from the figure that the approximation deteriorates as μ increases to 0.2 and beyond. In fact, at $\mu = 0.3$, the analytical approximation is not even close to the numerical solution in either form or magnitude. For $\mu > 0.3$, the analytical approximation is not shown.

3.5.2   Center Manifold Reduction

At a Hopf bifurcation point, the fixed point typically has nonempty stable and center manifolds and an empty unstable manifold, as illustrated in Figure 3.5.2. The basic idea underlying this approach is to restrict attention to the dynamics on the center manifold W^c,

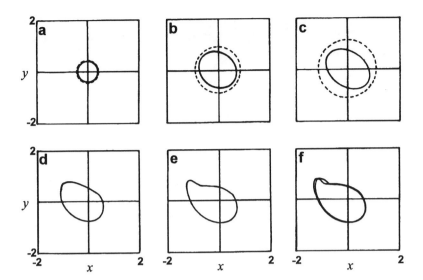

Figure 3.5.1: Periodic solutions of (3.4.1)–(3.4.3): (a) $\mu = 0.05$, (b) $\mu = 0.2$, (c) $\mu = 0.3$, (d) $\mu = 0.35$, (e) $\mu = 0.44$, and (d) $\mu = 0.445$. Reprinted with permission from Nayfeh and Balachandran (1990a).

thereby reducing the order of the dynamical system while retaining the essential features of the dynamic behavior near the bifurcation point. The procedure used in this section is identical to that described in Section 2.3.4.

Example 3.27. We reduce the system (3.4.1)–(3.4.3) to its normal form in the vicinity of the Hopf bifurcation that takes place at $(0, 0, 0, 0)$. To capture the dependence of the center manifold on the parameter μ, we use the suspension trick; that is, we augment (3.4.1)–(3.4.3) with the equation

$$\dot{\mu} = 0 \qquad (3.5.27)$$

In the augmented system, the terms μx and μy are treated as nonlinear terms.

At the bifurcation point, the linear part of (3.4.1)–(3.4.3) and (3.5.27) is in normal form. The $x - y - \mu$ space is the center eigenspace, while the z axis is the stable eigenspace. Noting that the local

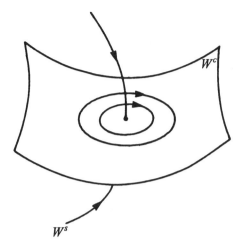

Figure 3.5.2: Illustration of stable and center manifolds of a fixed point at Hopf bifurcation.

center manifold is tangent to the center eigenspace at the origin (i.e., $\partial z/\partial x = \partial z/\partial y = \partial z/\partial \mu = 0$ at $(x, y, \mu) = (0, 0, 0)$), we approximate the manifold by the following polynomial:

$$z(x, y, \mu) = b_1 x^2 + b_2 xy + b_3 y^2 + \cdots \tag{3.5.28}$$

We have assumed that x and y are of the same order and μ is $\mathcal{O}(x^2)$. Equation (3.5.28) constrains the center manifold to be three–dimensional in the four–dimensional $x - y - z - \mu$ space.

We first choose the constants b_i in (3.5.28) such that (3.4.3) is satisfied and then use (3.5.28) to obtain the projection of the system (3.4.1)–(3.4.3) onto the center manifold. To this end, we substitute (3.5.28) into (3.4.1) and obtain

$$\dot{x} = \mu x - y - b_1 x^3 - b_2 x^2 y - b_3 xy^2 + \cdots \tag{3.5.29}$$

Then, substituting (3.5.28) into (3.4.3) yields

$$\begin{aligned} 2b_1 x\dot{x} + b_2 \dot{x}y + b_2 x\dot{y} + 2b_3 y\dot{y} \\ = (-b_1 x^2 - b_2 xy - b_3 y^2) + y^2 + \cdots \end{aligned} \tag{3.5.30}$$

Substituting for $\dot{x}$ and $\dot{y}$ from (3.5.29) and (3.4.2) into (3.5.30) and keeping up to second-order terms only, we obtain

$$
\begin{aligned}
-2b_1 xy - b_2 y^2 + b_2 x^2 + 2b_3 xy \\
= -b_1 x^2 - b_2 xy - b_3 y^2 + y^2 + \cdots
\end{aligned}
\tag{3.5.31}
$$

Equating the coefficients of x^2, xy, and y^2 on both sides of (3.5.31), we arrive at

$$
b_2 + b_1 = 0, \quad 2b_1 - b_2 - 2b_3 = 0, \quad b_2 - b_3 = -1
$$

Hence,

$$
b_1 = \frac{2}{5}, \quad b_2 = -\frac{2}{5}, \quad \text{and } b_3 = \frac{3}{5}
$$

Substituting for the b_i in (3.5.29) results in

$$
\dot{x} = \mu x - y - \frac{2}{5}x^3 + \frac{2}{5}x^2 y - \frac{3}{5}xy^2 + \cdots
\tag{3.5.32}
$$

Thus, the four-dimensional system (3.4.1)–(3.4.3) and (3.5.27) is reduced to the three-dimensional system (3.5.32), (3.4.2), and (3.5.27) in the vicinity of the bifurcation point.

Periodic solutions of (3.5.32) and (3.4.2) near the origin can be constructed using either the method of multiple scales or the method of normal forms (Nayfeh, 1973, 1981, 1993). Since we used the method of multiple scales to construct periodic solutions of the three-dimensional system in the preceding section, we use the method of normal forms in this section. To apply the method of normal forms, we find it convenient to combine the two first-order real-valued equations (3.4.2) and (3.5.32) into a single complex-valued equation using the transformation

$$
\zeta = x + iy \quad \text{and} \quad \bar{\zeta} = x - iy
\tag{3.5.33}
$$

Consequently,

$$
x = \frac{1}{2}(\zeta + \bar{\zeta}) \quad \text{and} \quad y = -\frac{1}{2}i(\zeta - \bar{\zeta})
\tag{3.5.34}
$$

Using this transformation, we combine (3.4.2) and (3.5.32) into

$$
\begin{aligned}
\dot{\zeta} = i\zeta + \mu\zeta - \frac{1}{40}\Big[(-1 + 2i)\zeta^3 \\
+ (9 + 2i)\zeta^2\bar{\zeta} + (9 - 2i)\zeta\bar{\zeta}^2 - (1 + 2i)\bar{\zeta}^3\Big]
\end{aligned}
\tag{3.5.35}
$$

To determine the normal form of (3.5.35), we introduce the near–identity transformation

$$\zeta = \xi + h(\xi, \bar{\xi}) \tag{3.5.36}$$

and choose h so that the transformed equation takes on the simplest possible form. Substituting (3.5.36) into (3.5.35) yields

$$\begin{aligned}
\dot{\xi} = i\xi + ih &- \frac{\partial h}{\partial \xi}\dot{\xi} - \frac{\partial h}{\partial \bar{\xi}}\dot{\bar{\xi}} + \mu\xi - \frac{1}{40}\Big[(-1 + 2i)\xi^3 \\
&+ (9 + 2i)\xi^2\bar{\xi} + (9 - 2i)\xi\bar{\xi}^2 - (1 + 2i)\bar{\xi}^3\Big] + \cdots
\end{aligned} \tag{3.5.37}$$

It follows from (3.5.37) that, to the first approximation,

$$\dot{\xi} = i\xi \tag{3.5.38}$$

Using (3.5.38) to eliminate $\dot{\xi}$ and $\dot{\bar{\xi}}$ from the right–hand side of (3.5.37), we obtain

$$\begin{aligned}
\dot{\xi} = i\xi + ih &- i\frac{\partial h}{\partial \xi}\xi + i\frac{\partial h}{\partial \bar{\xi}}\bar{\xi} + \mu\xi \\
&- \frac{1}{40}\Big[(-1 + 2i)\xi^3 + (9 + 2i)\xi^2\bar{\xi} + (9 - 2i)\xi\bar{\xi}^2 \\
&- (1 + 2i)\,\bar{\xi}^3\Big]
\end{aligned} \tag{3.5.39}$$

The form of the terms on the right–hand side of (3.5.39) suggests seeking h in the form

$$h = d_1\xi + d_2\xi^3 + d_3\xi^2\bar{\xi} + d_4\xi\bar{\xi}^2 + d_5\bar{\xi}^3 \tag{3.5.40}$$

Substituting (3.5.40) into (3.5.39) yields

$$\begin{aligned}
\dot{\xi} = i\xi + \mu\xi &- \frac{9 + 2i}{40}\xi^2\bar{\xi} - \left(2id_2 - \frac{1 - 2i}{40}\right)\xi^3 \\
&+ \left(2id_4 - \frac{9 - 2i}{40}\right)\xi\bar{\xi}^2 + \left(4id_5 + \frac{1 + 2i}{40}\right)\bar{\xi}^3
\end{aligned} \tag{3.5.41}$$

It follows from (3.5.41) that we can choose d_2, d_4, and d_5 so as to eliminate the terms proportional to ξ^3, $\xi\bar{\xi}^2$, and $\bar{\xi}^3$, thereby producing the normal form

$$\dot{\xi} = i\xi + \mu\xi - \frac{9+2i}{40}\xi^2\bar{\xi} \qquad (3.5.42)$$

To compare the present solution obtained by using the method of normal forms with that obtained in the preceding section by using the method of multiple scales, we let

$$\xi = iae^{i(t+\beta)} \qquad (3.5.43)$$

where a and β are functions of t. Substituting (3.5.43) into (3.5.42) and separating real and imaginary parts, we obtain

$$\dot{a} = \mu a - \frac{9}{40}a^3 \qquad (3.5.44)$$

$$\dot{\beta} = -\frac{1}{20}a^2 \qquad (3.5.45)$$

in agreement with (3.5.21) and (3.5.22) obtained by using the method of multiple scales.

Hassard, Kazarinoff, and Wan (1981) present a FORTRAN code called BIFOR2 for constructing periodic solutions through center manifold reduction in Hopf bifurcation problems.

3.5.3 General Case

In this section, we use the method of multiple scales to construct periodic solutions of the general system

$$\dot{\mathbf{x}} = \mathbf{F}(\mathbf{x}; \alpha) \qquad (3.5.46)$$

near a Hopf bifurcation point $(\mathbf{x}_c; \alpha_c)$. First, we introduce the transformation

$$\mathbf{x} - \mathbf{x}_c = \epsilon\mathbf{y} \quad \text{and} \quad \alpha - \alpha_c = \epsilon^2\mu \qquad (3.5.47)$$

where ϵ is a small nondimensional quantity that is used as a bookkeeping device. Substituting (3.5.47) into (3.5.46), expanding the right-hand side in a Taylor series for small $\|\mathbf{y}\|$ and small $|\mu|$, and using the fact that $\mathbf{F}(\mathbf{x}_c; \alpha_c) = 0$, we obtain

$$\dot{\mathbf{y}} = J\mathbf{y} + \epsilon\mathbf{Q}(\mathbf{y}, \mathbf{y}) + \epsilon^2\mathbf{C}(\mathbf{y}, \mathbf{y}, \mathbf{y}) + \epsilon^2\mu B\mathbf{y} + \cdots \qquad (3.5.48)$$

where J is the $n \times n$ Jacobian matrix of $\mathbf{F}$ evaluated at $(\mathbf{x}_c; \alpha_c)$, B is an $n \times n$ constant matrix, $\mathbf{Q}(\mathbf{y}, \mathbf{y})$ is generated by a vector–valued symmetric bilinear form $\mathbf{Q}(\mathbf{u}, \mathbf{v})$, and $\mathbf{C}(\mathbf{y}, \mathbf{y}, \mathbf{y})$ is generated by a vector–valued symmetric trilinear form $\mathbf{C}(\mathbf{u}, \mathbf{v}, \mathbf{w})$. At $\alpha = \alpha_c$, the Jacobian matrix J has a pair of purely imaginary eigenvalues $\pm i\omega$, with all of the remaining eigenvalues being in the open left–half of the complex plane.

Using the method of multiple scales, we seek an expansion in the form

$$\mathbf{y} = \mathbf{y}_1(T_0, T_2) + \epsilon \mathbf{y}_2(T_0, T_2) + \epsilon^2 \mathbf{y}_3(T_0, T_2) + \cdots \tag{3.5.49}$$

where $T_0 = t$ and $T_2 = \epsilon^2 t$. Substituting (3.5.49) into (3.5.48) and equating coefficients of like powers of ϵ, we obtain

$$D_0 \mathbf{y}_1 - J\mathbf{y}_1 = 0 \tag{3.5.50}$$
$$D_0 \mathbf{y}_2 - J\mathbf{y}_2 = \mathbf{Q}(\mathbf{y}_1, \mathbf{y}_1) \tag{3.5.51}$$
$$D_0 \mathbf{y}_3 - J\mathbf{y}_3 = -D_2 \mathbf{y}_1 + \mu B \mathbf{y}_1 + 2\mathbf{Q}(\mathbf{y}_1, \mathbf{y}_2) + \mathbf{C}(\mathbf{y}_1, \mathbf{y}_1, \mathbf{y}_1) \tag{3.5.52}$$

The nondecaying solution of (3.5.50) can be expressed as

$$\mathbf{y}_1 = A(T_2)\mathbf{p}e^{i\omega T_0} + \bar{A}(T_2)\bar{\mathbf{p}}e^{-i\omega T_0} \tag{3.5.53}$$

where $\mathbf{p}$ is the right eigenvector of J corresponding to the eigenvalue $i\omega$; that is,

$$J\mathbf{p} = i\omega\mathbf{p} \tag{3.5.54}$$

We let $\mathbf{q}$ be the left eigenvector of J corresponding to $i\omega$; that is,

$$\mathbf{q}^T J = i\omega\mathbf{q}^T \tag{3.5.55}$$

We assume that $\mathbf{p}$ and $\mathbf{q}$ have been normalized so that

$$\mathbf{q}^T \mathbf{p} = 1 \tag{3.5.56}$$

Substituting (3.5.53) into (3.5.51) yields

$$D_0 \mathbf{y}_2 - J\mathbf{y}_2 = \mathbf{Q}(\mathbf{p}, \bar{\mathbf{p}})A\bar{A} + \mathbf{Q}(\mathbf{p}, \mathbf{p})A^2 e^{2i\omega T_0} + cc \tag{3.5.57}$$

The solution of (3.5.57) can be expressed as

$$\mathbf{y}_2 = 2\mathbf{z}_0 A\bar{A} + 2\mathbf{z}_2 A^2 e^{2i\omega T_0} + cc \qquad (3.5.58)$$

where

$$J\mathbf{z}_0 = -\frac{1}{2}\mathbf{Q}(\mathbf{p},\bar{\mathbf{p}}) \qquad (3.5.59)$$

and

$$(2i\omega - J)\mathbf{z}_2 = \frac{1}{2}\mathbf{Q}(\mathbf{p},\mathbf{p}) \qquad (3.5.60)$$

Substituting (3.5.53) and (3.5.58) into (3.5.52) yields

$$\begin{aligned}
D_0\mathbf{y}_3 - J\mathbf{y}_3 = &-A'\mathbf{p}e^{i\omega T_0} + \mu B\mathbf{p}Ae^{i\omega T_0} \\
&+ [8\mathbf{Q}(\mathbf{p},\mathbf{z}_0) + 4\mathbf{Q}(\bar{\mathbf{p}},\mathbf{z}_2) + 3\mathbf{C}(\mathbf{p},\mathbf{p},\bar{\mathbf{p}})]\,A^2\bar{A}e^{i\omega T_0} \quad (3.5.61) \\
&+ cc + \text{NST}
\end{aligned}$$

where $A' = D_2 A$ and NST stands for terms that do not produce secular terms. Eliminating the terms that lead to secular terms from (3.5.61), we have

$$A' = \mu\beta_1 A + 4\beta_2 A^2\bar{A} \qquad (3.5.62)$$

where

$$\beta_1 = \mathbf{q}^T B\mathbf{p} \qquad (3.5.63)$$

$$\beta_2 = 2\mathbf{q}^T\mathbf{Q}(\mathbf{p},\mathbf{z}_0) + \mathbf{q}^T\mathbf{Q}(\bar{\mathbf{p}},\mathbf{z}_2) + \frac{3}{4}\mathbf{q}^T\mathbf{C}(\mathbf{p},\mathbf{p},\bar{\mathbf{p}}) \qquad (3.5.64)$$

Letting $A = \frac{1}{2}a\exp(i\theta)$ and separating real and imaginary parts in (3.5.62) yields

$$a' = \mu\beta_{1r}a + \beta_{2r}a^3 \qquad (3.5.65)$$

$$a\theta' = \mu\beta_{1i}a + \beta_{2i}a^3 \qquad (3.5.66)$$

where $\beta_1 = \beta_{1r} + i\beta_{1i}$ and $\beta_2 = \beta_{2r} + i\beta_{2i}$. Equivalent results were first obtained by Howard (1979) and Abed and Fu (1986).

3.6 EXERCISES

3.1. Consider the periodically forced linear system:

$$\ddot{x} + \omega^2 x = F\cos(\Omega t)$$

(a) Determine the solution of this system subject to the initial conditions $x(0) = x_0$ and $\dot{x}(0) = 0$.

(b) Discuss the nature of this solution in each of the following cases: (i) $\Omega = \omega$ and (ii) $\Omega \neq \omega$.

3.2. A planar autonomous system is described by

$$\dot{x} = y - x^3 + x$$
$$\dot{y} = -x - y^3 + y$$

(a) Apply Bendixson's criterion and discuss if periodic solutions are possible.

(b) Verify that there is a stable limit cycle in the plane.

3.3. The following system is called the **Mathieu oscillator** (e.g., Nayfeh and Mook, 1979, Chapter 5):

$$\ddot{u} + 2\mu\dot{u} + (\delta + 2\epsilon \cos 2t)\, u = 0$$

(a) Use the transformation $u = ve^{-\mu t}$ and obtain the standard form

$$\ddot{v} + \left(\delta - \mu^2 + 2\epsilon \cos 2t\right) v = 0$$

Such equations with periodic coefficients are called **Hill's equations**.

(b) Use Floquet theory to study the stability of periodic solutions in the following cases: (i) $\mu = 0$, $\delta = 1.0$, and $\epsilon = 0.1$; (ii) $\mu = 0.2$, $\delta = 1.0$, and $\epsilon = 0.1$; and (iii) $\mu = 0.2$, $\delta = 1.0$, and $\epsilon = 0.8$.

3.4. The van der Pol oscillator is described by

$$\dot{x}_1 = x_2$$
$$\dot{x}_2 = -\mu(x_1^2 - 1)x_2 - x_1$$

(a) Apply Bendixson's criterion and discuss if periodic solutions are possible in the following cases: (i) $\mu = 0$ and (ii) $\mu \neq 0$.

(b) Use the method of multiple scales to construct analytically a periodic solution for small $|\mu|$. For $\mu = 0.05$, conduct a stability analysis by using Floquet theory.

(c) Use the shooting method of Section 6.5.2 to determine the periodic solutions in the following cases: (i) $\mu = -1$ and (ii) $\mu = 1$. Conduct a stability analysis using Floquet theory.

The van der Pol oscillator has been extensively studied because of its relevance to many applications. A partial list of examples includes lasers (Lamb, 1964), Q machines (Lashinsky, 1969), arc discharge (Keen and Fletcher, 1970), beam–plasma systems (Deneef and Lashinsky, 1973; Nakamura, 1971), oil–film journal bearings (Jain and Srinivasan, 1975), flutter of plates and shells (Fung, 1955; Dowell, 1975; Holmes, 1977; Nayfeh and Mook, 1979), vehicle dynamics (Beaman and Hedrick, 1980; Cooperrider, 1980), and electrical activity in gastrointestinal tracts of humans and animals (Linkens, 1974, 1976).

3.5. A planar system is described by

$$\dot{x}_1 = x_2$$
$$\dot{x}_2 = -x_1 + x_1^2$$

Construct the phase portrait for this system and discuss if there are any limit cycles.

3.6. A planar system is described by

$$\dot{x}_1 = 1 - x_1 x_2$$
$$\dot{x}_2 = x_1$$

Show that there are no equilibrium points. Construct the phase portrait for this system and discuss if there are any limit cycles.

3.7. Use the methods of multiple scales and normal forms to simplify the system in the vicinity of the origin

$$\dot{u} = v$$
$$\dot{v} = -u + \alpha u^2 + \beta uv$$

3.8. Consider the system

$$\dot{x}_1 = -x_2 + x_1 \left[\lambda + \mu(x_1^2 + x_2^2) - (x_1^2 + x_2^2)^2 \right]$$
$$\dot{x}_2 = x_1 + x_2 \left[\lambda + \mu(x_1^2 + x_2^2) - (x_1^2 + x_2^2)^2 \right]$$

Transform these equations into polar coordinates and discuss the bifurcations of periodic orbits as a function of λ and μ.

3.9. Consider the system

$$\dot{x} = -y + xz$$
$$\dot{y} = x + yz$$
$$\dot{z} = -z - (x^2 + y^2) + z^2$$

Show that the center manifold of the origin is

$$z = -x^2 - y^2 + \cdots$$

Then, show that this system can be reduced near the origin to

$$\dot{x} = -y - x(x^2 + y^2)$$
$$\dot{y} = x - y(x^2 + y^2)$$

Use polar coordinates to determine the exact solution of the reduced system.

3.10. Show that none of the following systems has a limit cycle:

(a)

$$\dot{x}_1 = x_2$$
$$\dot{x}_2 = g(x_1) + ax_2, \quad a \neq 0$$

(b)

$$\dot{x}_1 = -x_1 + x_1^3 + x_1 x_2^2$$
$$\dot{x}_2 = -x_2 + x_2^3 + x_1^2 x_2$$

(c)

$$\dot{x}_1 = 1 - x_1 x_2$$
$$\dot{x}_2 = x_1$$

3.11. Consider the following system (Nemytskii and Stepanov, 1960, p. 24):

$$\dot{x} = -y + x(1 - x^2 - y^2)/(x^2 + y^2)^{1/2}$$
$$\dot{y} = x + y(1 - x^2 - y^2)/(x^2 + y^2)^{1/2}$$

Show that $x(t) = \cos(t + \theta_0), y(t) = \sin(t + \theta_0)$ represents a stable limit cycle of the system.

3.12. Consider the following system (Nemytskii and Stepanov, 1960, p. 25):

$$\dot{x} = -y + x(1 - x^2 - y^2)^2$$
$$\dot{y} = x + y(1 - x^2 - y^2)^2$$

Show that $x(t) = \cos(t + \theta_0), y(t) = \sin(t + \theta_0)$ represents a semistable limit cycle of the system; that is, it is a periodic solution that is stable on one side and unstable on the other (and thus is unstable).

3.13. Consider the following system (Nemytskii and Stepanov, 1960, p. 26):

$$\dot{x} = -y + xf\left[(x^2 + y^2)^{1/2}\right],$$
$$\dot{y} = x + yf[(x^2 + y^2)^{1/2}]$$

Here, f is defined as $f(r) = (r^2 - 1) \sin \left[1/(r^2 - 1) \right]$ for $r \neq 1$ and $f(\pm 1) = 0$, then limit cycles are given by $x(t) = r_j \cos(t + \theta_0), y(t) = r_j \sin(t + \theta_0)$ for $j = 0, 1, \cdots$, where $r_0 = 1$ and $r_j = (1 + 1/j\pi)^{1/2}$ for $j = 1, 2, \cdots$. Which of these limit cycles are stable?

3.14. Given that $x = r \cos \theta, y = r \sin \theta$ for $r \geq 0$, and

$$\dot{x} = -y + x \left[1 - f(x, y)/a \right]$$
$$\dot{y} = x + y \left[1 - f(x, y)/a \right]$$

for $a > 0$ and f is a continuous function, show that

$$\dot{r} = \left[1 - f(r \cos \theta, r \sin \theta)/a \right] r, \quad \dot{\theta} = 1$$

Deduce that the trivial solution is stable if $a < f(0, 0)$ and unstable if $0 < f(0, 0) < a$. Show that, if $0 < a < f(x, y)$ for all x, y, then the solution $(x, y) \to (0, 0)$ as $t \to \infty$ for all initial conditions.

Verify that, if $f(x, y) = g(r)$ and $a = g(r_0)$ for some positive differentiable function g and some positive constant r_0, then there exist solutions of the form

$$x(t) = r_0 \cos(t + \theta_0), \quad y(t) = r_0 \sin(t + \theta_0)$$

Show that the orbit of these periodic solutions is stable if $g'(r_0) = 0$ and unstable if $g'(r_0) < 0$.

Further, given that $g(r) = 1 + 1/\left[1 + (r - 1)^2 \right]$, sketch the bifurcation diagram in the first quadrant of the (a, r_0) plane, and sketch the phase portraits in the (x, y) plane for each of the qualitatively different cases that arise (Drazin, 1992, pp. 41-42).

3.15. Consider the following system (Nayfeh and Mook, 1979):

$$\ddot{x} + x - \epsilon x^3 = 0$$

show that there are oscillations of amplitude a with period

$$T = \frac{4K \left[\epsilon a^2/(2 - \epsilon a^2) \right]}{(1 - \frac{1}{2}\epsilon a^2)^{1/2}}$$

where K is the complete elliptic integral of the first kind defined by

$$K(m) = \int_0^{\pi/2} \frac{d\phi}{(1 - m\sin^2\phi)^{1/2}} \quad \text{for } m < 1$$

Deduce that

$$T = 2\pi \left[1 + \frac{3}{8}\epsilon a^2 + O(\epsilon^2 a^4) \right] \quad \text{as } \epsilon a^2 \to 0$$

3.16. Consider the equation (Drazin, 1992, pp. 66-67)

$$\dot{z} = (a - |z|^2)z + \epsilon$$

where z is a complex function of the real variable t but a and ϵ are real constants. Expressing $z = re^{i\theta}$ for non–negative modulus r and real phase θ, show that

$$\dot{r} = (a - r^2)r + \epsilon\cos\theta, \quad \dot{\theta} = -\frac{\epsilon\sin\theta}{r}$$

Investigate the equilibrium solutions and their stability. Sketch the bifurcation curves in the (a, r) plane for fixed $\epsilon > 0$, $\epsilon = 0$, and $\epsilon < 0$.

3.17. Consider the following periodically forced oscillator:

$$\dot{x}_1 = x_2$$
$$\dot{x}_2 = -x_1 - 0.4x_2 - x_1^3 + \cos(1.05t)$$

Use the shooting method of Section 6.5.2 to determine the periodic solution of this system and examine its stability by using Floquet theory.

3.18. Consider the following autonomous system:

$$\dot{x} = \mu x - 3\omega y + \alpha x \left(\frac{x^2}{9} + y^2 \right)$$
$$\dot{y} = \frac{\omega}{3}x + \mu y + \alpha y \left(\frac{x^2}{9} + y^2 \right)$$

Determine the periodic solution of this system, construct a Poincaré map, and carry out a stability analysis for each of the following cases: (i) $\mu = 1$, $\alpha = -1$, and $\omega = 1$; and (ii) $\mu = -1$, $\alpha = 1$, and $\omega = 1$.

3.19. Consider the following autonomous system:

$$\dot{x} = x - y - x\left(x^2 + y^2\right)$$
$$\dot{y} = x + y - y\left(x^2 + y^2\right)$$
$$\dot{z} = \gamma z$$

(a) Verify that $(\cos t, \sin t, 0)$ is a periodic solution of this system.

(b) Construct a Poincaré map and carry out a stability analysis for each of the following cases: (i) $\gamma = 1$ and (ii) $\gamma = -1$.

3.20. Consider the planar autonomous system:

$$\dot{x} = y$$
$$\dot{y} = -x - by\left(y^2 + x^2 - a^2\right)$$

Determine the periodic solution of this system, construct a Poincaré map, and carry out a stability analysis for each of the following cases: (i) $b = -1$ and (ii) $b = 1$.

3.21. Shaw and Rand (1989) used the following system to study forced oscillations of an inverted pendulum constrained to oscillate between two closely spaced rigid barriers:

$$\ddot{x} + 2\mu\dot{x} - x = F\cos(\omega t) \quad |x| < 1$$
$$\dot{x} \to -r\dot{x} \quad\quad\quad\quad |x| = 1$$

where x describes the position of the pendulum; the locations of the rigid barriers are $x = 1$ and $x = -1$; μ is a measure of the friction; $r \leq 1$ provides a measure of energy loss during impact; F is the amplitude of forcing; and ω is the forcing frequency.

(a) Rewrite this system as a third–order autonomous system by using the variables x, $y = \dot{x}$, and $\theta = \omega t (\mathrm{mod} 2\pi)$.

(b) Discuss the motions for which the following section can be used as a Poincaré section:

$$\Sigma = \{(x, y, \theta) \in I \times \mathcal{R} \times S \mid x = 1, y < 0\}$$

The interval $I = [-1, 1]$.

(c) Examine the periodic motions of period $2\pi/\omega$ for the following cases: (i) $r = 1.0$, $\mu = 0.1$, $F = 0.6$, and $\omega = 4.0$; and (ii) $r = 1.0$, $\mu = 0.1$, $F = 2.3$, and $\omega = 1.95$. Explore how these motions can be studied by using the Poincaré map associated with the section defined in part (b).

3.22. Consider the two–dimensional system

$$\dot{x} = \mu x - \omega y + \alpha_{11} x^3 + \alpha_{12} x^2 y + \alpha_{13} x y^2 + \alpha_{14} y^3$$
$$\dot{y} = \omega x - \mu y + \alpha_{21} x^3 + \alpha_{22} x^2 y + \alpha_{23} x y^2 + \alpha_{24} y^3$$

when $\mu \ll 1$. Use the methods of multiple scales and normal forms to simplify this system. Under what conditions is the Hopf bifurcation subcritical, and under what conditions is it supercritical?

3.23. Consider the following autonomous system:

$$\dot{x} = \mu x - y - x(x^2 + y^2) - yz$$
$$\dot{y} = x + \mu y - y(x^2 + y^2) + xz$$
$$\dot{z} = -z + x^2 + y^2$$

where x, y, and z are the states and μ is the control parameter. Construct an analytical approximation for the bifurcating periodic solutions in the vicinity of the Hopf bifurcation point $(x, y, z, \mu) = (0, 0, 0, 0)$ using (i) the method of multiple scales and (ii) a combination of center manifold reduction and the method of normal forms.

3.24. Consider the system

$$\dot{x} = \mu x - \omega y - a_1 x z - a_2 y z$$
$$\dot{y} = \mu y + \omega x + a_3 x z + a_4 y z$$
$$\dot{z} = -\tau z + a_5 x^2 + a_6 y^2 + a_7 x y + a_8 x z + a_9 y z$$

where the a_i, μ, ω, and τ are constants and τ is positive and away from zero. Use the method of multiple scales and a combination of center manifold reduction and the method of normal forms to construct periodic solutions of this system for small μ.

3.25. Consider the system

$$\ddot{u} + \omega^2 u = (\epsilon - \alpha z)\dot{u}$$
$$\dot{z} + \tau z = u^2$$

ω and α are constants, τ is a positive constant that is away from zero, and ϵ is a small positive parameter. Use the method of multiple scales and show that, to the first approximation,

$$u \approx a \cos(\omega t + \beta)$$

where

$$\dot{a} = \frac{1}{2}\epsilon a - \frac{\alpha(\tau^2 + 8\omega^2)}{8\tau(\tau^2 + 4\omega^2)}a^3$$
$$\dot{\beta} = -\frac{\alpha\omega}{4(\tau^2 + 4\omega^2)}a^2$$

3.26. Consider the system in the preceding exercise. Let $u = x_1$ and $\dot{u} = x_2$ and rewrite the system as

$$\dot{x}_1 = x_2$$
$$\dot{x}_2 = -\omega^2 x_1 + (\epsilon - \alpha z)x_2$$
$$\dot{z} = -\tau z + x_1^2$$

Show that the center manifold of the origin is given by

$$z = \frac{1}{\tau(\tau^2 + 4\omega^2)} \left[(\tau^2 + 2\omega^2)x_1^2 - 2\tau x_1 x_2 + 2x_2^2 \right] + \cdots$$

Then, show that the dynamics on the manifold is governed by

$$\dot{x}_1 = x_2$$
$$\dot{x}_2 = -\omega^2 x_1 + \epsilon x_2 - \frac{\alpha}{\tau(\tau^2 + 4\omega^2)} \left[(\tau^2 + 2\omega^2)x_1^2 - 2\tau x_1 x_2 + 2x_2^2 \right] x_2$$
$$+ \cdots$$

Use the method of multiple scales and normal forms to simplify the dynamics on the center manifold and compare the results with those obtained in Exercise 3.25.

3.27. Consider the system in Exercise 3.25. Let $u = \zeta + \bar{\zeta}$ and $\dot{u} = i\omega(\zeta - \bar{\zeta})$ and transform the system into

$$\dot{\zeta} = i\omega\zeta + \frac{1}{2}\epsilon(\zeta - \bar{\zeta}) - \frac{1}{2}\alpha(\zeta - \bar{\zeta})z$$
$$\dot{z} = -\tau z + \zeta^2 + 2\zeta\bar{\zeta} + \bar{\zeta}^2$$

Show that the center manifold of the origin can be approximated by

$$z = \frac{\zeta^2}{\tau + 2i\omega} + \frac{2\zeta\bar{\zeta}}{\tau} + \frac{\bar{\zeta}^2}{\tau - 2i\omega} + \cdots$$

Then, show that the dynamics on the center manifold is governed by

$$\dot{\zeta} = i\omega\zeta + \frac{1}{2}\epsilon(\zeta - \bar{\zeta}) - \frac{1}{2}\alpha(\zeta - \bar{\zeta}) \left[\frac{\zeta^2}{\tau + 2i\omega} + \frac{2\zeta\bar{\zeta}}{\tau} + \frac{\bar{\zeta}^2}{\tau - 2i\omega} \right] + \cdots$$

Use the method of normal forms to simplify the dynamics on the center manifold into

$$\dot{\xi} = i\omega\xi + \frac{1}{2}\epsilon\xi - \frac{1}{2}\alpha \left[\frac{2}{\tau} - \frac{1}{\tau + 2i\omega} \right] \xi^2\bar{\xi} + \cdots$$

Let $\xi = \frac{1}{2}a\exp(i\beta)$ and then determine u and the equations governing a and β. Compare the result with that obtained in Exercise 3.25.

3.28. Consider the following set of coupled van der Pol oscillators used by Linkens (1974) to study electrical activity in gastrointestinal tracts of humans and animals:

$$\ddot{x}_1 + \omega_1^2(x_1 + \gamma x_2) - \epsilon\left[1 - (x_1 + \gamma x_2)^2\right]\dot{x}_1 = 0$$
$$\ddot{x}_2 + \omega_2^2(x_2 + \gamma x_1) - \epsilon\left[1 - (x_2 + \gamma x_1)^2\right]\dot{x}_2 = 0$$

For $\omega_2 \simeq \omega_1$ and $\epsilon = 0.1$, analytically construct the phase–locked solutions of this system and examine their stability.

3.29. Consider the following periodically forced oscillator:

$$\ddot{x} + 0.05\dot{x} + x + 0.005x^3 = 0.1\cos t + 0.3\cos(0.115t)$$

(a) Construct a first approximation for the periodic solution of this system by using the method of harmonic balance. You will need to assume a solution of the form

$$x \simeq p_1\cos t + q_1\sin t + p_2\cos(0.115t) + q_2\sin(0.115t)$$

(b) Use the shooting method of Section 6.5.2 to determine this periodic solution. (You are likely to experience difficulties because of the large period of the solution.)

Chapter 4

QUASIPERIODIC SOLUTIONS

A quasiperiodic solution is a dynamic solution characterized by two or more incommensurate frequencies. Two frequencies ω_1 and ω_2 are said to be **incommensurate** if ω_1/ω_2 is an irrational number. Extending this notion, we say that m frequencies $\omega_1, \omega_2, \cdots, \omega_m$ are incommensurate if the equation

$$n_1\omega_1 + n_2\omega_2 + \cdots + n_m\omega_m = 0$$

is satisfied only when each of the n_i is zero, where the n_i are integers. A quasiperiodic solution is called a k–**period quasiperiodic solution** if it is characterized by k incommensurate frequencies $\omega_1, \omega_2, \cdots, \omega_k$.

In general, a k–period quasiperiodic function has the form

$$x = x(\omega_1 t, \omega_2 t, \cdots, \omega_k t)$$

where x is periodic of period one separately in its k arguments and ω_1, $\omega_2, \cdots$, and ω_k are k incommensurate frequencies. The function x can be represented by a multiple Fourier series of the form

$$x = \sum_{n_1, n_2, \cdots, n_k}^{\infty} a_{n_1 \, n_2 \, \cdots \, n_k} e^{i(\mathbf{n}^T \boldsymbol{\omega})t}$$

where

$$\boldsymbol{\omega} = \begin{bmatrix} \omega_1 & \omega_2 & \cdots & \omega_k \end{bmatrix}^T \quad ; \quad \mathbf{n} = \begin{bmatrix} n_1 & n_2 & \cdots & n_k \end{bmatrix}^T$$

and the n_i are integers. Therefore, the spectrum of x consists of spikes at $\mid n_1\omega_1 + n_2\omega_2 + \cdots + n_k\omega_k \mid$, where some of these frequency components may have zero amplitude. Thus, although the waveform of a quasiperiodic signal may look complex because of the presence of many sinusoids in it, calculating its spectrum would reveal its simplicity. In principle, one can use the spectrum to distinguish a quasiperiodic function from a periodic function in that the spikes in the spectrum of a quasiperiodic function are not spaced at integer multiples of a particular frequency. However, in practice, due to the difficulty of determining whether the ratio of two measured values is rational or irrational, a spectrum that appears to be that of a quasiperiodic function may be that of a periodic function with a very long period.

K–period quasiperiodic functions fall under the class of **almost periodic functions**, which are defined as follows (Hale, 1963, Chapter 12). Let $\mathbf{F}(x, t)$ represent a function where $\mathbf{x} \in \mathcal{R}^n$ and $t \in \mathcal{R}^1$. Further, let D represent a region in the $(n + 1)$–dimensional space $(\mathbf{x}, t)$. Then, in D, the function $\mathbf{F}$ is said to be almost periodic in t uniformly in $\mathbf{x}$ if $\mathbf{F}$ is C^0 in D and if, for any $\delta < 0$, there exists a number $T(\delta) > 0$ such that the condition

$$\| \mathbf{F}(\mathbf{x}, t + \tau) - \mathbf{F}(\mathbf{x}, t) \| \leq \delta \text{ for some } \tau \in [0, \ T(\delta)]$$

is satisfied for all $(\mathbf{x}, t)$ in D. The above condition may be viewed as a type of recurrence property for orbits described by almost periodic functions. There are existence and stability theorems for almost periodic solutions of nonlinear systems (Hale, 1963, Part III; Urabe, 1974). These theorems can be used to ascertain analytically the existence of quasiperiodic solutions.

Example 4.1. We consider the system

$$\dot{x}_1 = x_2$$
$$\dot{x}_2 = -\omega^2 x_1 + F \cos (\Omega t)$$

When $\omega^2 = 8$, $F = 10$, and $\Omega = 2$, the solution of this system is

$$x_1 = a \cos(2\sqrt{2}t) + b \sin(2\sqrt{2}t) + 2.5 \cos(2t)$$
$$x_2 = 2\sqrt{2} \left[-a \sin(2\sqrt{2}t) + b \cos(2\sqrt{2}t) \right] - 5.0 \sin(2t)$$

where the constants a and b are determined by the initial condition (x_{10}, x_{20}). When a and/or b are nonzero, the solution is characterized by the two incommensurate frequencies $2\sqrt{2}$ and 2. Hence, the solution (x_1, x_2) is two–period quasiperiodic.

In Section 4.1, we discuss Poincaré maps of quasiperiodic solutions. In Section 4.2, we discuss the circle map. In Section 4.3, we describe methods for constructing quasiperiodic solutions. In Section 4.4, we discuss the stability of quasiperiodic solutions. Finally, in Section 4.5, we discuss synchronized or mode–locked solutions.

4.1 POINCARÉ MAPS

Like other solutions of an autonomous system, quasiperiodic solutions are invariant under time shifts. On the other hand, quasiperiodic solutions of nonautonomous systems are not invariant under time shifts. However, in both autonomous and nonautonomous systems, a quasiperiodic orbit does not close on itself.

Example 4.2. We consider the dissipative autonomous system (3.4.9)–(3.4.14). The asymptotic solution of this system is two–period quasiperiodic when $\Gamma = 0.625$, $\mu_1 = 0.5$, $\mu_2 = 0.5$, $\mu_3 = 0.1$, $\nu_1 = 0.375$, $\nu_2 = 0.25$, $\nu_3 = 0.5$, and $F = 0.537$. The orbit of this solution, which is shown in Figure 3.4.15a, does not close on itself. On an associated two–sided Poincaré section, the discrete points collect on two loops, as seen in Figure 3.4.15b. Here, the quasiperiodic solution arises as a consequence of a Neimark bifurcation of a periodic solution. We note that a two–period quasiperiodic solution can also arise due to a codimension–two bifurcation of a fixed point. In this case, Hopf and static bifurcations occur simultaneously (Langford, 1979, 1983; Spirig, 1983).

Example 4.3. For $\omega^2 = 8$, $\mu = 0$, $F = 10$, and $\Omega = 2$, the solution of (1.2.2) and (1.2.3) initiated from (3.5, 0) is given by

$$x_1 = \cos(2\sqrt{2}t) + 2.5\cos(2t)$$

$$x_2 = -2\sqrt{2}\sin(2\sqrt{2}t) - 5.0\sin(2t)$$

To construct a Poincaré section of this quasiperiodic orbit, we start at $t = 0$ and sample the orbit at intervals of $2\pi/\Omega = \pi$. Then, at $t_n = n\pi$, we collect the discrete points

$$x_{1n} = \cos(2\sqrt{2}n\pi) + 2.5$$
$$x_{2n} = -2\sqrt{2}\sin(2\sqrt{2}n\pi) \qquad (4.1.1)$$

From (4.1.1), we find that the discrete points fall on the closed curve

$$(x_{1n} - 2.5)^2 + \frac{x_{2n}^2}{8} = 1 \qquad (4.1.2)$$

Here, it is possible to define explicitly the Poincaré map $\mathbf{P}$. Before we define this map, we use the following coordinate transformation

$$x_{1n} = 2.5 + \cos\theta_n \text{ and } x_{2n} = -2\sqrt{2}\sin\theta_n$$

to reduce (4.1.1) to

$$\theta_n = 2\sqrt{2}n\pi \pmod{2\pi}$$

This enables us to define the Poincaré map as

$$\theta_{n+1} = \theta_n + 2\sqrt{2}\pi \qquad (4.1.3)$$

Equation (4.1.3), which maps the circumference of a circle onto itself, does not have any fixed points or periodic points. If the map (4.1.3) did, then $\theta_{n+j} = \theta_j \pmod{2\pi}$ will admit a solution. Then, it follows from (4.1.3) that $2\sqrt{2}j\pi = 0 \pmod{2\pi}$ or $j\sqrt{2} = 0 \pmod 1$, which is not true because $\sqrt{2}$ is an irrational number and j is a nonzero integer. All the iterates of (4.1.3) are densely packed on a circle.

Example 4.4. We consider the nonautonomous system

$$\dot{x}_1 = x_2$$
$$\dot{x}_2 = -x_1 - 2x_2 - x^3 + \cos t + \cos\sqrt{2}t \qquad (4.1.4)$$

After the transients die out, the solution of this system settles on a

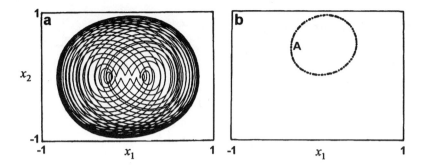

Figure 4.1.1: Two–period quasiperiodic attractor of (4.1.4): (a) orbit and (b) Poincaré section.

two–period quasiperiodic orbit characterized by the incommensurate frequencies 1 and $\sqrt{2}$. A portion of this quasiperiodic orbit obtained through numerical integrations is shown in Figure 4.1.1a. To construct a Poincaré section, we rewrite (4.1.4) as the following four–dimensional system:

$$
\begin{aligned}
\dot{x}_1 &= x_2 \\
\dot{x}_2 &= -x_1 - 2x_2 - x_1^3 + \cos\theta_1 + \cos\theta_2 \\
\dot{\theta}_1 &= 1 \\
\dot{\theta}_2 &= \sqrt{2}
\end{aligned}
\qquad (4.1.5)
$$

Then, we define the global section

$$
\Sigma = \left\{ (x_1, x_2, \theta_1, \theta_2) \in \mathcal{R}^1 \times \mathcal{R}^1 \times S^1 \times S^1 \mid \theta_1 = \theta_{10} = 0 \right\} \qquad (4.1.6)
$$

shown in Figure 4.1.1b. Again, the discrete points fall on a closed curve.

In general, a two–period quasiperiodic orbit can be better visualized on the surface of a torus (see Chapter 1) in a three–dimensional space. Before we deal with a quasiperiodic orbit, let us start with the periodic orbit shown on the surface of a torus in Figure 4.1.2a. Here, the angular coordinates θ_1 and θ_2 are defined by

$$
\theta_1 = \omega_1 t \ (\text{mod } 2\pi) \quad \text{and} \quad \theta_2 = \omega_2 t \ (\text{mod } 2\pi) \qquad (4.1.7)
$$

where $\omega_2/\omega_1 = 4$. The coordinates θ_1 and θ_2 correspond to the large and small diameters of the torus, respectively. The periodic orbit unfolds

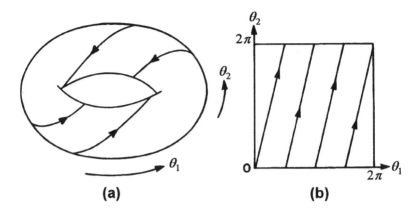

Figure 4.1.2: (a) Illustration of periodic solution of (4.1.7) on the surface of a torus when $\omega_2 = 4\omega_1$ and (b) development of the torus. The arrows indicate directions of evolution for positive times.

along a helix on the surface of the two–torus. When this torus is cut and developed, we obtain the $\theta_2 - \theta_1$ plane shown in Figure 4.1.2b. From (4.1.7), we obtain

$$\frac{d\theta_2}{d\theta_1} = \frac{\omega_2}{\omega_1}$$

which is the equation for a straight line in the $\theta_2 - \theta_1$ plane. On this plane, when any of the angular coordinates reaches a value of 2π it falls back to 0 before the evolution continues. The finite number of parallel lines on the plane indicates that the corresponding orbit closes on itself or repeats and is periodic. To understand why these lines are parallel, let us recall from Section 1.2 that two trajectories of a deterministic system initiated from two different initial conditions cannot intersect each other. If the lines on the developed sheet are not parallel, two trajectories initiated from two different initial conditions will intersect each other. For this reason, the lines on the developed sheet have to be parallel to each other.

If the frequencies ω_1 and ω_2 in (4.1.7) are incommensurate, then the corresponding orbit is two–period quasiperiodic. Like a periodic orbit, this orbit also unfolds along a helix on the surface of a torus. However, this orbit does not close on itself and meanders over the whole surface of the torus. Consequently, when the torus is developed into a plane sheet, there is a dense set of parallel lines after a finite time and an infinite number of parallel lines as $t \to \infty$.

To define a Poincaré section for the quasiperiodic orbit, we choose the plane $\theta_1 = \theta_{10}$, as shown in Figure 4.1.3. Starting at the time $t_0 = \theta_{10}/\omega_1$, we sample the quasiperiodic orbit at intervals of $2\pi/\omega_1$. If we use $\hat{\theta}_k$ to denote θ_2 at the kth intersection of the orbit with the section, it follows from (4.1.7) that

$$\hat{\theta}_k = \omega_2 \left(t_0 + k\frac{2\pi}{\omega_1} \right) \ (\text{mod } 2\pi)$$

Therefore, the associated Poincaré map $\mathbf{P}$ is given by

$$\hat{\theta}_{k+1} = \hat{\theta}_k + 2\pi\frac{\omega_2}{\omega_1} \tag{4.1.8}$$

This map, which maps a circle onto itself, does not have any fixed points or periodic points when ω_1 and ω_2 are incommensurate. Furthermore, the iterates of this map are densely packed on a circle. There is an infinite number of intersections with the Poincaré section because the quasiperiodic orbit does not close on itself. Although the closed curve is densely packed with intersection points, successive intersections with

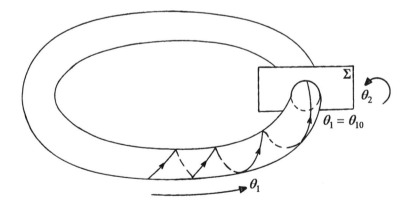

Figure 4.1.3: Portion of a two–period quasiperiodic orbit on a torus and the chosen Poincaré section.

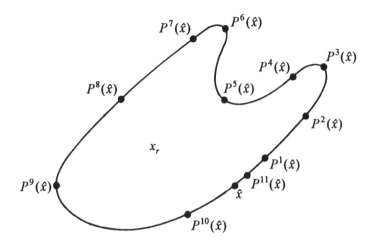

Figure 4.1.4: Poincaré section of a two–period quasiperiodic orbit. The iterates $\mathbf{P}^{10}(\hat{\mathbf{x}})$ and $\mathbf{P}^{11}(\hat{\mathbf{x}})$ bracket the initial point $\hat{\mathbf{x}}$.

the Poincaré section do not sequentially traverse the closed curve.

As noted earlier, when the frequencies ω_1 and ω_2 in (4.1.7) are commensurate, the corresponding orbit on the torus closes on itself and is periodic. If the frequencies ω_1 and ω_2 are equal, (4.1.8) has a fixed point. On the other hand, if the ratio ω_2/ω_1 is a rational number p/q with p and q relatively prime, (4.1.8) has a period–q point.

4.1.1   Winding Time and Rotation Number

In general, the discrete points on a Poincaré section of a two–period quasiperiodic orbit fall on a closed curve, which is not necessarily circular, as depicted in Figure 4.1.4. On the two–dimensional section, $\hat{\mathbf{x}}$ represents the initial point and $\mathbf{P}^i(\hat{\mathbf{x}})$ represents the discrete point obtained after the ith iterate of the Poincaré map $\mathbf{P}$. On the Poincaré section, the tenth and eleventh iterates bracket $\hat{\mathbf{x}}$ after we go once around the closed loop. Let the i_{k-1}th and i_kth iterates bracket $\hat{\mathbf{x}}$ after we go k times around the closed loop. Then, the limit

$$T_w = \lim_{k \to \infty} \frac{i_k}{k} \qquad\qquad (4.1.9)$$

is called the **winding time** (e.g., Kaas–Petersen, 1987). So, T_w represents the average number of iterates of **P** required to get back to $\hat{\mathbf{x}}$. This time is a real number for a two–period quasiperiodic orbit and an integer for a periodic orbit. For the quasiperiodic orbit of Figure 4.1.4, T_w is close to 11.

The inverse of the winding time is called the **rotation** or **winding number** ρ; that is,

$$\rho = \frac{1}{T_w} \tag{4.1.10}$$

An alternate definition for the rotation number is as follows. In Figure 4.1.4, let us choose a reference point $\mathbf{x}_r$ inside the closed loop and denote the angle between the vectors $\mathbf{P}^i(\hat{\mathbf{x}}) - \mathbf{x}_r$ and $\mathbf{P}^{i-1}(\hat{\mathbf{x}}) - \mathbf{x}_r$ by α_i. Then, assuming that $0 < \alpha_i < 2\pi$, the rotation number is defined as (e.g., Arnold, 1988; Kaas–Petersen, 1987)

$$\rho = \frac{1}{2\pi} \lim_{k \to \infty} \sum_{i=1}^{k} \frac{\alpha_i}{k} \tag{4.1.11}$$

There is a rigorous body of theory underlying the statements made in the context of (4.1.7) and (4.1.8). Some of the related theorems are due to Poincaré, Denjoy, Arnold, Moser, and Herman (Arnold, 1988, Chapter 3). According to Arnold, a Poincaré map describing a transformation from a circle onto itself can be written in the form

$$P(\theta) = \theta + g(\theta) \; ; \; g(\theta + 2\pi) = g(\theta) \text{ and } g'(\theta) > -1 \tag{4.1.12}$$

The condition $g'(\theta) > -1$ ensures that $P'(\theta) > 0$, as a result of which P is an orientation preserving map.

Example 4.5. For $\omega^2 = 8$, $\mu = 0$, $F = 10$, and $\Omega = 2$, (4.1.3) represents the Poincaré map for the quasiperiodic orbit of (1.2.2) and (1.2.3) initiated from $(3.5, 0)$. We note that this map is in the form (4.1.12). By using (4.1.11), we find that the associated rotation number is $\sqrt{2}$.

Example 4.6. We consider the system (4.1.4) for the section shown in Figure 4.1.1b. The rotation number is $\sqrt{2}$, and the winding time is

$\sqrt{2}/2$. Therefore, if one starts from the point labeled A on the orbit, it will take on the average $\sqrt{2}/2$ iterates of the Poincaré map to get back to this point.

From the above examples, it is clear that if we have a two–period quasiperiodic solution characterized by the incommensurate frequencies ω_1 and ω_2, the rotation number is either ω_1/ω_2 or ω_2/ω_1, depending on the chosen Poincaré section.

4.1.2   Second–Order Poincaré Map

As discussed earlier in this section, a two–period quasiperiodic orbit is reduced to a closed loop of densely filled points on a Poincaré section. By constructing another section, one can further reduce this closed loop of points to a single point (Kaas–Petersen, 1987; Parker and Chua, 1989). This point is a fixed point of what is called a **second–order Poincaré map**, which is a transformation from the current intersection to the next intersection on the second Poincaré section. Because Lorenz (1984) was one of the first to use second Poincaré sections, these sections are sometimes called **Lorenz sections** (e.g., Abraham and Shaw, 1992, Chapter 9).

In Figure 4.1.5, we illustrate a Poincaré section Σ_1 of a three–dimensional two–period quasiperiodic orbit. A portion of the quasiperiodic orbit is also shown. The intersection points lie on the closed loop Γ_1. The section Σ_1 is spanned by the vectors e_1 and e_2, while the second Poincaré section Σ_2 is oriented along the vector e_2. The points close to the intersection of Γ_1 with Σ_2 lie on the curve Γ_2. At the intersection point, the vector e_1 is tangential to Γ_2 while the vector e_2 is normal to this curve. The intersection of Γ_1 with Σ_2 is the fixed point of the map associated with Σ_2.

To extend the concept illustrated in Figure 4.1.5 to higher dimensions, let the section Σ_1 be n–dimensional. Further, let the n vectors $e_1, e_2, \cdots, e_n$ span Σ_1. Then, we define the second section Σ_2 such that $\Sigma_2 \subset \Sigma_1$. An example of such a section is the space spanned by the $(n-1)$ vectors $e_2, e_3, \cdots, e_n$.

Let us suppose that the orbit associated with Figure 4.1.5 is described by (4.1.7). Then, constructing a second Poincaré section for this orbit amounts to determining the locations on the orbit where

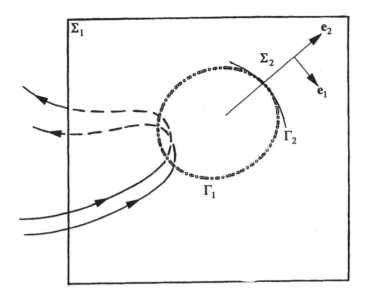

Figure 4.1.5: Illustration of first and second Poincaré sections for a two–period quasiperiodic orbit evolving in a three–dimensional space.

θ_1 and θ_2 simultaneously assume prechosen values. However, θ_1 and θ_2 do not simultaneously assume prechosen values for quasiperiodic orbits. Hence, approximations have to be made in determining the intersections with the second Poincaré section. For further information on construction of second Poincaré sections, we refer the reader to Kaas–Petersen (1987) and Parker and Chua (1989).

4.1.3 Comments

In this section, we explained how a two–period quasiperiodic solution can be studied on the surface of a two–torus. Extending this notion, one can study an n–period quasiperiodic solution on the surface of an n–torus. Each point on the n–torus is described by a set of n angular coordinates. Further, there are $(n-1)$ rotation numbers associated with an n–period quasiperiodic solution.

4.2 CIRCLE MAP

The **circle map**, also known as the **sine map**, is given by

$$x_{n+1} = F(x_n) = x_n + \Omega - \frac{K}{2\pi} \sin(2\pi x_n) \,(\text{mod } 1) \qquad (4.2.1)$$

This map is a special case of the two–dimensional **standard map** (e.g., Lichtenberg and Lieberman, 1992, Chapter 7). When the state x_n is restricted to the interval $[0, 1)$ by using the mod 1 function, (4.2.1) maps the interval $[0, 1)$ onto itself. In most studies, Ω is restricted to the interval $[0, 1]$. The circle map is linear when $K = 0$ and nonlinear when $K \neq 0$. At $K = 0$, (4.2.1) reduces to (4.1.8) if we set $x_n = \theta_n/2\pi$ and identify Ω as ω_2/ω_1.

For $0 \leq K < 1$, we have $F(x + 1) = F(x)$ and $F'(x) > 0$. Hence, the circle map is an orientation preserving diffeomorphism. At $K = 1$, this map is a homeomorphism. For $K > 1$, this map is noninvertible because it is not one–to–one. There is a great deal of interest in the range $0 \leq K < 1$ because (4.2.1) satisfies the requirements of the Poincaré map (4.1.12). Using (4.1.11), we find that the rotation or winding number for the circle map is given by

$$
\begin{aligned}
\rho &= \lim_{n \to \infty} \frac{(x_1 - x_0) + (x_2 - x_1) + \cdots + (x_n - x_{n-1})}{n} \\
&= \lim_{n \to \infty} \frac{x_n - x_0}{n}
\end{aligned}
\qquad (4.2.2)
$$

where the x_i are the values obtained by suspending the mod 1 action. For $K = 0$, the rotation number is Ω. For $K \neq 0$, the rotation number is a function of Ω and K. In practice, one uses a finite value of n. This value needs to be sufficiently large ($n > 500$ is suggested) to obtain accurate values for ρ.

For $0 \leq K < 1$, the results of many theorems (Arnold, 1988, Chapter 3) are applicable to the circle map. On the basis of these theorems, the following statements can be made. For invertible maps, the rotation number ρ is independent of the initial value x_0. When ρ is a rational number, (4.2.1) has periodic points. On the other hand, when ρ is an irrational number, the iterates of (4.2.1) are densely packed on a circle and the associated dynamics is quasiperiodic. The winding number is a continuous function of the map $F(x)$ and hence also of any continuous parameters characterizing F.

Regions of Ω in which ρ is rational are called **mode–locking** or **phase–locking** or **frequency–locking regions**. The rotation number $\rho = p/q$, where p and q are prime integers, if and only if (Arrowsmith and Place, 1990)

$$F^q(x) - (x + p) = 0 \tag{4.2.3}$$

for some $x \in \mathcal{R}$ as shown below. If (4.2.3) is satisfied for some $x = x_0$, then

$$F^q(x_0) = x_0 + p \tag{4.2.4}$$

and

$$F^{mq}(x_0) = x_0 + mp \tag{4.2.5}$$

If we let $n = mq$, then

$$\rho = \lim_{n \to \infty} \frac{F^n(x_0) - x_0}{n} = \lim_{m \to \infty} \frac{F^{mq}(x_0) - x_0}{mq} = \frac{p}{q} \tag{4.2.6}$$

Next, we show that, if (4.2.3) is not satisfied, then $\rho \neq p/q$. To this end, we note that

$$F^k(x) = x + k\Omega - \frac{K}{2\pi} F_k(K, \Omega, x) \tag{4.2.7}$$

where

$$F_k(K, \Omega, x) = \sum_{l=0}^{k-1} \sin[2\pi F^l(x)] \tag{4.2.8}$$

and

$$\sin[2\pi F^k(x + 1)] = \sin[2\pi F^k(x)] \tag{4.2.9}$$

for any k. Letting $\Omega = p/q + \beta$, we find from (4.2.7)–(4.2.9) that

$$F^q(x) = x + p + G(K, \frac{p}{q} + \beta, x) \tag{4.2.10}$$

where

$$G(K, \frac{p}{q} + \beta, x) = q\beta + F^q(K, \frac{p}{q} + \beta, x) \tag{4.2.11}$$

We observe that, because F^q is periodic in x, then G is periodic in x, and attains its maximum and minimum on $[0, 1]$, and hence is bounded. If (4.2.3) is not satisfied for some $x \in \mathcal{R}$, then

$$G\left(K, \frac{p}{q} + \beta, x\right) \neq 0$$

for all $x \in \mathcal{R}$, and hence G is bounded away from zero. Therefore,

$$\left| \rho - \frac{p}{q} \right| \geq \min_{x \in \mathcal{R}} \left| \frac{G\left(K, \frac{p}{q} + \beta, x\right)}{q} \right| \qquad (4.2.12)$$

Consequently, $\rho = p/q$ if and only if (4.2.3) is satisfied.

Equation (4.2.3) can be used to determine the boundaries of the Arnold tongues. For small K, one can use a perturbation scheme, as in Exercises 4.10–4.13, but, for a general value of K, one may have to use a numerical scheme.

Example 4.7. We use (4.2.3) to determine the boundaries of the Arnold tongues when $q = 1$ and $p = 0$ or 1. When $q = 1$, it follows from (4.2.1) and (4.2.3) that $\rho = p$ if and only if

$$x + \Omega - \frac{K}{2\pi} \sin 2\pi x = x + p \qquad (4.2.13)$$

for some $x \in \mathcal{R}$. Consequently,

$$\Omega = p + \frac{K}{2\pi} \sin 2\pi x \qquad (4.2.14)$$

for some $x \in [0, 1]$. Equation (4.2.14) is satisfied for some $x \in [0, 1]$ provided that $\Omega - p \leq \pm K/2\pi$. Therefore, $\rho = 0$ in the wedge $\Omega = \pm K/2\pi$ and $\rho = 1$ in the wedge $\Omega = 1 \pm K/2\pi$.

Example 4.8. We use (4.2.3) to determine the boundaries of the Arnold tongue when $\rho = \frac{1}{2}$. When $p = 1$ and $q = 2$, it follows from (4.2.3) that $\rho = \frac{1}{2}$ if and only if

$$F^2(x) = x + 1 \qquad (4.2.15)$$

for some $x \in [0,1]$. Moreover, it follows from (4.2.1) that

$$F^2(x) = x + 2\Omega - \epsilon \sin 2\pi x - \epsilon \sin[2\pi(x + \Omega - \epsilon \sin 2\pi x)] \quad (4.2.16)$$

where $\epsilon = K/2\pi$. Combining (4.2.15) and (4.2.16) and letting $\Omega = \frac{1}{2} + \beta$, we conclude that $\rho = \frac{1}{2}$ if and only if

$$2\beta - \epsilon \sin 2\pi x - \epsilon \sin\left[2\pi\left(x + \frac{1}{2} + \beta - \epsilon \sin 2\pi x\right)\right] = 0 \quad (4.2.17)$$

or

$$2\beta - \epsilon \sin 2\pi x + \epsilon \sin\left[2\pi\left(x + \beta - \epsilon \sin 2\pi x\right)\right] = 0 \quad (4.2.18)$$

Expanding the last term in (4.2.18) for small ϵ and β yields

$$2\beta - \epsilon \sin 2\pi x + \epsilon \sin 2\pi x + 2\pi\epsilon(\beta - \epsilon \sin 2\pi x)\cos 2\pi x + O(\epsilon^3, \epsilon\beta^2, \epsilon^2\beta) = 0$$

or

$$\beta - \frac{1}{2}\epsilon^2\pi \sin 4\pi x + \pi\epsilon\beta \cos 2\pi x + \cdots = 0$$

Hence,

$$\beta = \frac{1}{2}\epsilon^2\pi \sin 4\pi x + O(\epsilon^3) \quad (4.2.19)$$

Equation (4.2.19) is satisfied for some $x \in [0,1]$ provided that

$$|\beta| \le \frac{1}{2}\epsilon^2\pi \quad (4.2.20)$$

Consequently, $\rho = \frac{1}{2}$ in the Arnold tongue given by

$$\Omega = \frac{1}{2} \pm \frac{1}{2}\epsilon^2\pi + \cdots$$

or

$$\Omega = \frac{1}{2} \pm \frac{K^2}{8\pi} + \cdots$$

For $0 < K < 1$, we schematically depict the mode–locking regions in Figure 4.2.1 following Jensen, Bak, and Bohr (1983, 1984). These regions are known as **Arnold's tongues**. Each mode–locked region appears as a distorted triangle with its apex at a rational number on the Ω axis. Within each region, the rotation number is this constant

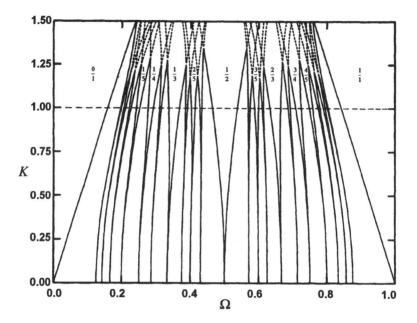

Figure 4.2.1: A few dominant Arnold tongues. Reprinted with permission from Jensen, Bak, and Bohr (1984).

rational number. When K is close to zero, all of the mode–locking regions are quite small. Hence, for a random value of Ω, the probability that the winding number is rational is almost zero. As the strength of the nonlinearity K increases, the width of each tongue increases. For $K \simeq 1$, the probability that the rotation number is rational for a random value of Ω is almost one. At $K = 1$, the set of mode–locking regions is fractal. For $K > 1$, the mode–locking motions overlap, implying the coexistence of different periodic oscillations. Moreover, the rotation number is no longer unique because it depends on the initial condition; that is, x_0 in (4.2.2).

The graph of ρ versus Ω has a very curious appearance. It resembles a staircase with nonuniform horizontal steps (plateaus), as is evident from Figure 4.2.2. Each time ρ takes on the value of a rational number, the graph of $\rho(\Omega)$ has a horizontal step because it is constant in an open interval of Ω. Therefore, in each of these steps, ρ assumes the

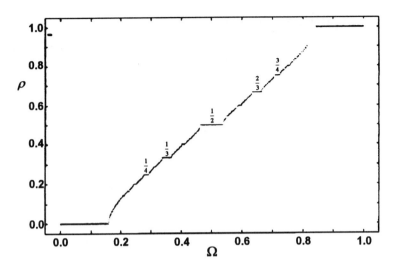

Figure 4.2.2: The devil's staircase for the circle map at $K = 1$.

value of a particular rational number, the corresponding motion is periodic and, hence, there is mode locking. In between two steps, ρ varies continuously with Ω, assuming values of all irrational numbers between the two rational numbers corresponding to these steps, and the corresponding motions are quasiperiodic. (For each irrational number, there is a unique value of Ω.) If a portion of Figure 4.2.2 is magnified, we see a similar structure at the new scale with the presence of more horizontal steps becoming evident. Because there are an infinite number of rational numbers distributed between any two rational numbers, a similar structure is observed at all scales. The graph $\rho(\Omega)$ is said to be **self–similar**. This interesting structure in the $\rho - \Omega$ plane is known as **devil's staircase**.

The concept of **Farey tree** from arithmetics (Allen, 1983; Schroeder, 1986) is useful for understanding the structure of the mode–locking regions in Figures 4.2.1 and 4.2.2. This tree provides a hierarchical arrangement of rational numbers on the unit interval. According to this arrangement, between two mode–locked regions with rotation numbers p_1/q_1 and p_2/q_2, there is a mode–locked region with the rotation number

$\hat{p}/\hat{q} = (p_1 + p_2)/(q_1 + q_2)$, where

$$\frac{p_1}{q_1} < \frac{\hat{p}}{\hat{q}} < \frac{p_2}{q_2}$$

For example, if we consider the mode–locking regions corresponding to the ratios $\frac{1}{3}$ and $\frac{1}{2}$, we find that there is a mode–locking region corresponding to the ratio $\frac{2}{5}$ in between these two regions. Between the regions $\frac{1}{3}$ and $\frac{2}{5}$ there is a mode–locking region corresponding to $\frac{3}{8}$, and between the regions $\frac{1}{2}$ and $\frac{2}{5}$ there is a mode–locking region corresponding to $\frac{3}{7}$. In the context of Figure 4.2.2, continuing the process, we find that there is an infinite number of steps between any two steps, giving rise to the devil's staircase description. Because there is an infinite number of steps on a bounded interval, the widths of most of these steps must be infinitesimal. The steps having non–negligible widths correspond to simple rational numbers; that is, rational numbers p/q, where the integers p and q are small, as in $\frac{1}{1}, \frac{1}{2}, \frac{1}{3}$, etc.

In many applications, two or more oscillators with independent natural frequencies influence each other in such a way as to produce a **synchronization** of these oscillators. In the synchronized or frequency-locked state, the oscillators exhibit periodic oscillations. To understand such frequency lockings and transitions from periodic to quasiperiodic oscillations, circle maps have been frequently used. The study of Choi and Noah (1992a) on a rotating machinery provides one example. An extensive list of other examples can be found in the review article of Glazier and Libchaber (1988). Other relevant references include Franceschini (1983), Bohr, Bak, and Jensen (1984), Jensen, Kadanoff, Libchaber, Procaccia, and Stavans (1985), and Stavans, Heslot, and Libchaber (1985).

4.3   CONSTRUCTIONS

Quasiperiodic solutions of continuous–time systems have been constructed by using a number of methods. These include the method of averaging, the method of multiple scales, the spectral balance method, and the Poincaré map method.

4.3.1   Method of Multiple Scales

We illustrate this method by constructing quasiperiodic solutions of the following weakly nonlinear system (Hale, 1963, Chapter 18):

Example 4.9. We consider the system

$$\ddot{x}_1 + \omega_1^2 x_1 = \epsilon(1 - x_1^2 - \delta x_2^2)\dot{x}_1 \qquad (4.3.1)$$

$$\ddot{x}_2 + \omega_2^2 x_2 = \epsilon(1 - \alpha x_1^2 - x_2^2)\dot{x}_2 \qquad (4.3.2)$$

When $\epsilon = 0$, we have a set of linear oscillators. When $\epsilon \neq 0$ and $\delta = \alpha = 0$, we have a set of uncoupled van der Pol oscillators. When $\epsilon \neq 0$, $\alpha \neq 0$, and $\delta \neq 0$, (4.3.1) and (4.3.2) represent two coupled van der Pol oscillators.

Here, we consider the case where ϵ is small and positive, $\alpha \neq 0$, $\delta \neq 0$, and the frequencies ω_1 and ω_2 are away from each other. First, we expand the states according to

$$x_j = x_{j0}(T_0, T_1) + \epsilon x_{j1}(T_0, T_1) + O(\epsilon^2) \quad \text{for } j = 1, 2 \qquad (4.3.3)$$

where $T_n = \epsilon^n t$. The transformation of the time derivative is given by (3.5.4). Substituting (4.3.3) into (4.3.1) and (4.3.2), using (3.5.4), and equating coefficients of like powers of ϵ, we obtain the following hierarchy of equations:

$O(\epsilon^0)$:

$$D_0^2 x_{10} + \omega_1^2 x_{10} = 0 \qquad (4.3.4)$$

$$D_0^2 x_{20} + \omega_2^2 x_{20} = 0 \qquad (4.3.5)$$

$O(\epsilon^1)$:

$$D_0^2 x_{11} + \omega_1^2 x_{11} = -2D_0 D_1 x_{10} + (1 - x_{10}^2 - \delta x_{20}^2)D_0 x_{10} \qquad (4.3.6)$$

$$D_0^2 x_{21} + \omega_2^2 x_{21} = -2D_0 D_1 x_{20} + (1 - \alpha x_{10}^2 - x_{20}^2)D_0 x_{20} \qquad (4.3.7)$$

The solutions of (4.3.4) and (4.3.5) can be expressed as

$$x_{10} = A_1(T_1)e^{i\omega_1 T_0} + cc \qquad (4.3.8)$$

$$x_{20} = A_2(T_1)e^{i\omega_2 T_0} + cc \qquad (4.3.9)$$

where cc stands for the complex conjugate of the preceding terms and the A_i are complex–valued functions, which will be determined by imposing the solvability conditions at the next level of approximation.

Substituting (4.3.8) and (4.3.9) into (4.3.6) and (4.3.7), recalling that ω_2 is away from ω_1, and eliminating the terms that produce secular terms, we obtain

$$-2A_1' + A_1 - A_1^2 \bar{A}_1 - 2\delta A_2 \bar{A}_2 A_1 = 0 \qquad (4.3.10)$$
$$-2A_2' + A_2 - A_2^2 \bar{A}_2 - 2\alpha A_1 \bar{A}_1 A_2 = 0 \qquad (4.3.11)$$

where the prime denotes the derivative with respect to T_1. Substituting $A_j = \frac{1}{2} a_j \exp(i\beta_j)$ into (4.3.10) and (4.3.11) and separating real and imaginary parts, we obtain

$$a_1' = \frac{1}{2} a_1 \left(1 - \frac{1}{4} a_1^2 - \frac{1}{2} \delta a_2^2 \right) \qquad (4.3.12)$$

$$a_2' = \frac{1}{2} a_2 \left(1 - \frac{1}{2} \alpha a_1^2 - \frac{1}{4} a_2^2 \right) \qquad (4.3.13)$$

$$\beta_1' = 0 \qquad (4.3.14)$$

$$\beta_2' = 0 \qquad (4.3.15)$$

To obtain the first approximation, we substitute (4.3.8) and (4.3.9) into (4.3.3) after expressing the A_j in polar form. The result is

$$x_1 \simeq a_1 \cos(\omega_1 t + \beta_1) \quad \text{and} \quad x_2 \simeq a_2 \cos(\omega_2 t + \beta_2) \qquad (4.3.16)$$

where the a_j and β_j are described by (4.3.12)–(4.3.15). These equations are equivalent to those obtained by Hale (1963, Chapter 18) using the method of averaging.

When ω_1 and ω_2 are incommensurate, a nontrivial fixed point of (4.3.12) and (4.3.13) corresponds to the quasiperiodic solution (4.3.16) of (4.3.1) and (4.3.2). This fixed point is given by

$$a_{10} = 2\sqrt{\frac{1 - 2\delta}{1 - 4\alpha\delta}} \quad \text{and} \quad a_{20} = 2\sqrt{\frac{1 - 2\alpha}{1 - 4\alpha\delta}} \qquad (4.3.17)$$

We note that real solutions of (4.3.17) exist only in the following two regions of the $\alpha - \delta$ plane: (i) $\alpha < \frac{1}{2}$ and $\delta < \frac{1}{2}$ and (ii) $\alpha > \frac{1}{2}$ and

$\delta > \frac{1}{2}$. The stability of the fixed point (a_{10}, a_{20}) can be determined by examining the eigenvalues of the Jacobian matrix associated with (4.3.12) and (4.3.13); that is, the eigenvalues of

$$A = \frac{-1}{1 - 4\alpha\delta} \begin{bmatrix} 1 - 2\delta & 2\delta\sqrt{(1 - 2\alpha)(1 - 2\delta)} \\ 2\alpha\sqrt{(1 - 2\alpha)(1 - 2\delta)} & 1 - 2\alpha \end{bmatrix} \tag{4.3.18}$$

The eigenvalues are given by

$$\lambda_{1,2} = \frac{1 - \alpha - \delta}{4\alpha\delta - 1}$$

$$\pm \frac{[(1 - \alpha - \delta)^2 - (1 - 4\alpha\delta)(1 - 2\delta)(1 - 2\alpha)]^{1/2}}{4\alpha\delta - 1} \tag{4.3.19}$$

Both of the eigenvalues have negative real parts in region (i) of the $\alpha - \delta$ plane. On the other hand, one of the eigenvalues is positive and the other is negative in region (ii). Hence, the fixed point is stable in region (i) and unstable in region (ii). Consequently, the quasiperiodic solutions of (4.3.1) and (4.3.2) are stable in region (i) and unstable in region (ii).

The quasiperiodic solution (4.3.16) corresponding to the fixed point (4.3.17) can be visualized on the surface of a two–torus. For this torus, the diameters are $2a_{10}$ and $2a_{20}$, and the angular coordinates are $\theta_1 = \omega_1 t + \beta_{10}$ and $\theta_2 = \omega_2 t + \beta_{20}$, respectively.

The method described in this section can be used to construct quasiperiodic solutions of many weakly nonlinear systems. The studies of Neu (1979), Rand and Holmes (1980), and Storti and Rand (1982) provide a few examples. Gilsinn (1993) used a series expansion to construct a higher–order approximation for a quasiperiodic solution of a weakly nonlinear system.

4.3.2 Spectral Balance Method

We note that the method of harmonic balance can be used to obtain periodic solutions of a continuous–time system. In this method, the problem of determining a periodic solution of a continuous–time system is converted into the problem of determining a solution of an algebraic system of equations. The solution of the algebraic system provides the amplitudes of the sinusoidal components that make up the periodic solution.

An extension of the method of harmonic balance, called the **spectral balance method,** can be used to construct quasiperiodic solutions of a continuous–time system. In the spectral balance method, the solution is represented by a multidimensional Fourier series, and the problem of determining a quasiperiodic solution is converted into the problem of determining a solution of an algebraic system of equations (Chua and Ushida, 1981; Ushida and Chua, 1984). The solution of this system provides the amplitudes associated with the different spectral peaks in the assumed series. We outline the spectral balance method below by using an example.

Example 4.10. To construct an approximate two–period quasiperiodic solution of (4.3.1) and (4.3.2), we expand the variables in the multiple Fourier series

$$\mathbf{x} = \sum_{n_1,n_2}^{\infty} \mathbf{a}_{n_1 n_2} e^{i(n_1 \omega_1 + n_2 \omega_2)t}$$

where $\mathbf{a}$ is a complex–valued vector function. If we only keep the terms that correspond to $n_1, n_2 = 0, \pm 1$, the components of $\mathbf{x}$ are approximated by

$$x_j \simeq X_{j0} + X_{j1} \cos(\omega_j t) + X_{j2} \sin(\omega_j t), \quad j = 1, 2 \qquad (4.3.20)$$

where the real quantities X_{j0}, X_{j1}, and X_{j2} are the unknowns that need to be determined.

Substituting (4.3.20) into (4.3.1) and (4.3.2) and matching the coefficients of the different spectral peaks on both sides of the equations, we obtain the following system of algebraic equations to determine the spectral coefficients X_{j0}, X_{j1}, and X_{j2}:

$$X_{10} = 0$$
$$X_{20} = 0$$
$$X_{11}\left[1 - X_{10}^2 - \frac{1}{4}(X_{11}^2 + X_{12}^2) - \delta X_{20}^2 - \frac{\delta}{2}(X_{21}^2 + X_{22}^2)\right] = 0$$

$$X_{12}\left[1 - X_{10}^2 - \frac{1}{4}(X_{11}^2 + X_{12}^2) - \delta X_{20}^2 - \frac{\delta}{2}(X_{21}^2 + X_{22}^2)\right] = 0$$

$$X_{21}\left[1 - \alpha X_{10}^2 - \frac{\alpha}{2}(X_{11}^2 + X_{12}^2) - X_{20}^2 - \frac{1}{4}(X_{21}^2 + X_{22}^2)\right] = 0$$

$$X_{22}\left[1 - \alpha X_{10}^2 - \frac{\alpha}{2}(X_{11}^2 + X_{12}^2) - X_{20}^2 - \frac{1}{4}(X_{21}^2 + X_{22}^2)\right] = 0$$

Solving these equations, we obtain

$$X_{10} = X_{20} = 0 \tag{4.3.21}$$

$$X_{11} = X_{12} = \sqrt{\frac{2 - 4\delta}{1 - 4\alpha\delta}} \quad \text{and} \quad X_{21} = X_{22} = \sqrt{\frac{2 - 4\alpha}{1 - 4\alpha\delta}} \tag{4.3.22}$$

Substituting (4.3.21) and (4.3.22) into (4.3.20), we find that the results agree with those obtained by using the method of multiple scales. However, unlike the method of multiple scales, the spectral balance method does not provide the transient solution.

In larger systems, a Newton–Raphson method may be used to solve the algebraic system of equations for the spectral coefficients. Furthermore, for systems with arbitrary nonlinearities, it will be necessary to compute forward and inverse Fourier transforms of two–period quasiperiodic functions. An efficient and accurate algorithm for this purpose has been proposed by Kundert, Sorkin, and Sangiovanni–Vincentelli (1983).

4.3.3   Poincaré Map Method

In one version of this method, Kevrekidis, Aris, Schmidt, and Pelikan (1985) reduced the problem of determining a two–period quasiperiodic solution of a continuous–time system to the problem of determining the closed curve on an associated Poincaré section. The scheme proposed by Kevrekidis et al. is only applicable to periodically forced systems.

In another version, Kaas–Petersen (1985a, 1985b, 1987) reduced the problem of determining quasiperiodic solutions to the problem of finding a fixed point of a second–order Poincaré map. His approach is applicable to all continuous–time systems. Kaas–Petersen (1987) also described a continuation scheme for two–period quasiperiodic solutions based on the second–order Poincaré map. In this scheme, an analytical approximation is used to initiate the continuation scheme. For periodically forced systems, Choi and Noah (1992a) proposed a modified scheme and successfully applied it to a rotating machinery problem. For quasiperiodically forced systems, Ling (1991) improved on Kaas–Petersen's method and developed a shooting scheme for constructing solutions.

The reader is referred to the above–mentioned references for more information on Poincaré map methods.

4.4 STABILITY

When one constructs a quasiperiodic solution of a weakly nonlinear system by using either the method of averaging or the method of multiple scales, one can obtain information on the stability of the quasiperiodic solution by studying the stability of the corresponding fixed point of the averaged equations. According to the theorems provided by Hale (1963, Part III), a quasiperiodic solution is stable (unstable) if the corresponding hyperbolic fixed point is stable (unstable).

When one constructs a quasiperiodic solution by using the method proposed by Kaas–Petersen, the stability of the constructed solution depends on the stability of the fixed point of the second–order Poincaré map. Iooss and Joseph (1980, Chapter X) discuss how the stability of a two–period quasiperiodic solution can be determined through an associated Poincaré map. The concept of Lyapunov exponents, discussed in Section 7.8, is useful in ascertaining if an asymptotic state is a (stable) quasiperiodic solution.

Next, we consider the structural stability of the system (4.1.8). When the corresponding rotation number ρ is irrational, we have a two–period quasiperiodic flow. This two–period quasiperiodic flow is replaced by a periodic flow if a perturbation added to (4.1.8) makes ρ a rational number. According to a theorem due to Arnold (1988, Chapter 3), (4.1.8) is structurally stable if and only if ρ is a rational number and the corresponding periodic solutions are hyperbolic. This example gives the impression that the realization of a quasiperiodic solution is an unlikely prospect. However, this is not so. Using the circle map as an example, Arnold (1988) showed that it is possible to realize two–period quasiperiodic flows.

The structural stability of systems associated with n–period quasi-periodic flows for $n \geq 3$ was studied theoretically by Ruelle, Takens, and Newhouse (Ruelle and Takens, 1971; Newhouse, Ruelle, and Takens, 1978) and numerically by Grebogi, Ott, and Yorke (1985). The results of Ruelle, Takens, and Newhouse indicate that three- and four-period quasiperiodic flows can be destroyed by providing appropriate perturbations to the considered system.

4.5 SYNCHRONIZATION

In many coupled oscillators, as a parameter is varied the periodic so-lution that exists during a synchronized or mode–locked state loses stability, giving rise to a quasiperiodic solution. This loss of synchro-nism is of interest in many applications (e.g., Choi and Noah, 1992a; Linkens, 1974, 1976; Storti and Rand, 1982). Linkens used the method of harmonic balance, while Storti and Rand used the method of multiple scales.

Example 4.11. We consider the case where ω_2 is close to ω_1 in (4.3.1) and (4.3.2). When $\epsilon = 0$, the set of linear oscillators is characterized by the frequencies ω_1 and ω_2. However, when ϵ is small and positive, it may be possible for the two oscillators to be synchronized to one basic frequency due to the nonlinear coupling. This is investigated in the multiple–scale analysis that follows.

We seek an expansion of the form (4.3.3) for the states x_1 and x_2. Further, to describe quantitatively the nearness of ω_2 to ω_1, we introduce the detuning parameter σ defined as

$$\omega_2 = \omega_1 + \epsilon\sigma \tag{4.5.1}$$

Substituting (4.3.3) into (4.3.1) and (4.3.2) and equating coefficients of like powers of ϵ, we obtain (4.3.4)–(4.3.7).

The solution of (4.3.4) and (4.3.5) is given by (4.3.8) and (4.3.9). Next, substituting (4.3.8) and (4.3.9) into (4.3.6) and (4.3.7), using (4.5.1), and eliminating the terms that produce secular terms in x_{11} and x_{21}, we obtain

$$2A_1' = A_1 - A_1^2 \bar{A}_1 - 2\delta A_2 \bar{A}_2 A_1 + \delta A_2^2 \bar{A}_1 e^{2i\sigma T_1} \qquad (4.5.2)$$

$$2A_2' = A_2 - 2\alpha A_1 \bar{A}_1 A_2 - A_2^2 \bar{A}_2 + \alpha A_1^2 \bar{A}_2 e^{-2i\sigma T_1} \qquad (4.5.3)$$

Expressing the A_j in the polar form $A_j = \frac{1}{2} a_j \exp(i\beta_j)$ and separating real and imaginary parts in (4.5.2) and (4.5.3) yields

$$a_1' = \frac{1}{2}a_1 \left(1 - \frac{1}{4}a_1^2 - \frac{1}{2}\delta a_2^2\right) + \frac{1}{8}\delta a_1 a_2^2 \cos\gamma \qquad (4.5.4)$$

$$a_2' = \frac{1}{2}a_2 \left(1 - \frac{1}{2}\alpha a_1^2 - \frac{1}{4}a_2^2\right) + \frac{1}{8}\alpha a_1^2 a_2 \cos\gamma \qquad (4.5.5)$$

$$a_1\beta_1' = \frac{1}{8}\delta a_1 a_2^2 \sin\gamma \qquad (4.5.6)$$

$$a_2\beta_2' = -\frac{1}{8}\alpha a_1^2 a_2 \sin\gamma \qquad (4.5.7)$$

where

$$\gamma = 2\sigma T_1 + 2\beta_2 - 2\beta_1 \qquad (4.5.8)$$

Eliminating β_1 and β_2 from (4.5.6)–(4.5.8), we arrive at

$$a_1 a_2 \gamma' = a_1 a_2 \left[2\sigma - \frac{1}{4}(\alpha a_1^2 + \delta a_2^2)\sin\gamma\right] \qquad (4.5.9)$$

Substituting (4.3.8) and (4.3.9) into (4.3.3), expressing the A_j in polar form, and using (4.5.1) and (4.5.8), we obtain to the first approximation

$$x_1 \approx a_1 \cos(\omega_1 t + \beta_1) \qquad (4.5.10)$$

$$x_2 \approx a_2 \cos(\omega_1 t + \beta_1 + \frac{1}{2}\gamma) \qquad (4.5.11)$$

where a_1, a_2, and γ are given by (4.5.4), (4.5.5), and (4.5.9).

The fixed points $(a_{10}, a_{20}, \gamma_0)$ of (4.5.4), (4.5.5), and (4.5.9), which correspond to $a_1' = 0$, $a_2' = 0$, and $\gamma' = 0$, can be classified into four types: (1) $a_{10} = 0$ and $a_{20} = 0$; (2) $a_{10} = 2$ and $a_{20} = 0$; (3) $a_{10} = 0$ and $a_{20} = 2$; and (4) $a_{10} \neq 0$ and $a_{20} \neq 0$. The first type of fixed points corresponds to trivial solutions of (4.3.1) and (4.3.2), while the second, third, and fourth types of fixed points correspond to periodic solutions of (4.3.1) and (4.3.2). However, only the fourth type of fixed points corresponds to synchronized solutions; that is, solutions where x_1 and x_2 are synchronized. These fixed points are solutions of

$$\frac{1}{4}a_{10}^2 + \frac{1}{2}\delta a_{20}^2 - \frac{1}{4}\delta a_{20}^2 \cos\gamma_0 = 1 \qquad (4.5.12)$$

$$\frac{1}{2}\alpha a_{10}^2 + \frac{1}{4}a_{20}^2 - \frac{1}{4}\alpha a_{10}^2 \cos\gamma_0 = 1 \qquad (4.5.13)$$

$$\frac{1}{4}(\alpha a_{10}^2 + \delta a_{20}^2)\sin\gamma_0 = 2\sigma \qquad (4.5.14)$$

From (4.5.6), (4.5.10), and (4.5.11), we deduce that the synchronization frequency is

$$\omega_1 + \epsilon\beta_1' = \omega_1 + \frac{1}{8}\epsilon\delta a_{20}^2 \sin\gamma_0$$

Next, a stability analysis along the lines of Nayfeh and Mook (1979, Section 7.5.3) is conducted. We superimpose a disturbance $(\hat{a}_1, \hat{a}_2, \hat{\gamma})$ on the fixed point $(a_{10}, a_{20}, \gamma_0)$; carry out the corresponding substitutions in (4.5.4), (4.5.5), and (4.5.9); linearize in the disturbance terms; and obtain

$$\begin{aligned}
\hat{a}_1' &= \left(\frac{1}{2} - \frac{3}{8}a_{10}^2 - \frac{1}{4}\delta a_{20}^2 + \frac{1}{8}\delta a_{20}^2 \cos\gamma_0\right)\hat{a}_1 \\
&\quad + \left(-\frac{1}{2}\delta a_{10}a_{20} + \frac{1}{4}\delta a_{10}a_{20} \cos\gamma_0\right)\hat{a}_2 \qquad (4.5.15) \\
&\quad + \left(-\frac{1}{8}\delta a_{10}a_{20}^2 \sin\gamma_0\right)\hat{\gamma}
\end{aligned}$$

$$\begin{aligned}
\hat{a}_2' &= \left(-\frac{1}{2}\alpha a_{10}a_{20} + \frac{1}{4}\alpha a_{10}a_{20} \cos\gamma_0\right)\hat{a}_1 \\
&\quad + \left(-\frac{1}{8}\alpha a_{10}^2 a_{20} \sin\gamma_0\right)\hat{\gamma} \qquad (4.5.16) \\
&\quad + \left(\frac{1}{2} - \frac{3}{8}a_{20}^2 - \frac{1}{4}\alpha a_{10}^2 + \frac{1}{8}\alpha a_{10}^2 \cos\gamma_0\right)\hat{a}_2
\end{aligned}$$

$$
\begin{aligned}
a_{10}a_{20}\hat{\gamma}' = \Big\{ & \Big[2\sigma - \frac{1}{4}(\alpha a_{10}^2 + \delta a_{20}^2)\sin\gamma_0\Big]a_{20} \\
& - \frac{1}{2}\alpha a_{10}^2 a_{20}\sin\gamma_0\Big\}\hat{a}_1 \\
+ \Big\{ & \Big[2\sigma - \frac{1}{4}(\alpha a_{10}^2 + \delta a_{20}^2)\sin\gamma_0\Big]a_{10} \\
& - \frac{1}{2}\delta a_{10}a_{20}^2\sin\gamma_0\Big\}\hat{a}_2 \\
- \Big[& \frac{1}{4}a_{10}a_{20}(\alpha a_{10}^2 + \delta a_{20}^2)\cos\gamma_0\Big]\hat{\gamma}
\end{aligned}
\qquad (4.5.17)
$$

For the first type of fixed points, we substitute $a_{10} = 0$ and $a_{20} = 0$ into (4.5.15)–(4.5.17). Then, (4.5.15) and (4.5.16) reduce to

$$
\hat{a}_1' = \frac{1}{2}\hat{a}_1 \quad \text{and} \quad \hat{a}_2' = \frac{1}{2}\hat{a}_2
\qquad (4.5.18)
$$

while (4.5.17) reduces to an identity. It is clear from (4.5.18) that a fixed point of the first type is always unstable. Consequently, the trivial solution of (4.3.1) and (4.3.2) is unstable too.

For the second type of fixed points, we substitute $a_{10} = 2$ and $a_{20} = 0$ into (4.5.15)–(4.5.17) and obtain

$$
\hat{a}_1' = -\hat{a}_1
\qquad (4.5.19)
$$

$$
\hat{a}_2' = \left(\frac{1}{2} - \alpha + \frac{1}{2}\alpha\cos\gamma_0\right)\hat{a}_2
\qquad (4.5.20)
$$

$$
\left(2\sigma - \alpha\sin\gamma_0\right)\hat{a}_2 = 0
\qquad (4.5.21)
$$

Equation (4.5.21) is always satisfied if

$$
\sin\gamma_0 = \frac{2\sigma}{\alpha}
\qquad (4.5.22)
$$

Consequently, the second type of fixed points is possible only if $|\sigma/\alpha| \leq \frac{1}{2}$. We infer from (4.5.19) and (4.5.20) that a fixed point of the second type is stable if

$$
\lambda = \frac{1}{2} - \alpha + \frac{1}{2}\alpha\cos\gamma_0 < 0
\qquad (4.5.23)
$$

where

$$\cos \gamma_0 = \pm\sqrt{1 - \frac{4\sigma^2}{\alpha^2}} \qquad (4.5.24)$$

For the third type of fixed points, we substitute $a_{10} = 0$ and $a_{20} = 2$ into (4.5.15)–(4.5.17) and obtain

$$\hat{a}_1' = \left(\frac{1}{2} - \delta + \frac{1}{2}\delta \cos \gamma_0\right)\hat{a}_1 \qquad (4.5.25)$$

$$\hat{a}_2' = -\hat{a}_2 \qquad (4.5.26)$$

$$\left(2\sigma - \delta \sin \gamma_0\right)\hat{a}_1 = 0 \qquad (4.5.27)$$

Equation (4.5.27) is always satisfied if

$$\sin \gamma_0 = \frac{2\sigma}{\delta} \qquad (4.5.28)$$

Therefore, the third type of fixed points is possible only if $|\delta/\alpha| \le \frac{1}{2}$. We infer from (4.5.25) and (4.5.26) that a fixed point of the third type is stable if

$$\lambda = \frac{1}{2} - \delta + \frac{1}{2}\delta \cos \gamma_0 < 0 \qquad (4.5.29)$$

where

$$\cos \gamma_0 = \pm\sqrt{1 - \frac{4\sigma^2}{\delta^2}} \qquad (4.5.30)$$

To determine the stability of a fixed point of the fourth type, we need to determine the eigenvalues of the Jacobian matrix

$$A = -\frac{1}{8}\begin{bmatrix} 2a_{10}^2 & 2\delta a_{10}a_{20}(2 - \cos \gamma_0) & \delta a_{10}a_{20}^2 \sin \gamma_0 \\ 2\alpha a_{10}a_{20}(2 - \cos \gamma_0) & 2a_{20}^2 & \alpha a_{10}^2 a_{20} \sin \gamma_0 \\ 4\alpha a_{10} \sin \gamma_0 & 4\delta a_{20} \sin \gamma_0 & 2(\alpha a_{10}^2 + \delta a_{20}^2)\cos \gamma_0 \end{bmatrix} \qquad (4.5.31)$$

If $(a_{10}, a_{20}, \gamma_0)$ is stable (unstable), the corresponding synchronized state of (4.3.1) and (4.3.2) is stable (unstable).

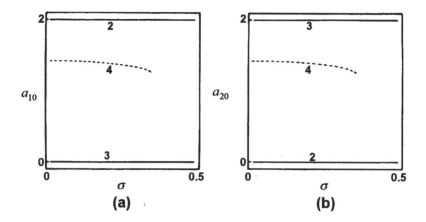

Figure 4.5.1: Response curves for (4.5.4), (4.5.5), and (4.5.9).

In Figure 4.5.1, we show the bifurcation diagram when σ is used as a control parameter while $\alpha = \delta = 1$. The branches of stable and unstable fixed points of (4.5.4), (4.5.5), and (4.5.9) are represented by solid and broken lines, respectively. The different curves are symmetric with respect to the vertical axis. The fixed points of the fourth type, which were numerically found to exist for $\sigma < \sigma_c \simeq 0.353553$, are saddle points. Consequently, the corresponding synchronized solutions of (4.3.1) and (4.3.2) are of the saddle type and unstable. The fixed points of the second and third types that exist for $\mid \sigma \mid \leq 0.5$ are all stable, and the corresponding periodic solutions of (4.3.1) and (4.3.2) are stable. For a value of $\sigma < \sigma_c$, the stable manifold of the fixed point of the fourth type is two–dimensional, while the unstable manifold is one–dimensional. In the (a_1, a_2, γ) space, the stable manifold separates the basins of attraction of the fixed points $(a_{10}, a_{20}) = (0, 2)$ and $(a_{10}, a_{20}) = (2, 0)$. Numerical simulations conducted with (4.3.1) and (4.3.2) for $\epsilon = 0.001$, $\omega_1 = 1$, $\delta = 1$, and $\alpha = 1$ confirm the presence of unstable synchronized solutions and other stable periodic solutions in agreement with the predictions of the perturbation analysis.

In another case, we set $\sigma = 0.35$ and $\alpha = 1$ and varied δ in the range $1 \leq \delta < 5$. The fixed points of the fourth type are found to be unstable, and the corresponding synchronized solutions of (4.3.1) and (4.3.2) are unstable as well. To gain a deeper understanding of the solutions of

(4.3.1) and (4.3.2) requires further simulations and global analyses.

Example 4.12. We consider a primary resonance excitation of the Rayleigh oscillator; that is,

$$\ddot{u} + \omega_0^2 u = \epsilon\left(\dot{u} - \frac{1}{3}\dot{u}^3\right) + \epsilon F \cos \Omega t \qquad (4.5.32)$$

where ϵ is a small positive parameter and $\Omega \approx \omega_0$. To describe the nearness of Ω to ω_0, we introduce the detuning parameter σ defined by

$$\Omega = \omega_0 + \epsilon\sigma \qquad (4.5.33)$$

We use the method of multiple scales and seek a first–order uniform expansion of (4.5.32) in the form

$$u(t; \epsilon) = u_0(T_0, T_1) + \epsilon u_1(T_0, T_1) + \cdots \qquad (4.5.34)$$

where $T_0 = t$ and $T_1 = \epsilon t$. Substituting (4.5.34) and (3.5.4) into (4.5.32) and equating coefficients of like powers of ϵ, we obtain

$O(\epsilon^0)$:

$$D_0^2 u_0 + \omega_0^2 u_0 = 0 \qquad (4.5.35)$$

$O(\epsilon)$:

$$D_0^2 u_1 + \omega_0^2 u_1 = -2D_0 D_1 u_0 + D_0 u_0 - \frac{1}{3}(D_0 u_0)^3 + F \cos \Omega T_0 \quad (4.5.36)$$

where $D_n = \partial/\partial T_n$.

The solution of (4.5.35) can be expressed as

$$u_0 = A(T_1)e^{i\omega_0 T_0} + \bar{A}(T_1)e^{-i\omega_0 T_0} \qquad (4.5.37)$$

where the function $A(T_1)$ is determined by eliminating the secular terms from u_1.

Substituting (4.5.37) into (4.5.36) yields

$$D_0^2 u_1 + \omega_0^2 u_1 = -i\omega_0 \left(2A' - A + \omega_0^2 A^2 \bar{A} \right) e^{i\omega_0 T_0}$$
$$+ \frac{1}{3} i\omega_0^3 A^3 e^{3i\omega_0 T_0} + \frac{1}{2} F e^{i\Omega T_0} + cc \tag{4.5.38}$$

where cc stands for the complex conjugate of the preceding terms, and
the prime indicates the derivative with respect to T_1. Using (4.5.33) in
eliminating the terms that lead to secular terms in u_1 from (4.5.38), we
obtain

$$2A' = A - \omega_0^2 A^2 \bar{A} - \frac{iF}{2\omega_0} e^{i\sigma T_1} \tag{4.5.39}$$

Expressing A in the polar form $A = \frac{1}{2} a \exp[i(\beta + \sigma T_1)]$, where a and β
are functions of T_1, and separating real and imaginary parts in (4.5.39),
we have

$$a' = \frac{1}{2} a - \frac{1}{8} \omega_0^2 a^3 - \frac{F}{2\omega_0} \sin \beta \tag{4.5.40}$$

$$a\beta' = -\sigma a - \frac{F}{2\omega_0} \cos \beta \tag{4.5.41}$$

Substituting the polar form of A into (4.5.37) and then substituting the
result into (4.5.34), we find that, to the first approximation, u is given
by

$$u = a \cos(\omega_0 t + \epsilon \sigma t + \beta) + \cdots$$

or

$$u = a \cos(\Omega t + \beta) + \cdots \tag{4.5.42}$$

on account of (4.5.33), where a and β are defined by (4.5.40) and
(4.5.41).

In the case of free oscillations, $F = 0$ and (4.5.40) and (4.5.41)
reduce to

$$a' = \frac{1}{2} a - \frac{1}{8} \omega_0^2 a^3 \tag{4.5.43}$$

$$a\beta' = -\sigma a \tag{4.5.44}$$

For nontrivial solutions, $a \neq 0$, and it follows from (4.5.44) that

$$\beta = -\sigma T_1 + \beta_0 = -\epsilon \sigma t + \beta_0 \tag{4.5.45}$$

where β_0 is a constant. Substituting (4.5.45) into (4.5.42), we find that, to the first approximation, the free oscillations of (4.5.32) are given by

$$u = a\cos(\omega_0 t + \beta_0) + \cdots \tag{4.5.46}$$

where a is given by (4.5.43), which has the normal form of a supercritical pitchfork bifurcation. Using separation of variables, we find that the solution of (4.5.43) is

$$a = \left[\left(\frac{1}{\hat{a}_0^2} - \frac{1}{4}\omega_0^2\right)e^{-\epsilon t} + \frac{1}{4}\omega_0^2\right]^{-1/2}$$

where $\hat{a}_0 \neq 0$ is the initial value of a. Consequently, to the first approximation, the free oscillations of (4.5.32) are given by

$$u = \left[\left(\frac{1}{\hat{a}_0^2} - \frac{1}{4}\omega_0^2\right)e^{-\epsilon t} + \frac{1}{4}\omega_0^2\right]^{-1/2}\cos(\omega_0 t + \beta_0) + \cdots \tag{4.5.47}$$

It follows from (4.5.47) that, as $t \to \infty$,

$$u \to \frac{2}{\omega_0}\cos(\omega_0 t + \beta_0) + \cdots \tag{4.5.48}$$

for all values of $\hat{a}_0 \neq 0$. In other words, when $F = 0$, (4.5.32) represents a **self–excited oscillator**.

When $F \neq 0$, one would expect the response to contain the basic frequencies Ω and ω_0; that is, the excitation frequency and the self–excited frequency. As shown below, this is so when Ω is not close to ω_0 (i.e., σ is not small). In this case, if Ω and ω_0 are not commensurate, the response is two–period quasiperiodic. However, when σ is small, the response is periodic and is characterized by the frequency Ω. In other words, the free–oscillation component of the response is **entrained** by the forced component, resulting in a synchronized response. Such periodic responses correspond to the fixed points (a_0, β_0) of (4.5.40) and (4.5.41). Setting $a' = 0$ and $\beta' = 0$ in (4.5.40) and (4.5.41), we find that their fixed points are given by

$$\frac{1}{2}a_0 - \frac{1}{8}\omega_0^2 a_0^3 = \frac{F}{2\omega_0}\sin\beta_0 \tag{4.5.49}$$

$$-\sigma a_0 = \frac{F}{2\omega_0}\cos\beta_0 \tag{4.5.50}$$

Squaring and adding (4.5.49) and (4.5.50) yield the frequency–response equation

$$\rho(1 - \rho)^2 + 4\sigma^2\rho = \frac{1}{4}F^2 , \quad \rho = \frac{1}{4}\omega_0^2 a_0^2 \tag{4.5.51}$$

The stability of the fixed points depends on the eigenvalues of the Jacobian matrix of (4.5.40) and (4.5.41); that is, the eigenvalues of

$$A = \begin{bmatrix} \frac{1}{2} - \frac{3}{8}\omega_0^2 a_0^2 & -\frac{F}{2\omega_0}\cos\beta_0 \\ \frac{F}{2\omega_0 a_0^2}\cos\beta_0 & \frac{F}{2\omega_0 a_0}\sin\beta_0 \end{bmatrix} \tag{4.5.52}$$

They are given by

$$\lambda^2 - (1 - 2\rho)\lambda + \Lambda = 0 \tag{4.5.53}$$

where

$$\Lambda = \frac{1}{4}(1 - 4\rho + 3\rho^2) + \sigma^2 \tag{4.5.54}$$

When $\Lambda < 0$, the roots of (4.5.53) are real and have different signs; hence, the fixed point is a saddle. When $\Lambda = 0$, one of the eigenvalues is zero and hence the fixed point is nonhyperbolic. When $\Lambda > 0$,

$$\lambda = \frac{1}{2} - \rho \pm \sqrt{\frac{1}{4}\rho^2 - \sigma^2} \tag{4.5.55}$$

Hence, the fixed point is a stable node if $\rho^2 \geq 4\sigma^2$ and $\rho > \frac{1}{2}$ and is an unstable node if $\rho^2 \geq 4\sigma^2$ and $\rho < \frac{1}{2}$. On the other hand, the fixed point is a focus if $\rho^2 < 4\sigma^2$, which is stable if $\rho > \frac{1}{2}$ and unstable if $\rho < \frac{1}{2}$. When $\rho^2 < 4\sigma^2$ and $\rho = \frac{1}{2}$, the two eigenvalues are purely imaginary and the fixed point is a center.

In Figure 4.5.2, we display the regions of the different fixed points in the $\rho - \sigma$ plane. The broken lines $\Lambda = 0$ and $\rho = \frac{1}{2}$ represent the loci of the nonhyperbolic fixed points, while the solid line $\rho = 2\sigma$ has been included to demarcate the different regions. Next, we examine if these nonhyperbolic fixed points are bifurcation points when σ is used as a control parameter. By using (4.5.51) and (4.5.53), we find that

$$\frac{d\rho}{d\sigma} = -\frac{8\sigma\rho}{1 + 3\rho^2 - 4\rho + 4\sigma^2} = -\frac{2\sigma\rho}{\Lambda} \tag{4.5.56}$$

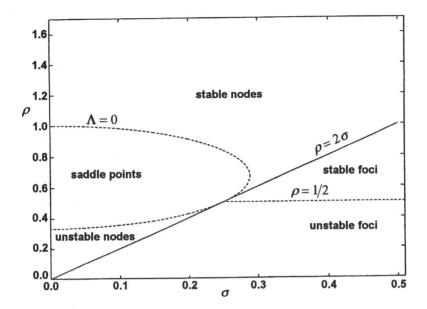

Figure 4.5.2: Classification of fixed points of (4.5.40) and (4.5.41).

From (4.5.56), we deduce that the curve $\Lambda = 0$ is the locus of tangent bifurcation points. When $\rho^2 < 4\sigma^2$, it follows from (4.5.55) that

$$\lambda_r = \frac{1}{2} - \rho \quad \text{and} \quad \lambda_i = \sqrt{\sigma^2 - \frac{1}{4}\rho^2}$$

Therefore,

$$\frac{d\lambda_r}{d\sigma} = \frac{d\lambda_r}{d\rho}\frac{d\rho}{d\sigma} = -\frac{d\rho}{d\sigma} = \frac{2\sigma\rho}{\Lambda}$$

which, at $\rho = \frac{1}{2}$, becomes

$$\frac{d\lambda_r}{d\sigma} = -\frac{16\sigma}{16\sigma^2 - 1} \neq 0$$

Consequently, $\rho = \frac{1}{2}$ represents the locus of Hopf bifurcation points. The point $(\sigma, \rho) = (0.25, 0.5)$, where the loci of tangent and Hopf bifurcation points meet, represents a codimension–two bifurcation point because both roots of (4.5.53) are zero.

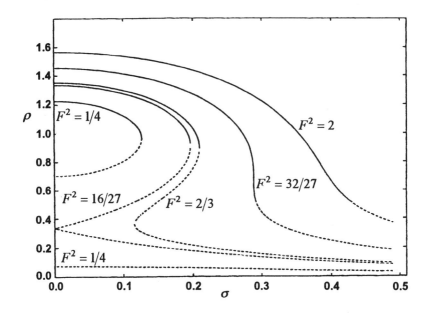

Figure 4.5.3: Response curves for (4.5.40) and (4.5.41).

In Figure 4.5.3, we show a family of frequency–response curves. These curves are symmetric with respect to the ρ–axis. The solid and broken lines represent stable and unstable fixed points, respectively. For free oscillations, we set $F = 0$ and find that the response curve is made up of the lines $\rho = 0$ and $\rho = 1$. As F is increased, the frequency–response curve consists of two branches: an unstable branch running near the σ–axis and a closed oval formed by stable and unstable branches. This oval can be approximated by an ellipse whose center is at $(\sigma, \rho) = (0, 1)$. This structure can be seen for $F^2 = \frac{1}{4}$ in the figure. As F is increased further, the oval expands while the branch near the σ–axis moves away from the axis. When F^2 reaches the critical value $\frac{16}{27}$, there is a **double point** or **cusp** at $(\sigma, \rho) = (0, \frac{1}{3})$. For $F^2 = \frac{2}{3}$, we note that the response curve has **two turning points**. At one of them, an unstable branch and a stable branch meet. Consequently, this turning point is a saddle–node bifurcation point. At the other turning point, two unstable branches meet. This turning point is also a bifurcation point because locally there are no solutions on one side

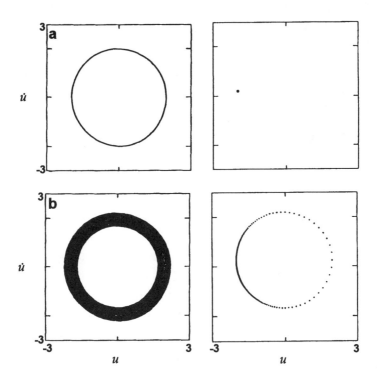

Figure 4.5.4: Orbits and Poincaré sections of attractors of (4.5.32): (a) $\omega_0 = 1$, $\epsilon = 0.1$, $F^2 = \frac{1}{4}$, $\sigma = 0.120$ and (b) $\omega_0 = 1$, $\epsilon = 0.1$, $F^2 = \frac{1}{4}$, $\sigma = 0.122$.

of the bifurcation point. However, the associated bifurcation is not a saddle–node bifurcation. As F is increased further, the response curve continues to be multivalued until F^2 exceeds the second critical value $\frac{32}{27}$, beyond which the frequency–response curve becomes single–valued. Further, there is a Hopf bifurcation point on this response curve. The frequency–response curve for $F^2 = 1$, which is not shown in Figure 4.5.3, passes through the degenerate point $(\sigma, \rho) = (0.25, 0.5)$.

For $F^2 = \frac{1}{4}$, let us start from the point $(\sigma, \rho) = (0, 1.2258)$ on the stable branch of fixed points and gradually increase σ. Then, the fixed point of (4.5.40) and (4.5.41) experiences a saddle–node bifurcation

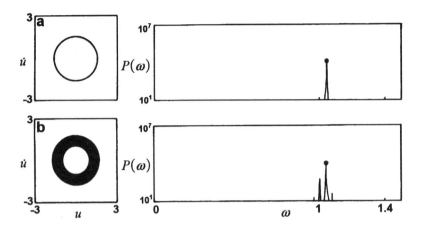

Figure 4.5.5: Orbits and power spectra for attractors of (4.5.32): (a) $\omega_0 = 1$, $\epsilon = 0.1$, $F^2 = 2$, $\sigma = 0.41$; and (b) $\omega_0 = 1$, $\epsilon = 0.1$, $F^2 = 2$, $\sigma = 0.45$.

at the critical value $\sigma_c \simeq 0.126$. The corresponding phase–locked solution of (4.5.32) experiences a cyclic–fold bifurcation. To examine the consequence of this bifurcation, we conducted numerical simulations of (4.5.32) for $\omega_0 = 1$, $\epsilon = 0.1$, and $F^2 = \frac{1}{4}$. For these parameter values, a cyclic–fold bifurcation occurs at about $\sigma_c \simeq 0.121$. In Figures 4.5.4a and 4.5.4b, we show the attractors realized at $\sigma = 0.120$ and $\sigma = 0.122$, respectively. The corresponding Poincaré sections obtained by using the excitation frequency as the clock frequency are also shown. The attractor preceding the bifurcation is periodic, while the attractor subsequent to the bifurcation is two–period quasiperiodic. Consequently, synchronism is lost due to a cyclic–fold bifurcation.

For $F^2 = 2$, when σ is gradually increased from zero, the perturbation analysis predicts that a loss of synchronism will occur at $\sigma_c \simeq 0.43$. Numerical simulations conducted for $\epsilon = 0.1$, $\omega_0 = 1$, and $F^2 = 2$ confirm this prediction and reveal that the phase–locked solution loses stability due to a supercritical Neimark bifurcation. In Figures 4.5.5a and 4.5.5b, we show the periodic and two–period quasiperiodic attractors realized at $\sigma = 0.41$ and $\sigma = 0.45$, respectively. The corresponding power spectra are also shown. In Figure 4.5.5b, there is a dominant peak at $\omega = 1.045$ with uniformly spaced sidebands surrounding it. This structure is indicative of a **modulated motion**, as discussed in

Section 7.5. Here, the basic frequencies are 1.045 and the modulation frequency. The latter frequency, which is given by the sideband spacing, is produced by the Neimark bifurcation.

In the context of forced oscillations of the van der Pol oscillator, the classic analytical studies of Cartwright and Littlewood (1945) and Levinson (1949) are worth noting. Further, Nayfeh and Mook (1979, Chapter 4) conducted an extensive study of synchronized oscillations of the forced Rayleigh oscillator for different excitation conditions. A partial list of other studies on the forced van der Pol oscillator includes Holmes and Rand (1978), Levi (1981), Shaw (1981), Guckenheimer and Holmes (1983), Abraham and Scott (1985), Abraham and Simo (1986), and Thompson and Stewart (1986).

4.6 EXERCISES

4.1. Consider the three–dimensional system (Langford, 1985)

$$\dot{x}_1 = (\lambda - b)x_1 - cx_2 + x_1x_3 + dx_1(1 - x_3^2)$$
$$\dot{x}_2 = cx_1 + (\lambda - b)x_2 + x_2x_3 + dx_2(1 - x_3^2)$$
$$\dot{x}_3 = \lambda x_3 - (x_1^2 + x_2^2 + x_3^2)$$

when $b = 3.0, c = 0.25$, and $d = 0.2$.

(a) Show that for $\lambda > 0$ and small, there is an asymptotically stable equilibrium point, with a positive x_3 near the origin.

(b) Show that at $\lambda \approx 1.68$ the equilibrium point undergoes a super-critical Hopf bifurcation, resulting in the birth of a stable limit cycle.

(c) Show that for $\lambda > 2.0$ the limit cycle undergoes a secondary Hopf bifurcation, resulting in the birth of a two–period quasiperiodic solution.

4.2. Consider the following set of coupled van der Pol oscillators

$$\ddot{u}_1 + \omega_1^2 u_1 = \epsilon(\alpha_1 - \alpha_2 u_1^2)\dot{u}_1 + \epsilon\alpha_3 u_2$$

$$\ddot{u}_2 + \omega_2^2 u_2 = \epsilon(\alpha_1 - \alpha_2 u_2^2)\dot{u}_2 + \epsilon\alpha_4 u_4$$

where $\omega_2 = \omega_1 + \epsilon\sigma$ and ϵ is a small parameter. Show that to the first approximation

$$u_1 \approx A_1(T_1)e^{i\omega_1 T_0} + cc \text{ and } u_2 \approx A_2(T_1)e^{i\omega_2 T_0} + cc$$

where

$$2i\omega_1 A_1' = i\omega_1(\alpha_1 - \alpha_2 A_1 \bar{A}_1)A_1 + \alpha_3 A_2 e^{i\sigma T_1}$$
$$2i\omega_2 A_2' = i\omega_2(\alpha_1 - \alpha_2 A_2 \bar{A}_2)A_2 + \alpha_4 A_1 e^{-i\sigma T_1}$$

Use the modulation equations to determine the synchronized solutions and their stability.

4.3. Consider the following set of coupled van der Pol oscillators used by Linkens (1974) to study electrical activity in gastrointestinal tracts of humans and animals:

$$\ddot{x}_1 + \omega_1^2(x_1 + \gamma x_2) - \epsilon\left[1 - (x_1 + \gamma x_2)^2\right]\dot{x}_1 = 0$$
$$\ddot{x}_2 + \omega_2^2(x_2 + \gamma x_1) - \epsilon\left[1 - (x_2 + \gamma x_1)^2\right]\dot{x}_2 = 0$$

For $\omega_2 \simeq \omega_1$ and $\epsilon = 0.1$, analytically construct the phase–locked solutions of this system and examine their stability.

4.4. Consider the sinusoidally forced Rayleigh oscillator

$$\ddot{u} + \omega^2 u = \epsilon\left(1 - \frac{1}{3}\dot{u}^2\right)\dot{u} + K\cos(\Omega t)$$

Analytically construct the synchronized solutions and determine their stability when $\Omega \approx 3\omega$ and when $\Omega \approx \frac{1}{3}\omega$.

4.5. The response of a van der Pol oscillator with delayed amplitude limiting to a sinusoidal excitation is governed by

$$\ddot{u} + \omega^2 u = 2\epsilon\left[(1 - z)\dot{u} - \dot{z}u\right] - 2\epsilon k\Omega\sin(\Omega t\tau)$$
$$\dot{z} + z = u^2$$

where ω, ϵ, Ω, and τ are constants with τ being positive and away from zero. For the case of primary resonance (i.e., $\Omega = \omega + \epsilon\sigma$,) show that to the first approximation (Nayfeh, 1968)

$$u \approx a \cos(\omega t + \beta)$$

where

$$\dot{a} = \epsilon(1 - \frac{1}{4}\alpha_r a^2)a + \epsilon k \cos(\epsilon\sigma t - \beta)$$

$$\dot{\beta} = -\frac{1}{4}\epsilon\alpha_i a^2 + \frac{\epsilon k}{a}\sin(\epsilon\sigma t - \beta)$$

and $\alpha = \alpha_r + i\alpha_i$ is a function of ω and τ. Use the modulation equations to determine the synchronized motions and their stability.

4.6. Consider the periodically forced van der Pol oscillator:

$$\dot{x}_1 = x_2$$
$$\dot{x}_2 = -(x_1^2 - 1)x_2 - x_1 + \cos(\sqrt{3}t)$$

Numerically verify that the solution of this system settles on a two-period quasiperiodic orbit after transients die out. Construct a Poincaré section for this orbit, and determine the corresponding rotation number.

4.7. Consider the following quasiperiodically forced oscillator:

$$\dot{x}_1 = x_2$$
$$\dot{x}_2 = -x_1 - x_1^3 - 0.4x_2 + \left[1 + \cos\left(\sqrt{2}t\right)\right]\cos t$$

Numerically verify that the solution of this system settles on a two-period quasiperiodic orbit after transients die out. Construct a Poincaré section for this orbit, and determine the corresponding rotation number.

4.8. A damped and driven pendulum is governed by

$$\ddot{\theta} + 2\mu\dot{\theta} + \sin\theta = F$$

Use central–difference schemes to approximate the time derivatives as in Exercise 2.58 and examine the resulting system in the context of the circle map (4.2.1).

4.9. Consider the following linearization obtained near a Hopf bifurcation point of a two–dimensional map:

$$\begin{bmatrix} x_{k+1} \\ y_{k+1} \end{bmatrix} = \begin{bmatrix} \mu & -\omega \\ \omega & \mu \end{bmatrix} \begin{bmatrix} x_k \\ y_k \end{bmatrix}$$

where $\sqrt{\mu^2 + \omega^2} = 1$. Note that this map is an orientation preserving map.

(a) Show that a circle of a given radius is invariant under the action of this map.

(b) Discuss when periodic and nonperiodic solutions of this map are possible.

4.10. Consider the circle map (4.2.1). Show that the boundaries of the Arnold tongue within which $\rho = \frac{1}{3}$ is given by

$$\Omega = \frac{1}{3} \pm \frac{\sqrt{3}\pi}{6} \epsilon^2 \pm \frac{\sqrt{7}\pi}{6} \epsilon^3 + \cdots$$

where $\epsilon = K/2\pi$.
Hint: It follows from (4.2.3) that $\rho = \frac{1}{3}$ if and only if

$$F^3(x) - (x + 1) = 0$$

It follows from (4.2.7) and (4.2.8) that

$$F^3(x) = x + 3\Omega - \epsilon \sin 2\pi x - \epsilon \sin[2\pi(x + \Omega - \epsilon \sin 2\pi x)]$$
$$-\epsilon \sin\{2\pi [x + 2\Omega - \epsilon \sin 2\pi x - \epsilon \sin 2\pi(x + \Omega - \epsilon \sin 2\pi x)]\}$$

Let $\Omega = \frac{1}{3} + \beta$ and show that

$$3\beta - \epsilon \sin 2\pi x - \epsilon \sin \left[2\pi \left(x + \frac{1}{3} + \beta - \epsilon \sin 2\pi x \right) \right]$$

$$-\epsilon \sin \left\{ 2\pi \left[x + \frac{2}{3} + 2\beta - \epsilon \sin 2\pi x \right. \right.$$

$$\left. \left. - \epsilon \sin 2\pi \left(x + \frac{1}{3} + \beta - \epsilon \sin 2\pi x \right) \right] \right\} = 0$$

Let $\beta = \epsilon^2 \beta_2(x) + \epsilon^3 \beta_3(x) + \cdots$, expand for small ϵ, and determine $\beta_2(x)$ and $\beta_3(x)$. Find the infimum and supremum of $\beta_2(x)$ and $\beta_3(x)$ to determine the boundaries of the tongue.

4.11. If

$$\eta = \exp \left[\frac{2\pi i p}{q} \right]$$

where p and q are positive integers with no common factors, show that

$$\sum_{k=1}^{q-1} \eta^k = -1$$

Then, show that

$$\sum_{k=1}^{q-1} \sin \left[2\pi \left(x + \frac{kp}{q} \right) \right] = -\sin 2\pi x$$

4.12. If

$$\eta = \exp \left[\frac{2\pi i p}{q} \right]$$

where p and q are positive integers without common factors and $q \geq 3$, show that

$$\sum_{k=1}^{q-1} \eta^k (\eta^k - 1) = 0$$

$$\sum_{k=1}^{q-1} \eta^{-k} (\eta^k - 1) = q$$

4.13. Consider the circle map (4.2.1). Show that the boundaries of the Arnold tongue within which $\rho = p/q$ is given by

$$\Omega = \frac{p}{q} + \frac{\pi \sin(2\pi p/q)}{2(1 - \cos 2\pi p/q)}\epsilon^2 + \cdots$$

where $\epsilon = K/2\pi$. Then, show that

$$\Omega = \frac{1}{4} + \frac{\pi}{2}\epsilon^2 + \cdots$$

$$\Omega = \frac{3}{4} - \frac{\pi}{2}\epsilon^2 + \cdots$$

$$\Omega = \frac{1}{3} + \frac{\sqrt{3}\pi}{6}\epsilon^2 + \cdots$$

$$\Omega = \frac{2}{3} - \frac{\sqrt{3}\pi}{6}\epsilon^2 + \cdots$$

Hint: Show that

$$F^q(x) = x + p + \beta q - \epsilon \sum_{k=0}^{q-1} \sin[2\pi F^k(x)]$$

where $\Omega = \frac{p}{q} + \beta$. Then, show that

$$F^q(x) = x + p + \beta q - \epsilon \sum_{k=0}^{q-1} \sin \left\{ 2\pi \left(x + \frac{kp}{q} + k\beta \right) \right.$$
$$\left. - \epsilon \sum_{l=0}^{k-1} \sin[2\pi F^l(x)] \right\}$$

Expand for small ϵ and β and obtain

$$F^q(x) = x + p + \beta q - \epsilon \sin 2\pi x - \epsilon \sum_{k=1}^{q-1} \sin \left[2\pi \left(x + \frac{kp}{q} \right) \right]$$
$$- 2\epsilon\pi\beta \sum_{k=1}^{q-1} k \cos \left[2\pi \left(x + \frac{kp}{q} \right) \right]$$
$$+ 2\pi\epsilon^2 \sum_{k=1}^{q-1} \cos \left[2\pi \left(x + \frac{kp}{q} \right) \right] \sum_{l=0}^{k-1} \sin \left[2\pi \left(x + \frac{lp}{q} \right) \right]$$
$$+ O(\epsilon^3, \beta^2, \epsilon^2\beta)$$

Then use Exercise 4.11 to show that

$$\beta = -\frac{2\pi\epsilon^2}{q} \sum_{k=1}^{q-1} \cos\left[2\pi\left(x + \frac{kp}{q}\right)\right] \sum_{l=0}^{k-1} \sin\left[2\pi\left(x + \frac{kp}{q}\right)\right]$$

Finally, use Exercise 4.12 to determine the summations.

Chapter 5

CHAOS

In this chapter, we explore **chaotic solutions** of maps and continuous–time systems. These solutions are also bounded like equilibrium, periodic, and quasiperiodic solutions. There is no precise definition for a chaotic solution because it cannot be represented through standard mathematical functions. However, a chaotic solution is an aperiodic solution, which is endowed with some special identifiable characteristics. From a practical point of view, **chaos** can be defined as a bounded steady–state behavior that is not an equilibrium solution or a periodic solution or a quasiperiodic solution. The attractor associated with a chaotic motion in state space is not a simple geometrical object like a finite number of points, a closed curve, or a torus. In fact, it is not even a smooth surface; that is, it is not a manifold. Chaotic attractors are complicated geometrical objects that possess fractal dimensions.

In contrast with the spectra of periodic and quasiperiodic attractors, which consist of a number of sharp spikes, the spectrum of a chaotic signal has a continuous broadband character. In addition to the broadband component, the spectrum of a chaotic signal often contains spikes that indicate the predominant frequencies of the signal. A chaotic motion is the superposition of a very large number of unstable periodic motions. Thus, a chaotic system may dwell for a brief time on a motion that is very nearly periodic and then may change to another periodic motion with a period that is k times that of the preceding motion. This constant evolution from one periodic motion to another produces a long–time impression of randomness while showing short–

term glimpses of order.

Chaotic systems are also characterized by sensitivity to initial conditions; that is, tiny differences in the input can be quickly amplified to create overwhelming differences in the output. This is the so–called **butterfly effect**. A small perturbation created by the wings of a butterfly today in Beijing, China, can produce a torrential rainstorm next month in California.

In Section 5.1, we discuss chaotic solutions of maps, and in Section 5.2 we discuss chaotic solutions of continuous–time systems. In Sections 5.3–5.5 we discuss the period–doubling, intermittency, and quasiperiodic routes to chaos. In Section 5.6 we discuss crises, in Section 5.7 we discuss the Melnikov theory of homoclinic and heteroclinic tangles, and in Section 5.8 we discuss bifurcations of homoclinic orbits.

5.1 MAPS

We have discussed so far three classes of solutions of a map, namely, fixed–points, periodic points, and dense sets of points on closed loops. The orbit of a fixed point is the fixed point itself, while the orbit of a period–k point is a set of k discrete points. When the iterates of a map are densely packed on a closed loop, the orbit of each point on this loop is the closed loop. Hence, in the above three classes, the associated orbits are well defined and in a sense regular. Another class of solutions of a map is what is called **chaotic solutions**. These solutions are characterized by erratic orbits and some special features as illustrated by the examples given below.

Example 5.1. For $0 < \alpha \leq 1$, the one–dimensional noninvertible map (1.1.4) maps the unit interval onto itself. In Figure 5.1.1, the orbits of the attractors of (1.1.4) for three different values of α are shown. It is clear that there is a period–one solution at $\alpha = 0.7$ and two period–two solutions at $\alpha = 0.8$. The orbit of Figure 5.1.1c, which is irregular, corresponds to an **aperiodic** or a **chaotic attractor**.

To determine how the chaotic solution arises, we examine the bifurcations experienced by the solutions of (1.1.4) when α is used as a

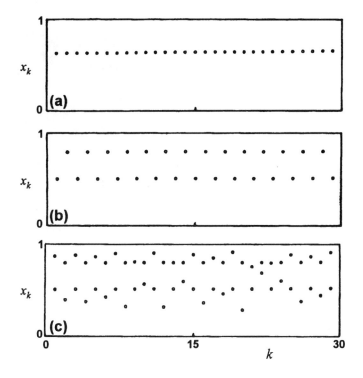

Figure 5.1.1: Orbits of the logistic map: (a) $\alpha = 0.7$, (b) $\alpha = 0.8$, and (c) $\alpha = 0.92$.

control parameter. As α is gradually increased from 0.7, the nontrivial fixed point of (1.1.4) experiences a period–doubling bifurcation at $\alpha_1 = 0.75$, giving rise to two stable period–two solutions. As α is increased further, each period–two solution experiences a period–doubling bifurcation at $\alpha_2 \simeq 0.8623795$. This results in the birth of four stable period–four solutions. Each of these periodic solutions experiences a period–doubling bifurcation at $\alpha_3 \simeq 0.8860225$. As α is increased further, period–doubling bifurcations occur at $\alpha_4 \simeq 0.8911018$, and so on. With each successive bifurcation, the period of the realized attractor increases. Eventually, the sequence culminates in an infinite–period or aperiodic attractor at $\alpha_\infty \simeq 0.8924864$. The point α_∞ is an accumulation point of the period–doubling sequence. This aperiodic solution continues for $\alpha > \alpha_\infty$. The sequence of period–

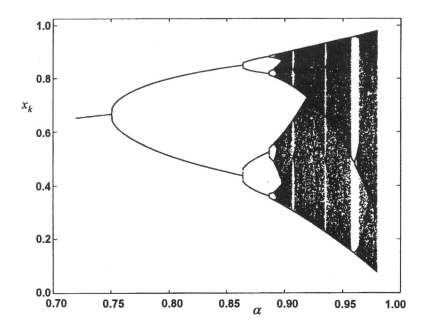

Figure 5.1.2: Bifurcation diagram for the logistic map.

doubling bifurcations leading up to the chaotic solution is bifurcation illustrated in Figure 5.1.2. In this numerically constructed diagram, the unstable solutions are not shown.

For a smooth single–hump map $F(x)$ with continuous first derivative and nonvanishing second derivative at its extremum (i.e., has a quadratic maximum), such as (1.1.4), Feigenbaum (1978) showed that the sequence of period–doubling control–parameter values scales according to the law

$$\lim_{k \to \infty} \frac{\alpha_k - \alpha_{k-1}}{\alpha_{k+1} - \alpha_k} = \delta = 4.66292016 \cdots \qquad (5.1.1)$$

The important result is that δ is the same for all period–doubling sequences associated with smooth maps having a quadratic maximum; that is, δ is universal. This universal constant is called the **Feigenbaum number**.

In Figure 5.1.2, we note that we do not have aperiodic solutions for all $\alpha \geq \alpha_\infty$. In fact, the presence of period–six attractors in a window about $\alpha = 0.91$, period–five attractors in a window about $\alpha = 0.93$, and period–three attractors in a window about $\alpha = 0.96$ are quite clear. The window of the period–three attractors is the largest. A finer resolution will reveal the presence of other **periodic windows**. According to related analyses, periodic windows corresponding to the periods 6, 12, 5, 3, 6, 12, 8, 7, and 5 are expected to occur in that order, with the period–three window being the largest (Metropolis, Stein, and Stein, 1973; May, 1976; Ott, 1981, 1993). Furthermore, it can be numerically verified that the period–three solutions arise through a saddle–node bifurcation at $\alpha \simeq 0.957$ and exist in the interval $0.957 < \alpha < 0.9625$. As α is increased beyond 0.9625, the period–three solutions experience period–doubling bifurcations, as seen in Figure 5.1.2.

Alligood, Yorke, and Yorke (1987) investigated the conditions under which period–doubling cascades occur. The **period–doubling sequence to chaos** has also been observed in many continuous–time systems. A partial list includes the studies of Huberman and Crutchfield (1979), Crutchfield, Farmer, Packard, Shaw, Jones, and Donnelly (1980), Linsay (1981), D'Humieres, Beasley, Huberman, and Libchaber (1982), Novak and Frelich (1982), Testa, Perez, and Jeffries (1982), Swinney (1983), Bergé, Pomeau, and Vidal (1984), Räty, Isomäki, and von Boehm (1984), Thompson and Stewart (1986), Moon (1987), Seydel (1988), Zavodney and Nayfeh (1988), Jackson (1989), Nayfeh and Sanchez (1989), Parker and Chua (1989), Zavodney, Nayfeh, and Sanchez (1989), Nayfeh, Hamdan, and Nayfeh (1990), and Kim and Noah (1991).

Next, we consider two points close to the period–two attractor at $\alpha = 0.8$, which is shown in Figure 5.1.1b. The separation d_0 between these points is 0.1. For the evolutions initiated from these points, the separation d_k varies with the iterate number k, as shown in Figure 5.1.3a. On the average, the decay of d_k to zero follows

$$d_k = d_0 e^{\gamma k} \tag{5.1.2}$$

where γ represents the average slope in the $\log d_k$ versus k graph. Here, γ is obviously negative. The exponent γ, which is determined through (5.1.2), represents the **Lyapunov exponent** of the one–dimensional map (1.1.4). (A detailed discussion of Lyapunov exponents is given in

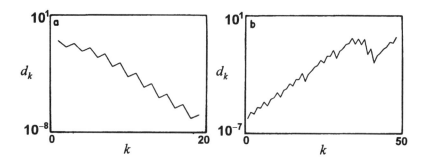

Figure 5.1.3: Sensitivity to initial conditions: (a) $\alpha = 0.8$ and (b) $\alpha = 0.92$.

Section 7.8.) We note that the Lyapunov exponent associated with a periodic orbit of a one–dimensional map is negative.

To construct Figure 5.1.3b, we considered the initial points 0.875660 and 0.875661 close to the chaotic orbit of Figure 5.1.1c at $\alpha = 0.92$. (Note that the initial conditions are chosen close to the attractor, so that the corresponding transient phases are as short as possible.) In this case, $d_0 = 10^{-6}$. The separation d_k increases exponentially for $k < 35$ before leveling off. This leveling off occurs because d_k cannot grow any larger than the size of the unit interval $[0, 1]$. In the intermediate regime, d_k is modeled by (5.1.2), with γ being positive. The exponential separation seen in Figure 5.1.3b between iterates initiated from two close initial conditions is often described as **sensitivity to initial conditions**.

To understand why such exponential separations occur, we illustrate the action of the logistic map for $\alpha \simeq 1$ in Figure 5.1.4. We note that $F(x = 0) = 0$, $F(x = \frac{1}{2}) = \alpha$, and $F(x = 1) = 0$. The interval $[0, \frac{1}{2}]$ is mapped to an interval nearly twice its size; that is $[0, \alpha]$. The interval $[\frac{1}{2}, 1]$ is also mapped to an interval $[0, \alpha]$ nearly twice its size. However, the orientation is reversed. Hence, the action of the map can be interpreted as a two–step process. In the first step, the unit interval is uniformly **stretched** to an interval nearly twice its size. In the second step, the stretched interval is **folded** so that the resulting segment is confined to the unit interval, as illustrated in Figure 5.1.4. The stretching action is responsible for exponential divergence of adjacent orbits. This action alone will lead to an unbounded object. Hence,

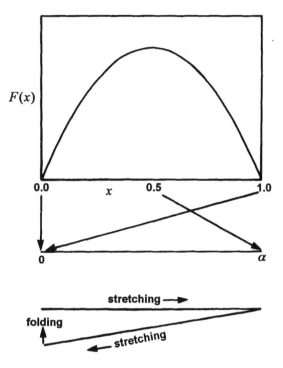

Figure 5.1.4: Illustration of the action of the logistic map when $\alpha \simeq 1$.

the folding action is necessary to keep the attractor bounded. We note that the folding action causes the map to be a two–to–one map and, consequently, a noninvertible map. In other words, it is always possible to go forward from x_n to x_{n+1} but not possible to return from x_{n+1} to its precedent. Because stretching and folding actions are necessary to create a chaotic solution, a one–dimensional invertible map cannot have this solution.

Chaotic solutions are known to occur in one–dimensional and higher noninvertible maps and in two–dimensional and higher invertible maps (Collet and Eckmann, 1980; Ott, 1981, 1993; Eckmann and Ruelle, 1985; Ruelle, 1989a). Further, one or more positive Lyapunov exponents are a distinguishing characteristic of a chaotic solution. An attractor with two or more positive Lyapunov exponents is called **hyperchaos** (Rössler, 1979a). Due to these positive exponents, there is sensitivity to initial conditions and, consequently, orbits initiated from two nearby points on a chaotic attractor move away from each other at an exponential rate until the separation reaches the size of the attractor. From a practical standpoint, this is an important point to reckon with. Because of finite precision and noise, there is always some uncertainty in specifying an initial condition. If this initial condition is located on a chaotic orbit, long–time prediction of the evolution initiated from this point is not possible because the uncertainty gets amplified due to the positive Lyapunov exponents. Hence, despite the fact that a chaotic solution is governed by a deterministic system, finite precision and noise limit one in predicting the observed evolution. Of course, in theory, with infinite precision, a chaotic solution is deterministic and predictable.

Example 5.2. The Hénon map (1.1.5) and (1.1.6) is a two–dimensional invertible map for $\beta \neq 0$. When $\mid \beta \mid < 1$, this map is dissipative and hence contracts areas in the $x - -y$ plane. We choose $\alpha = 1.4$ and $\beta = 0.3$. For these values, $d = \text{Det} [D_{\mathbf{x}}\mathbf{F}] = -0.3$. Because the determinant d is negative, the map is orientation reversing, and because $\mid d \mid < 1$, the map is dissipative. Numerical calculations show that the forward iterates of this map are attracted to the object shown in Figure 5.1.5a. We show 15,000 iterates on the attractor obtained after initiation from $(x, y) = (0.185, 0.191)$. There is a banded structure

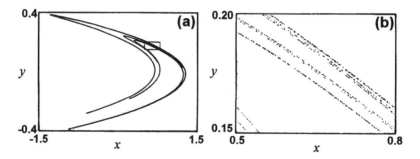

Figure 5.1.5: (a) Attractor of the Hénon map for $\alpha = 1.4$ and $\beta = 0.3$ and (b) enlargement of box shown in a.

with the iterates falling on lines separated by gaps. In Figure 5.1.5b, which is an enlargement of the box shown in Figure 5.1.5a, we see the same structure. When a region of Figure 5.1.5b is enlarged, a similar structure is again observed. Objects such as the attractor of Figure 5.1.5a are examples of **fractal objects** (Mandelbrot, 1977, 1983). The structure of a fractal object is invariant with respect to different scales. Attractors that have fractal geometry are called **strange attractors**.

Next, we choose the points $(0.6000, 0.1940)$ and $(0.599999, 0.194000)$ close to the attractor shown in Figure 5.1.5, initiate evolutions from these points, and obtain the graph shown in Figure 5.1.6. We see that the separation d_k grows exponentially from the initial separation $d_0 = 10^{-6}$ before leveling off at the size of the attractor. For $10 < k < 30$, the growth of d_k follows

$$d_k \simeq d_0 e^{\gamma_1 k}$$

where γ_1 is positive. This exponent characterizes the stretching action of the map. Due to sensitive dependence on initial conditions, the attractor of Figure 5.1.5 is a chaotic attractor. As in this case, in many cases a chaotic attractor is also a strange attractor. Therefore, in the literature, the words **strange** and **chaotic** are often used interchangeably.

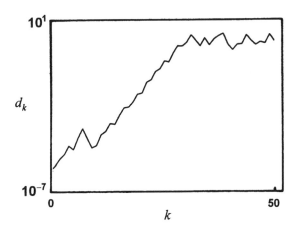

Figure 5.1.6: Sensitivity to initial conditions on the Hénon attractor realized for $\alpha = 1.4$ and $\beta = 0.3$.

Because the Hénon map is two–dimensional, there are two Lyapunov exponents γ_1 and γ_2 associated with an orbit of this map (see Section 7.8). These exponents govern the local growth or decay of an area in the $x - y$ plane. Since the map is dissipative, the iterates of (1.1.5) and (1.1.6) are attracted to a subset of the $x - y$ space. The contracting action of the map is described by γ_2, which is negative. In numerical simulations, the contracting action is observable during the transients leading up to the attractor, as shown in Figure 5.1.7. In one iteration, the rectangle $ABCD$ in Figure 5.1.7a is stretched in the horizontal direction, contracted in the vertical direction, and folded into the horseshoe shown in Figure 5.1.7b. Thus, each iterate of the Hénon map is characterized by stretching, folding, and contracting actions. Repeated stretching, folding, and contracting actions result in the attractor shown in Figure 5.1.5a.

Example 5.3. We consider the two–dimensional dissipative **horseshoe map**, which is due to Smale (1967, 1980). The first iteration of this map, which maps the plane onto itself, is illustrated in Figure 5.1.8a. During this iteration, the unit square $ABCD$ is stretched by a factor greater than two along the vertical direction and contracted by

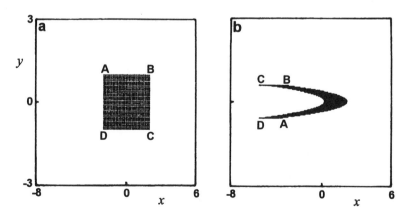

Figure 5.1.7: Illustration of the action of the Hénon map when $\alpha = 1.4$ and $\beta = 0.3$.

a factor larger than two along the horizontal direction into a rectangle. This rectangle is then folded into a horseshoe. Only the rectangular regions V_0 and V_1 of the horseshoe remain inside the unit square. Because a portion of the horseshoe is outside the unit square, the square is not mapped onto itself. (If the stretching factor is less than two, the horseshoe will lie inside the unit square, and the square is mapped onto itself.) Furthermore, since the map is dissipative, the area of the horseshoe is less than one. There are two Lyapunov exponents γ_1 and γ_2 associated with an orbit of the horseshoe map. The exponent γ_1 is positive and associated with the stretching action, while the exponent γ_2 is negative and associated with the contracting action.

Repeated applications of the horseshoe map to the unit square leads to a fractal object. To illustrate this point, in Figure 5.1.8b we show the cross sections along the line FF' for the first two iterations of the

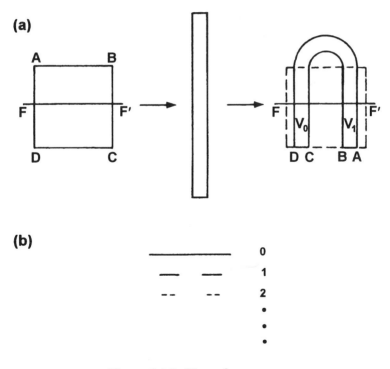

Figure 5.1.8: Horseshoe map.

mapping. Prior to the first iteration, the cross section is a line of unit length. After the first iteration, the cross section consists of two segments with a gap separating them. Following the second iteration, the cross section consists of four segments, which are again separated by gaps. Thus, after the third iteration, we obtain eight segments with gaps between consecutive segments and so on. The fractal object obtained after n iterations can be constructed as follows. To obtain the cross section at the end of the first iteration, we delete portions from the middle and the ends of the unit line to obtain two segments. For the second iteration, we repeat this procedure with each of the two segments obtained at the end of the first iteration to obtain four segments. Thus, repeated applications of this procedure results in a fractal object, which is an example of a **Cantor set**.

The solutions and dynamics of the horseshoe map have been extensively studied and documented (Smale, 1967, 1980; Guckenheimer and Holmes, 1983; Wiggins, 1990). In many rigorous studies, the concept of chaos is built on the basis of this map. In these studies, to prove that the considered continuous–time system exhibits chaos, one shows that an associated Poincaré map exhibits a dynamics qualitatively similar to that of the horseshoe map.

5.2   CONTINUOUS–TIME SYSTEMS

Chaotic solutions represent another class of dynamic solutions besides periodic and quasiperiodic solutions. The orbits of periodic and quasiperiodic solutions are regular, while the orbit of a chaotic solution is erratic. As in the case of maps, there is sensitivity to initial conditions on a chaotic attractor of a continuous–time system. (Poincaré was aware of this sensitivity to initial conditions.) In other words, minute differences in input can quickly be amplified to create overwhelming differences in output. This is usually referred to as the **butterfly effect**; that is, a small perturbation created by the wings of a butterfly today in Beijing, China, can produce a torrential rainstorm next month in California.

Example 5.4. We consider the following three–dimensional system (Rössler, 1976a):

$$\dot{x} = -(y + z) \tag{5.2.1}$$

$$\dot{y} = x + ay \tag{5.2.2}$$

$$\dot{z} = b + (x - c)z \tag{5.2.3}$$

This system, which is characterized by the states x, y, and z, has a quadratic nonlinearity. We set $a = 0.55$, $b = 2$, and $c = 4$ and numerically integrated these equations from $(0.1, 0.1, 0.1)$. After the transients had died out, the solution of this system settled on the attractor shown in Figure 5.2.1a. To examine the cross section of this attractor, we chose the two–sided Poincaré section $z = 4$. The discernible fractal structure in Figure 5.2.1b is indicative of a strange

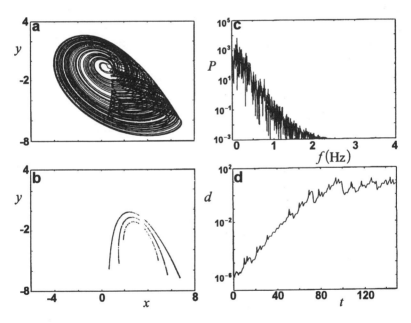

Figure 5.2.1: Attractor of the Rössler equations: (a) two–dimensional projection, (b) two–sided Poincaré section, (c) power spectrum of x, and (d) sensitivity to initial conditions on the attractor.

attractor. The power spectrum associated with the state x, which is displayed in Figure 5.2.1c, has a **broadband character**. In Figure 5.2.1d, we display how the separation d between evolutions initiated from two points separated by $d_0 = 10^{-6}$ varies with time. Both of the initial points are located close to the attractor. The exponential growth of separation for $t < 90$ is clearly noticeable. The separation saturates at the size of the attractor for $t \geq 90$. From Figure 5.2.1, we conclude that there is sensitivity to initial conditions on the attractor. Hence, this attractor is chaotic. In this case, the attractor is both strange and chaotic. The broadband character observed in the power spectrum is another distinguishing characteristic of a chaotic solution.

There are n Lyapunov exponents associated with an orbit of an n-dimensional autonomous system. As explained in Section 7.8, these exponents provide a measure of the orbital divergence in the vicinity of an orbit. Furthermore, one of the exponents associated with a periodic, quasiperiodic, or chaotic attractor is always zero. This corresponds to a direction tangent to the orbit. In the case of an orbit of the Rössler system, there are three Lyapunov exponents. From Figure 5.2.1d, it is clear that one of these exponents γ_1 is positive, indicating a stretching action in the formation of the attractor. Because there is dissipation, one of the other exponents, say γ_2, needs to be negative to account for the accompanying contracting action. The third exponent γ_3 is zero. To illustrate the stretching and folding actions, in Figure 5.2.2, we show the three-dimensional attractor. Because of a noticeable funnel-shaped structure in Figures 5.2.1a and 5.2.2, Abraham and Shaw (1992, Chapter 9) call this attractor the **Rössler funnel**.

For an orbit of a dissipative system, the sum of the Lyapunov exponents is always negative (see Section 7.8). Hence, for a chaotic orbit of an autonomous system, one of the Lyapunov exponents is zero, the sum of the Lyapunov exponents is negative, and one or more of the Lyapunov exponents are positive. This means that chaotic solutions can only occur in three-dimensional and higher autonomous systems. Moreover, for three-dimensional dissipative systems, only one of the Lyapunov exponents can be positive. However, for four- and higher-dimensional dissipative systems, two or more Lyapunov exponents can be positive and the corresponding motions are called **hyperchaos** (Rössler, 1979a). Kapitaniak and Steeb (1991) studied

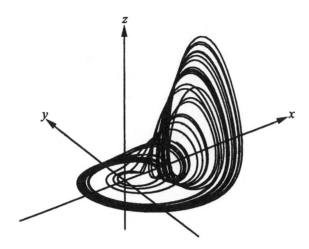

Figure 5.2.2: Three–dimensional plot of the Rössler funnel.

numerically harmonically excited two coupled van der Pol oscillators and found attractors with two positive Lyapunov exponents. In planar autonomous systems, the Poincaré–Bendixson theorem (Coddington and Levinson, 1955; Hale, 1969) rules out the possibility of irregular solutions. However, as illustrated in the next example, chaotic solutions can occur in two–dimensional nonautonomous systems.

Example 5.5. We consider the following parametrically excited nonlinear oscillator (Zavodney, Nayfeh, and Sanchez, 1989):

$$\ddot{x} + \omega_0^2 x + \epsilon \left[2\mu\dot{x} + \delta x^2 + g \cos\left(\Omega t\right)x \right] + \epsilon^2 \alpha x^3 = 0 \qquad (5.2.4)$$

Here, ϵ is a small positive parameter and $\Omega = 2\omega_0 + \epsilon\sigma$. This two–dimensional nonautonomous system can be rewritten as a three–dimensional autonomous system. Further, when $\mu > 0$, the system is dissipative. For $\epsilon = 0.1$, $g = 0$, $\omega_0 = 1$, $\delta = 5.0$, and $\alpha = 4.0$, the system has two stable and one unstable equilibrium positions. Because of the presence of two stable equilibrium positions in the unforced case, this system is referred to as a **two–well potential** system.

If we set $\omega_0 = 1$, $\Omega = 2.0$, and $\mu = 1$ and use g as a control parameter, one of the periodic solutions of this system experiences a sequence of period–doubling bifurcations, culminating in an aperiodic attractor. The periodic attractor realized at $g = 4.0$ is illustrated in Figure 5.2.3a. In Figure 5.2.3b, we show the periodic attractor realized at $g = 4.7$ subsequent to the first period–doubling bifurcation. In Figure 5.2.3c, we display the attractor realized at $g = 5.15$ following the second period–doubling bifurcation. The aperiodic attractor realized at $g = 5.40$, after a sequence of period–doubling bifurcations, is shown in Figure 5.2.3d. The appearance of subharmonics in the power spectrum following each period–doubling bifurcation is clearly noticeable. The power spectrum associated with the aperiodic attractor has a distinctive broadband character.

To construct a Poincaré section for an orbit of the oscillator, we used one–half of the excitation frequency as the sampling frequency. Hence, we obtain one point on this section for the attractor realized before the first period–doubling bifurcation, two points for the attractor realized after the first period–doubling bifurcation, and so on. We illustrate the sequence of period–doubling bifurcations in Figure 5.2.4. In this numerically constructed bifurcation diagram, the discrete points on the Poincaré section of the attractor realized at each value of g are displayed.

To verify that the attractor realized at $g = 5.40$ is chaotic, one needs to show sensitivity to initial conditions on this attractor. To this end, we chose two points separated by $d_0 = 10^{-6}$ close to the attractor and initiated evolutions from them. The variation of the separation d with time t is illustrated in Figure 5.2.5. The separation clearly grows exponentially in the range $110 < t < 310$ before leveling off at the size of the attractor. Consequently, there is a positive Lyapunov exponent γ_1 associated with the aperiodic orbit at $g = 5.40$. The other Lyapunov exponent γ_2 is negative to account for the dissipation.

We note that, in the parameter range $4 < g < 6$, another periodic solution of the oscillator also experiences a sequence of period–doubling bifurcations. Chaotic solutions of two–well potential systems have been extensively studied (e.g., Holmes, 1979; Holmes and Moon, 1983; Moon and Li, 1985; Dowell and Pezeshki, 1988; Zavodney and Nayfeh, 1988; Szemplińska–Stupnicka, Plaut and Hsieh, 1989; Abou–Rayan, Nayfeh, Mook, and Nayfeh, 1993).

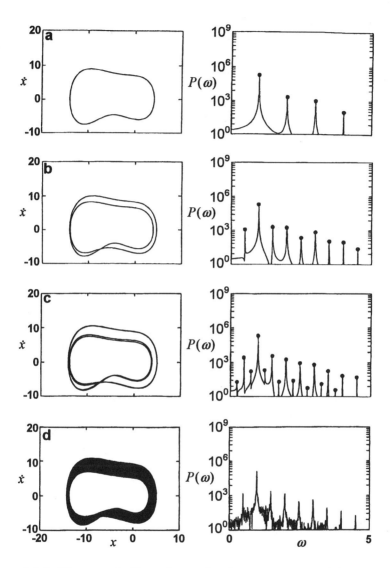

Figure 5.2.3: Phase portraits and associated power spectra: (a) $g = 4.0$, (b) $g = 4.7$, (c) $g = 5.15$, and (d) $g = 5.40$.

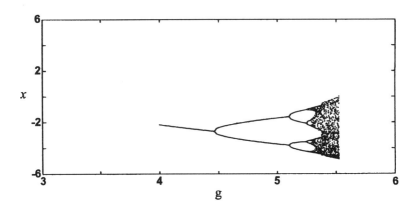

Figure 5.2.4: Illustration of bifurcations on the Poincaré section.

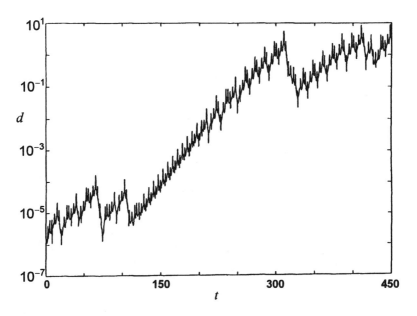

Figure 5.2.5: Sensitivity to initial conditions on the attractor realized at $g = 5.4$.

From the preceding sections, it is clear that a strange attractor has a fractal geometry while a chaotic attractor is characterized by one or more positive Lyapunov exponents. In a general situation, to determine whether an attractor is fractal or not, one needs to compute what is called the **dimension** of an attractor. The dimension of an attractor determines the minimum number of essential variables required to describe the dynamics on the attractor. When the considered attractor is fractal, the dimension is not an integer but a real number. In Section 7.9, we provide some definitions for dimension and discuss methods to compute it.

In the examples we have discussed, the strange attractors are chaotic too. However, this is not always the case. Strange and nonchaotic attractors have been found in quasiperiodically forced systems (Grebogi, Ott, and York, 1985; Awrejcewicz and Rheinhardt, 1990; Brindley and Kapitaniak, 1991a). In these cases, the attractors have a fractal structure but are not associated with any positive Lyapunov exponents. Further, they have some unique spectral characteristics (Romeiras and Ott, 1987). Besides the studies mentioned above, other relevant studies include those of Romeiras, Bondeson, Ott, Antonsen, and Grebogi (1987), Ding, Grebogi, and Ott (1989), Brindley and Kapitaniak (1991b), Brindley, Kapitaniak, and El Naschie (1991), and Kapitaniak (1991).

5.3 PERIOD–DOUBLING SCENARIO

Currently, there are many fairly well–understood transitions to chaotic solutions. Some of these transitions, such as the period–doubling sequence, are associated with local bifurcations, while the other transitions are associated with global bifurcations.

In the period–doubling scenario, as a control parameter is gradually varied, we proceed from a periodic solution to a chaotic solution via a sequence of period–doubling bifurcations. The period–doubling sequence to chaos was first found in the context of one–dimensional noninvertible maps (e.g., Feigenbaum, 1978; Collet and Eckmann, 1980). As discussed in Sections 5.1 and 5.2, it is now known to occur in two–dimensional and higher invertible maps and three–dimensional and higher continuous–time systems.

The period–doubling sequence to chaos has also been observed in many experimental studies. A partial list includes the studies of Maurer and Libchaber (1979), Giglio, Musazzi, and Perine (1981), and Libchaber, Fauve, and Laroche (1982) on the Rayleigh–Bénard convection; Linsay (1981), Testa, Perez, and Jeffries (1982), Van Buskirk and Jeffries (1985), and Matsumoto, Chua, and Tokunaga (1987) on electrical circuits; Jorgensen and Rutherford (1983) on chemical reactions; and Levin, Pompe, Wilke, and Koch (1985) on pendulums. In some of these studies, the appearance of subharmonics in the power spectrum is used to discern the occurrence of a period–doubling bifurcation. In many numerical and experimental studies, the control–parameter values corresponding to the period doubling bifurcations scale according to (5.1.1). On the basis of the first four bifurcations, Giglio, Musazzi, and Perine (1981) obtained $\delta \simeq 4.3$; Linsay (1981) obtained $\delta \simeq 4.5$; and Libchaber, Fauve, and Laroche (1982) obtained $\delta \simeq 4.4$.

In the absence of noise, an infinite number of period–doubling bifurcations takes place in the transition to chaos (e.g., Feigenbaum, 1978; Eckmann, 1981). However, in practice, noise is always present. In the presence of noise, some of the higher period–doubling bifurcations are suppressed, resulting in a finite sequence of bifurcations. The influence of noise on the period–doubling scenario has been investigated in the studies of Crutchfield and Huberman (1980), Crutchfield, Nauenberg, and Rudnick (1981), Eckmann (1981), Mayer–Kress and Haken (1981), and Shraiman, Wayne, and Martin (1981).

5.4 INTERMITTENCY MECHANISMS

A second route to chaos observed frequently in physical experiments is the onset via intermittency. **Intermittency** in fluid mechanics refers to the state in which the laminar flow is interrupted by turbulent outbreaks or bursts at irregular intervals. In fact, spatial–temporal intermittency is a well–known phenomenon that can be observed in interior and exterior boundary layers, pipe flow, flow between rotating cylinders, and fully developed turbulent flows. In this book, we focus on temporal intermittency, which can occur in low–order dynamical systems.

Next, we describe the main features of this route. For values of a control parameter r less than a critical value r_i, the dynamical system has an attracting limit cycle. Thus, the system oscillates in a regular fashion and is stable to small perturbations. An example is given in Figure 5.4.1a, which shows the time trace of the velocity measured by Bergé, Dubois, Manneville, and Pomeau (1980) in the Rayleigh–Bénard convection at the Rayleigh number $R/R_c = 270$. As r slightly exceeds the threshold value r_i (the intermittency threshold), the system response consists of long stretches of oscillations (laminar phases) that appear to be regular and closely resemble the oscillatory behavior for $r < r_i$, but this regular behavior is intermittently interrupted by chaotic outbreaks (turbulent bursts) at irregular intervals. An example is

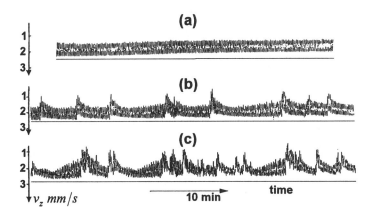

Figure 5.4.1: Experimentally observed intermittency in Rayleigh–Bénard convection: Prandtl number $= 130$, aspect ratios $\Gamma_1 = 2.0$ and $\Gamma_2 = 1.2$. The turbulent bursts occur with increasing frequency as the Rayleigh number is increased. Reprinted with permission from Bergé, Dubois, Manneville, and Pomeau (1980).

shown in Figure 5.4.1b for the velocity measured by Bergé, Dubois, Manneville, and Pomeau (1980) in the Rayleigh–Bénard convection at $R/R_c = 300$. With increasing r, the laminar phases between two consecutive bursts become shorter and shorter and more and more difficult to recognize, as shown in Figure 5.4.1c for the Rayleigh–Bénard experiment at $R/R_c = 335$. As r is increased further, eventually the laminar phases disappear and the response becomes fully irregular (chaotic).

As aforedescribed, in the intermittency mechanism, as a control parameter r exceeds the intermittency threshold r_i , the system response explodes into a larger attractor with the old periodic attractor being a subset of the new chaotic attractor. Thus, as a result of the bifurcation, a periodic orbit is replaced with chaos rather than with a nearby stable periodic orbit. This is implied by the fact that, during the bursts, the trajectory goes "far" away from the vicinity of the periodic orbit that exists for $r < r_i$. Three types of generic bifurcations meet these requirements, namely, cyclic–fold, subcritical Hopf, and subcritical period-doubling bifurcations. Consequently, Manneville and Pomeau (1979) and Pomeau and Manneville (1980) labeled the intermittency mechanisms associated with these bifurcations as **type I**, **type II**, and **type III** intermittency, respectively. The loss of stability of a periodic orbit via one of the aforementioned three generic bifurcations is not sufficient for intermittency to occur. The other necessary condition is the existence of a global "relaminarization" mechanism that repeatedly reinjects the trajectory in the neighborhood of the original periodic orbit (ghost or phantom orbit). Otherwise, the trajectory will never revisit the ghost orbit.

In one–dimensional (noninvertible) maps, chaos through types I and III intermittencies can occur. However, chaos through type II intermittency is not possible. Chaos through all three intermittency mechanisms is possible in two–dimensional or higher invertible maps and three–dimensional or higher continuous systems. By using one-dimensional maps as prototypes, Bergé, Pomeau, and Vidal (1984, Chapter IX) provide a detailed illustration of types I, II, and III intermittencies. Experimental reports of type I intermittency include those of Bergé, Dubois, Manneville, and Pomeau (1980) in the Rayleigh–Bénard convection; Pomeau, Roux, Rossi, Bachelart, and Vidal (1981) and Roux, Rossi, Bachelart, and Vidal (1981) in the Belousov–Zhabotin-sky reaction; Jeffries and Perez (1982) in electronic oscillators; Mullin and Darbyshire (1989) and Price and Mullin (1991) in the Taylor–Couette flow; and Tang, Li, and Weiss (1992) in a coherently pumped laser. In analog and digital computer simulations, Ben–Jacob, Gold-hirsch, Imry, and Fishman (1982), Seifert (1983), and Yeh and Kao (1983) observed type I intermittency in Josephson junctions. Stein-tuch and Schmidt (1988) found type I intermittency in their analysis of ammonia oxidation on a platinum wire.

Huang and Kim (1987), Herzel, Plath, and Svensson (1991), and Ringuet, Roże, and Gouesbet (1993) reported experimental observations of type II intermittency in an electronic oscillator, the oxidation of methanol on zeolithe–supported palladium, and a hydrodynamic system, respectively. Aubry, et al. (1988) identified a regular form of type II intermittency in the wall region of a turbulent boundary layer. Experimental reports of type III intermittency include those of Dubois, Rubio, and Bergé (1983) in the Rayleigh–Bénard convection; Baier, Wegmann, and Hudson (1989) and Kreisberg, McCormick, and Swinney (1991) in the Belousov–Zhabotinsky reaction; Pujol, Arjona, and Corbalán (1993) in laser systems; and Richter, Peinke, Clauss, Rau, and Parisi (1991) in a semiconductor system. Theoretical reports of type III intermittency include those of Yang and Sethna (1991) in parametrically excited nearly square plates, Malasoma, Lamarque, and Jezequel (1994) in a parametrically excited single–degree–of–freedom system with quadratic and cubic nonlinearities, and Paidoussis and Botez (1995) in a three–degree–of–freedom articulated cylinder system subjected to annular axial flow and impacting on the outer pipe.

Recently, Price and Mullin (1991) experimentally observed a new type of intermittency mechanism in a variant of the Taylor–Couette flow problem. The main features of their observations are the extreme regularity of bursting and the presence of a hysteretic transition between singly periodic and intermittent flows in some ranges of the control parameters. In contrast, the preceding intermittency mechanisms are characterized by nonhysteretic transitions and irregular reinjection mechanisms. In another recent work, Platt, Spiegel, and Tresser (1993) defined a process called **On–Off Intermittency**. In this process, there is aperiodic switching between a static state and irregular bursts of oscillations. In contrast, in all of the preceding cases, switching occurred between regular or periodic oscillations and irregular oscillations. In the context of an n–dimensional system of equations, the number of state variables k that exhibit on–off intermittency can be less than n. On–off intermittency has been observed in the work of Malasoma, Lamarque, and Jezequel (1994), who considered a parametrically excited single–degree–of–freedom system (Example 5.8, below).

5.4.1 Type I Intermittency

Next, we discuss a theoretical model for the interpretation of type I intermittency. An experimental observation of this type is shown in Figure 5.4.1. At the bifurcation point, one of the Floquet multipliers associated with the periodic solution leaves the unit circle through $+1$. Similarly, one of the eigenvalues of the corresponding fixed point on the Poincaré map exits the unit circle through $+1$. The manifold corresponding to this eigenvalue contains the essential information relating to the intermittency because there is dissipation along all of the other directions. Therefore, we consider the one–dimensional map

$$x_{n+1} = P(x_n; \epsilon) \qquad (5.4.1)$$

associated with this manifold, where ϵ is a control parameter that passes through zero at the onset of intermittency. This map can be approximated in the bifurcation region as

$$x_{n+1} = x_n + x_n^2 + \epsilon + \cdots \qquad (5.4.2)$$

When $\epsilon < 0$, it is clear from Figure 5.4.2a that the return map intersects the identity map $x_{n+1} = x_n$ in two points: a stable fixed point x_s and an unstable fixed point x_u. These fixed points correspond to coexisting stable and unstable limit cycles of the dynamical system. An orbit initiated in the basin of attraction of the stable solution converges to it as time increases. Iterations started from $x > x_u$ diverge rapidly from x_u, whereas iterations started from $x < x_u$ converge to x_s, as indicated by the arrows.

As ϵ is increased to zero, the two fixed points approach each other and coalesce at the fixed point x^*, as shown in Figure 5.4.2b. Because

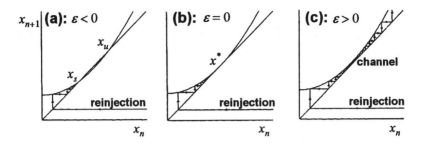

Figure 5.4.2: A set of one–dimensional iterative maps illustrating the type I intermittent transition. (a) Stable and unstable fixed points coexist. (b) Fixed points coalesce, threshold for onset of intermittency. (c) Fixed points disappear, "ghost" of limit cycle with reinjection bursts.

the return map is tangent to the identity map $x_{n+1} = x_n$, the associated bifurcation is called **tangent bifurcation**. Iterations started from $x > x^*$ diverge rapidly from x^*, whereas iterations started from $x < x^*$ converge to x^*.

When $\epsilon > 0$, the return map does not intersect the identity map $x_{n+1} = x_n$, and there are no fixed points in the neighborhood of this part of the map, as shown in Figure 5.4.2c. Instead, for small positive values of ϵ, a narrow corridor or channel or sluice emerges between the return map and the identity map. Whenever any trajectory approaches the channel closely, it will drift slowly through the channel, with the successive iterates accumulating in the narrowest part of the channel. The time taken to traverse the channel depends on the channel width, which in turn depends on the value of ϵ. The narrower the channel is, the longer is the laminar time; that is, the longer is the time during which the orbit is trapped in the channel. As aforementioned, the tangent bifurcation is a necessary condition for intermittency. The other necessary condition is the reinjection via a global mechanism of the escaped trajectory to the vicinity of the channel, as shown in Figure 5.4.2c. While outside the channel, the orbit oscillates irregularly (chaotically) until it is reinjected into the channel when by chance the chaotic orbit approaches the channel. Bergé, Pomeau, and Vidal (1984) discuss reinjection mechanisms in the context of (a) flow on a torus, (b) baker's transformation, and (c) Smale's transformation.

To calculate the time taken by the orbit to traverse the channel, we assume that ϵ is a small positive number so that the steps in x with successive iterates in the channel are very small; that is, $x_{n+1} - x_n$ is small. This allows us to approximate $x_{n+1} - x_n$ by dx/dn and then replace the difference equation (5.4.2) with the differential equation

$$\frac{dx}{dn} = \epsilon + x^2 \qquad (5.4.3)$$

Thus, for an orbit that is reinjected into the channel at $x = x_0 << -\epsilon$, the number of iterations needed to traverse the channel is given by

$$n = \int_{x_0}^{x_f} \frac{dx}{\epsilon + x^2} = \frac{1}{\sqrt{\epsilon}} \left[\arctan\left(\frac{x_f}{\sqrt{\epsilon}}\right) - \arctan\left(\frac{x_0}{\sqrt{\epsilon}}\right) \right] \qquad (5.4.4)$$

As $x_0 \to -\infty$ and $x_f \to \infty$, $n = \pi/\sqrt{\epsilon}$. Therefore, the number of iterations and hence the approximate time necessary to traverse the channel is $O(\epsilon^{-1/2})$. This means that the bursts of irregular oscillations will be separated by approximately periodic intervals of duration $O(\epsilon^{-1/2})$. The relaminarization comes about through reinjection into the channel. An alternate method, based on renormalization group analysis, is used by Hu and Rudnick (1986) to determine the channel transit time.

Example 5.6. Following Manneville and Pomeau (1979), we consider the three–dimensional Lorenz system to illustrate type I intermittency. This system is given by

$$\dot{x} = \sigma(y - x) \qquad (5.4.5)$$
$$\dot{y} = \rho x - y - xz \qquad (5.4.6)$$
$$\dot{z} = -\beta z + xy \qquad (5.4.7)$$

We set $\sigma = 10$ and $\beta = \frac{8}{3}$ and use ρ as the control parameter. At $\rho = 166.0$, this system has a periodic attractor. The corresponding oscillations in the state z are depicted in Figure 5.4.3a. As ρ is gradually increased, a cyclic–fold bifurcation occurs at $\rho_c \simeq 166.06$. Subsequently, chaotic oscillations follow for $\rho > \rho_c$. In Figure 5.4.3b, finite intervals of regular oscillations are interrupted by intermittent

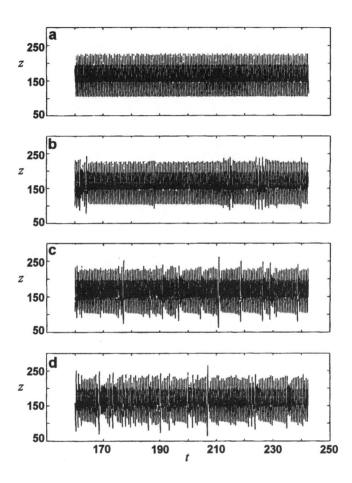

Figure 5.4.3: Transition to chaos through intermittency in the Lorenz equations for $\sigma = 10$ and $\beta = \frac{8}{3}$. The time histories of the state z are shown for the following cases: (a) $\rho = 166.0$, (b) $\rho = 166.1$, (c) $\rho = 166.5$, and (d) $\rho = 167.0$.

bursts of irregular oscillations. The regular oscillations are similar to those seen in Figure 5.4.3a. The bursts of irregular oscillations are more prominent in Figures 5.4.3c and 5.4.3d, with the duration of regular oscillations decreasing in size and the bursts becoming more frequent as ρ is increased.

Example 5.7. We use the logistic map

$$x_{n+1} = F(x_n) = \alpha x_n (1 - x_n) \tag{5.4.8}$$

to illustrate that chaos through intermittency can also occur in maps. The period–three orbits of (5.4.8) are solutions of

$$x = F^3(x) \tag{5.4.9}$$

These solutions first appear at

$$\alpha_c = \frac{1}{4}(1 + 2\sqrt{2}) \approx 0.9571068$$

For $\alpha \geq \alpha_c$, in addition to the period–one solutions

$$x = 0 \text{ and } x = 1 - \frac{1}{4}\alpha^{-1} \tag{5.4.10}$$

there are six other solutions. These solutions correspond to an unstable period–three orbit and a stable period–three orbit. The stable period–three orbit realized at $\alpha = 0.959$ is shown in Figure 5.4.4a. At the critical point $\alpha = \alpha_c$, the stable and unstable period–three orbits coincide, and the map $F^3(x)$ is tangent to the identity map. Therefore, $\alpha = \alpha_c$ corresponds to a tangent bifurcation point. For α just below α_c, the only real solutions of (5.4.9) are those given by (5.4.10). Near the tangent points, which occur at $\alpha = \alpha_c$, there are three narrow channels between the map $F^3(x)$ and the identity map. An orbit initiated in one of these channels spends a long time nearly trapped in this channel before it eventually escapes from this channel. After coming out of the channel, the orbit oscillates chaotically until it again enters one of the three channels. The processes of entrapment (laminar phase), escape, irregular oscillations (bursts), and reinjection into one of the three channels continues indefinitely. This is an example of type I intermittency. In Figure 5.4.4b, we show a time history exhibiting this type of intermittency at $\alpha = 0.956$.

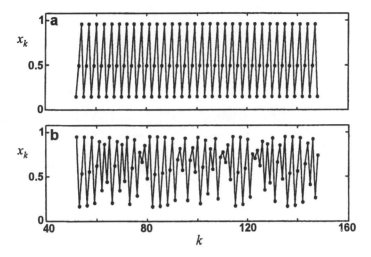

Figure 5.4.4: Transition to chaos through intermittency in the logistic map. The orbits are shown for the following cases: (a) $\alpha = 0.959$ and (b) $\alpha = 0.956$. The discrete points that make up each orbit are connected by straight lines.

5.4.2 Type III Intermittency

Type III intermittency is associated with a subcritical period–doubling or flip bifurcation. An experimental example is shown in Figure 5.4.5. It represents the time variation of the horizontal temperature gradient near the threshold of type III intermittency in the Rayleigh–Bénard convection, as measured by Dubois, Rubio, and Bergé (1983) at the Rayleigh number $R/R_c = 416.5$. For a Rayleigh number below the threshold intermittency value R_i, we have a stable limit cycle. As R exceeds R_i, the behavior is quite close to a periodic (laminar) one most of the time. However, it is interrupted from time to time by turbulent bursts. It is clear from Figure 5.4.5 that the amplitude of the subharmonic progressively increases while the amplitude of the fundamental harmonic progressively decreases with each successive oscillation. When the amplitude of the subharmonic exceeds a critical value, a kind of final catastrophe occurs, interrupting the laminar phase and marking the beginning of the turbulent burst. Subsequently, a global

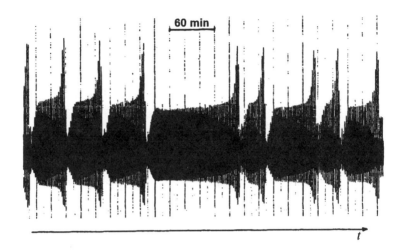

Figure 5.4.5: Type III intermittency in the Rayleigh–Bénard convection at $R/R_c \approx 416.5$. Shown is the time trace of the horizontal temperature gradient. Reprinted with permission from Dubois, Rubio, and Bergé (1983).

mechanism, such as baker's transformation or Smale's transformation, reinjects the orbit near the ghost or phantom (former) periodic attractor, and the process continues indefinitely.

To estimate the time of the laminar phase, we first note that this intermittency is associated with a Floquet multiplier or an eigenvalue of a fixed point of the Poincaré return map exiting the unit circle through -1. Next, we consider the manifold corresponding to this eigenvalue because it contains the essential information relating to the intermittency because there is dissipation along all of the other directions. Thus, the map associated with this manifold has the form of the one–dimensional map (5.4.1), which can be approximated near the intermittency threshold as

$$x_{n+1} = -(1 + \epsilon)x_n + \alpha_1 x_n^2 + \alpha_2 x_n^3 + \cdots \qquad (5.4.11)$$

where the constants α_1 and α_2 depend on the dynamical system under consideration. Here, $x = 0$ is a fixed point of this map for all ϵ. For slightly positive values of ϵ the fixed point $x = 0$ is unstable, whereas for slightly negative values of ϵ the fixed point $x = 0$ is stable. To analyze the mapping (5.4.11), we use its second iterate; that is,

$$x_{n+2} = -(1+\epsilon)\left[-(1+\epsilon)x_n + \alpha_1 x_n^2 + \alpha_2 x_n^3\right]$$
$$+\alpha_1\left[(1+\epsilon)^2 x_n^2 - 2\alpha_1(1+\epsilon)x_n^3\right] - \alpha_2(1+\epsilon)^3 x_n^3 + \cdots$$

Since we are interested in small $\mid x \mid$ and $\mid \epsilon \mid$, we can simplify the second return map to

$$x_{n+2} = (1+2\epsilon)x_n + \alpha x_n^3 + \cdots \qquad (5.4.12)$$

where $\alpha = -2(\alpha_2 + \alpha_1^2)$.

For small values of ϵ, the steps in x with successive iterates in the laminar phase are very small; that is, $x_{n+2} - x_n$ is small. Hence, we approximate $x_{n+2} - x_n$ by dx/dn and then replace the difference equation (5.4.12) with the differential equation

$$\frac{dx}{dn} = 2\epsilon x + \alpha x^3 \qquad (5.4.13)$$

which is the normal form of a pitchfork bifurcation. The bifurcation is supercritical if α is negative and subcritical if α is positive. Consequently, type III intermittency is possible only when $\alpha > 0$, so that (5.4.13) does not have stable fixed–point solutions for $\epsilon > 0$ and it has a single stable fixed–point solution for $\epsilon < 0$.

Scaling x by $\sqrt{2\epsilon/\alpha}u$ and n by $k(2\epsilon)^{-1}$, we rewrite (5.4.13) as

$$\frac{du}{dk} = u(1 + u^2) \qquad (5.4.14)$$

where

$$x = u\sqrt{\frac{2\epsilon}{\alpha}} \quad \text{and} \quad n = \frac{k}{2\epsilon} \qquad (5.4.15)$$

Equation (5.4.14) is a universal form. The index $n = k/2\epsilon$, which is the count of the number of iterations at the fundamental frequency, is a measure of time. The scaling indicates that the duration of the laminar phase during type III intermittency is $O(\epsilon^{-1})$ as opposed to $O(\epsilon^{-1/2})$ during type I intermittency.

The solution of (5.4.14) can be expressed as

$$2k = \ln \frac{u^2(1 + u_0^2)}{u_0^2(1 + u^2)} \tag{5.4.16}$$

where $u = u_0$ when $k = 0$. As $x \to \infty, u \to \infty$ and (5.4.16) becomes

$$2k = \ln \frac{1 + u_0^2}{u_0^2} \quad \text{so that} \quad n = \frac{1}{4\epsilon} \ln \frac{1 + u_0^2}{u_0^2} \tag{5.4.17}$$

Example 5.8. Following Malasoma, Lamarque, and Jezequel (1994), we consider the parametrically excited single–degree–of–freedom system

$$\ddot{x} + 0.2\dot{x} + x + 1.5x^2 + 0.5x^3 + Fx \cos \omega t = 0 \tag{5.4.18}$$

In the bifurcation analyses, ω is used as a control parameter while F is held fixed. The trivial fixed point $(x, \dot{x}) = (0, 0)$ is a solution of the system for all ω. Malasoma, Lamarque, and Jezequel (1994) numerically constructed Poincaré sections for orbits of this system by using the excitation frequency ω. For each orbit, the discrete points $[x(nT), \dot{x}(nT)]$ were collected at time intervals of $T = 2\pi/\omega$. The bifurcation diagram shown in Figure 5.4.6 was generated by keeping F fixed at 0.85 and incrementing the control parameter ω in steps of $\Delta\omega = 0.0002$. In this diagram, the values $\dot{x}(nT)$ corresponding to the attractor realized at each value of ω are plotted. (For each value of ω, the first 3,000 points on the Poincaré section were discarded, and the next 1,000 points were collected.)

It is clear from Figure 5.4.6 that, as ω is increased past $\omega_c = 1.6067$, there is an abrupt transition from the trivial point attractor to an aperiodic attractor. A subcritical Hopf bifurcation of the trivial fixed point occurs at $\omega_H \approx 1.60670376591$. Therefore, the local bifurcation associated with the transition from the trivial solution to the aperiodic solution appears to be a subcritical Hopf bifurcation. Associated with this bifurcation is an intermittent transition to chaos as evident from Figure 5.4.7, where the time history of the velocity is shown for a particular ω. We see small (pseudoperiodic) oscillations of increasing amplitude with random interruptions in the form of

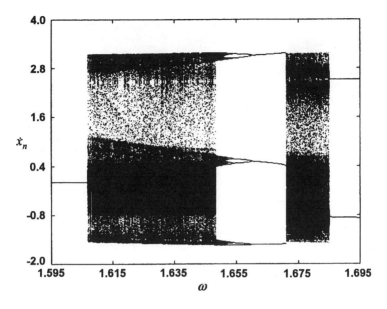

Figure 5.4.6: Illustration of bifurcations on the Poincaré section when $F = 0.85$. Reprinted with permission from Malasoma, Lamarque, and Jezequel (1994).

irregular or chaotic–like bursts. As the frequency is increased further from the intermittency threshold ω_c, the pseudoperiodic oscillations are interrupted more frequently, indicating a fully chaotic state. As the frequency is decreased toward ω_c, the average duration between the intermittent bursts becomes longer and longer and the amplitude of the pseudoperiodic oscillations becomes smaller and smaller. As ω is decreased below ω_c, the intermittent bursts disappear, and the system response returns to the equilibrium state $(x, \dot{x}) = (0, 0)$. The results in Figure 5.4.7 are indicative of on–off intermittency.

As ω is increased past $\omega_F \approx 1.685481$, a subcritical period–doubling bifurcation occurs. As a consequence of this local bifurcation, there is the possibility of a transition of type III intermittency from a periodic solution to a chaotic solution. The time history shown in Figure 5.4.8 is indicative of this transition. The amplitude of one harmonic associated with the period–two motion increases while the amplitude of the other harmonic decreases. The pseudoperiodic oscillations (laminar

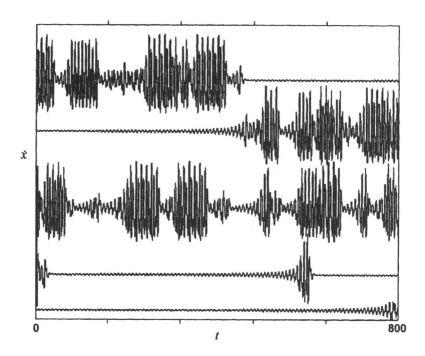

Figure 5.4.7: Intermittent transition from a fixed point to chaos associated with a subcritical Hopf bifurcation in the parametrically excited single–degree–of–freedom system (5.4.18) for $F = 0.85$ and $\omega = 1.6068$. Reprinted with permission from Malasoma, Lamarque, and Jezequel (1994).

stretches) are interrupted by chaotic–like bursts. When ω is increased toward ω_F, the durations of the laminar stretches become longer and longer and the chaotic bursts disappear for $\omega > \omega_F$.

In Figure 5.4.9, we show a bifurcation diagram of the system (5.4.18) for $F = 1.0$, determined with a control–parameter increment $\Delta\omega = 0.0001$. As ω is increased past $\omega_H \approx 1.528781$, the system undergoes a supercritical Hopf bifurcation, resulting in the birth of two stable small-amplitude period–two limit cycles. They are destroyed in a cyclic–fold bifurcation at $\omega_s \approx 1.5293183$. The postbifurcation state is a chaotic attractor. The results displayed in Figure 5.4.10 indicate that a type I intermittency is associated with this local bifurcation.

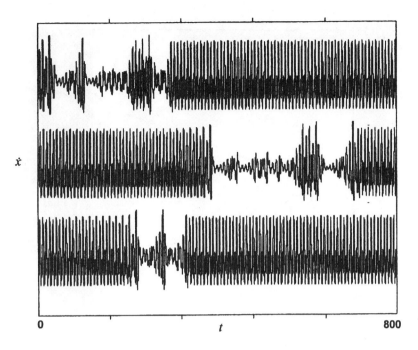

$\dot{x}$

0   t 800

Figure 5.4.8: Type III intermittency in the response of the parametrically excited single–degree–of–freedom system (5.4.18) for $F = 0.85$ and $\omega = 1.68525$. Reprinted with permission from Malasoma, Lamarque, and Jezequel (1994).

5.4.3   Type II Intermittency

When two complex–conjugate Floquet multipliers or two complex–conjugate eigenvalues of the Poincaré return map exit the unit circle away from the real axis, a Hopf bifurcation of the fixed point of the map occurs. If this bifurcation is subcritical, there is a possibility of type II intermittency. An experimental observation of this intermittency is provided in Figure 5.4.11. Shown is the time history of the temperature measured by Herzel, Plath, and Svensson (1991) during the oxidation of methanol on zeolithe–supported palladium catalyst. The time series consists of laminar stretches of pseudoquasiperiodic oscillations with interruptions in the form of chaotic bursts, characteristic of type II intermittency.

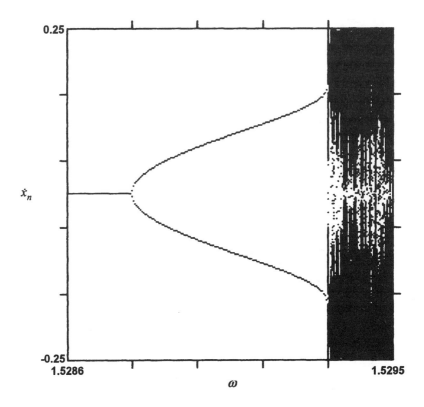

Figure 5.4.9: Illustration of bifurcations on the Poincaré section for $F = 1.0$. Reprinted with permission from Malasoma, Lamarque, and Jezequel (1994).

To determine the duration of the laminar stretches associated with this intermittency, we consider the dynamics on the manifold corresponding to the pair of complex–conjugate eigenvalues leaving the unit circle. This manifold contains the essential information relevant to the considered intermittency because of the dissipation in all of the other directions. The two–dimensional map governing this dynamics is given by

$$r_{n+1} = (1 + \epsilon)r_n + \alpha r_n^3 \tag{5.4.19}$$

$$\theta_{n+1} = \theta_n + \omega + \beta r_n^2 \tag{5.4.20}$$

where α, β, and ω are real constants that depend on the dynamical

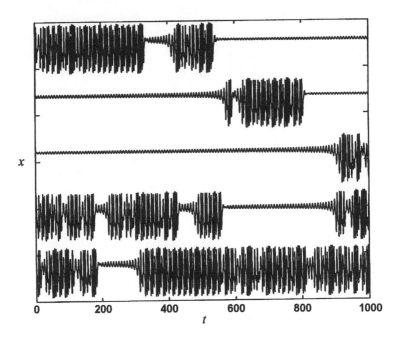

Figure 5.4.10: A time series indicating type I intermittency in the response of the parametrically excited single–degree–of–freedom system (4.14.18) for $F = 1.0$ and $\omega = 1.53$. Reprinted with permission from Malasoma, Lamarque, and Jezequel (1994).

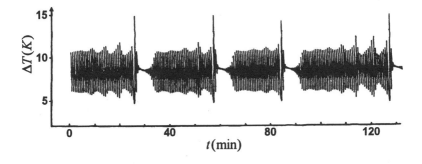

Figure 5.4.11: A time series indicating type II intermittency measured during the oxidation of methanol on zeolithe supported Pd. Reprinted with permission from Herzel, Plath, and Svensson (1991).

system under consideration, and ϵ is a control parameter that passes through zero at the onset of the intermittency. For small values of ϵ, the steps in r with successive iterates in the laminar phase are small; that is, $r_{n+1} - r_n$ is small. This allows us to approximate $r_{n+1} - r_n$ by $\dot{r}$ and then replace the difference equation (5.4.19) with the differential equation

$$\dot{r} = \epsilon r + \alpha r^3 \qquad (5.4.21)$$

For type II intermittency to occur, the bifurcation must be subcritical; that is, α must be positive. For $\epsilon < 0$, there are three fixed–point solutions of (5.4.21): the trivial solution, which is stable, and

$$r = \pm\sqrt{-\epsilon/\alpha}$$

which are unstable. For $\epsilon > 0$, there is only one fixed–point solution of (5.4.21), which is unstable. Thus, as the control parameter ϵ is increased through zero, the system undergoes a subcritical Hopf bifurcation in which two unstable limit cycles of the corresponding continuous system are destroyed. Introducing the scaling

$$r = u\sqrt{\frac{\epsilon}{\alpha}} \quad \text{and} \quad t = \frac{\tau}{\epsilon} \qquad (5.4.22)$$

we transform (5.4.21) into the universal form

$$\frac{du}{d\tau} = u(1 + u^2) \qquad (5.4.23)$$

The scaling suggests that the duration of the laminar stretches is $O(\epsilon^{-1})$ as in type III intermittency as opposed to $O(\epsilon^{-1/2})$ for type I intermittency.

5.5   QUASIPERIODIC ROUTES

At the present time, there are many analytically and numerically established quasiperiodic transitions to chaos. Some of these transitions have also been experimentally observed.

5.5.1 Ruelle–Takens Scenario

Motivated to seek an explanation for hydrodynamic turbulence (see Landau and Lifschitz, 1959), Landau (1944) proposed one of the first scenarios for transition to an aperiodic motion (turbulence). In this scenario, first, a Hopf bifurcation of an equilibrium solution occurs, leading to the birth of a periodic attractor. Subsequently, this periodic attractor undergoes a secondary Hopf bifurcation, leading to the birth of a two–period quasiperiodic attractor. This attractor undergoes another Hopf bifurcation, resulting in a three–period quasiperiodic attractor. Thus, a sequence of Hopf bifurcations takes place with each bifurcation, adding a fundamental frequency. Hence, after the $(k-1)$th Hopf bifurcation, the state x_i associated with the motion is described by

$$x_i = f_i(\omega_1 t, \omega_2 t, \cdots, \omega_k t) \tag{5.5.1}$$

where the ω_j are incommensurate and f_i is periodic with period one in its k arguments. As k increases, the motion becomes more irregular or "turbulent." So, in the **Landau scenario,** an infinite sequence of Hopf bifurcations is required in the transition to a turbulent motion, which is characterized by the presence of an infinite number of incommensurate frequencies. Therefore, a large number of states is required to characterize a turbulent motion resulting through the Landau scenario. It is to be noted that Hopf (1948) advocated a similar scenario with mathematical rigor. The irregular motion at the culmination of the Landau scenario can be called chaotic if there is sensitivity to initial conditions. However, as pointed out by Ruelle (1989a), an irregular motion with just an infinite number of incommensurate frequencies is not characterized by sensitivity to initial conditions.

In 1971, Ruelle and Takens demonstrated with mathematical rigor that a quasiperiodic flow on an N–torus, where $N \geq 4$, can be perturbed under fairly general conditions to produce a robust chaotic attractor. This theory was further extended in the work of Newhouse, Ruelle, and Takens (1978). An arbitrary perturbation of a quasiperiodic flow on an N–torus, where $N \geq 3$, may lead to (a) the persistence of the quasiperiodicity with N incommensurate frequencies, (b) frequency locking during which the associated orbit is either periodic or quasiperi-

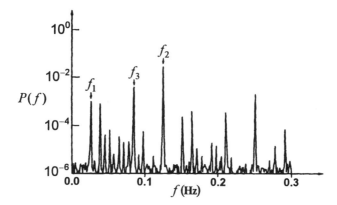

Figure 5.5.1: Power spectrum associated with a three–period quasiperiodic attractor. The incommensurate frequencies have been marked. Reprinted with permission from Gollub and Benson (1980).

odic with less than N incommensurate frequencies, or (c) chaos. The theoretical results of Ruelle and Takens (1971) and Newhouse, Ruelle, and Takens (1978) form the basis for the **Ruelle–Takens scenario**, which is in remarkable contrast to the scenario proposed by Landau and Hopf. In the Ruelle–Takens scenario, a finite number of successive Hopf bifurcations can lead from an equilibrium solution to a chaotic attractor. Specifically, this attractor can be observed after the birth of a three–period quasiperiodic attractor. At the present time, there are no theoretical means to predict the critical values at which three–period quasiperiodic and chaotic attractors, respectively, appear. In contrast with the Landau scenario, a small number of states are sufficient to characterize a complex aperiodic attractor. This means that chaos is possible in finite–dimensional systems.

Many experimentally observed transitions to chaos have been interpreted by using the Ruelle–Takens scenario. Gollub and Swinney (1975) and Swinney and Gollub (1978) presented power spectra to illustrate the transitions observed in the Taylor–Couette flow and the Rayleigh–Bénard convection experiments. This quasiperiodic transition to chaos was also observed in the Rayleigh–Bénard convection experiment of Gollub and Benson (1980). In Figure 5.5.1, the power spectrum associated with the three–period quasiperiodic attractor observed in the transition is shown. Three-period quasiperiodic motions were also observed in the experiments of Gorman, Reith, and Swinney (1980) and Tavakol and Tworkowski (1984). The transition to chaos observed in the experiments of Martin, Leber, and Martienssen (1984) can be interpreted in terms of the Ruelle–Takens scenario. A good experimental illustration of this scenario is provided by the study of Cumming and Linsay (1988) with a nonlinear electronic oscillator. The influence of noise on the Ruelle–Takens scenario has been addressed by Eckmann (1981) and the references mentioned therein.

While the studies mentioned in the previous paragraph illustrate that a chaotic motion can be observed subsequent to the observation of a three–period quasiperiodic motion, there are other studies that indicate that an observation of a three–period quasiperiodic motion does not always mean that a chaotic motion is to follow. In this regard, we note the numerical studies of Grebogi, Ott, and Yorke (1983c, 1985) with maps. Through extensive studies, they showed that a "weak" to "moderate" perturbation of a three–period quasiperiodic flow rarely leads to a strange attractor. This perturbation often results in either a quasiperiodic or a periodic flow. However, as the strength of the perturbation increases, the likelihood of realizing a chaotic attractor also increases. Walden, Kolodner, Passner, and Surko (1984) illustrated through a Rayleigh–Bénard experiment that stable four–period and five–period quasiperiodic oscillations can be observed in experiments.

5.5.2 Torus Breakdown

In the preceding scenario, a three–period quasiperiodic attractor can be perturbed to produce a chaotic attractor. In contrast, here, a chaotic attractor appears following the appearance of a two–period quasiperiodic attractor. In each of these routes, first a point attractor loses stability due to a supercritical Hopf bifurcation, resulting in a periodic attractor. Subsequently, this periodic attractor experiences a supercritical secondary Hopf bifurcation, resulting in a two–period quasiperiodic attractor (two–torus). On an associated Poincaré section, the intersection points densely fill up a closed loop. The associated evolution of the continuous system can be characterized in terms of two oscillatory modes with incommensurate frequencies. As the considered control parameter is further varied, a rich variety of bifurcations of the torus can take place. The postbifurcation state can be one of the following: (1) a complex periodic attractor (phase–locked oscillations or **mixed–mode oscillations**); (2) a nonstrange attractor whose corresponding orbit exhibits the feature of intermittent excursions similar to those discussed in the context of intermittency mechanisms in Section 5.4; and (3) a chaotic attractor.

If the coupling between the two oscillatory modes is strong enough, the quasiperiodic attractor loses stability, resulting in a phase–locked attractor. On the considered Poincaré section, the closed loop is broken up, resulting in a discrete number of points. In this case, we have synchronization. As discussed in Section 4.5, synchronization can be studied for weak nonlinearities by using a perturbation method, such as the method of multiple scales, method of averaging, or method of normal forms. As the control parameter is further varied, the transition from the periodic attractor to a chaotic attractor can take place through either a period–doubling scenario or an intermittency scenario.

Let us suppose that the quasiperiodic flow is characterized by the frequencies ω_2 and ω_1. An application of a suitable small perturbation to this flow can result in the replacement of this flow by a flow asymptotic to attracting periodic orbits with a rational frequency ratio p/q. As discussed in Section 4.5, this phenomenon is called **frequency locking** or **synchronization** because the frequency ratio ω_2/ω_1 locks into a rational number p/q. When ω_2/ω_1 is close to p/q, an interesting nonstrange attractor can be the postbifurcation state. An orbit on this attractor spends a long time near the ghost or phantom of the phase–locking orbit, from which it occasionally unlocks.

A third way the system may evolve as the control parameter is varied is the destruction of the torus and the emergence of a chaotic attractor. This transition to chaos through a two–period quasiperiodic attractor is often described as **chaos via torus breakdown**.

To understand the routes described in this section, we consider a Poincaré section associated with the dissipative flow on the two–torus. The corresponding dissipative map is two–dimensional, which can be put in the form

$$\theta_{i+1} = P_1(\theta_i, r_i) \ (\mathrm{mod}\ 2\pi) \tag{5.5.2}$$

$$r_{i+1} = P_2(\theta_i, r_i) \qquad (5.5.3)$$

where θ and r are angular and radial coordinates, respectively, and θ is restricted to the interval $[0, 2\pi)$ by using the mod 2π operation. We consider a location close to the secondary Hopf bifurcation point. Furthermore, we assume that dissipation occurs due to strong contraction in the radial direction and that, after transients die out, there is a simple attracting closed curve on the Poincaré section. In this case, we have $r_i = g(\theta_i)$, and the two–dimensional Poincaré map reduces to

$$\theta_{i+1} = P_1[\theta_i, g(\theta_i)] = F(\theta_i) \qquad (5.5.4)$$

which resembles the circle map of Section 4.2. For a quasiperiodic attractor, the rotation number associated with (5.5.4) is irrational. As discussed earlier, one way in which the system can evolve as the control parameter is varied beyond the secondary Hopf bifurcation point is the destruction of the torus and the emergence of chaos. In this scenario, the closed curve in the Poincaré section deforms, then develops wrinkles, becomes fractal, and finally breaks down. Thus, the transition to chaos from the two–period quasiperiodic attractor occurs through the destruction of the closed curve in the Poincaré section. Although the Poincaré map associated with the evolution is an invertible map, the reduced map, such as (5.5.4), becomes noninvertible after wrinkles develop in the Poincaré section. When the reduced map becomes noninvertible, the rotation number is not well defined. This means that the associated evolution need not be either quasiperiodic or periodic; that is, chaos is possible. Therefore, the loss of invertibility of the reduced map is a landmark for torus breakdown and the ensuing emergence of chaos.

For illustration, we consider the circle map described by

$$x_{n+1} = F(x_n) = x_n + \Omega - \frac{K}{2\pi} \sin(2\pi x_n) \qquad (5.5.5)$$

where x_n is restricted to the interval $[0, 1)$ by using the mod 1 operation. The parameter K determines the strength of the nonlinearity of the map. There are two parameters in this equation, namely, K and Ω. In Figure 5.5.2, we depict how the nature of this one–dimensional map changes with respect to K when Ω is held constant at 0.2 (Jensen, Bak,

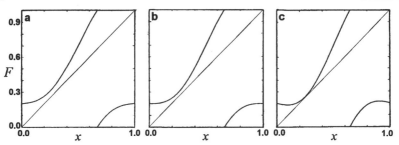

Figure 5.5.2: Circle map for $\Omega = 0.2$: (a) $K = 0.9$, (b) $K = 1.0$, and (c) $K = 1.2$. The map is invertible in (a) and (b) and noninvertible in (c).

and Bohr, 1984). For $K = 0.9$, the map is invertible. Furthermore, numerical simulations show that forward iterates of this map settle on a quasiperiodic orbit; that is, the winding number is irrational. At $K = 1$, there is an inflection point at $x = 0$. (A Taylor series expansion of the trigonometric term in the vicinity of $x = 0$ indicates that this inflection is due to a cubic nonlinearity.) The circle map is still invertible at $K = 1$. However, in this case, numerical simulations show that forward iterations of this map settle on a phase–locked orbit; that is, the corresponding winding number is rational. As K is increased beyond 1, the circle map becomes noninvertible as illustrated by Figure 5.5.2c. (The horizontal line $F = 0.2$ intersects the graph at two locations, indicating that the map is no longer one–to–one.) As a consequence, the rotation number is no longer well defined, which means that chaos is possible. Numerical simulations indicate that forward iterates of this map settle on an irregular orbit, indicating chaos.

From the extensive studies that have been carried out (e.g., Jensen et al., 1984), it is now known that the line $K = 1$ separates two regions in the $K - \Omega$ plane, as shown in Figure 4.2.1. For $K < 1$, as Ω varies at a fixed K, the Arnold tongues are separated and the map displays both periodic and quasiperiodic motions. As $K \to 1$, the widths of the Arnold tongues increase; that is, the rational intervals and hence the mode–locking regions increase in size. At the critical value $K = 1$, Jensen et al. (1984) found that the set of rational intervals is fractal. Consequently, the different mode–locking intervals form a Cantor set whose dimension is about 0.87. Hence, on $K = 1$, the probability that the winding number is rational for a random choice of

Ω is almost 1. For $K > 1$, the mode–locking regions overlap, which implies that many different periodic motions can coexist for given (K, Ω), depending on the initial conditions. Moreover, the rotation number, which characterizes the periodicity or quasiperiodicity of the considered motion, becomes undefined. An undefined rotation number means that chaos is possible. For the circle map, which is invertible for $K \leq 1$ and noninvertible for $K > 1$, the line $K = 1$ represents the critical boundary. In the case of other maps, the critical boundary may not be a straight line. The study of Bohr, Bak, and Jensen (1984) serves as a good example. They constructed Poincaré maps for a periodically forced, damped oscillator and reduced them to one–dimensional maps. The numerically found critical boundaries for these one–dimensional maps are well-defined curves in the two–parameter plane. The study of Aronson, McGehee, Kevrekidis, and Aris (1986) is illustrative of how complex the critical boundary can be in higher–dimensional maps. The critical boundary is not well defined in many higher–dimensional maps.

In the transition to chaos, the sequence of bifurcations that takes place depends on how the parameters are varied in the two–parameter plane, the structure of the mode–locking regions, and the critical boundary. To fix ideas, we consider a hypothetical one–dimensional map with the two parameters K and Ω. Three mode–locking regions and the critical boundary of this map are depicted in Figure 5.5.3. This map is invertible below the critical curve and noninvertible above it. Along route 1, Ω is held fixed at an irrational number while K is gradually varied. Along this route, we proceed from quasiperiodic motions to phase–locked motions to chaos. The phase–locked motions occur at the location where the critical boundary is crossed. In some cases, because of the fractal nature of the critical boundary, it is possible to go directly from quasiperiodic motions to chaos without phase locking. Both K and Ω need to be varied to follow routes 2 and 3. In the case of route 2, we go from phase–locked motions to chaos. On the other hand, in the case of route 3, we alternate between quasiperiodic and phase–locked motions before the emergence of chaos. In another system, it may be possible to realize routes 2 and 3 by varying a single parameter such as K. The realization depends on the structure of the phase–locking regions and the critical boundary for the considered map.

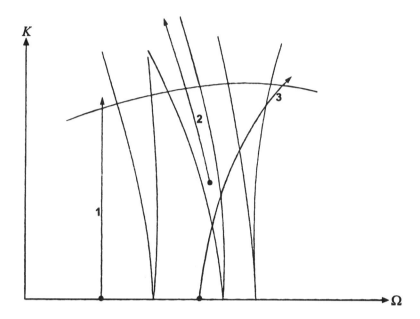

Figure 5.5.3: Three possible routes to chaos in the $K - \Omega$ plane.

Curry and Yorke (1977) considered a dissipative two–dimensional map and numerically illustrated that phase locking occurs before the emergence of the chaotic attractor. A detailed exposition of the work of Curry and Yorke is provided in Chapter VII of the book by Bergé, Pomeau, and Vidal (1984). Aronson, Chory, Hall, and McGehee (1982) also conducted interesting numerical studies with maps in the plane. The scaling behavior in the route to chaos via torus breakdown was investigated by Feigenbaum, Kadanoff, and Shenker (1982), Rand, Ostlund, Sethna, and Siggia (1982), Shenker (1982), and Ostlund, Rand, Sethna, and Siggia (1983).

Sano and Sawada (1983) considered a fourth–order autonomous system of differential equations with cubic nonlinearities and examined bifurcations by using a single control parameter. They numerically illustrated a quasiperiodic transition to chaos by using Poincaré maps, which were reduced to one–dimensional maps in terms of an angular coordinate. Sano and Sawada showed that the one–dimensional map is invertible prior to the emergence of chaos and noninvertible after the emergence of chaos, thus linking the destruction of the two–torus with the noninvertibility of the one–dimensional map. Matsumoto, Chua, and Tokunaga (1987) considered a three–dimensional autonomous system of differential equations representing a nonlinear electrical circuit. Through detailed analysis and experimental and numerical simulations, they illustrated chaos through two different quasiperiodic routes. In one of the routes, as the control parameter was varied, quasiperiodic and phase–locked oscillations appeared and disappeared alternately before chaos emerged through a cyclic–fold bifurcation of a phase–locked solution. In another route, as the control parameter was varied, quasiperiodic and phase–locked oscillations appeared and disappeared alternately before chaos emerged through a period–doubling sequence from a phase–locked oscillation. Choi and Noah (1992a) also observed this transition to chaos in the study of a rotating machinery. Furthermore, above criticality, Guevara, Glass, and Shrier (1981), Perez and Glass (1982), Glass, Guevara, Shrier, and Perez (1983), Schell, Fraser, and Kapral (1983), Mackay and Tresser (1984), and Bélair and Glass (1985) observed that the phase–locked oscillations within a mode–locking region experience period–doubling bifurcations leading to chaos.

The theorems of Afraimovich and Shilnikov (1983a) address the destruction of two–tori in $\mathcal{R}^N$ for $N \geq 3$ in the presence of two or more control parameters.

Experimental observations of chaos via torus breakdown include those of Fenstermacher, Swinney, and Gollub (1979), Roux (1983), Bergé, Pomeau, and Vidal (1984, Chapter VII), Anishchenko, Letchford, and Safonova (1985), Argoul and Roux (1985), Stavans, Heslot, and Libchaber (1985), Argoul, Arneodo, Richetti, and Roux (1987), Basset and Hudson (1989), and Xu and Schell (1990).

Example 5.9. We consider, after Steinmetz and Larter (1991) and Larter, Olsen, Steinmetz, and Geest (1993), the following four–dimensional system:

$$\dot{A} = -k_1 ABX - k_3 ABY + k_7 - k_{-7}A \qquad (5.5.6)$$

$$\dot{B} = -k_1 ABX - k_3 ABY + k_8 \qquad (5.5.7)$$

$$\dot{X} = k_1 ABX - 2k_2 X^2 + 2k_3 ABY - k_4 X + k_6 \qquad (5.5.8)$$

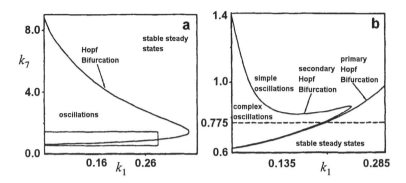

Figure 5.5.4: Bifurcation diagram for the peroxidase–oxidase reaction The primary Hopf bifurcation shown in (a) separates the region of oscillatory solutions from the region of stable steady states. The region in (a) enclosed in the box is enlarged in (b), and both of the Hopf bifurcations are indicated along with the three regions they define in (b). The dashed line in (b) is the one parameter path chosen for studying the transition to chaos. Reprinted with permission from Steinmetz and Larter (1991).

$$\dot{Y} = -k_3 ABY + 2k_2 X^2 - k_5 Y \qquad (5.5.9)$$

This system represents a model of the peroxidase-oxidase reaction. Steinmetz and Larter (1991) performed a bifurcation analysis by letting $k_2 = 1250$, $k_3 = 0.046875$, $k_4 = 20$, $k_5 = 1.104$, $k_6 = 0.001$, $k_{-7} = 0.1175$, and $k_8 = 0.5$. In the bifurcation set shown in Figure 5.5.4a, the loci of the primary Hopf bifurcation (PHB) and secondary Hopf bifurcation (SHB) points are displayed in the $k_7 - k_1$ plane. The PHB curve separates equilibrium solutions from dynamic solutions, which may be periodic, quasiperiodic, or chaotic. By examining the stability of the periodic solutions just to the left of the PHB curve, Steinmetz and Larter determined the SHB curve, which is shown together with an enlargement of the PHB curve in Figure 5.5.4b. Thus, between the two Hopf bifurcation curves, the solutions are periodic, and immediately to the left of the SBH curve the solutions are two–period quasiperiodic. The corresponding oscillations occur on a two–torus attractor. An example is shown in Figure 5.5.5.

Fixing k_7 at 0.775 and decreasing k_1 slowly from a value to the right of the PHB curve in Figure 5.5.4b, one finds that the solution

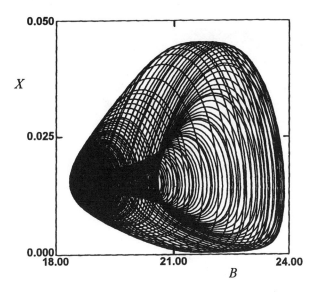

Figure 5.5.5: Quasiperiodicity on a two–torus in the peroxidase–oxidase reaction: $k_1 = 0.2$ and $k_7 = 0.775$. Reprinted with permission from Larter et al. (1993).

is constant. As k_1 is decreased to cross the PHB curve, the solutions become periodic. As k_1 is decreased below $k_{SHB} = 0.206$, the solutions become two–period quasiperiodic. The two–torus realized at $k_1 = 0.2$ is shown in Figure 5.5.5. Immediately to the left of the SHB curve, at each k_1, a Poincaré section provides a cross section of the associated two–torus. The intersections with a Poincaré section fill up a closed curve uniformly and densely. In Figure 5.5.6a, we show a Poincaré section of the two–torus attractor realized at $k_1 = 0.205$. The cross section of the two–torus attractor is elliptical. As k_1 is decreased further, the elliptical section grows without distortion. As k_1 is substantially decreased below k_{SHB}, the ellipse gets distorted through flattening at one end and bulging at the other end, as shown in Figure 5.5.6b at $k_1 = 0.170$. This distortion results in the wrinkles seen in Figure 5.5.6b. Moreover, the intersections with the Poincaré section cover the closed curve nonuniformly. The highly wrinkled torus is associated with the development of an inflection point in the (reduced) circle map

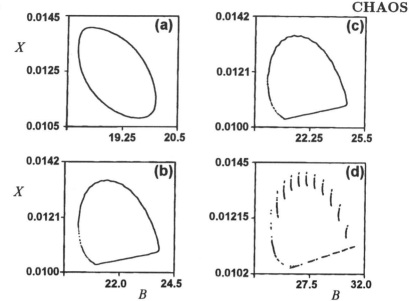

Figure 5.5.6: The four stages of the torus in the peroxidase–oxidase reaction. The Poincaré sections are shown for (a) the smooth, undistorted torus at $k_1 = 0.205$; (b) the wrinkled torus at $k_1 = 0.170$; (c) the fractal torus at $k_1 = 0.1634$; and (d) the broken torus at $k_1 = 0.1178$. Reprinted with permission from Steinmetz and Larter (1991).

constructed from the Poincaré section (Fig. 5.5.7c). The presence of the inflection point means that the inverse of the map may be multivalued, which in turn means that the rotation number is undefined. Therefore, chaos is possible. As k_1 is decreased further, the torus gets further distorted. The coverage of the closed curve of the points of intersection with the Poincaré section becomes increasingly more nonuniform, and then the torus becomes fractal, as shown in Figure 5.5.6c at $k_1 = 0.1634$. The fractal structure of the Poincaré map emphasizes the landmark of folding and stretching of trajectories on the two–torus. The fractal torus is associated with the development of a region of negative slope in the circle map, as seen in Figure 5.5.7d. The negative slope makes the one–dimensional map noninvertible and, consequently, chaos is possible. As k_1 is decreased further, the torus is destroyed, as illustrated in Figure 5.5.6d at $k_1 = 0.1178$. In fact, the

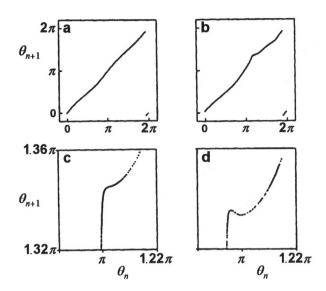

Figure 5.5.7: Circle maps illustrating the transition from quasiperiodicity to chaos. The corresponding values of k_1 are (a) 0.2057, (b) 0.190, (c) 0.170, and (d) 0.1634. Reprinted with permission from Steinmetz and Larter (1991).

graph is a section through a strange attractor. Therefore, chaos has emerged as a result of the destruction of the torus.

The transition from a fractal torus to a broken torus can be characterized through Poincaré sections as discussed in the previous paragraph. There is an alternative way to characterize this transition. Thus far, the discussed states of the peroxidase–oxidase reaction beyond the SHB curve are either quasiperiodic or chaotic. For $k_1 < k_{SHB}$, Larter and Steinmetz (1991) also found periodic states alternating between quasiperiodic and chaotic states. These periodic states can be used to characterize the transition from a fractal torus to a broken torus. The time traces and Poincaré sections for three such states are depicted in Figure 5.5.8. These complex periodic states cannot be characterized in terms of a single frequency as simple periodic oscillations can be. The finite number of points in the Poincaré sections shown in Figures 5.5.8d–f confirm the periodicity of these states. The section shown in Figure 5.5.8d corresponds to a phase–locked motion on a smooth two–

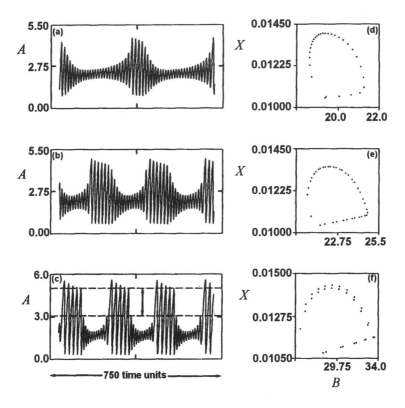

Figure 5.5.8: Phase–locked and mixed–mode oscillations in a model of the peroxidase–oxidase reaction. The corresponding k_1 values are (a) 0.193762205, (b) 0.1621, and (c) 0.1033. Reprinted with permission from Larter and Steinmetz (1991).

torus. The corresponding reduced map is invertible. The periodic oscillations associated with Figures 5.5.8b and 5.5.8e correspond to a phase–locked motion on a wrinkled torus. At the corresponding value of k_1, the reduced map is not invertible. The periodic oscillations in Figure 5.5.8c occur at a value of k_1 at which the torus is broken. The corresponding Poincaré section in Figure 5.5.8f is indicative of phase–locking on a broken torus. In this case, the periodic oscillations are quite complex. One can clearly distinguish between the large– and small–amplitude oscillations in Figure 5.5.8c. However, no such clear distinction can be made for the oscillatory states occurring on a wrinkled or fractal torus. Periodic oscillations, such as those seen in Figure 5.5.8c, where there is a clear disparity between small–amplitude and large–amplitude oscillations, are called **mixed–mode oscillations**. The appearance of these oscillations marks the transition from the fractal torus to the broken torus.

In the literature, the following two notations have been commonly used to characterize mixed–mode oscillations: L^S and P_L^S (e.g., Marek and Schreiber, 1991; Field and Györgyi, 1993). In both of these notations, L is the number of large–amplitude oscillations and S is the number of small–amplitude oscillations per cycle. Thus, the periodic state in Figure 5.5.8c can be denoted by either 5^{10} or P_5^{10}. The rotation number of this state is $(S + L)^{-1}$ or $\frac{1}{15}$.

The rotation number is also helpful in characterizing the transitions from the smooth torus to the wrinkled torus; from the wrinkled torus to the fractal torus; and from the fractal torus to the broken torus. In Figure 5.5.9, we display the variation of the rotation number with respect to the parameter $k_1' = k_{SHB} - k_1$. The interval corresponding to the smooth or undistorted torus is too small to be discerned. The

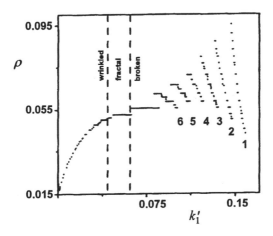

Figure 5.5.9: The rotation number as a function of $k_1' = k_{SHB} - k_1$. The number of large oscillations in each of the cascades on the broken torus is indicated. Reprinted with permission from Larter and Steinmetz (1991).

transition from this torus to the wrinkled one corresponds to the minimum that occurs at a very small value of k_1'. Furthermore, the transition from the wrinkled torus to the fractal torus cannot be easily identified from the variation of the rotation number with k_1'. However, as discussed earlier, this transition can be easily identified from the variation of the reduced or circle map with k_1' because it corresponds to the development of an inflection point in this map. The transition from the fractal torus to the broken torus can be easily identified from the variation of the rotation number with k_1', as seen in Figure 5.5.9. This transition is marked by the appearance of the six devil staircases, which are labeled with the sequence numbers 6, 5, 4, 3, 2, and 1. These numbers represent the number of large–amplitude oscillations L per cycle in each of the cascades of phase–locking states on the broken torus. Each staircase is referred to as an L–**sequence**. In Figure 5.5.10, the 5–sequence staircase is illustrated. Many of the steps in the staircases are associated with very complex periodic states of the form

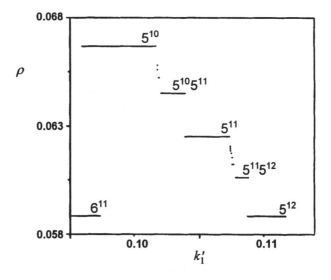

Figure 5.5.10: Variation of the rotation number ρ with k_1' for the 5–sequence staircase The primary and secondary states are labeled. Reprinted with permission from Steinmetz and Larter (1991).

$(P_L^S)^m (P_L^{S+1})^n$. These complex states are referred to as **concatenated states**. They are formed by combining the two primary states P_L^S and P_L^{S+1}. The rotation number for such a complex state is given by

$$\rho = \frac{m + n}{m(L + S) + n(L + S + 1)} \qquad (5.5.10)$$

The steps within each staircase follow a Farey tree construction, as noted by Steinmetz and Larter (1991) and Larter, Olsen, Steinmetz, and Geest (1993). Maselko and Swinney (1986, 1987) observed mixed–mode oscillations in the Belousov–Zhabotinsky reaction, which they explained in terms of a Farey tree construction. Petrov, Scott, and Showalter (1992) examined the development of mixed–mode oscillations in a chemical system.

The mixed–mode oscillations may become chaotic via, for example, a period–doubling scenario or an intermittency scenario. The resulting chaotic states are random mixtures of nearby periodic states. The chaotic state that is a mixture of P_L^S, and P_L^{S+1} is denoted by $C_L^{S,S+1}$, where C stands for chaos. In Figures 5.5.11b–d we display the three periodic states P_1^0, P_1^1 and P_1^2, and in Figure 5.5.11e we display the chaotic state $C_1^{2,3}$ observed in an experiment on the Belousov–Zhabotinsky reaction by Turner, Roux, McCormick, and Swinney (1981). Each chaotic regime can contain many subintervals that are periodic, and chaotic states exist between periodic orbits, as seen in Figure 5.511a. The result is what is called an **alternating periodic–chaotic sequence**. Such sequences are prevalent in chemical reactions (e.g., Swinney, 1983; Marek and Schreiber, 1991; Field and Györgyi, 1993).

5.5.3 Torus Doubling

A second scenario by which a two–period quasiperiodic motion bifurcates into chaos is torus doubling. In this scenario, a fixed–point solution loses stability as a control parameter is varied via a supercritical Hopf bifurcation, leading to the birth of a stable periodic solution characterized by the frequency ω_1. There is a point attractor in a corresponding Poincaré section. As the control parameter is further varied,

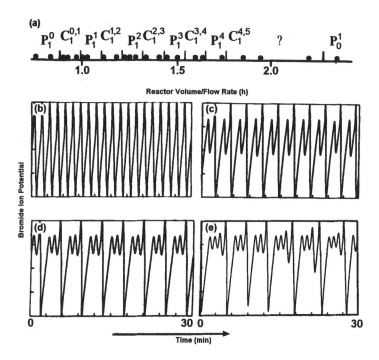

Figure 5.5.11: (a) An alternating periodic–chaotic sequence in the Belousov–Zhabotinsky reaction. (b)–(d) Time series for the states P_1^0, P_1^1, and P_1^2. (e) A time series for the third chaotic state $C_1^{2,3}$, where the number of small–amplitude oscillations following each large–amplitude oscillation is either two or three, but is unpredictable. Reprinted with permission from Turner, Roux, McCormick, and Swinney (1981).

the periodic solution loses stability via a supercritical secondary Hopf bifurcation, producing a second (incommensurate) frequency ω_2. The resulting two–period quasiperiodic attractor is a two–torus. The intersections with a Poincaré section of this attractor densely fill a closed curve when ω_2/ω_1 is irrational. As the control parameter is further varied, the two–torus attractor undergoes a cascade of period–doubling bifurcations in which the period $2\pi/\omega_2$ is doubled in each bifurcation. After each period–doubling bifurcation, the postbifurcation state is a new torus that forms two loops around the original torus.

There are two possibilities: The cascade is either complete or incomplete. In the latter case, the system undergoes a transition to chaos, as in the preceding section, on a deformed torus, as observed by Anishchenko, Letchford, and Safonova (1985) and Bassett and Hudson (1989). When the cascade is complete, chaos emerges at the end of the sequence, and the system response is a chaotically modulated motion. An experimental example is provided in Figure 5.5.12. These results, which show a sequence of period–doubling tori culminating in chaos, were obtained by Balachandran and Nayfeh (1991). They excited a two–beam two–mass structure (see Fig. 7.9.4), near the primary resonance of the second vibration mode at a constant excitation level of 30 mili g rms. The response was monitored with two strain gauges, which were mounted along the axes of the horizontal and vertical beams, as shown in Figure 7.9.4. The results shown in Figure 5.5.12 are Poincaré sections of the motion obtained by stroboscoping the strain–gauge outputs at the excitation frequency. The intersections in Figure 5.5.12a uniformly fill up, a closed curve, indicating a smooth or undistorted two–torus. We see two loops of intersection points in Figure 5.5.12b. Each of these loops is similar to the single loop seen in Figure 5.5.12a. It appears that two asymmetric attractors (the one

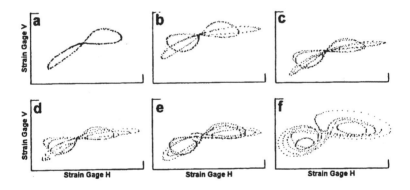

Figure 5.5.12: Torus doublings observed in an experiment with a two–beam two–mass structure. Shown are Poincaré sections of the motion. Reprinted with permission from Balachandran and Nayfeh (1991).

associated with Fig. 5.5.12a and its reflected version) merged to form the attractor associated with Figure 5.5.12b. The attractor associated with Figure 5.5.12b undergoes a sequence of torus–doubling bifurcations that result in the creation of a chaotic state. Poincaré sections of the attractor after the first, second, and third bifurcations are shown in Figures 5.5.12c–e, respectively. Figure 5.5.12f corresponds to the chaotic attractor observed at the culmination of the scenario. The power spectra obtained in the experiments are also illustrative of this sequence (Balachandran and Nayfeh, 1991).

In the experiments of Anishchenko, Letchford, and Safonova (1985), the two–torus attractor realized after the first torus–doubling bifurcation was destroyed by the subsequent bifurcations, resulting in chaos. In other experiments with a nonlinear circuit, Anishchenko, Astakhov, Letchford, and Safonova (1983a) observed a complete torus–doubling sequence to chaos. Torus doubling was also observed in the string experiments of Molteno and Tufillaro (1990) and Molteno (1993). Both incomplete and complete sequences of torus doublings have also been observed in many analytical and numerical studies. A partial list includes the studies of Arneodo, Coullet, and Spiegel (1983), Franceschini (1983), Kaneko (1984), Miles (1984), Tousi and Bajaj (1985), Nayfeh and Zavodney (1986), Gu and Sethna (1987), Nayfeh (1987a,b, 1988), Nayfeh and Nayfeh (1990), Nayfeh and Raouf (1987), Streit, Bajaj, and Krousgrill (1988), Johnson and Bajaj (1989), Umeki and Kambe (1989), Bajaj and Johnson (1990, 1992), Bajaj and Tousi (1990), Miles and Henderson (1990), Pai and Nayfeh (1990), Raouf and Nayfeh (1990a,b) Nayfeh, Raouf, and Nayfeh (1991), Restuccio, Krousgrill, and Bajaj (1991), and Steindl and Troger (1991).

5.6   CRISES

The term **crisis** was introduced by Grebogi, Ott, and Yorke (1983a) to describe certain sudden qualitative changes in the chaotic dynamics of dissipative dynamical systems as a control parameter is varied. A **crisis** occurs when a chaotic attractor comes into contact with an unstable periodic solution. It is to be noted that when a chaotic attractor comes into contact with the stable manifold of either an unstable periodic solution or a saddle point, due to the nature of the stable manifold, it comes into contact with the unstable periodic motion or the saddle point.

Grebogi, Ott, Romeiras, and Yorke (1987) distinguished three types of crises, according to the nature of the discontinuity induced in the chaotic attractor. In the first type, the chaotic attractor is suddenly destroyed as the control parameter α passes through its critical crisis value α_c; this is called a **boundary** or an **exterior crisis**. In the postbifurcation state, the motion is transiently chaotic before it tends to either a bounded motion (i.e., fixed point, periodic solution, quasiperiodic solution, or chaotic solution) or an unbounded solution. A boundary crisis is an example of what is called a **blue sky catastrophe** or **dangerous bifurcation**. This catastrophe refers to the sudden disappearance of an attractor from the state space of a system (Abraham, 1985; Thompson and Stewart, 1986; Abraham and Shaw, 1992).

In the second type, the size of the chaotic attractor suddenly increases as α is varied through α_c; this is called an **interior crisis**. During this crisis, the chaotic attractor collides with an unstable equilibrium or periodic solution that is in the interior of the basin of attraction. In the third type, two or more chaotic attractors of a system with symmetries merge to form one chaotic attractor as α is varied through α_c; this is called an **attractor merging crisis**. The new chaotic attractor can be larger in size than the union of the chaotic attractors before the crisis. As the parameter α is varied in the other direction, the inverse of these crises occurs; that is, a sudden creation, shrinking, or splitting of a chaotic attractor occurs. The latter two crises are sometimes called **explosive bifurcations** (Thompson, Stewart, and Ueda, 1994). In each of these crises, the postbifurcation state is characterized by a certain temporal behavior. This behavior has been quantified in terms of time scales by Grebogi, Ott, Romeiras, and Yorke (1987).

Crises have been observed in many experimental and numerical studies (e.g., Rössler, 1976b; Simó, 1979; Huberman and Crutchfield, 1979; Ueda, 1980a; Grebogi, Ott, and Yorke, 1983a; Jeffries and Perez,

1983; Brorson, Dewey, and Linsay, 1983; Ikezi, deGrassie, and Jensen, 1983; Gaspard and Nicolis, 1983; Rollins and Hunt, 1984; Gwinn and Westervelt, 1985; Iansiti, Hu, Westervelt, and Tinkham, 1985; Dangoisse, Glorieux, and Hennequin, 1986; Ishii, Fujisaka, and Inoue, 1986; Carroll, Pecora, and Rachford, 1987; Gu and Sethna, 1987; Streit, Bajaj, and Krousgrill, 1988; Bajaj, 1991; Bajaj and Johnson, 1992; Wang, Abed, and Hamdan, 1994; and Nayfeh and Chin, 1994).

Next, we give several examples of crises in discrete– and continuous– time systems.

Example 5.10. Following Grebogi, Ott, and Yorke (1983a), we consider the one–dimensional quadratic map

$$x_{n+1} = a - x_n^2 \tag{5.6.1}$$

Its bifurcation diagram is shown in Figure 5.6.1. The fixed points of this map are given by

$$x = a - x^2 \tag{5.6.2}$$

or

$$x_s = -\frac{1}{2} + \sqrt{\frac{1}{4} + a} \ \text{ and } \ x^* = -\frac{1}{2} - \sqrt{\frac{1}{4} + a} \tag{5.6.3}$$

Hence, when $a < -\frac{1}{4}$, there are no fixed points, and all orbits initiated from $|x| < \infty$ tend to $-\infty$ as n tends to ∞. As a is increased from a value less than $-\frac{1}{4}$, a tangent bifurcation occurs at $a = -\frac{1}{4}$ at which a stable and an unstable fixed point are created. The eigenvalue of the Jacobian of the map is given by

$$\lambda = -2x$$

For the fixed point x_s,

$$\lambda = 1 - \sqrt{1 + 4a} \tag{5.6.4}$$

and hence it is stable for $-\frac{1}{4} < a < \frac{3}{4}$. On the other hand, for the fixed point x^*,

$$\lambda = 1 + \sqrt{1 + 4a} \tag{5.6.5}$$

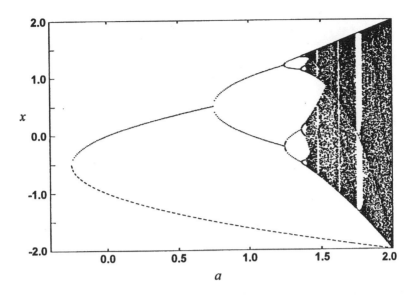

Figure 5.6.1: Bifurcation diagram for the quadratic map.

and hence it is unstable for all $a > -\frac{1}{4}$. The evolutions for all initial conditions greater than $|\ x^*\ |$ diverge to $-\infty$ and for all initial conditions less than $|\ x^*\ |$ converge to x_s. In other words, the basin of attraction of the periodic orbit x_s is $|\ x\ | \leq -x^*; x^*$ is plotted as a dashed curve in Figure 5.6.1.

As a is increased past $\frac{3}{4}$, the stable fixed point undergoes a period–doubling bifurcation because λ exits the unit circle through -1, resulting in the creation of two period–two orbits and an unstable period–one orbit. Further increases in a lead to a whole cascade of period doublings. This sequence has a finite accumulation point at $a = a_\infty = 1.40095$. At this point, there is an infinite number of unstable periodic orbits because each period–doubling bifurcation gives rise to an unstable periodic orbit. As a is increased beyond a_∞, a chaotic attractor emerges. The basin of attraction is again $|\ x\ | \leq -x^*$. There are many small intervals of a values between a_∞ and 2 for which the unique stable solution is periodic, repeating exactly after m iterates. In other words, there are many small windows of periodic solutions, as seen in Figure 5.6.1. As $a \to 2$, the periodic windows become narrower and chaotic

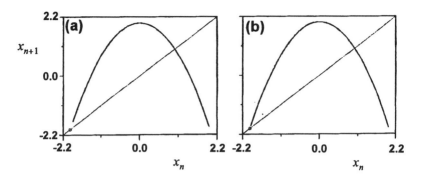

Figure 5.6.2: Return maps of the quadratic map: (a) a $= 1.97$ and (b) a $=$ 1.99.

solutions predominate. The basin of attraction of the periodic and chaotic orbits for $-\frac{1}{4} < a < 2$ is given by $\mid x \mid \leq -x^*$.

As a is increased past 2, the attracting chaotic orbit is destroyed, and all initial conditions tend to orbits that approach $-\infty$. In other words, the chaotic attractor experiences a crisis or a blue sky catastrophe at $a = 2$. It is clear from Figure 5.6.1 that as $a \to 2$ the size of the chaotic attractor increases, and at $a = 2$ it touches the unstable fixed point x^*. This can be seen more clearly in Figure 5.6.2, where we show the return maps for $a = 1.97$ and 1.99. The small circle on the identity map $x_{n+1} = x_n$ is the unstable point x^*. The heavy dots are 1,000 long–term iterates computed from (5.6.1) after discarding the first 500 iterates. Comparison of parts a and b of Figure 5.6.2 shows that the left edge of the chaotic attractor moves closer to the unstable point x^* as $a \to 2$. For the one–dimensional map under consideration, the left edge of a chaotic attractor is the second image of $x = 0$; that is, $x = a - a^2$. Hence, the distance between the left edge of the chaotic attractor at a and the unstable fixed point x^* at a is (Stewart and Lansbury, 1992)

$$d = a - a^2 + \frac{1}{2} + \sqrt{\frac{1}{4} + a} \qquad (5.6.6)$$

which is a continuous function of a. It is easily seen that $d = 0$ when $a = 2$. Thus, the destruction of the chaotic orbit coincides with its collision with the unstable fixed point x^* on its basin boundary. As a

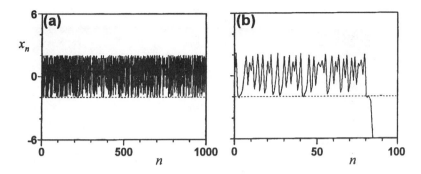

Figure 5.6.3: Time series x_n for the quadratic map near the boundary crisis: (a) a = 1.99 and (b) a = 2.01.

result, both the chaotic attractor and its basin are destroyed. This is an example of a boundary or an exterior crisis (Grebogi, Ott, and Yorke, 1983a) or a blue sky catastrophe (Abraham, 1985; Abraham and Shaw, 1992).

Beyond $a = 2$, the chaotic attractor is unstable to perturbations, and for all initial conditions the iterates ultimately diverge to $-\infty$. The result is a discontinuous or catastrophic disappearance or death of the chaotic attractor. However, it leaves behind a definite signature called **transient chaos**. An example is shown in Figure 5.6.3b. For values of a slightly larger than 2, typical orbits started with initial conditions in the region formerly occupied by the destroyed chaotic attractor appear to bounce around in this region (ghost or phantom of the chaotic attractor) in a chaotic fashion, which is indistinguishable from the behavior (Fig. 5.6.3a) for values of a slightly less than 2. This behavior may extend for a possibly long time, depending on the initial conditions, but the orbit will eventually move away from the region of the ghost attractor and diverge to $-\infty$. This phenomenon, where the initial dynamics of the system appears to be chaotic, is called **transient chaos** (Yorke and Yorke, 1979; Parker and Chua, 1987). The time length of a chaotic transient depends sensitively on the initial conditions. However, looking at a very large ensemble of initial conditions in the basin of the ghost or phantom attractor close to α_c, one finds that the time length of chaotic transients is given by the exponential probability density (Grebogi et

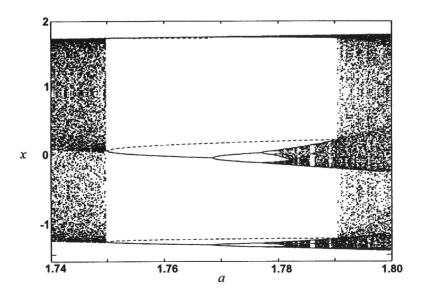

Figure 5.6.4: Enlargement of the bifurcation diagram in the region of the period–three window.

al., 1987)

$$P(\tau) \sim \exp\left(-\frac{\tau}{\tau_0}\right) \tag{5.6.7}$$

where τ_0 is the characteristic transient life time.

As a is decreased through 2 (i.e., the control parameter α is varied through α_c in the other direction), the crisis creates rather than destroys a chaotic attractor.

During a second type of crisis (namely, **interior crisis**) in the quadratic map, the chaotic attractor collides with an unstable orbit within its basin of attraction. To illustrate this, we enlarge the period–three window in Figure 5.6.1. The result is shown in Figure 5.6.4. Period–three orbits are solutions of

$$x = f^3(x; a) \quad \text{where} \quad f(x; a) = a - x^2 \tag{5.6.8}$$

There are two obvious solutions of (5.6.8), namely, the period–one solutions given by (5.6.3). In this range of a, both of them are unstable. The other solutions can be obtained graphically or numerically. In

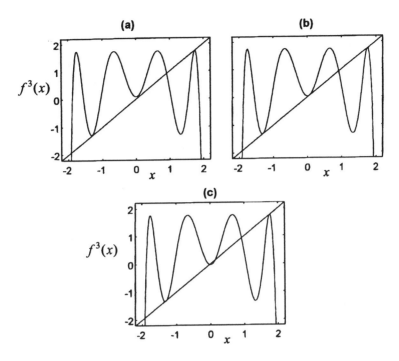

Figure 5.6.5: Third iterated maps of the quadratic map: (a) $a = 1.74$, (b) $a = 1.75$, and (c) $a = 1.76$.

Figure 5.6.5, we plot the functions $f^3(x)$, the third return map, and x for three values of a: (a) at the left of the window where a $= 1.74$, (b) at the start of the window where a $= 1.75$, and (c) inside the window where a $= 1.76$. At the left of the window, the curve $f^3(x)$ intersects the curve $y = x$ twice: once at x_s and the other at x^*, both of which are unstable at this value of a. At the start of the window, the curve $f^3(x)$ is tangent to the curve $y = x$ at three values of x. These three points are the cyclic steady–state values at the start of the window. As discussed earlier, this type of bifurcation is called **tangent bifurcation**. For values of a slightly larger than 1.75, one can expect intermittent chaos, as discussed in Section 5.4. For a slightly larger than 1.75, in addition to the two isolated intersections found for $a < 1.75$, the curve $f^3(x)$ intersects the curve x at three pairs of values of x. The slopes of $f^3(x)$ in the neighborhood of three of the latter six points (one from each close pair) are sufficiently steep that the trajectory wanders away from them. In other words, the eigenvalues of the Jacobian of $f^3(x)$ at these points are outside the unit circle, and hence they are unstable. The dashed curves in Figure 5.6.4 denote the unstable period–three orbit created at the tangent bifurcation. On the other hand, the eigenvalues of the Jacobian of $f^3(x)$ at the other three points are inside the unit circle, and hence they are stable. Consequently, the cyclic behavior initiated by the tangent bifurcation continues to be stable.

As a increases past 1.7692, the stable period–three orbit undergoes a sequence of period–doubling bifurcations, resulting in chaotic bands at $a \approx 1.7806$. We note that, for a range of values of a less than a critical value $a_3^* \simeq 1.790327492$, the chaotic attractor lies within three distinct bands, but that, as a increases beyond a_3^*, the three chaotic bands widen to form a single chaotic band. This widening coincides with the three chaotic bands simultaneously colliding with the unstable period–three orbit (dashed curves in Fig. 5.6.4) created at $a = 1.75$, the location of the tangent bifurcation that initiated the period–three window. This bifurcation is called an **interior crisis** by Grebogi, Ott, and Yorke (1983a) because the sudden change in the chaotic attractor is the result of its collision with an unstable orbit within its basin boundary ($|x| \leq -x^*$). Such an interior crisis was first documented in a differential equation by Ueda (1980a,b, 1991).

As a is decreased below a_3^*, the single chaotic attractor splits into three chaotic bands. Interior crises are examples of explosive bifurcations.

An interior crisis is accompanied by what Grebogi, Ott, and Yorke (1983a) call a **crisis–induced intermittency**. For a value of a slightly larger than a_3^*, the orbit on the attractor spends long stretches of time in the region to which the old (ghost) attractor was confined before the crisis. Following these stretches of time, the attractor bursts from the ghost region and bounces around in the new enlarged region. It then returns back to the ghost region for another stretch of time and then bursts again, and so on. In Figure 5.6.6a, we plot the time series obtained for every third iterate of the map for $a = 1.7903 < a_3^*$ just before the crisis. At this value of a, the orbit cycles through the three chaotic bands. Because every third iterate is plotted, the orbit in Figure 5.6.6a is confined to one of the three bands. In Figures 5.6.6b and 5.6.6c,

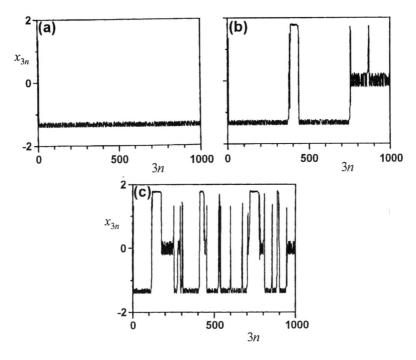

Figure 5.6.6: Time series x_{3n} for the quadratic map near the crisis terminating the period–three window: (a) $a = 1.7903$, (b) $a = 1.79033$, and (c) $a = 1.7904$.

we plot the orbits for the two values 1.79033 and 1.7904 of a above a_3^*. When a is slightly larger than a_3^*, it is clear from Figure 5.6.6b that for long stretches of time the orbit remains in one band and occasionally bursts out of it, but the orbit returns back to this band or to one of the other bands. As a increases further away from a_3^*, as is the case in Figure 5.6.6c, the time stretch during which the orbit remains in one band decreases and the frequency of the bursts increases.

Similar interior crises, resulting in the widening of a chaotic attractor, are associated with other tangent bifurcations. In Figure 5.6.7, we plot the functions $f^5(x)$ and x for three values of a near the start of the period–five window: (a) at the left of the window where $a = 1.615$, (b) at the start of the window where $a = 1.624399$, and (c) inside the window where $a = 1.64$. At the left of the window, the curve $f^5(x)$

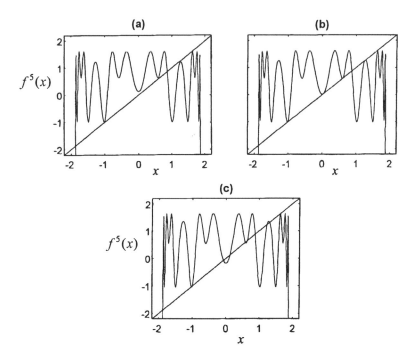

Figure 5.6.7: Fifth iterated maps of the quadratic map: (a) $a = 1.615$, (b) $a = 1.624399$, and (c) $a = 1.64$.

crosses the curve x at the two isolated points x^* and x_s, which are un–stable at this value of a. At the start of the window, the curve $f^5(x)$ is tangent to the curve x at five values of x. Again, these five points are the cyclic steady–state values at the start of the window. This is a tangent bifurcation, which is also associated with intermittent chaos, as discussed in Section 5.4. Inside the window (Fig. 5.6.7c), in addition to the two isolated crossings that are unstable, the curve $f^5(x)$ crosses the curve x at five pairs of values of x. Five of these points (one from each close pair) correspond to an unstable period–five orbit denoted by a broken curve in the enlarged bifurcation diagram in Figure 5.6.8, and the other five points correspond to a stable period–five orbit.

It follows from Figure 5.6.8 that, for a range of values of a less than the critical value $a_5^* \approx 1.633359$, the chaotic attractor lies within five distinct bands. As a increases beyond a_5^*, the five chaotic bands suddenly widen to form a single band. This sudden change coincides

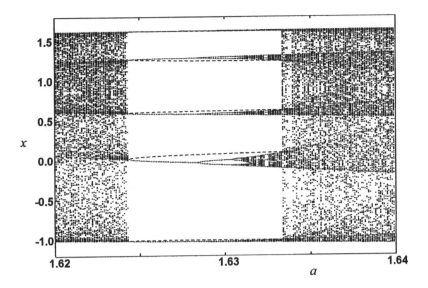

Figure 5.6.8: Enlargement of the bifurcation diagram in the region of the period–five window.

precisely with the collision of the five chaotic bands with the unstable period–five orbit, which lies within the basin of attraction of the five–band attractor. As a decreases through a_5^*, the single chaotic attractor undergoes a **reverse interior crisis**, resulting in the splitting of the attractor into five chaotic bands.

Again, the interior crisis at the end of the period–five window is associated with a crisis–induced intermittency. In Figure 5.6.9a, we plot the time series obtained for every fifth iterate just before the crisis. The attractor in this case cycles through five bands, and, because the fifth iterate is plotted, the trajectory is confined to one band. For $a = 1.63336$, which is slightly larger than a_5^*, the trajectory remains in one band for a long stretch of time before it bursts out of it and then the orbit returns back to this band or to one of the other four bands. As a is increased further, the bursting frequency increases, as is evident in Figure 5.6.9c at $a = 1.63337$.

The third type of crisis present in the dynamics of the quadratic map is what Grebogi, Ott, and Yorke (1987) call **attractor merging**

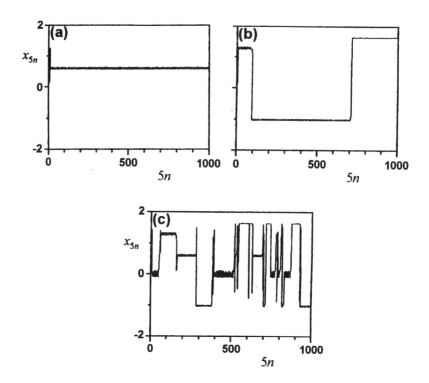

Figure 5.6.9: Time series x_{5n} for the quadratic map near the crisis terminating the period–five window: (a) $a = 1.63335$, (b) $a = 1.63336$, and (c) $a = 1.63337$.

crisis. To describe this crisis, we show in Figure 5.6.10 an enlargement of the bifurcation diagram in Figure 5.6.1. For a slightly less than $a_3 \approx 1.4070$, there are eight chaotic bands, while for a slightly larger than a_3, there are four chaotic bands. For $a < a_3$, an orbit initiated in one of these bands will return to that band after eight iterates. Thus, a band can be thought of as an attractor of $f^8(x)$, the eighth–iterated map. Similarly, the band with which it merges at a_3 can be thought of as an attractor of $f^8(x)$. Thus, at $a = a_3$, we have simultaneous crises of these two bands in which they collide with the unstable period–four orbit between these bands. In Figure 5.6.11, we show the eight–band attractor at $a = 1.403$ and the unstable period–four orbit denoted by the crosses. As a is increased past a_3, the bands collide with the

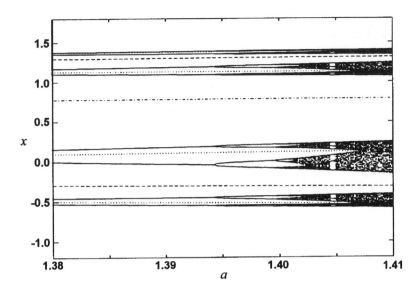

Figure 5.6.10: Enlargement of the bifurcation diagram of the quadratic map illustrating the attractor merging crisis.

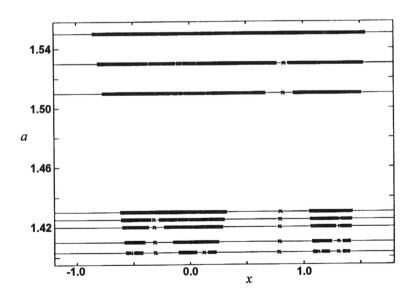

Figure 5.6.11: Illustration of attractor merging crises.

unstable period–four orbit, resulting in their merger. The resulting four–band chaotic attractor at $a = 1.41$ is shown.

The four–band chaotic attractor increases in size as a is increased past 1.41 and gets closer to the unstable period–two orbit marked by the crosses in Figure 5.6.11. At $a = a_2 \approx 1.4297$, the bands collide with the unstable periodic orbit between them, resulting in their merger. The resulting two–band chaotic attractor at $a = 1.43$ is shown.

Again, as a is increased beyond 1.43, the two bands increase in size and move closer to the unstable period–one orbit marked by the cross in Figure 5.6.11. At $a = a_1 \approx 1.5425$, the two bands collide with the unstable periodic orbit between them, resulting in their merger. The resulting single–band attractor at $a = 1.55$ is also shown in Figure 5.6.11.

Example 5.11. Following Abraham and Stewart (1986), we consider the velocity–forced van der Pol oscillator

$$\dot{x} = 0.7y + 10x(0.1 - y^2) \tag{5.6.9}$$

$$\dot{y} = -x + 0.25 \sin{(1.5t)} + C \tag{5.6.10}$$

In Figure 5.6.12a, we show the Poincaré section at a driving angle π for $C = 0.08$. At this value of C, the system has a chaotic (Birkhoff–Shaw) attractor and a saddle limit cycle. The Poincaré section of this limit cycle is the saddle point represented by the asterisk symbol *. The stable manifold of the saddle limit cycle is a smooth two–dimensional surface whose Poincaré section is the one–dimensional stable manifold of the saddle point, as shown in Figure 5.6.12a. The chaotic attractor is represented by 1,000 return points computed from a single trajectory. The stable manifold of the saddle limit cycle forms the basin boundary of the chaotic attractor. As C is increased, the saddle limit cycle and its stable manifold move closer to the chaotic attractor. Moreover, the stable manifold develops a finger that also moves closer to the chaotic attractor, as seen in the Poincaré section in Figure 5.6.12b at $C = 0.09$. As C is increased further to a critical value C_c, the chaotic attractor collides with the stable manifold of the saddle limit cycle (its basin boundary). As a result, the chaotic attractor and its basin of attraction

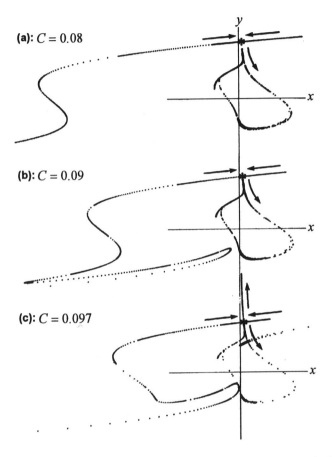

(a): $C = 0.08$

(b): $C = 0.09$

(c): $C = 0.097$

Figure 5.6.12: Poincaré sections of the orbits of the asymmetrically forced van der Pol equations (5.6.9) and (5.6.10) showing a blue sky disappearance of a Birkhoff–Shaw chaotic attractor by collision with an unstable periodic orbit (Thompson and Stewart, 1986).

are destroyed in a boundary crisis. For $C > C_c$, the chaotic attractor no longer exists but is replaced with a chaotic transient. In Figure 5.6.12c, we show the Poincaré section at $C = 0.097$, which is slightly larger than the crisis value C_c. If one starts a trajectory with an initial condition that is contained in the basin of attraction of the chaotic attractor that exists for $C < C_c$, one will find that this trajectory typically moves toward the region in the phase space of the $C < C_c$ attractor (ghost attractor), bounces around in this region chaotically for a period of time that depends on the initial conditions and the closeness of C to C_c, suddenly leaves this region, and moves off toward infinity, a chaotic transient. In Figure 5.6.13a, we show the time history of the chaotic attractor at $C = 0.096$, and in Figure 5.6.13b we show the chaotic transient at $C = 0.097$.

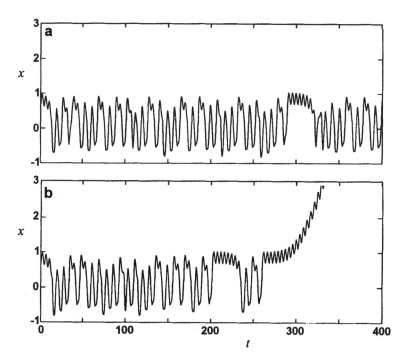

Figure 5.6.13: Typical temporal evolutions of $x(t)$: (a) $C = 0.096$ and (b) $C = 0.097$.

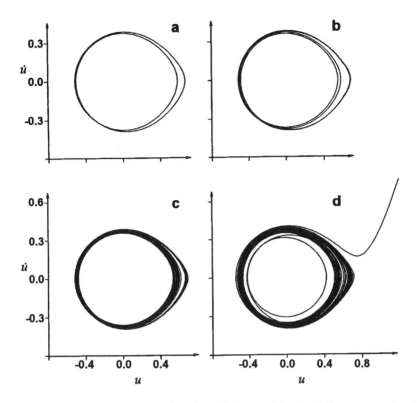

Figure 5.6.14: Phase portraits for the solutions of (5.6.11) demonstrating (a) period doubling at $\Omega = 0.6200$, (b) period quadrupling at $\Omega = 0.6130$, (c) chaotic motion at $\Omega = 0.6117$, and (d) unbounded motion at $\Omega = 0.6116$.

Example 5.12. We consider, after Nayfeh and Khdeir (1986a,b), the roll response of a ship in beam seas modeled by

$$\ddot{\theta} + (0.7037)^2(\theta - 0.598\theta^3 - 0.939\theta^5) + 0.0455\dot{\theta} + 0.2\dot{\theta}^3$$
$$= (0.7037)^2(\theta_s - 0.598\theta_s^3 - 0.939\theta_s^5) + 0.15\cos\Omega t \quad (5.6.11)$$

when $\theta_s = 0.13963$ radians. Starting from a value of $\Omega = 0.8$, one finds that the response is periodic with the period $T = 2\pi/\Omega$. As Ω is decreased, the period–one limit cycle deforms and increases in size. As Ω is decreased below approximately 0.626, the motion undergoes a sequence of period–doubling bifurcations, culminating in chaos at

$\Omega \approx 0.6117$. In Figures 5.6.14a–c, we show the phase portrait $u - \dot{u}$, where $u = \theta - \theta_s$, of the period–two, period–four, and chaotic motions for $\Omega = 0.6200, 0.6130$, and 0.6117. As Ω is decreased below a critical value Ω_c, which is very close to 0.6117, the chaotic attractor collides with the boundary of its basin of attraction. As a result, the chaotic attractor and its boundary are destroyed. Considering a trajectory with the same initial conditions as in Figure 5.6.14c and setting $\Omega = 0.6116$, which is slightly below Ω_c, we obtain the orbit shown in Figure 5.6.14d. The presence of a chaotic transient is clear. The associated orbit is initially attracted to the phase–space region formerly occupied by the chaotic attractor for $\Omega < \Omega_c$. It bounces around in this region apparently for a long time, but suddenly moves away from this region and approaches a distant solution, which is infinity in this case.

Example 5.13. We follow Nayfeh, Hamdan, and Nayfeh (1990) and consider the response of a single–machine quasi–infinite busbar power system modeled by

$$\ddot{\theta} + 0.2025\dot{\theta} = 1 - 1.969(1 + 0.1 \cos \Omega t) \sin(\theta - 0.1 \sin \Omega t) \quad (5.6.12)$$

Starting from a value of Ω larger than 9, we find that the response is periodic, having the period $T = 2\pi/\Omega$. As Ω is decreased, the period–one orbit deforms, increases in size, and then undergoes a sequence of period–doubling bifurcations, culminating in chaos. The phase portraits of the period–two, period–four, and chaotic orbits found at $\Omega = 8.4, 8.28$, and 8.274 are shown in Figures 5.6.15a–c. As Ω is decreased to Ω_c, which is slightly less than 8.274, the chaotic attractor collides with the boundary of its basin of attraction. As a result, the attractor and its basin are destroyed. For an $\Omega < \Omega_c$, such as 8.26, the orbit bounces around chaotically in the phase–space region formerly occupied by the chaotic attractor for $\Omega > \Omega_c$, but eventually moves away from this region and approaches infinity.

Example 5.14. Following Thompson and Stewart (1986), we consider the velocity–forced van der Pol oscillator

$$\dot{x} = y - B \sin(1.9t) \quad (5.6.13)$$

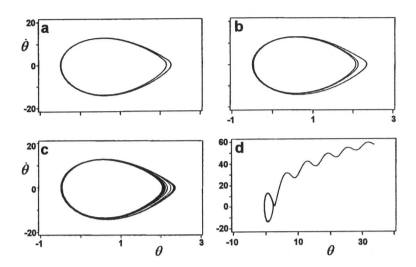

Figure 5.6.15: Phase portraits of (5.6.12): (a) $\Omega = 8.4$, (b) $\Omega = 8.28$, (c) $\Omega = 8.274$, and (d) $\Omega = 8.26$. Reprinted with permission from Nayfeh, Hamdan, and Nayfeh (1990).

$$\dot{y} = -x + (1 - x^2)y \qquad\qquad (5.6.14)$$

In Figure 5.6.16a, we show the Poincaré sections of the attractors at $t = 0$ and $B = 1.0$. The attractors, which are two in number, are in the form of Rössler bands. The associated intersections of a trajectory with the Poincaré section jump alternatively between pairs of opposite bands. The Poincaré section is also transverse to an unstable period–two orbit of the saddle type. In Figure 5.6.16, we mark each intersection of this saddle limit cycle with the Poincaré section by the asterisk symbol *. On this section, between a pair of adjacent pieces, there is a saddle point. The stable manifold of the saddle limit cycle is very close to the banded attractors. As B is increased to a critical value $B_c \approx 1.02$, the two attractors simultaneously collide with the stable manifold of the saddle limit cycle. As a consequence of this crisis, the two chaotic attractors merge to form a single larger attractor of the Birkhoff–Shaw type, as shown in Figure 5.6.16b at $B = 1.03$.

Example 5.15. Following Ishii, Fujisaka, and Inoue (1986), we

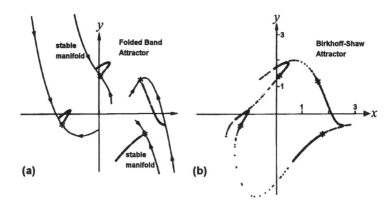

Figure 5.6.16: Poincaré sections illustrating a crisis in equations (5.6.13) and (5.6.14). There are banded chaotic attractors preceding the crisis and a Birkhoff–Shaw bagel following the crisis (Thompson and Stewart, 1986).

consider the response of a damped single–degree–of–freedom system to a harmonic excitation; that is,

$$\ddot{x} + \dot{x} - 10x + 100x^3 = F \sin(4.5t) \qquad (5.6.15)$$

The potential energy of this system has a maximum at $x = 0$ (saddle) and two minima at $x = \pm\sqrt{0.1}$ (centers). In other words, the potential has two wells. In the presence of damping, $x = 0$ remains a saddle, but the centers become stable foci. For values of F slightly less than $F_c \approx 0.8495$, there are two symmetrically disposed chaotic attractors, one confined to each well. One of the symmetric attractors is shown in Figure 5.6.17a for $F = 0.8492$. As F is increased, the two symmetric attractors enlarge, and at $F = F_c$ both touch simultaneously the stable manifolds of the two period–three saddle limit cycles, which form the boundary separating their basins of attraction. In other words, they collide simultaneously with the saddle limit cycles on their basin boundary. For $F > F_c$, the two chaotic attractors experience an attractor merging crisis and become a single attractor, subsuming the two single attractors, which exist for $F < F_c$, as shown in Figure 5.6.17b. The orbit following the crisis switches intermittently from one well to the other. In Figure 5.6.18a, we show the time history of one of the symmetric attractors for $F = 0.849 < F_c$. Clearly, it is confined

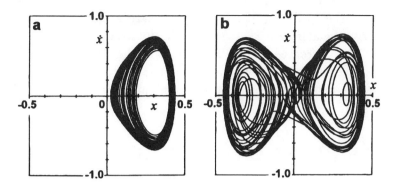

Figure 5.6.17: Typical phase space portraits of (5.6.15) for (a) $F = 0.8492 <$ F_c and (b) $F = 0.865 > F_c$. For $F = 0.8492$, there is another chaotic attractor for $x < 0$ statistically the same as in (a), which is realized for different initial conditions. Reprinted with permission from Ishii, Fujiska, and Inoue (1986).

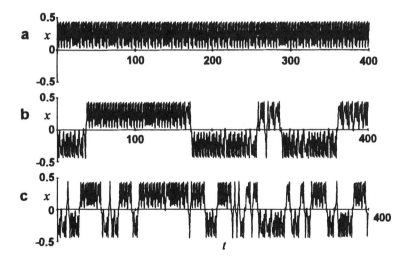

Figure 5.6.18: Typical temporal evolutions $x(t)$ for (a) $F = 0.849$, (b) $F = 0.853$, and (c) $F = 0.865$. Reprinted with permission from Ishii, Fujiska, and Inoue (1986).

to the right well. In Figure 5.6.18b, we show the time history of the attractor for $F = 0.853$, which is slightly larger than the crisis value F_c. Clearly, the attractor switches chaotically from the right to the left well. As F is increased to 0.865, which is further from F_c than in the previous case, the frequency of switching of the attractor from one well to the other increases, as is evident from comparing parts b and c of Figure 5.6.18.

5.7   MELNIKOV THEORY

In Section 2.1.3, we defined homoclinic and heteroclinic orbits to saddle points. These definitions can be generalized to orbits that are homoclinic or heteroclinic to general invariant sets I_i, such as limit cycles. If the orbit of a point p in phase space of a mapping or a flow approaches an invariant set I as t approaches $\pm\infty$, then the orbit of p is said to be **homoclinic** to I. On the other hand, if the orbit of a point p in phase space of a mapping or a flow approaches an invariant set I_1 as t approaches $+\infty$ and approaches another invariant set I_2 as t approaches $-\infty$, then the orbit of p is said to be **heteroclinic** to I_1 and I_2.

5.7.1   Homoclinic Tangles

To describe homoclinic tangles, we consider the system

$$\ddot{x} - x + x^3 + 0.125\dot{x} = F\cos\Omega t \qquad (5.7.1)$$

where F is a constant. The potential energy of the system has two minima and one maximum, a two–well potential. The phase portrait of the undamped and unforced system is shown in Figure 5.7.1a. There is a saddle S at $(x, \dot{x}) = (0, 0)$, corresponding to the maximum of the potential energy, and two centers C_1 and C_2, corresponding to the two minima of the potential energy, at $(x, \dot{x}) = (\pm 1, 0)$. The orbit labeled Γ approaches the saddle as t approaches $\pm\infty$, and hence Γ is homoclinic to the saddle point. The stable and unstable manifolds W^s and W^u of the saddle intersect nontransversely (they are tangent to each other).

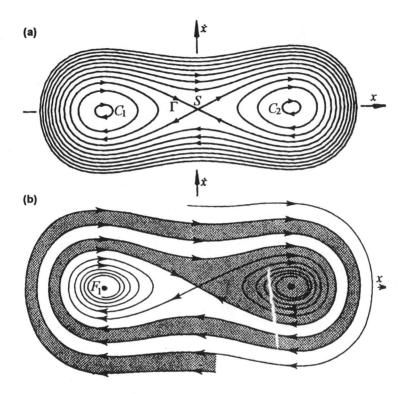

Figure 5.7.1: Undamped and damped phase trajectories for the nonlinear oscillator (5.7.1) when $F = 0$. There are two stable equilibrium states and one unstable equilibrium state.

Including the damping term destroys the homoclinic orbit, as shown in Figure 5.7.1b. The saddle remains a saddle S, but the centers C_1 and C_2 become the stable foci F_1 and F_2, respectively. Now, the stable and unstable manifolds W^s and W^u of the saddle do not intersect, and the stable manifold of S divides the phase space into two regions. The region to the right of W^s is the basin of attraction of the right focus F_2 (motions confined to the right well of the potential), whereas the region to the left of W^s is the basin of attraction of the left focus F_1 (motions confined to the left well of the potential).

If the system is driven by a weak periodic excitation (i.e., a small F), the same diagram in Figure 5.7.1b can be regarded as the

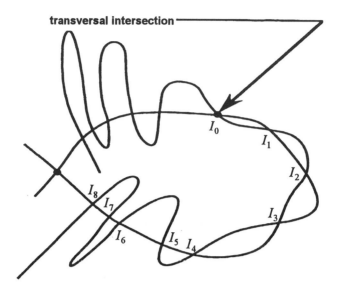

Figure 5.7.2: Illustration of the homoclinic tangle occurring at a hyperbolic saddle point. The tangle forms with an infinite number of intersections $I_0, I_1, I_2, \cdots$. Two nearby points may be mapped far apart, as is the case on a chaotic attractor.

Poincaré section of the three–dimensional phase space $(x, \dot{x}, t)$, except that each line should be regarded as a sequence of dots corresponding to successive intersections of a trajectory with the Poincaré section because an orbit of a Poincaré map is a sequence of points. As the excitation amplitude F is increased with Ω being kept fixed, the unstable and stable manifolds move closer to each other, and at a critical value F_c of F they may touch or intersect at a point I_0, as shown in Figure 5.7.2. We note that the actual trajectories in the (x_1, x_2, t) space do not intersect each other, but the stable and unstable manifolds can intersect each other in the Poincaré section because they are not trajectories but sequences of points. Because the stable manifold W^s is invariant, every point I_0 on W^s is mapped into a new point $I_1 = P(I_0)$ on W^s, where P is the Poincaré map. Similarly, because I_0 is also on W^u, which is also invariant, $I_{-1} = P^{-1}(I_0)$ must also be on W^u. Hence, I_{-1} represents another intersection of W^s and W^u closer to the saddle. Continuing this reasoning, we conclude that if the stable and unstable manifolds intersect once they must intersect infinitely many times at the points $I_n = P^n(I_0)$ and $I_{-n} = P^{-n}(I_0)$ for $n = 1, 2, 3, \cdots$. Each point of transversal intersection I_i is called a **transversal homoclinic point,** and the orbit of this intersection point under the Poincaré map produces a so–called (transversal) **homoclinic orbit**.

The intersection of the stable and unstable manifolds of the saddle has far–reaching consequences on the complexity of the dynamics, as can be seen in Figure 5.7.2. As the unstable manifold approaches the saddle point, the loops between adjacent homoclinic points are stretched parallel to the local unstable manifold W^u and contracted parallel to the local stable manifold W^s. The fate of the stable manifold is similar in reverse iterations, resulting in the **homoclinic tangle** shown in Figure 5.7.2. Because of the strong bending of the manifolds near the saddle point, a small parallelogram of the plane near I_1 will suffer stretching and folding much like that seen in the context of the horseshoe map. (The **Smale–Birkhoff Homoclinic Theorem** [e.g., Guckenheimer and Holmes, 1983; Wiggins, 1988, 1990] implies that the dynamics near a transverse homoclinic point of a diffeomorphism is similar to that of a horseshoe map.) As a result, two points that are initially close together will be found far apart after a few iterations, resulting in unpredictability or sensitive dependence on initial conditions, which is a hallmark of chaos. Therefore, a transverse intersection of W^s with W^u implies chaos–like behavior in a neighborhood of the intersection. Moreover, the boundary of respective regions of initial conditions in phase space that result in trajectories tending to the left or right well (i.e., basin boundary) becomes fractally divided in a neighborhood of the phase space surrounding the transverse intersections (i.e., transverse homoclinic points).

5.7.2 Heteroclinic Tangles

To describe heteroclinic tangles, we consider the system

$$\ddot{x} + x - x^3 + 0.4\dot{x} = F \cos \Omega t \qquad (5.7.2)$$

The phase portrait of the undamped unforced system is shown in Figure 5.7.3. The potential energy of the system has two maxima and one

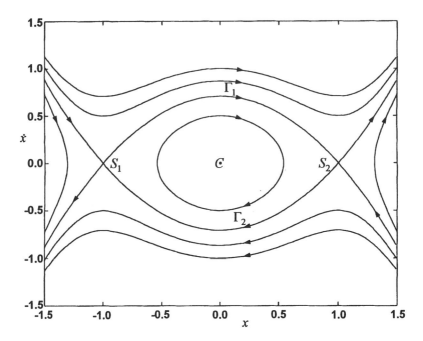

Figure 5.7.3: Phase portrait for the oscillator $\ddot{x} + x - x^3 = 0$.

minimum, a one–well potential. There is a center C at $(x, \dot{x}) = (0, 0)$,
corresponding to the minimum of the potential energy, and two saddle
points S_1 and S_2 at $(x, \dot{x}) = (\pm 1, 0)$, corresponding to the two maxima
of the potential energy. The orbit that leads to S_1 as $t \to -\infty$
(its unstable manifold) and to S_2 as $t \to \infty$ (its stable manifold)
is a **heteroclinic half–orbit**; it is denoted by Γ_1. They intersect
nontransversely. Similarly, the orbit that leads to S_1 as $t \to \infty$ and to
S_2 as $t \to -\infty$ is another heteroclinic half–orbit; it is denoted by Γ_2.
The union of Γ_1 and Γ_2 is called a **heteroclinic orbit**. Including the
damping term destroys the heteroclinic half–orbits and transforms the
center into a stable focus F and leaves the saddles as saddles. The stable
and unstable manifolds W^s and W^u of the saddles do not intersect. The
stable manifolds of the two saddle points divide the phase space into
three regions. Evolutions initiated in the regions to the right of the
stable manifold of S_2 and to the left of the stable manifold of S_1 are
attracted to infinity. The region bounded by S_1 and S_2 is the basin of

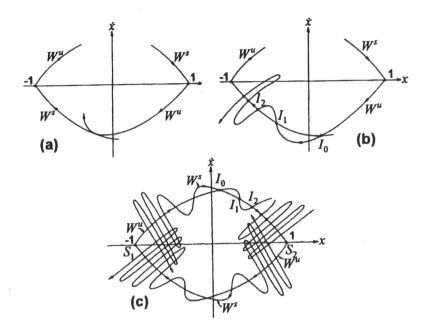

Figure 5.7.4: The formation of a heteroclinic tangle in the Poincaré map of (5.7.2). (a) The unstable and stable orbits barely touch, signaling the beginning of chaos. (b) and (c) The tangle forms with an infinite number of intersections $I_0, I_1, I_2, \cdots$. Two nearby points may be mapped far apart, as is the case on a chaotic attractor.

attraction of the focus F.

In the presence of a weak excitation (i.e., small F), the diagram in Figure 5.7.3 may be regarded as a Poincaré section of the three–dimensional trajectory, except that each line should be regarded as a sequence of dots corresponding to successive intersections of a trajectory with the Poincaré section. As the excitation amplitude is increased while the excitation frequency is kept fixed, the stable and unstable manifolds of the two saddle points move closer to each other. As F is increased to a critical value F_c, the two manifolds approach each other and touch (Fig. 5.7.4a) or even intersect (Fig. 5.7.4b) at some point I_0. Due to the invariant property of W^s and W^u, once they intersect once, they must intersect infinitely many times because

$$\text{if } I_0 \in W^s \cap W^u \text{ then } P^m(I_0) \in W^S \cap W^u \qquad (5.7.3)$$

for any integer m. Each point of transversal intersection is called a **transversal heteroclinic point,** and the orbit of this intersection point under the Poincaré map produces a (transversal) **heteroclinic orbit.** Consequently, the manifolds oscillate wildly because the images of heteroclinic points are heteroclinic points.

As the unstable manifold of S_1 approaches S_2, the loops between adjacent heteroclinic points are stretched parallel to the local unstable manifold W^u and contracted parallel to the local stable manifold W^s, as shown in Figure 5.7.4c. The resulting configuration is called a **heteroclinic tangle.** Because of the strong folding and stretching near the saddle points, a small parallelogram near I_0 will suffer stretching and folding similar to that seen in the context of a horseshoe–type map. Once again, horseshoe–type maps lead to unpredictability or sensitivity to initial conditions, which is a hallmark of chaos. Moreover, the basin boundary separating initial conditions that lead to bounded (safe) and unbounded (unsafe) motions becomes fractally divided in a neighborhood of the phase space surrounding the transverse intersections (i.e., transverse heteroclinic points).

In Figure 5.7.5, we show, after Nayfeh and Sanchez (1989), a series of basin–boundary metamorphoses as the level of F is increased for $\Omega = 0.8$. We note that as F increases from 0.3, the once smooth basin boundary develops fingers (parts a and b) following the scenario described by Grebogi, Ott, and Yorke (1986). The white region represents the basin of attraction of bounded solutions, and the dark region represents the set of initial conditions that take the system to an unbounded solution. The entanglement of the two regions becomes complicated and possibly fractal as F increases, as shown in parts c–e, and the basin of attraction of bounded solutions fades away, as shown in part f. In the studies of Soliman and Thompson (1989) and Thompson and Soliman (1990), the erosion of a basin of attraction has been quantified in terms of a measure.

The disappearance of the basin of attraction of bounded solutions is associated with the point of escape from the potential well, as no initial

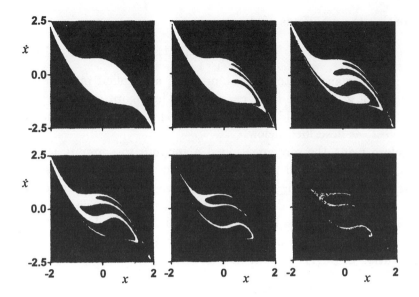

Figure 5.7.5: Basin boundary metamorphoses for $\Omega = 0.8$: F increases from left to right and from top to bottom in the sequence (0.30, 0.32, 0.33, 0.345, 0.38, 0.394) labeled a to f. Reprinted with permission from Nayfeh and Sanchez (1989).

conditions can take the system to a solution other than infinity. Each of the basins in Figure 5.7.5 was numerically generated by examining the outcomes of numerical integrations initiated from a grid of 500×500 initial conditions. This type of identification of the proper initial conditions is not generally feasible, but perhaps other techniques such as cell–to–cell mapping (Hsu, 1980, 1981, 1987, 1992) or boundary mapping would be more appropriate. In any case, the computational effort is considerable.

5.7.3 Numerical Prediction of Manifold Intersections

The intersection of the stable and unstable manifolds of the saddle points in a Poincaré map may be directly observed by numerically computing the manifolds of the system

$$\dot{\mathbf{x}} = \mathbf{f}(\mathbf{x}) + \epsilon\mathbf{g}(\mathbf{x}, t) \tag{5.7.4}$$

where $\mathbf{f}$ is a planar vector field; ϵ provides an explicit measure of the smallness of $\mathbf{g}$; and $\mathbf{g}(\mathbf{x}, t + T) = \mathbf{g}(\mathbf{x}, t)$. The algorithm of Parker and Chua (1989, Chapter 6) is applied to (5.7.4) to generate the manifolds of interest, as described below.

Before using the aforementioned algorithm to compute the manifolds, we need to find approximations to the saddle points of the Poincaré map describing (5.7.4). For future reference, the exact locations of the saddles are denoted by $\mathbf{x}^*$. As a first step, we characterize the local behavior of the Poincaré map in the neighborhood of the saddle point $\mathbf{x}_c^*$ of the associated conservative system. Thus, we express the solution of (5.7.4) near $\mathbf{x}_c^*$ as

$$\mathbf{x}_\epsilon(t) = \mathbf{x}_c^* + \epsilon\mathbf{u}(t) \tag{5.7.5}$$

Substituting (5.7.5) into (5.7.4) and equating the coefficients of ϵ on both sides, we obtain

$$\dot{\mathbf{u}} = D\mathbf{f}(\mathbf{x}_c^*)\mathbf{u} + \mathbf{g}(\mathbf{x}_c^*, t) + \cdots \tag{5.7.6}$$

Equation (5.7.6) is a system of linear first–order nonhomogeneous differential equations with constant coefficients that may be solved analytically using an integrating factor. If we subsequently realize that a fixed point of the Poincaré map requires that $\mathbf{u}(\theta) = \mathbf{u}(\theta + T)$, then the perturbed saddle can be approximated according to (Li and Moon, 1990a) as

$$\mathbf{x}_\epsilon - \mathbf{x}_c^* = \epsilon\left[I - e^{D\mathbf{f}(\mathbf{x}_c^*)^T}\right]^{-1}\int_\theta^{\theta+T} e^{D\mathbf{f}(\mathbf{x}_c^*)(\theta+T-\tau)}\mathbf{g}(\mathbf{x}_c^*, \tau)d\tau + \cdots \tag{5.7.7}$$

where θ specifies the Poincaré section, T is the period of the excitation, and I is the identity matrix. The above expression may be computed numerically. In some cases, this approximation may be inadequate. A more accurate approximation can be obtained by using a modified Hooke and Jeeves algorithm.

To compute the manifolds given four saddle–point approximations in regions I, II, III, and IV, as shown in Figure 5.7.6, we apply an

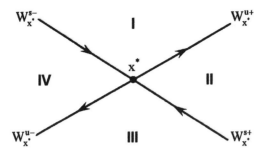

Figure 5.7.6: Phase space near a saddle point.

algorithm presented by Parker and Chua (1989). The basic idea of the algorithm is quite simple. First, the map $\mathbf{P}(\mathbf{x})$ is applied to one of the saddle–point approximations, which is denoted by $\mathbf{x}_s$. The resulting point marks the start of one of the half–manifolds. Second, the map is applied to a succession of points $(\mathbf{x}_i, i = 1, 2, 3, \cdots)$ starting near $\mathbf{x}_s$ to create the half–manifold. This procedure is repeated to create all the half–manifolds of interest.

A major limitation makes this simple technique ineffective without some modification. An arbitrary $\mathbf{P}(\mathbf{x}_{i+1})$ might not be "close enough" to $\mathbf{P}(\mathbf{x}_i)$, meaning that the approximation of the manifold will be crude. To overcome this problem, Parker and Chua suggest that the distance between $\mathbf{P}(\mathbf{x}_i)$ and $\mathbf{P}(\mathbf{x}_{i+1})$ be computed. If the distance is greater than some tolerance, then the corresponding points $\mathbf{x}_i$ and $\mathbf{x}_{i+1}$ are interpolated to obtain a new value $\mathbf{x}_j$, to which $\mathbf{P}$ is applied, and the resulting point is tested for proximity to $\mathbf{P}(\mathbf{x}_i)$. If this distance is again too large, further interpolations are executed as necessary to satisfy the tolerance. A noteworthy feature of this algorithm is that an arbitrary iterate $\mathbf{P}(\mathbf{x}_i)$ is only accepted as part of the manifold when a $\mathbf{P}(\mathbf{x}_j)$ close enough to $\mathbf{P}(\mathbf{x}_i)$ is determined. The iteration procedure and the refinement just described are illustrated in Figure 5.7.7.

At this point, we note again the need to approximate the saddle point accurately. Earlier, it was noted that the linearized approximation may be inadequate. To appreciate why, we recall that $\mathbf{P}(\mathbf{x}_s)$, not $\mathbf{x}_s$, is the first point of the manifold generated. As a result, both the approximation and its first iterate must be "close" to the actual saddle

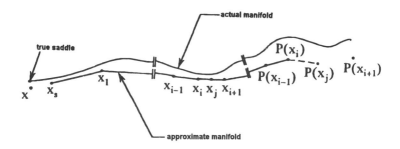

Figure 5.7.7: Parker and Chua's algorithm: A manifold is computed by applying the Poincaré map **P** to successive points on the manifold, beginning with the saddle–point estimate x_s. If two computed points are too far apart, an intermediate point is calculated by interpolating the two points whose iterates were those computed points and then applying **P**. Here, for example, $P(x_{i+1})$ is too far from $P(x_i)$, so x_i and x_{i+1} are interpolated to obtain x_j and thus $P(x_j)$. Since $P(x_j)$ is close enough to $P(x_i)$, $P(x_i)$ is accepted as part of the manifold approximation. The dashed line indicates that acceptance of $P(x_j)$ hinges on the proximity of $P(x_{i+1})$.

point x^*. Otherwise, a substantial portion of the manifold is omitted, possibly including the segment containing the transverse intersections sought. Yet, due to the sensitive nature of this Poincaré map, the first iterates of points reasonably close to x^*, such as x_ϵ, tend to be quite distant from either point.

5.7.4   Analytical Prediction of Manifold Intersections

Melnikov (1963) developed a global analysis technique that yields a condition on the occurrence of a heteroclinic (or homoclinic) bifurcation. Such a bifurcation is said to have occurred if a heteroclinic (homoclinic) set is either created or destroyed as a parameter is varied. As discussed in the preceding two sections, the occurrence of these bifurcations can lead to major changes in the basins of attraction of qualitatively different types of motion, the onset of chaos, and the mixing or intermingling between the safe and unsafe regions. The Melnikov criterion helps in an indirect way in ascertaining the values of the different parameters for which the heteroclinic or homoclinic bifurcations occur and, hence, for which chaos is to be expected or for which the basins of bounded and unbounded regions are likely to be intermingled. To provide a flavor for applications, we mention that this intermingling is not desirable in the context of the capsizing of a ship or the synchronization of a power system.

Next, we derive the so–called Melnikov function for the system (5.7.4) when $\epsilon << 1$ and the unperturbed system is Hamiltonian; that is,

$$\dot{x}_1 = f_1(x_1, x_2) = \frac{\partial H}{\partial x_2} \tag{5.7.8}$$

$$\dot{x}_2 = f_2(x_1, x_2) = -\frac{\partial H}{\partial x_1} \tag{5.7.9}$$

where H is a scalar–valued smooth function, the Hamiltonian. Thus, $H(x_1, x_2)$ is constant on trajectories. The Melnikov function is a measure of the distance between stable and unstable manifolds when that distance is small (Guckenheimer and Holmes, 1983; Wiggins, 1988, 1990; Arrowsmith and Place, 1990).

The parameter ϵ is an explicit measure of the smallness of the perturbation $\mathbf{g}$ and may be equated to unity if $\mathbf{g}$ is sufficiently small. For the purposes of this derivation, the only restriction placed on $\mathbf{f}$ and $\mathbf{g}$, other than the smoothness characteristics and that $\mathbf{g}(\mathbf{x}, t + T) = \mathbf{g}(\mathbf{x}, t)$, is that the two–dimensional state space of the unperturbed system $\dot{\mathbf{x}} = \mathbf{f}(\mathbf{x})$ contains at least one homoclinic orbit in two–dimensional phase space. The case of heteroclinic orbits can be treated similarly.

At this point, two additional parameters are defined in order to describe fully the three–dimensional manifold structures associated with both the unperturbed and perturbed continuous–time systems. First, the symbol θ denotes the particular Poincaré section chosen; it may vary in the interval spanning the period of $\mathbf{g}$, namely, $(0, T]$. Second, $t \in (-\infty, +\infty)$ is defined as the time required for a chosen point on the discrete–time manifold in section θ to be reached by the trajectory beginning at a given initial condition. Although θ and t are both values of time, θ refers only to the Poincaré section chosen, while t is the elapsed time describing a particular trajectory. As an example,

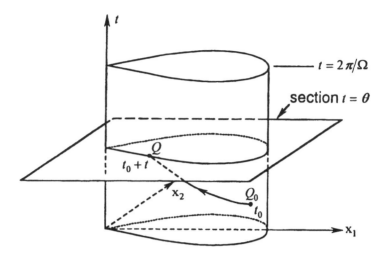

Figure 5.7.8: A generic unperturbed homoclinic invariant–manifold structure in three–dimensional phase space is illustrated. The arbitrary point Q on the surface is specified by selecting a particular Poincaré section θ and the time t required to travel from the reference point Q_0 to the point Q.

we consider the unperturbed continuous–time manifolds illustrated in Figure 5.7.8; any point on the surface can be located by specifying θ and t.

Unlike the unperturbed system, for which all Poincaré sections are identical, the perturbed system is described by Poincaré sections that are distinct for each value of θ in the interval spanning the period of the forcing function **g**. Any value of θ outside this interval may be identified with a θ value that is inside the interval and that is also an integral multiple of T distant from the original value. This identification is possible because any point located on one of the two sections will eventually appear on the other. Thus, for any perturbed system, only the Poincaré sections defined for an interval of the forcing period are unique.

We will not focus on the three–dimensional continuous–time manifolds, but instead on the discrete–time manifolds found in particular two–dimensional Poincaré sections. Such a section is illustrated in Figure 5.7.9, which contains schematics of both the homoclinic orbit of the

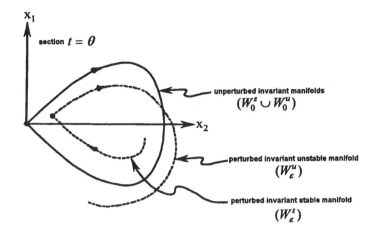

Figure 5.7.9: The invariant manifolds for both the unperturbed system (solid curve) and the damped, forced system (broken curves) are pictured. Note that the stable and unstable manifolds coincide for the unperturbed case, creating a homoclinic orbit.

unperturbed system and the broken manifolds that result when damping or forcing is added. The saddle is assumed to be at the origin so that $\mathbf{f}(\mathbf{0}) = \mathbf{0}$.

A distance function that characterizes the proximity of the perturbed manifolds is now constructed. During the upcoming development, it may be helpful to refer to the illustration in Figure 5.7.10. First, we consider the unperturbed homoclinic orbit in the Poincaré section defined by an arbitrary value θ. Then, we draw a perpendicular line L to the orbit at an arbitrary but fixed location $\mathbf{x}_c$, defined to be the value of the time–varying function $\mathbf{x}_c(t - \theta)$ when $t = \theta$. The quantity $\mathbf{x}_c(t - \theta)$ is a vector that traces out the unperturbed homoclinic orbit as t varies. Next, the curves $\mathbf{x}^s(t; \theta)$ and $\mathbf{x}^u(t; \theta)$ are defined as the perturbed stable and unstable manifolds in the section θ that correspond to the homoclinic orbit. These manifolds intersect the line L at the respective locations A^s and A^u when $t = \theta$. While the curves may intersect L more than once, the points A^s and A^u are those closest to the point $\mathbf{x}_c$. If the time t is fixed at θ, the distance between the manifolds projected along L is simply the separation vec-

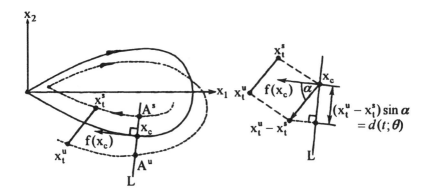

Figure 5.7.10: Construction of the time–dependent distance function $d(t;\theta)$ is shown. (a) The points A^s and A^u represent the points on each of the stable and unstable manifolds closest to the point $\mathbf{x}_c$. They are defined to occur at the elapsed time $t = \theta$. Two other arbitrary trajectories ($\mathbf{x}_t^s$ and $\mathbf{x}_t^u$) with $t > \theta$ are shown as well. Note that the vector $\mathbf{f}(\mathbf{x}_c)$ is the time derivative of $\mathbf{x}$ evaluated at $\mathbf{x}_c$ and is thus tangent to the curve at this point. (b) The projection of the separation between the two arbitrary trajectories along L is clearly the labeled quantity $(\mathbf{x}_t^u - \mathbf{x}_t^s)\sin\alpha$, hence the definition of $d(t;\theta)$.

tor $\mathbf{x}^u(\theta;\theta) - \mathbf{x}^s(\theta;\theta)$. If the time t is no longer constrained to be θ, the distance between the manifolds projected along the line L may be expressed as a function of time. We note that this distance is by definition minimized for the particular section and the chosen value of $\mathbf{x}_c$ if t is set equal to θ. However, we wish to consider all locations on every possible Poincaré section to determine conclusively whether transverse intersections occur in the perturbed system. To do this, we need to examine the function d, which depends on both the time t and the section θ and is defined as

$$d(t;\theta) \equiv \frac{\mathbf{f}[\mathbf{x}_c(t-\theta)] \wedge [\mathbf{x}^u(t;\theta) - \mathbf{x}^s(t;\theta)]}{|\mathbf{f}[\mathbf{x}_c(t-\theta)]|} \qquad (5.7.10)$$

The wedge symbol $\wedge$ denotes a vector cross product of which only the magnitude is significant. Its computation is identical to the standard cross product identified by the $\times$ symbol, except that the result is a scalar rather than a vector. With this definition in mind, the truth of (5.7.10) may be readily established through examination of the geometry presented in Figure 5.7.10b. Although our interest in the function $d(t; \theta)$ concerns only the value at the time $t = \theta$, since that is by definition when the manifolds approach most closely for the section θ, the definition of d as a time–dependent function is necessary to transform the crucial value $d(\theta; \theta)$ from a function of the perturbed manifolds (through the trajectories $\mathbf{x}^u$ and $\mathbf{x}^s$) to a function of only the unperturbed manifolds.

The existence of real zeros of the function $d(\theta; \theta)$ in the interval $(0, T]$ indicates conclusively that transverse intersections are predicted to occur. If two manifolds are to intersect, they must do so in a one–dimensional homoclinic trajectory, a curve that converges to the path traced by the saddle point of the Poincaré map as the section varies. Because every point on the stable (unstable) manifold will eventually approach the trajectory of the perturbed saddle point $\mathbf{x}^*$ as the system moves forward (backward) in time, any point on an intersection of the manifolds will approach the perturbed saddle trajectory as one moves either forward or backward in time. The implication is that an intersection point will travel roughly "around" the homoclinic orbit of the unperturbed system. Because only the Poincaré sections defined by one repetition of the interval $(0, T]$ are unique, all of the information about the journey of the intersection point in question around the homoclinic orbit must be contained in these sections. The conclusion is that the arbitrary choice $\mathbf{x}_c$ where we decided to draw the perpendicular line L is unimportant. If all nonredundant values of θ are considered, then any existing transverse intersections will pass through that section and be detected by our calculation of $d(\theta; \theta)$.

The next step in developing the Melnikov function consists of expressing the trajectories $\mathbf{x}^u$ and $\mathbf{x}^s$ in terms of only unperturbed quantities. To this end, we consider the function Δ_ϵ, which is related to the function $d(t; \theta)$ and is defined as

$$\Delta_\epsilon(t; \theta) \equiv \mathbf{f}[\mathbf{x}_c(t - \theta)] \wedge [\mathbf{x}^u(t; \theta) - \mathbf{x}^s(t; \theta)] \qquad (5.7.11)$$

Thus, we expand the quantities $\mathbf{x}^u$ and $\mathbf{x}^s$ in terms of ϵ as

$$\mathbf{x}^{u,s}(t; \theta) = \mathbf{x}_c(t - \theta) + \epsilon \eta^{u,s}(t; \theta) + O(\epsilon^2) \qquad (5.7.12)$$

Substituting this expansion into (5.7.4) and separating powers of ϵ, we obtain

Order ϵ^0

$$\dot{\mathbf{x}}_c(t - \theta) = \mathbf{f}[\mathbf{x}_c(t - \theta)] \qquad (5.7.13)$$

Order ϵ

$$\dot{\eta}^{u,s}(t;\theta) = D\mathbf{f}[\mathbf{x}_c(t - \theta)]\eta^{u,s}(t;\theta) + \mathbf{g}[\mathbf{x}_c(t - \theta), t] \qquad (5.7.14)$$

At the same time, we also substitute the expansions (5.7.12) into the expression (5.7.11) for Δ_ϵ and obtain

$$\begin{aligned} \Delta_\epsilon(t;\theta) &= \mathbf{f}[\mathbf{x}_c(t - \theta)] \wedge \epsilon\eta^u(t;\theta) \\ &\quad - \mathbf{f}[\mathbf{x}_c(t - \theta)] \wedge \epsilon\eta^s(t;\theta) + O(\epsilon^2) \\ &= \epsilon\Delta_\epsilon^u(t;\theta) - \epsilon\Delta_\epsilon^s(t;\theta) + O(\epsilon^2) \end{aligned} \qquad (5.7.15)$$

Differentiating $\Delta_\epsilon^u(t,\theta)$ with respect to t yields

$$\dot{\Delta}_\epsilon^u = D\mathbf{f}[\mathbf{x}_c(t - \theta)]\dot{\mathbf{x}}_c(t - \theta) \wedge \eta^u(t,\theta) + \mathbf{f}[\mathbf{x}_c(t - \theta)] \wedge \dot{\eta}^u(t,\theta) \quad (5.7.16)$$

Substituting for $\dot{\mathbf{x}}_c(t - \theta)$ and $\dot{\eta}^u(t,\theta)$ from (5.7.13) and (5.7.14) into (5.7.16) yields

$$\begin{aligned} \dot{\Delta}_\epsilon^u &= D\mathbf{f}[\mathbf{x}_c(t - \theta)]\mathbf{f}[\mathbf{x}_c(t - \theta)] \wedge \eta^u(t,\theta) \\ &\quad + \mathbf{f}[\mathbf{x}_c(t - \theta)] \wedge D\mathbf{f}[\mathbf{x}_c(t - \theta)]\eta^u(t,\theta) \\ &\quad + \mathbf{f}[\mathbf{x}_c(t - \theta)] \wedge \mathbf{g}[\mathbf{x}_c(t - \theta), t] \end{aligned} \qquad (5.7.17)$$

Next, we appeal to the identity

$$A\mathbf{x} \wedge \mathbf{y} + \mathbf{x} \wedge A\mathbf{y} = (\mathrm{Tr}A)(\mathbf{x} \wedge \mathbf{y}) \qquad (5.7.18)$$

where A is a 2×2 matrix and $\mathrm{Tr}A$ is the trace of A. This identity can be proven by direct computation. Using (5.7.18), we rewrite (5.7.17) as

$$\begin{aligned} \dot{\Delta}_\epsilon^u(t;\theta) &= \mathrm{Tr}\left\{D\mathbf{f}[\mathbf{x}_c(t - \theta)]\right\}\mathbf{f}[\mathbf{x}_c(t - \theta)] \wedge \eta^u(t;\theta) \\ &\quad + \mathbf{f}[\mathbf{x}_c(t - \theta)] \wedge \mathbf{g}[\mathbf{x}_c(t - \theta), t] \end{aligned} \qquad (5.7.19)$$

It follows from (5.7.8) and (5.7.9) that

$$Df = \begin{bmatrix} \dfrac{\partial^2 H}{\partial x_1 \partial x_2} & \dfrac{\partial^2 H}{\partial x_2^2} \\[2mm] -\dfrac{\partial^2 H}{\partial x_1^2} & -\dfrac{\partial^2 H}{\partial x_1 \partial x_2} \end{bmatrix} \tag{5.7.20}$$

Hence. $\mathrm{Tr}\,\{Df[\mathbf{x}_c(t-\theta)]\} \equiv 0$ and (5.7.19) reduces to

$$\dot{\Delta}_\epsilon^u = \mathbf{f}[\mathbf{x}_c(t-\theta)] \wedge \mathbf{g}[\mathbf{x}_c(t-\theta),t] \tag{5.7.21}$$

Following a similar procedure, we have

$$\dot{\Delta}_\epsilon^s = \mathbf{f}[\mathbf{x}_c(t-\theta)] \wedge \mathbf{g}[\mathbf{x}_c(t-\theta),t] \tag{5.7.22}$$

Integrating (5.7.21) and (5.7.22) from $-\infty$ to θ and θ to $+\infty$, respectively, we obtain

$$\Delta_\epsilon^u(\theta;\theta) - \Delta_\epsilon^u(-\infty;\theta) = \int_{-\infty}^{\theta} \mathbf{f}[\mathbf{x}_c(t-\theta)] \wedge \mathbf{g}[\mathbf{x}_c(t-\theta),t]\,dt \tag{5.7.23}$$

$$\Delta_\epsilon^s(+\infty;\theta) - \Delta_\epsilon^s(\theta;\theta) = \int_{\theta}^{+\infty} \mathbf{f}[\mathbf{x}_c(t-\theta)] \wedge \mathbf{g}[\mathbf{x}_c(t-\theta),t]\,dt \tag{5.7.24}$$

We note that

$$\Delta_\epsilon^u(-\infty,\theta) = \lim_{t\to-\infty} \{\mathbf{f}[\mathbf{x}_c(t-\theta)] \wedge \eta^u(t;\theta)\} \tag{5.7.25}$$

But $\lim_{t\to-\infty}[\mathbf{x}_c(t-\theta)] = \mathbf{0}$, the saddle. Since $\mathbf{f}(\mathbf{0}) = \mathbf{0}$ and $\eta^u(-\infty,\theta)$ is bounded, $\Delta_\epsilon^u(-\infty,\theta) = 0$. A similar reasoning gives $\Delta_\epsilon^s(\infty,\theta) = 0$. Adding (5.7.23) and (5.7.24) and recalling that $\Delta_\epsilon^u(-\infty;\theta) = 0$ and $\Delta_\epsilon^s(\infty;\theta) = 0$, we rewrite (5.7.15) as

$$\Delta_\epsilon(\theta;\theta) = \int_{-\infty}^{+\infty} \epsilon\mathbf{f}[\mathbf{x}_c(t-\theta)] \wedge \mathbf{g}[\mathbf{x}_c(t-\theta),t]\,dt + O(\epsilon^2) \tag{5.7.26}$$

This simplified result leads us to the definition of the Melnikov function, which is

$$M(\theta) = \int_{-\infty}^{+\infty} \mathbf{f}[\mathbf{x}_c(t-\theta)] \wedge \mathbf{g}[\mathbf{x}_c(t-\theta),t]\,dt \tag{5.7.27}$$

or equivalently

$$M(\theta) = \int_{-\infty}^{\infty} \mathbf{f}\left[\mathbf{x}_c(t)\right] \wedge \mathbf{g}\left[\mathbf{x}_c(t), t + \theta\right] dt \qquad (5.7.28)$$

The functional dependence in these equations is solely on the section level θ, where $0 < \theta \leq T$. We note from (5.7.10), (5.7.11), (5.7.27), and (5.7.28) that $M(\theta)$ is approximately proportional to $d(\theta; \theta)$. Let us suppose that $M(\theta)$ oscillates about zero and has simple zeros at the θ_i; that is, $M = 0$ and $dM/d\theta \neq 0$ at $\theta = \theta_i$. At these locations, the stable and unstable manifolds of the perturbed system intersect each other transversely.

5.7.5 Application of Melnikov's Method

In this section, we apply Melnikov's method to three examples.

Example 5.16. We construct the Melnikov function for the externally forced Duffing oscillator

$$\ddot{x} - x + x^3 = \epsilon[F\cos(\Omega t) - 2\mu\dot{x}] \qquad (5.7.29)$$

When $\epsilon = 0$, (5.7.29) reduces to

$$\ddot{x} - x + x^3 = 0 \qquad (5.7.30)$$

which has a saddle at $(x, \dot{x}) = (0, 0)$, two centers at $(x, \dot{x}) = (\pm 1, 0)$, and two homoclinic orbits, as shown in Figure 5.7.1. Letting $x = x_1$ and $\dot{x} = x_2$, we rewrite (5.7.30) as

$$\dot{x}_1 = x_2 = \frac{\partial H}{\partial x_2} \qquad (5.7.31)$$

$$\dot{x}_2 = x_1 - x_1^3 = -\frac{\partial H}{\partial x_1} \qquad (5.7.32)$$

Hence, the system is Hamiltonian with

$$H = \frac{1}{2}(x_2^2 - x_1^2 + \frac{1}{2}x_1^4) \qquad (5.7.33)$$

The level set of $H = 0$ consists of the two homoclinic orbits, and the saddle point $x_1 = x_2 = 0$.

To compute the homoclinic orbits, we solve $H = 0$ for x_2 in terms of x_1 and obtain

$$x_2 = \pm\sqrt{x_1^2 - \frac{1}{2}x_1^4} \tag{5.7.34}$$

which intersects the x_1 axis at $x_1 = \pm\sqrt{2}$, corresponding to the homoclinic orbits in the right– and left–half planes. In terms of x and $\dot{x}$, we rewrite (5.7.34) as

$$\dot{x} = \pm\sqrt{x^2 - \frac{1}{2}x^4} \tag{5.7.35}$$

Using separation of variables and considering the negative sign, we have

$$-t + c = \int^x \frac{dx}{\sqrt{x^2 - \frac{1}{2}x^4}} \tag{5.7.36}$$

where c is a constant. Letting $x = \sqrt{2}\operatorname{sech}(u)$ in (5.7.36), we obtain $u = -t + c$. Hence, $x = \sqrt{2}\operatorname{sech}(t - c)$. Choosing the time origin so that $x = \sqrt{2}$ at $t = 0$, we have $c = 0$ and hence $x = \sqrt{2}\operatorname{sech}(t)$. Consequently, the homoclinic orbit $\mathbf{x}_c = (x_1, x_2)^T$ in the right–half plane is given by

$$x_1 = \sqrt{2}\operatorname{sech}(t) \quad \text{and} \quad x_2 = -\sqrt{2}\operatorname{sech}(t)\tanh(t) \tag{5.7.37}$$

A similar calculation gives the homoclinic orbit in the left–half plane as

$$x_1 = -\sqrt{2}\operatorname{sech}(t) \quad \text{and} \quad x_2 = \sqrt{2}\operatorname{sech}(t)\tanh(t) \tag{5.7.38}$$

Using x_1 and x_2, we rewrite (5.7.29) as

$$\dot{x}_1 = x_2 \tag{5.7.39}$$

$$\dot{x}_2 = x_1 - x_1^3 + \epsilon[F\cos(\Omega t) - 2\mu x_2] \tag{5.7.40}$$

Hence,

$$\mathbf{f} = \left[x_2, \ x_1 - x_1^3\right]^T \quad \text{and} \quad \mathbf{g} = [0, \ F\cos(\Omega t) - 2\mu x_2]^T \tag{5.7.41}$$

Using (5.7.37) and substituting for $\mathbf{f}$ and $\mathbf{g}$ into (5.7.28) yields

$$M(\theta) = -\sqrt{2} \int_{-\infty}^{\infty} \operatorname{sech}(t) \tanh(t) \{ F \cos[\Omega(t + \theta)]$$
$$+ 2\sqrt{2}\mu \operatorname{sech}(t) \tanh(t) \} dt \qquad (5.7.42)$$

or

$$M(\theta) = -\sqrt{2}F \cos(\Omega\theta) \int_{-\infty}^{\infty} \operatorname{sech}(t) \tanh(t) \cos(\Omega t) dt$$
$$+ \sqrt{2}F \sin(\Omega\theta) \int_{-\infty}^{\infty} \operatorname{sech}(t) \tanh(t) \sin(\Omega t) dt \qquad (5.7.43)$$
$$- 4\mu \int_{-\infty}^{\infty} \operatorname{sech}^2(t) \tanh^2(t) dt$$

The first integral is zero because the integrand is odd, the second integral can be evaluated using the method of residues (see Exercise 5.12) to be $\pi\Omega \operatorname{sech}\left(\frac{1}{2}\Omega\pi\right)$, and the last integral can be easily evaluated to be $\frac{2}{3}$. Therefore,

$$M(\theta) = -\frac{8}{3}\mu + 2\sqrt{2}\pi\Omega F \sin(\Omega\theta) \operatorname{sech}\left(\frac{1}{2}\Omega\pi\right) \qquad (5.7.44)$$

Consequently, if

$$\frac{F}{\mu} > \frac{4 \cosh\left(\frac{1}{2}\Omega\pi\right)}{3\sqrt{2}\pi\Omega} \qquad (5.7.45)$$

then $M(\theta)$ has simple zeros, and hence transverse homoclinic points must occur. On the other hand, if the reverse of the inequality (5.7.45) is satisfied, $M(\theta)$ is bounded away from zero, and hence there are no homoclinic points. When the inequality (5.7.45) is replaced by an equality, $M(\theta)$ has a double zero at $\Omega\theta = \frac{1}{2}\pi$. This corresponds to W^u and W^s meeting tangentially rather than transversely.

In Figure 5.7.11a, we show the stable and unstable manifolds of the saddle point near $(0,0)$ for $\Omega = 1.0, \epsilon\mu = 0.125$, and $\epsilon F = 0.11$. In this case, the reverse of the inequality is satisfied, and hence $M(\theta)$ is bounded away from zero. Consequently, there are no homoclinic points according to the Melnikov criterion. This is confirmed in Figure 5.7.12a. As ϵF is increased to 0.19, the inequality is replaced with an equality, and hence the stable and unstable manifolds are predicted to meet tangentially. This is confirmed by the results in Figure 5.7.11b.

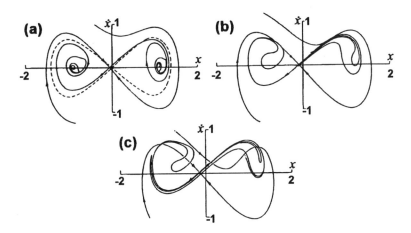

Figure 5.7.11: The stable and unstable manifolds of the saddle of (5.7.29):
(a) $\epsilon F = 0.11$; (b) $\epsilon F = 0.19$; and (c) $\epsilon F = 0.30$.

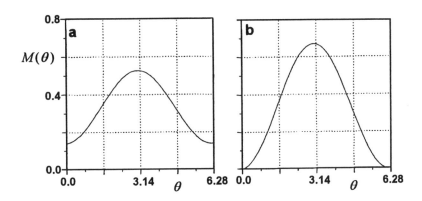

Figure 5.7.12: Melnikov functions for the Duffing equation: (a) $\epsilon F = 0.11$
and (b) $\epsilon F = 0.19$. The minimum of the Melnikov function is far away from
zero in (a) and very close to zero in (b).

As ϵF is increased past 0.19, the inequality is satisfied, and the stable and unstable manifolds are predicted to intersect transversely. Such a transversal intersection is shown in Figure 5.7.11c for $\epsilon F = 0.30$.

Example 5.17. We consider a parametrically excited single–degree–of–freedom system with a single–well potential. Specifically, we consider the system

$$\ddot{x} + \frac{1}{2}(x - x^3) = -\epsilon[Fx\cos(\Omega t) + 2\mu\dot{x}] \qquad (5.7.46)$$

Letting $x = x_1$ and $\dot{x} = x_2$, we rewrite (5.7.46) as

$$\dot{x}_1 = x_2 \qquad (5.7.47)$$

$$\dot{x}_2 = -\frac{1}{2}x_1 + \frac{1}{2}x_1^3 - \epsilon[Fx_1\cos(\Omega t) + 2\mu x_2] \qquad (5.7.48)$$

The unperturbed system is

$$\dot{x}_1 = x_2 = \frac{\partial H}{\partial x_2} \qquad (5.7.49)$$

$$\dot{x}_2 = -\frac{1}{2}x_1 + \frac{1}{2}x_1^3 = -\frac{\partial H}{\partial x_1} \qquad (5.7.50)$$

where

$$H = \frac{1}{2}\left(x_2^2 + \frac{1}{2}x_1^2 - \frac{1}{4}x_1^4\right) \qquad (5.7.51)$$

Hence, the unperturbed system is Hamiltonian. The level set of $H = \frac{1}{8}$ consists of the two saddles $(\pm 1, 0)$, the heteroclinic orbit, and the stable and unstable manifolds of the saddles, as shown in Figure 5.7.3.

To compute the heteroclinic orbit, we solve $H = \frac{1}{8}$ for x_2 in terms of x_1 and obtain

$$x_2 = \pm\sqrt{\frac{1}{4} - \frac{1}{2}x_1^2 + \frac{1}{4}x_1^4} = \pm\frac{1}{2}\left(1 - x_1^2\right) \qquad (5.7.52)$$

or

$$\dot{x} = \pm\frac{1}{2}(1 - x^2) \qquad (5.7.53)$$

where the plus and minus signs correspond to the Γ^+ and Γ^- half-orbits. Using separation of variables, we obtain

$$\pm 2t + c = \int^x \frac{dx}{1 - x^2} = \tanh^{-1} x \qquad (5.7.54)$$

Hence, $x = \pm \tanh(2t + c)$. Choosing the time origin so that $x = 0$ when $t = 0$, we have $c = 0$ and therefore $x = \pm \tanh(2t)$. Consequently, the heteroclinic orbit is given by

$$x_1 = \pm \tanh(2t) \quad \text{and} \quad x_2 = \pm 2 \operatorname{sech}^2(2t) \qquad (5.7.55)$$

Referring to (5.7.47) and (5.7.48), we conclude that

$$\mathbf{f} = \left[x_2, \ -\frac{1}{2}x_1 + \frac{1}{2}x_1^3 \right]^T \quad \text{and} \quad \mathbf{g} = [0, \ -Fx_1 \cos(\Omega t) - 2\mu x_2]^T \quad (5.7.56)$$

Substituting for $\mathbf{f}$ and $\mathbf{g}$ from (5.7.56) into (5.7.28), we have

$$M(\theta) = -\int_{-\infty}^{\infty} x_2 [Fx_1 \cos \Omega(t + \theta) + 2\mu x_2] dt \qquad (5.7.57)$$

which, upon using (5.7.55) with the plus signs, becomes

$$M(\theta) = -2F \cos(\Omega\theta) \int_{-\infty}^{\infty} \tanh(2t) \operatorname{sech}^2(2t) \cos(\Omega t) dt$$
$$+ 2F \sin(\Omega\theta) \int_{-\infty}^{\infty} \tanh(2t) \operatorname{sech}^2(2t) \sin(\Omega t) dt \quad (5.7.58)$$
$$- 8\mu \int_{-\infty}^{\infty} \operatorname{sech}^4(2t) dt$$

The first integral is zero because the integrand is odd, the second integral can be evaluated using the method of residues (see Exercise 5.14) to be $\Omega^2 \pi / 16 \sinh \left(\frac{1}{4}\Omega\pi \right)$, and the last integral can be evaluated by letting $u = \tanh 2t$ to be $\frac{2}{3}$. Hence,

$$M(\theta) = -\frac{16}{3}\mu + \frac{F\Omega^2 \pi \sin(\Omega\theta)}{8 \sinh \left(\frac{1}{4}\Omega\pi \right)} \qquad (5.7.59)$$

Consequently, if

$$\frac{F}{\mu} > \frac{128 \sinh \left(\frac{1}{4}\Omega\pi \right)}{3\Omega^2 \pi} \qquad (5.7.60)$$

then simple zeros of $M(\theta)$ and, hence, transverse heteroclinic points are possible. On the other hand, if the reverse of the inequality (5.7.60) is satisfied, $M(\theta)$ is bounded away from zero, and hence there are no heteroclinic points. When the inequality (5.7.60) is replaced with an equality, $M(\theta)$ has a double zero and the stable and unstable manifolds of the saddles S_1 and S_2 meet tangentially rather than transversely.

Example 5.18. Following Kreider (1992), we consider the behavior of a single–degree–of–freedom system with a three–well potential, linear and nonlinear damping, and a harmonic parametric excitation. Specifically, we consider

$$\ddot{\phi} + \phi - 1.9\phi^3 + 0.722\phi^5 + 0.2\dot{\phi} + 0.2\dot{\phi}^3 + h\phi\cos t = 0 \qquad (5.7.61)$$

The potential function of this system has three minima at $\phi = 0$ and $\phi \approx \pm 1.38$ and two maxima at $\phi \approx \pm 0.854$. Thus, the undamped and unforced system has two homoclinic orbits encircling the centers at $\phi \approx \pm 1.38$ and a heteroclinic orbit encircling the origin.

Li and Moon (1990a,b) calculated manifolds of two– and three–well potential oscillators and discussed the order in which homoclinic and heteroclinic bifurcations occur, as well as transient chaos and how to predict it. Falzarano (1990) examined manifold tangles as they pertain to a ship–roll equation with cubic (and quintic) nonlinearities. Falzarano, Shaw, and Troesh (1992) obtained closed–form expressions for the Melnikov function for the heteroclinic and homoclinic cases when the nonlinearity is cubic and discussed the application of lobe dynamics. Bikdash, Balachandran, and Nayfeh (1994) introduced the concept of the **Melnikov equivalent damping** as a global measure of the system damping and conducted a detailed analysis of this quantity, including a sensitivity analysis. Based on this new concept, they proposed a procedure in which the linear–plus–quadratic damping model can be approximated by a linear–plus–cubic damping model that yields the same Melnikov predictions and very similar steady–state and transient responses.

We let $\phi = \phi_1$ and $\dot{\phi} = \phi_2$ and rewrite (5.7.61) as a system of two first–order equations in the form

$$\dot{\phi}_1 = \phi_2 \qquad (5.7.62)$$

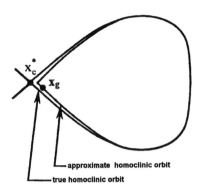

Figure 5.7.13: The homoclinic orbit is approximated by selecting an initial guess point x_g near the unperturbed saddle x_c^* and inside the homoclinic orbit and then integrating numerically until the curve closes on itself.

$$\dot{\phi}_2 = -\phi_1 + 1.9\phi_1^3 - 0.722\phi_1^5 - 0.2\phi_2 - 0.2\phi_2^3 - h\phi_1 \cos(t) \quad (5.7.63)$$

In this case,

$$\mathbf{f} = \left[\phi_2, \ -\phi_1 + 1.9\phi_1^3 - 0.722\phi_1^5\right]^T \quad (5.7.64)$$

and

$$\mathbf{g} = \left[0, \ -0.2\phi_2 - 0.2\phi_2^3 - h\phi_1 \cos(t)\right]^T \quad (5.7.65)$$

The undamped and unforced system is Hamiltonian with

$$H = \frac{1}{2}\phi_2^2 + \frac{1}{2}\phi_1^2 - \frac{1.9}{4}\phi_1^4 + \frac{0.722}{6}\phi_1^6 \quad (5.7.66)$$

Substituting for $\mathbf{f}$ and $\mathbf{g}$ from (5.7.64) and (5.7.65) into (5.7.28) yields the Melnikov function

$$M(\theta) = -\int_{-\infty}^{\infty} \left[0.2\phi_2^2 + 0.2\phi_2^4 + h\phi_1\phi_2 \cos\Omega(t - \theta)\right] dt \quad (5.7.67)$$

Next, we discuss homoclinic tangles, heteroclinic tangles, and so–called mixed tangles (they are described later). Whereas in the preceding two

Figure 5.7.14: A heteroclinic orbit has two possible approximations for use in computing the Melnikov function. The use of the upper inside curve predicts the intersections of the two upper curves, while the lower inside curve forecasts the intersections of the two lower curves. Note that, as for the homoclinic case, the beginning and end points of the approximate trajectories must lie near the corresponding saddle points.

examples we were able to determine the homoclinic and heteroclinic orbits of the unperturbed system analytically and hence compute the Melnikov function exactly, in the present example we compute these orbits numerically.

For homoclinic tangles, we approximately compute the homoclinic orbit by numerically integrating the unperturbed equation (5.7.61) from an initial guess point $\mathbf{x}_g$ inside the homoclinic curve and near the unperturbed saddle point $\mathbf{x}_c^*$ to some final point also near $\mathbf{x}_c^*$. This procedure is pictured schematically in Figure 5.7.13. Practically, the integral over $(-\infty, +\infty)$ is carried out by applying the trapezoidal rule (or another numerical integration technique) with an appropriate step size Δt between the two points established as the "beginning" and "end" of the orbit. We note that Δt is fixed implicitly by the calculation of the homoclinic orbit; the time step cannot be freely chosen in the evaluation of the integral, but must be selected during the calculation of the homoclinic orbit. In addition, although both of the two endpoints of the interval are theoretically identical to the unperturbed saddle $\mathbf{x}_c^*$, one needs to use approximations at either end of the trajectory to compute $M(\theta)$ because use of the exact saddle point will result in a degenerate trajectory consisting only of the saddle itself.

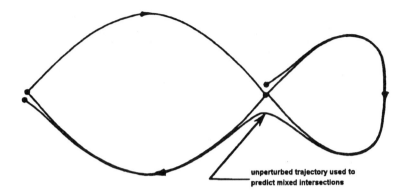

Figure 5.7.15: Pictured is the unperturbed trajectory used in the Melnikov integral when predicting intersections of one homoclinic manifold and one heteroclinic manifold in the perturbed map.

For heteroclinic tangles, we approximate the heteroclinic half–orbit in the upper half plane numerically proceeding from a point close to the unperturbed left saddle point to a point close to the unperturbed right saddle point, as shown in Figure 5.7.14. As might be intuitively expected, the use of the upper curve in Figure 5.7.14 in the Melnikov integral will lead to the detection of the intersection between the unstable manifold of the left saddle point and the stable manifold of the right saddle point. Analogously, the lower curve corresponds to intersections of the left stable and right unstable manifolds.

If it is possible to detect both homoclinic and heteroclinic intersections using the Melnikov function, what about finding mixed intersections? The occurrence of intersections of one heteroclinic manifold and one homoclinic manifold has been documented in the work of Falzarano (1990). Kreider (1992) used the Melnikov function to predict these intersections.

The key to computing the Melnikov function for predicting mixed intersections lies with the choice of unperturbed trajectory over which the computation is carried out. In the homoclinic case, the manifolds that intersect correspond to an unperturbed homoclinic orbit, the curve along which the Melnikov function is calculated. Similarly, in the heteroclinic case, the intersecting manifolds originate from a heteroclinic

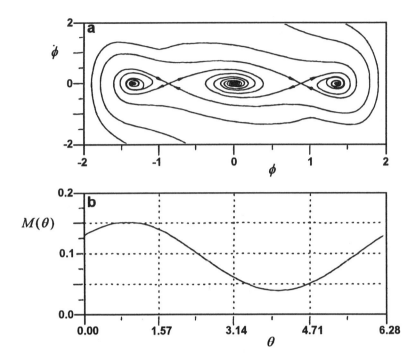

Figure 5.7.16: (a) The manifolds of the saddles of (5.7.61) for low forcing $h = 0.05$ and (b) the homoclinic Melnikov function. As expected, $M(\theta)$ does not approach zero.

half–orbit, which is used to compute $M(\theta)$. For detecting mixed tangles, then, it seems reasonable to use the unperturbed manifolds that later result in a mixed intersection when the perturbation is added. A trajectory that closely approximates a homoclinic orbit and a heteroclinic half–orbit fits this description; an example of such a trajectory is illustrated in Figure 5.7.15.

Next, we examine the effects of increasing the excitation amplitude on the invariant manifolds of the saddles of the Poincaré map for the model represented by (5.7.61). We begin with a low excitation level ($h = 0.05$). The Melnikov function displayed in Figure 5.7.16b does not have zeros. The corresponding manifold structure does not exhibit any intersections or other unusual characteristics, as seen from Figure

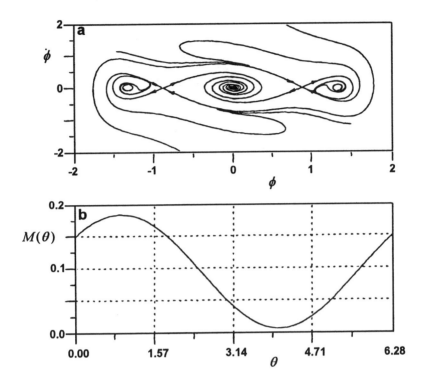

Figure 5.7.17: (a) The manifolds of the saddles of (5.7.61) for $h = 0.08$ and (b) the homoclinic Melnikov function. This function has a minimum close to zero as homoclinic intersections are imminent.

5.7.16a. The manifolds resemble the trajectories of a damped, unforced phase portrait. Fixed–point solutions occur near $\phi = 0$ and ± 1.38. These solutions are fixed points of the Poincaré map and therefore correspond to periodic solutions of the original continuous–time system.

When the forcing amplitude is increased to $h = 0.08$, the qualitative nature of the homoclinic manifolds changes, as is illustrated in Figure 5.7.17a. Rather than spiraling smoothly to fixed points as they did in the previous case, the unstable homoclinic manifolds bend and twist as they spiral inward. It is clear from Figure 5.7.17b that the homoclinic Melnikov function has a minimum close to zero.

The bending and twisting of the homoclinic manifolds act as a precursor to transverse intersections, which are first observed for an ex-

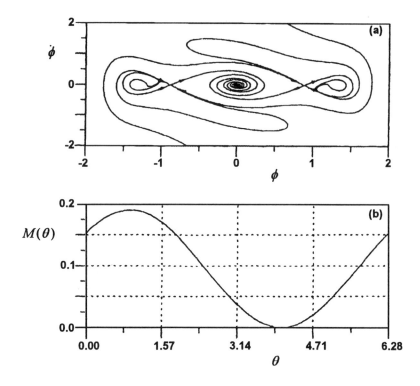

Figure 5.7.18: (a) The manifolds of the saddles of (5.7.61) for $h = 0.085$ and (b) the homoclinic Melnikov function. As homoclinic intersections begin to occur, the zeros of the Melnikov function are closely spaced.

citation level very near $h = 0.085$. As illustrated in Figure 5.7.18a, the intersections are pending at two different locations for each homoclinic structure: near the points $(\pm 0.2, \mp 1.1)$ and also near each of the saddles. We note that the minimum absolute value of M is very close to zero in Figure 5.7.18b. The four parts in Figures 5.7.17 and 5.7.18 indicate that Melnikov's method predicts the homoclinic tangles. Melnikov's theory is effective if the perturbations are small enough. The success of the method in this case tends to indicate that the perturbations are small enough. We recall that the significance of the onset of these intersections is the concurrent incidence of fractal regions in the transient basin. The heteroclinic manifolds in this case remain qualita-

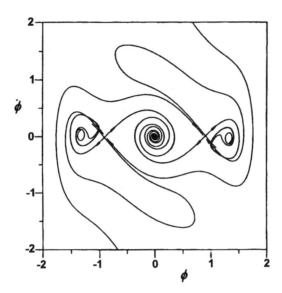

Figure 5.7.19: Well–developed homoclinic manifold intersections for $h = 0.10$. At larger forcing, the intersections of the manifolds is more easily discernible.

tively like those seen in the earlier cases, spiraling smoothly to a fixed point at the origin.

An excitation level of $h = 0.10$ is used to show more well–developed homoclinic manifold intersections. Several such crossings are clearly evident in Figure 5.7.19. By this time, it should be obvious that the Poincaré section for this system is symmetric about the origin, meaning that successive reflections of any point on the manifold structure across both coordinate axes generate a point also on the structure. (This behavior is due to the fact that [(5.7.61)] is an odd function of both ϕ and $\dot{\phi}$.) Consequently, any feature associated with one homoclinic manifold will also appear on the other. This symmetry will be taken for granted in the following development, in which only the left–most feature will be referenced.

Next, we consider mixed intersections. The results are summarized in Figure 5.7.20, which contains the Melnikov graphs and manifolds for $h = 0.215$. The technique seems to work quite well. As the

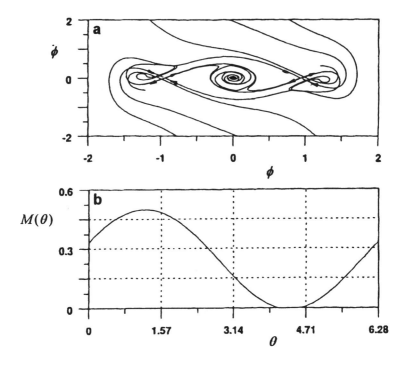

Figure 5.7.20: (a) The manifolds of the saddles of (5.7.61) for $h = 0.215$ and (b) a mixed Melnikov function. As mixed tangles develop, the minimum of the Melnikov function decreases slightly. The prediction in this case is slightly conservative.

homoclinic unstable manifold and heteroclinic stable manifold approach one another and then intersect ($h = 0.215$), the Melnikov function exhibits closely spaced roots in each case. Thus, Melnikov's method is slightly conservative, predicting intersections at a marginally lower forcing amplitude (just smaller than $h = 0.21$) than that at which tangles first occur (just larger than $h = 0.21$).

Further confirmation of Melnikov's method is sought by examining the heteroclinic intersections, which occur near $h = 0.34$. Once again, variation of M closely parallels the approach of the stable and unstable manifolds. Heteroclinic tangles are approaching but have yet to occur when $h = 0.32$. This behavior is predicted by the Melnikov function

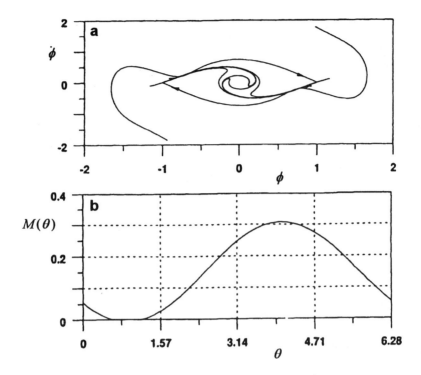

Figure 5.7.21: (a) The manifolds of the saddles of (5.7.61) for $h = 0.34$ and (b) the heteroclinic Melnikov function. When heteroclinic tangles are first established, the Melnikov function exhibits a minimum near zero.

pictured in Figure 5.7.21 along with the manifolds. In Figure 5.7.21a, the incidence of heteroclinic intersections is definitely established, and the plot of M in Figure 5.7.21b has two closely spaced roots, indicating a minimum near zero and thus good agreement. The intersections illustrated in Figure 5.7.21a indicate that the safe area surrounding the origin could be starting to diminish rapidly (e.g., Thompson and Soliman, 1990).

To summarize, we recall that homoclinic intersections were observed first at an excitation level of approximately 0.085, followed by mixed intersections at $h = 0.215$, and then heteroclinic intersections at $h = 0.34$. These different intersections are predicted by Melnikov's method.

5.7.6   Comments

In all the cases discussed thus far, the unperturbed system is not dissipative. There are studies in which a Melnikov analysis has been conducted for cases where the unperturbed system is dissipative (e.g., Salam, 1988). Melnikov's method has also been used to determine transverse homoclinic intersections in the context of averaged systems of equations (e.g., Holmes, 1980; Sanders, 1982; Guckenheimer and Holmes, 1983; Wiggins, 1988, 1990). Melnikov analyses have also been conducted in the context of multi–degree–of–freedom systems and infinite–degree–of–freedom systems (Holmes and Marsden, 1981, 1982; Wiggins, 1988). The analyses discussed in the previous sections are sometimes called **homoclinic** or **heteroclinic Melnikov analyses** to distinguish them from what are called **subharmonic Melnikov analyses**. In Chapter 3, we pointed out that there is a continuum of periodic orbits enclosed within homoclinic and heteroclinic orbits. Subharmonic Melnikov analyses (e.g., Wiggins, 1988, 1990) are useful in understanding the influence of perturbations on these periodic orbits.

5.8   BIFURCATIONS OF HOMOCLINIC ORBITS

In contrast with the preceding section, where we considered perturbations of homoclinic and heteroclinic orbits of two–dimensional Hamiltonian systems, in this section we consider bifurcations of orbits homoclinic or heteroclinic to saddles of autonomous systems. In three–dimensional and higher systems, the presence of a homoclinic orbit may imply the existence of chaotic behavior, horseshoes, and infinitely many nearby bifurcations, depending on the eigenvalues of the Jacobian matrix of the flow at the saddle point and on any symmetries that might be present in the system. In Section 5.8.1 we consider planar systems, in Section 5.8.2 we consider three–dimensional systems with the saddle having three purely real eigenvalues, in Section 5.8.3 we consider three–dimensional systems having orbits homoclinic to a saddle focus, and in Section 5.8.4 we discuss higher–dimensional systems.

5.8.1   Planar Systems

We follow Andronov, Leontovich, Gordon, and Maier (1971) and consider planar systems governed by

$$\dot{x} = \lambda_1 x + f_1(x, y; \epsilon) \tag{5.8.1}$$

$$\dot{y} = -\lambda_2 y + f_2(x, y; \epsilon) \tag{5.8.2}$$

where λ_1 and λ_2 are positive, f_1 and f_2 are $O(x^2+y^2)$, f_1 and f_2 are C^r where $r \geq 2$, and ϵ is a small dimensionless parameter. We assume that the system possesses an orbit homoclinic to the saddle $(x, y) = (0, 0)$ when $\epsilon = 0$ and that the homoclinic orbit is broken on both sides of $\epsilon = 0$. We note that, without loss of generality, the saddle has been transformed to the origin, and the local stable and unstable manifolds of the saddle have been used as coordinates.

In Figure 5.8.1b, we show the situation when $\epsilon = 0$; there is an orbit homoclinic to the saddle at the origin. In Figures 5.8.1a and 5.8.1c, we show the situation when the system is slightly perturbed so that the homoclinic orbit breaks in a transverse manner, resulting in the stable and unstable manifolds of the saddle missing each other. Without loss of generality, we assume that ϵ is normalized so that the unstable manifold passes just to the left of the stable manifold when $\epsilon < 0$ and to its right when $\epsilon > 0$, as shown in parts a and c, respectively.

To determine the nature of the orbit structure near the homoclinic orbit for $\epsilon \approx 0$, we construct a small box B ($\mid x \mid \leq a$ and $\mid y \mid \leq a$, where a is small) around the saddle, as shown in Figure 5.8.2. Because a is small, the flow can be assumed to be linear inside the box, and hence the equations governing the motion there are

$$\dot{x} = \lambda_1 x \quad \text{and} \quad \dot{y} = -\lambda_2 y \tag{5.8.3}$$

whose general solutions can be expressed as

$$x = c_1 e^{\lambda_1 t} \quad \text{and} \quad y = c_2 e^{-\lambda_2 t} \tag{5.8.4}$$

where c_1 and c_2 are constants. Starting a trajectory at (x_n, a) on the top edge of the box with $x_n > 0$, we have

$$x = x_n e^{\lambda_1 t} \quad \text{and} \quad y = a e^{-\lambda_2 t} \tag{5.8.5}$$

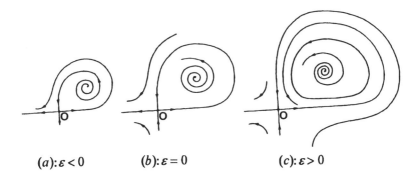

(a): $\varepsilon < 0$ (b): $\varepsilon = 0$ (c): $\varepsilon > 0$

Figure 5.8.1: Three phase portraits for a two–dimensional system. There is a homoclinic orbit in part b. After small perturbations of the system there may be no periodic orbits as in part a or one stable periodic orbit that passes close to the saddle point as in part c.

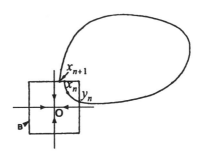

Figure 5.8.2: Analysis of the behavior near the homoclinic orbit in Figure 5.8.1. It is assumed that the behavior is linear in a small box B around O and that a trajectory started at the point x_n on the edge of B will next hit the edge of B at points y_n and x_{n+1}.

Thus, the time of flight T needed for this trajectory to reach the right edge of the box is given by

$$a = x_n e^{\lambda_1 T} \text{ or } T = \frac{1}{\lambda_1} \ln\left(\frac{a}{x_n}\right) \tag{5.8.6}$$

Consequently, this trajectory will emerge from the right edge of the box at

$$y_n = ae^{-\lambda_2 T} = ae^{-\delta \ln\left(\frac{a}{x_n}\right)} = a\left(\frac{x_n}{a}\right)^\delta = a^{1-\delta} x_n^\delta \tag{5.8.7}$$

where $\delta = \lambda_2/\lambda_1$.

Under the flow, a trajectory started at (a, y_n) can be described by

$$x = g(y; a, y_n, \epsilon) \tag{5.8.8}$$

where g is a smooth function of y, a, y_n, and ϵ. The homoclinic orbit can be obtained from (5.8.8) by setting $y_n = 0$ and $\epsilon = 0$; that is,

$$x = g(y; a, 0, 0) \tag{5.8.9}$$

because it emerges from the right edge of the box at $(a, 0)$. It reenters the box at its top edge at $(0, a)$. Hence,

$$g(a; a, 0, 0) = 0 \tag{5.8.10}$$

We note that the time taken by a trajectory started at (x_n, a) to pass through the box is very long compared with the time taken to get from (a, y_n) to the point (x_{n+1}, a) because the velocity of the flow is near zero within the box. Moreover, because of the smoothness of the flow with respect to ϵ and the initial conditions, we can approximate the trajectory (5.8.8) in terms of the homoclinic orbit by using a Taylor series expansion. The result is

$$x = g(y; a, y_n, \epsilon) = g(y; a, 0, 0)$$
$$+ \frac{\partial g}{\partial y_n}(y; a, 0, 0)y_n + \frac{\partial g}{\partial \epsilon}(y; a, 0, 0)\epsilon + \cdots \tag{5.8.11}$$

Because this trajectory reaches the top edge of the box at (x_{n+1}, a),

$$x_{n+1} = g(a; a, 0, 0) + \alpha y_n + \beta \epsilon \tag{5.8.12}$$

or

$$x_{n+1} = \alpha y_n + \beta \epsilon \qquad (5.8.13)$$

where

$$\alpha = \frac{\partial g}{\partial y_n}(a; a, 0, 0) \text{ and } \beta = \frac{\partial g}{\partial \epsilon}(a; a, 0, 0) \qquad (5.8.14)$$

The parameters α and β depend on properties of the global flow. Because trajectories cannot cross in the plane, α is positive. To determine the sign of β, we let $y_n = 0$ and obtain $x_{n+1} = \beta \epsilon$. In order that this expression be consistent with the three phase portraits in Figure 5.8.1, β must be positive.

Combining (5.8.7) and (5.8.13), we obtain the map

$$x_{n+1} = k x_n^\delta + \beta \epsilon, \text{ where } k = \alpha a^{1-\delta} \qquad (5.8.15)$$

which relates x_{n+1} to x_n in Figure 5.8.2. We reiterate that (5.8.15) is valid only for small ϵ and only for trajectories that pass close to the saddle. Moreover, the constants k and β depend on properties of the global flow, but they are positive. With this information, we can use this map to study the qualitative features of the flow provided that ϵ is small and we consider only trajectories that pass close to the saddle.

To study the dynamics generated by (5.8.15), we distinguish two cases: $\delta > 1$ and $\delta < 1$. When $\delta > 1$, the map has a fixed point x^* when $\epsilon > 0$ and no fixed points when $\epsilon < 0$, as shown in Figure 5.8.3. Provided ϵ is sufficiently small, x^* is small and the Jacobian $k\delta x^{*\delta-1}$ of the map is larger than zero but less than unity. Hence, the fixed point x^* is stable, and (x^*, a) lies on a stable periodic orbit of the flow. Consequently, when $\delta > 1$, the system (5.8.1) and (5.8.2) has a stable periodic orbit when $\epsilon > 0$ and no periodic orbits when $\epsilon < 0$.

When $\delta < 1$, the map has a fixed point at x^* when $\epsilon < 0$ and none when $\epsilon > 0$, as shown in Figure 5.8.4. For sufficiently small ϵ, x^* is small and the Jacobian $k\delta x^{*\delta-1}$ of the map at the fixed point is larger than unity. Hence, the fixed point is unstable and corresponds to an unstable periodic orbit. Consequently, when $\delta < 1$, the system has an unstable periodic orbit when $\epsilon < 0$ and no periodic orbits when $\epsilon > 0$.

The global bifurcation discussed in this section is an example of a blue sky catastrophe in which a limit cycle disappears through collision with a saddle equilibrium point. When $\delta > 1$, there is a stable periodic

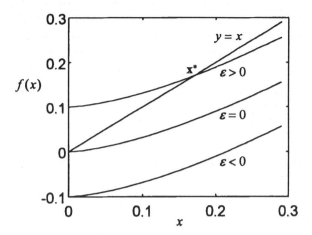

Figure 5.8.3: Return maps of (5.8.15) when $\delta > 1$, $k = 1.0, \beta = 1.0$, and $\delta = 1.5$. The three cases shown correspond to $\epsilon = -0.1, 0$, and 0.1. The identity map $y = x$ is also shown.

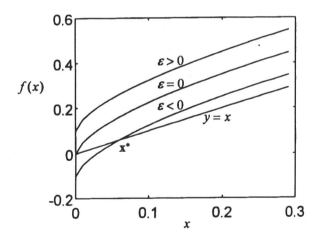

Figure 5.8.4: Return maps of (5.8.15) when $\delta < 1$, $k = 1.0, \beta = 1.0$, and $\delta = 0.65$. The three cases correspond to $\epsilon = -0.1, 0$, and 0.1. The identity map is also shown.

orbit for $\epsilon > 0$. As ϵ decreases, the limit cycle and the saddle move closer to each other, and at $\epsilon = 0$ the limit cycle and a branch of both the stable and unstable manifolds of the saddle point coincide, forming a homoclinic connection, which is doubly asymptotic to the saddle point. In other words, a trajectory starting on this orbit approaches the saddle as $t \to \pm\infty$. Thus, the homoclinic orbit can be thought of as a limit cycle with an infinite period. As ϵ decreases below zero, the limit cycle abruptly vanishes from the state space, hence, the name **blue sky catastrophe.** When $\delta < 1$, the blue sky catastrophe involves the abrupt disappearance of an unstable limit cycle through collision with a saddle equilibrium point.

When $\delta = 1$, the present first–order analysis fails. Andronov, Leontovich, Gordon, and Maier (1971) state that, in this case, multiple periodic solutions bifurcate from the homoclinic orbit and present results for special cases. Dangelmayr and Guckenheimer (1987) developed a method that can be used to treat this case.

The planar results discussed in this section are applicable to higher–dimensional systems without symmetry provided that the following conditions are satisfied. First, the system has an orbit homoclinic to a saddle point; that is, an orbit that approaches the saddle as $t \to \pm\infty$. Second, the eigenvalues of the Jacobian matrix associated with the saddle have a special structure. Out of all of the eigenvalues in the right–half of the complex plane, the closest to the imaginary axis is a real eigenvalue λ_1. And out of all of the eigenvalues in the left–half of the complex plane, the closest to the imaginary axis is a real eigenvalue $-\lambda_2$. Thus, the dynamics of the system near the homoclinic orbit can be reduced to that of a planar system because the eigenvalues $-\lambda_2$ and λ_1 determine how the homoclinic orbit approaches and leaves the saddle point.

Example 5.19. Following Diener (1984), we consider the system

$$\dot{x} = ky + \mu x(b - y^2) \qquad (5.8.16)$$
$$\dot{y} = -x + a \qquad (5.8.17)$$

where k, μ, a, and b are positive constants. The fixed points of this system are given by

$$x = a \text{ and } y = \frac{k}{2\mu a} \pm \sqrt{\frac{k^2}{4\mu^2 a^2} + b} \qquad (5.8.18)$$

The Jacobian of the system is

$$J = \begin{bmatrix} \mu(b - y^2) & k - 2\mu ay \\ -1 & 0 \end{bmatrix} \qquad (5.8.19)$$

whose eigenvalues are

$$\lambda^2 - \mu(b - y^2)\lambda + k - 2\mu ay = 0 \qquad (5.8.20)$$

or

$$\lambda^2 + \frac{ky}{a}\lambda \mp \sqrt{\frac{k^2}{4\mu^2 a^2} + b} = 0 \qquad (5.8.21)$$

Therefore, the fixed point corresponding to the positive sign in (5.8.18) is a saddle and the one corresponding to the negative sign is an unstable focus. Moreover, the sum of the eigenvalues is $-ky/a$, which is negative for the saddle because y is positive for it. Hence, $\lambda_1 - \lambda_2$ is negative, which implies that $\lambda_2 > \lambda_1$ or $\delta > 1$. When $a = a_c \approx 0.12$, the system has an orbit homoclinic to the saddle; this is shown in Figure 5.8.5b. Because $\delta > 1$, the system has a stable periodic orbit on one side of the homoclinicity condition (such an orbit is shown in Fig. 5.8.5a for $a = 0.1$) and no periodic orbits on the other side, as shown in Figure 5.8.5c for $a = 0.14$. In other words, as a increases past the critical value a_c at which a homoclinic connection exists, the stable limit cycle abruptly disappears from the state space in a blue sky catastrophe.

5.8.2 Orbits Homoclinic to a Saddle

In this section, we consider three–dimensional systems with symmetry and possessing orbits homoclinic to a saddle with three purely real eigenvalues. Transforming the saddle to the origin and using the eigenvectors of the saddle as coordinates, we write the equations governing the motions of these systems as

$$\dot{x} = -\lambda_1 x + f_1(x, y, z; \epsilon) \qquad (5.8.22)$$

$$\dot{y} = -\lambda_2 y + f_2(x, y, z; \epsilon) \qquad (5.8.23)$$

$$\dot{z} = \lambda_3 z + f_3(x, y, z; \epsilon) \qquad (5.8.24)$$

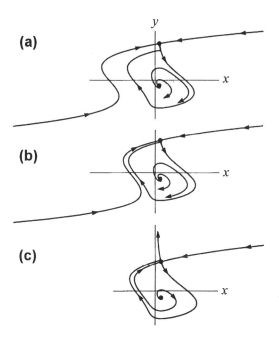

Figure 5.8.5: Phase portraits illustrating a blue sky catastrophe of a limit cycle in the acceleration–forced van der Pol equations (5.8.16) and (5.8.17) with DC bias (Thompson and Stewart, 1986): $k = 0.7, \mu = 10$, and $b = 0.1$.

where the λ_i are positive and the f_i are C^2 and vanish along with their first derivatives at $(x, y, z, \epsilon) = (0, 0, 0, 0)$.

When the system (5.8.22)–(5.8.24) has no symmetries, one can reduce this system to a planar system. Then, one can use a technique similar to that used in the preceding section to show that for $\epsilon \neq 0$ and sufficiently small, a periodic orbit bifurcates from the homoclinic orbit. For systems without symmetries, Wiggins (1988) proved that the periodic orbit is

(a) a sink for $\lambda_1 > \lambda_3$ and $\lambda_2 > \lambda_3$;

(b) a saddle for $\lambda_1 + \lambda_2 > \lambda_3, \lambda_1 < \lambda_3$ and/or $\lambda_2 < \lambda_3$;

(c) a source for $\lambda_1 + \lambda_2 < \lambda_3$.

In the presence of symmetries, we assume that $\lambda_1 > \lambda_2$ and let $\delta = \lambda_2/\lambda_3$. Again, one can use a technique similar to that used in the preceding section to prove the following theorem (Afraimovich, Bykov, and Shilnikov, 1977; Kaplan and Yorke, 1979c; Sparrow, 1982; Arneodo, Coullet, and Tresser, 1981; Wiggins, 1988). When $\delta < 1$, there is no recurrent behavior near the saddle on one side of the homoclinicity, and there exists an unstable strange invariant set containing a horseshoe on the other side. When $\delta > 1$, a stable symmetric periodic orbit exists on one side of the homoclinicity, and two stable but nonsymmetric orbits exist on the other side. All three periodic orbits have periods that tend to infinity as the homoclinicity is approached.

Example 5.20. We consider the Lorenz system

$$\dot{x} = -\sigma(x - y) \tag{5.8.25}$$

$$\dot{y} = \rho x - y - xz \tag{5.8.26}$$

$$\dot{z} = xy - \beta z \tag{5.8.27}$$

where σ, ρ, and b are positive constants. The fixed points of this system are

$$x = 0, \quad y = 0, \quad \text{and} \quad z = 0 \tag{5.8.28}$$

$$x = y = \pm\sqrt{\beta(\rho - 1)}, \quad \text{and} \quad z = \rho - 1 \tag{5.8.29}$$

Nontrivial solutions exist only when $\rho > 1$. The stability of these fixed points is determined by the eigenvalues of the Jacobian matrix

$$J = \begin{bmatrix} -\sigma & \sigma & 0 \\ \rho - z & -1 & -x \\ y & x & -\beta \end{bmatrix} \tag{5.8.30}$$

To study the stability of the trivial fixed point, we let $x = y = z = 0$ in J and find that the eigenvalues of J are

$$\lambda_1, \lambda_2 = -\frac{1}{2}(\sigma + 1) \pm \sqrt{\frac{1}{4}(\sigma + 1)^2 - \sigma(1 - \rho)}, \quad \lambda_3 = -\beta \tag{5.8.31}$$

or

$$\lambda_1, \lambda_2 = -\frac{1}{2}(\sigma + 1) \pm \sqrt{\frac{1}{4}(\sigma - 1)^2 + \sigma\rho}, \ \lambda_3 = -\beta \qquad (5.8.32)$$

Hence, when $\rho < 1$, all of the eigenvalues are purely real and negative, and the origin is a sink. When $\rho > 1$, one of the eigenvalues is positive and the other two eigenvalues are negative, and hence the origin is a saddle. Thus, at $\rho = 1$, we have a pitchfork bifurcation, where a stable fixed point loses stability and gives rise to two other stable fixed points.

To study the stability of the nontrivial fixed points, we substitute (5.8.29) into (5.8.30) and find that the eigenvalues of the Jacobian matrix are given by

$$\lambda^3 + (\sigma + \beta + 1)\lambda^2 + \beta(\sigma + \rho)\lambda + 2\sigma\beta(\rho - 1) = 0 \qquad (5.8.33)$$

According to the Routh–Hurwitz criterion, all three roots have negative real parts, and hence the nontrivial fixed points are stable if

$$(\sigma + \beta + 1)(\sigma + \rho) - 2\sigma(\rho - 1) > 0 \qquad (5.8.34)$$

or

$$\rho < \frac{\sigma(\sigma + \beta + 3)}{\sigma - \beta - 1} \qquad (5.8.35)$$

All of the roots of (5.8.33) are real if

$$4\left[\beta(\sigma + \rho) - \frac{1}{3}(\sigma + \beta + 1)^2\right]^3 + 27\left[2\sigma\beta(\rho - 1)\right.$$
$$\left. - \frac{1}{3}\beta(\sigma + \rho)(\sigma + \beta + \rho) + \frac{8}{27}(\sigma + \beta + 1)^3\right]^2 < 0 \qquad (5.8.36)$$

and one root is real and two roots are complex conjugates if the inequality is reversed.

Next, we fix the values of β and σ at $\frac{8}{3}$ and 10 used by Lorenz and examine the behavior of the unstable manifold of the origin as ρ is increased from zero. When $0 < \rho < 1$, the origin is globally stable, as shown in Figure 5.8.6a. As ρ is increased past unity, one of the eigenvalues of the origin becomes positive, with the other two eigenvalues remaining negative. Simultaneous with the loss of stability of the origin, the two nontrivial fixed points C_1 and C_2 are born with

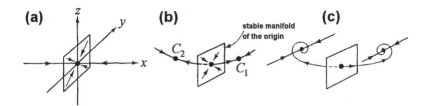

Figure 5.8.6: Evolution of the fixed points and the unstable manifold of the saddle of the Lorenz equations.

all of their eigenvalues being purely real and negative. Consequently, the unstable manifold of the origin is slightly curved, leading from the saddle directly to the attracting fixed points C_1 and C_2, as shown in Figure 5.8.6b. The stable manifold of the saddle separates the basin boundaries of the fixed points C_1 and C_2. As ρ is increased further, two of the eigenvalues of each of C_1 and C_2 approach each other, coalesce at $\rho = \rho_s \approx 1.346$, and become complex conjugates with negative real part past ρ_s. Hence, as ρ exceeds ρ_s, the unstable manifold of the origin spirals into C_1 and C_2, as shown in Figure 5.8.6c.

As ρ is increased past 10, the spiraling becomes more pronounced. The imaginary part of the complex eigenvalues increases in magnitude, while its real part becomes less negative. As a result, the unstable manifold of the saddle makes wider turns and settles more slowly to C_1 and C_2, as shown in Figure 5.8.7a. As ρ is increased further, the unstable and stable manifolds of the saddle get closer and closer to each other and at $\rho = \rho_h \approx 14.926$ intersect nontransversely to form a trajectory Γ that approaches the saddle for $t \to -\infty$ and for $t \to +\infty$. In other words, Γ is a homoclinic orbit. This is shown in Figure 5.8.7b. The eigenvalues of the saddle at $\rho = \rho_h$ are $-18.13, -2.67$, and 7.14. Thus, $\delta = 2.67/7.13 = 0.37 < 1$. Hence, according to the theorem, there is no recurrent behavior near the saddle on one side of the homoclinicity ($\rho < \rho_h$ in this case), and there is an unstable strange invariant set containing a horseshoe on the other side ($\rho > \rho_h$ in this case). However, as pointed out by Kaplan and Yorke (1979c) and Yorke and Yorke (1979), the resulting chaos is not attracting but transient. The nontrivial fixed points C_1 and C_2 continue to be the only attracting sets (see Fig. 5.8.7c) until ρ exceeds $\rho_c \approx 24.06$, where

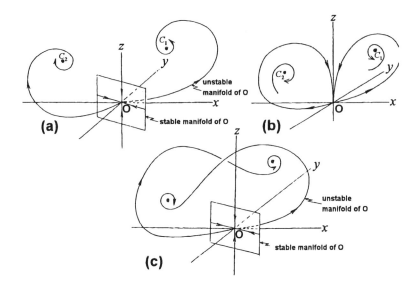

Figure 5.8.7: Behavior of the unstable manifold of the saddle of the Lorenz equations: (a) $\rho < \rho_h$, (b) $\rho = \rho_h$, and (c) $\rho > \rho_h$.

the transient chaos is converted into a chaotic attractor by a crisis. As ρ increases past $\rho = \rho_H \approx 24.74$, the nontrivial fixed points lose stability via a subcritical Hopf bifurcation. Therefore, in the small interval $\rho_c < \rho < \rho_H$, there are three attractors: the nontrivial fixed points C_1 and C_2 and a chaotic attractor.

5.8.3   Orbits Homoclinic to a Saddle Focus

In contrast with the preceding case in which the eigenvalues of the fixed point are purely real, we consider in this section three–dimensional systems with orbits homoclinic to a fixed point with one real eigenvalue and two complex–conjugate eigenvalues; that is, a saddle focus. Specifically, we consider after Shilnikov (1965, 1968, 1970) the three–dimensional system

$$\dot{x} = -\rho x - \omega y + f_1(x, y, z; \epsilon) \tag{5.8.37}$$
$$\dot{y} = \omega x - \rho y + f_2(x, y, z; \epsilon) \tag{5.8.38}$$
$$\dot{z} = \lambda z + f_3(x, y, z; \epsilon) \tag{5.8.39}$$

where ω, ρ, λ, and ϵ are constants and f_1, f_2, and f_3 are C^2 and vanish together with their first derivatives at $(x, y, z, \epsilon) = (0, 0, 0, 0)$. Here, ρ, ω, and λ are positive. We assume that the system (5.8.37)–(5.8.39) has a single orbit homoclinic to the origin (it is a saddle focus) when $\epsilon = 0$. We let $\delta = \rho/\lambda$. Shilnikov (1965, 1968, 1970) proved the following theorem. If $\delta > 1$, the system has a periodic orbit on one side of the homoclinicity and no recurrent behavior on the other side. The period of the orbit tends to infinity as $\epsilon \to 0$. If $\delta < 1$, then for $\epsilon = 0$ there is a countable infinity of unstable periodic orbits in a neighborhood of the homoclinic orbit. In fact, the system has a countable infinity of horseshoes. Moreover, Gaspard (1983) and Glendinning and Sparrow (1984) showed that, if $\delta < 1$ and $\epsilon \neq 0$, finitely many of the horseshoes persist and on one side of the homoclinicity there is a countable infinity of more complicated homoclinic orbits at parameter values $\epsilon_1, \epsilon_2, \ldots, \epsilon_n, \ldots$ tending to zero. This theorem also applies if both ρ and λ are negative by time reversal.

In the presence of the symmetry $(x, y, z) \to (-x, -y, -z)$, Glendinning (1984) showed that, for $\delta > 1$, each homoclinic bifurcation contributes three periodic orbits to the bifurcation diagram far from $\epsilon = 0$: two asymmetric orbits on one side of the homoclinicity and a symmetric orbit on the other side.

The three–dimensional results discussed in this section are applicable to higher–dimensional systems provided that the following conditions are satisfied. First, the system has an orbit homoclinic to a saddle–focus point. This orbit approaches the saddle focus as $t \to \pm\infty$. Second, the eigenvalues of the Jacobian matrix associated with the saddle focus have a special structure. Out of all of the eigenvalues in the right–half of the complex plane, the closest to the imaginary axis is a real eigenvalue λ. And out of all of the eigenvalues in the left–half of the complex plane, the closest to the imaginary axis is a pair of complex conjugate eigenvalues $-\rho \pm i\omega$, where $\omega \neq 0$. Consequently, the dynamics of the higher–dimensional system near the saddle focus can be reduced to the three–dimensional system (5.8.37)–(5.8.39) because the three mentioned eigenvalues determine how the homoclinic orbit approaches and leaves the saddle–focus point. By reversing time, the results also apply to the case in which the eigenvalues closest to the imaginary axis are $\rho \pm i\omega$ and $-\lambda$, where ρ and λ are positive.

In addition to the aforementioned references, we refer the reader to the following references: Arneodo, Coullet, and Tresser (1982), Gaspard, Kapral, and Nicolis (1984), Tresser (1984a,b), Wiggins (1988), Fowler (1990a,b), and Palis and Takens (1993). Homoclinic bifurcations in n dimensions are considered in detail by Wiggins (1988) and Fowler (1990a).

Example 5.21. We consider, after Chin, Nayfeh, and Mook (1993), the system

$$\dot{a}_1 = -\mu_1 a_1 - (\Lambda a_2 + \alpha a_2^3)\sin\gamma_1 - f a_2 \sin\gamma_2 \qquad (5.8.40)$$

$$\dot{a}_2 = -\mu_2 a_2 + \frac{1}{2}a_1 \sin\gamma_1 \qquad (5.8.41)$$

$$a_1\dot{\beta}_1 = (\Lambda a_2 + \alpha a_2^3)\cos\gamma_1 + f a_2 \cos\gamma_2 \qquad (5.8.42)$$

$$a_2\dot{\beta}_2 = \frac{1}{2}a_1 \cos\gamma_1 \qquad (5.8.43)$$

$$\gamma_1 = \beta_2 - \beta_1 \quad \text{and} \quad \gamma_2 = \sigma t - \beta_2 - \beta_1 \qquad (5.8.44)$$

where $\mu_1 = \mu_2 = 1.2334, \sigma = 4.0835, \Lambda = -11.4792, \alpha = 0.4972$, and $f = 0.0235F$. This system has an orbit homoclinic to a saddle focus when $F = F_h \approx 340.854$. The eigenvalues of the saddle focus are $0.4656, -1.2334 \pm 4.2036i$, and -2.9324. Hence, $\delta = 1.2334/0.4656 = 2.649 > 1$. Therefore, according to the Shilnikov theorem, the system has a stable limit cycle on one side of the homoclinicity and no recurrent behavior on the other side. Two–dimensional projections of the unstable manifolds of the saddle focus are shown in Figure 5.8.8. When $F = 330$, the unstable manifold leads to a limit cycle in one direction and to a sink in the other, as shown in Figure 5.8.8a. As F is increased, the limit cycle and saddle focus move closer to each other, as shown in Figure 5.8.8b at $F = 335$. When $F = F_h$, there is an orbit that is asymptotic to the saddle focus in forward and backward time; that is, an orbit homoclinic to the saddle focus has been formed, as shown in Figure 5.8.8c. As F is increases through F_h, there is no recurrent behavior nearby, and the unstable manifold of the saddle focus leads to the sink in both directions.

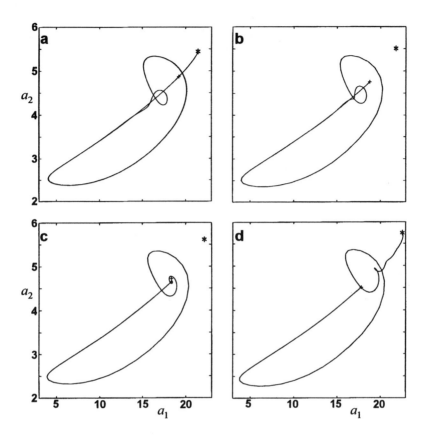

Figure 5.8.8: A two–dimensional projection of an unstable manifold onto the $a_2 - a_1$ plane for (a) $F = 330$ (both directions are included), (b) $F = 335$, (c) $F = 340.853$, and (d) $F = 350$: + denotes a saddle and * denotes a sink.

Example 5.22. We consider, after Nayfeh and Chin (1994), the system

$$\dot{a} = -a \left(\mu_2 + \frac{1}{4} f \sin \beta \right) \tag{5.8.45}$$

$$a\dot{\beta} = -a \left(\sigma + \frac{3}{4} \alpha_3 a^2 + \alpha_4 u_1^2 + \frac{1}{2} f \cos \beta \right) \tag{5.8.46}$$

$$\dot{u}_1 = v_1 \tag{5.8.47}$$

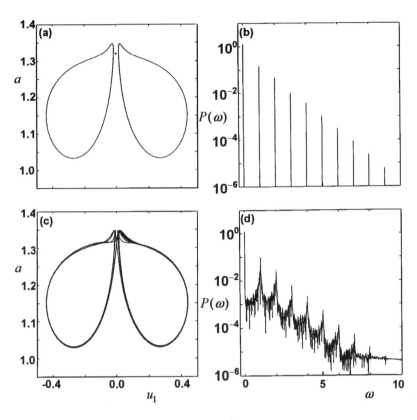

Figure 5.8.9: (a) A two–dimensional projection of the twin limit cycles onto the $a - u_1$ plane, (b) the FFT of a at $\sigma = -0.55799$, (c) a two–dimensional projection of the chaotic attractor onto the $a - u_1$ plane, and (d) the FFT of a at $\sigma = -0.558$. In (a) and (c), + denotes the saddle focus.

$$\dot{v}_1 = -\left(u_1 + 2\mu_1 v_1 + 4\alpha_1 u_1^3 + \frac{1}{2}\alpha_2 u_1 a^2\right) \qquad (5.8.48)$$

when $\mu_1 = 0.25, \mu_2 = 0.5, f = 2.5, \alpha_1 = \alpha_3 = 1, \alpha_2 = -2$, and $\alpha_4 = 4$. There is an orbit homoclinic to a saddle focus when $\sigma = \sigma_h \approx -0.55799$. The eigenvalues of the saddle focus at $\sigma = \sigma_h$ are $0.644, -0.5 \pm 0.855i$, and -1.148. Hence, $\delta = 0.5/0.644 = 0.776 < 1$. Therefore, according to the Shilnikov theorem, the trajectories of the system in the neighborhood of the homoclinic orbit form a homoclinic structure that contains a countable set of saddle periodic orbits, and hence a state trajectory wanders "randomly" between the unstable periodic trajectories. In other words, the system behavior is chaotic. In Figures 5.8.9a and 5.8.9b we show the two twin period orbits and their spectrum that exist at $\sigma = -0.55799$, just before the formation of the homoclinic orbit, and in Figures 5.8.9c and 5.8.9d we show the chaotic attractor and its spectrum at $\sigma = -0.558$, just after the formation of the homoclinic orbit. The broad spectrum of this attractor shown in Figure 5.8.9d and its Lyapunov exponents $(0.015, 0.0, -0.982, -1.195)$ confirm its chaotic nature.

5.8.4 Comments

Shilnikov (1967, 1970) also proved the following theorem. Let a four-dimensional system have an orbit homoclinic to a saddle focus when a parameter $\epsilon = 0$. Assume that the eigenvalues of the Jacobian matrix associated with the saddle focus are $\rho_1 \pm i\omega_1$ and $-\rho_2 \pm i\omega_2$, where ω_1 and $\omega_2 \neq 0$, ρ_1 and $\rho_2 > 0$, and $\rho_1 = \rho_2$. Then, for $\epsilon = 0$, there is a countable infinity of horseshoes in a neighborhood of the homoclinic orbit. At least thus far it appears that researchers have not encountered examples of natural systems exhibiting this homoclinicity.

We note that the theorems discussed in Sections 5.8.1–5.8.3 do not give a complete picture of the systems under consideration. The theorems are local in nature. Even if they predict complexity, such complexity may not be observed in numerical simulations or may be observed for parameter values far away from homoclinicity (Arneodo, Coullet, and Tresser, 1982; Glendinning and Sparrow, 1986), as is evident in the next example.

Example 5.23. We consider, after Arneodo et al. (1982), the piecewise linear system

$$\dot{x} = y \tag{5.8.49}$$

$$\dot{y} = z \tag{5.8.50}$$

$$\dot{z} = -y - \beta z + f(x) \tag{5.8.51}$$

where

$$f(x) = 1 + ax \quad \text{if} \quad x \leq 0 \tag{5.8.52}$$

$$f(x) = 1 - \mu x \quad \text{if } x \geq 0 \tag{5.8.53}$$

and $\beta, a,$ and μ are positive constants.

The system has two equilibria: $A = (-a^{-1}, 0, 0)$ in the left–half plane, and $B = (\mu^{-1}, 0, 0)$ in the right–half plane. The eigenvalues of the Jacobian matrix at A are given by

$$s^3 + \beta s^2 + s - a = 0 \tag{5.8.54}$$

We denote the three roots of (5.8.45) by λ and $\rho \pm i\omega$, where $\rho, \omega,$ and λ are real. Then,

$$\lambda + \rho + i\omega + \rho - i\omega = -\beta \tag{5.8.55}$$
$$\lambda(\rho + i\omega) + \lambda(\rho - i\omega) + (\rho + i\omega)(\rho - i\omega) = 1 \tag{5.8.56}$$
$$\lambda(\rho + i\omega)(\rho - i\omega) = a \tag{5.8.57}$$

Similarly, the eigenvalues of the Jacobian matrix at B are given by

$$s^3 + \beta s^2 + s + \mu = 0 \tag{5.8.58}$$

If we denote these eigenvalues by L and $R \pm i\Omega$, then

$$L + 2R = -\beta \tag{5.8.59}$$
$$2LR + R^2 + \Omega^2 = 1 \tag{5.8.60}$$
$$\lambda(R^2 + \Omega^2) = -\mu \tag{5.8.61}$$

For simplicity, we choose the parameters $\rho, \omega,$ and R instead of $\beta, a,$ and μ. Then, $\lambda, \beta, a, L, \Omega,$ and μ are given in terms of $\rho, \omega,$ and R as

$$\lambda = (1 - \rho^2 - \omega^2)/2\rho, \quad \beta = -(\lambda + 2\rho), \quad a = \lambda(\rho^2 + \omega^2) \tag{5.8.62}$$

$$L = -\beta - 2R, \quad \Omega^2 = 1 - R^2 - 2RL, \quad \mu = -L(R^2 + \Omega^2) \tag{5.8.63}$$

To illustrate the consequences of the Shilnikov theorem, we fix ρ and ω so that the fixed point A is a saddle focus with the desired value of $\delta = |\rho/\lambda|$. Then, we choose R so that the unstable manifold of A nontransversely intersects its stable manifold, thereby forming a homoclinic orbit to A. To accomplish this, we write the general solution of the system in the region $x \leq 0$ as

$$x = (c_1 \cos \omega t + c_2 \sin \omega t)e^{\rho t} + c_3 e^{\lambda t} - \frac{1}{a} \qquad (5.8.64)$$

$$y = \rho(c_1 \cos \omega t + c_2 \sin \omega t)e^{\rho t}$$
$$-\omega(c_1 \sin \omega t - c_2 \cos \omega t)e^{\rho t} + c_3 \lambda e^{\lambda t} \qquad (5.8.65)$$

$$z = (\rho^2 - \omega^2)(c_1 \cos \omega t + c_2 \sin \omega t)e^{\rho t}$$
$$-2\omega\rho(c_1 \sin \omega t - c_2 \cos \omega t)e^{\rho t} + c_3 \lambda^2 e^{\lambda t} \qquad (5.8.66)$$

Consequently, the unstable manifold of the saddle is given by

$$x = c_3 e^{\lambda t} - \frac{1}{a}, \quad y = c_3 \lambda e^{\lambda t}, \quad z = c_3 \lambda^2 e^{\lambda t} \qquad (5.8.67)$$

It intersects the plane $x = 0$ at the point M when $c_3 e^{\lambda t} = a^{-1}$, which corresponds to $y = \lambda a^{-1}$ and $z = \lambda^2 a^{-1}$. The stable manifold of A is given by

$$x = (c_1 \cos \omega t + c_2 \sin \omega t)e^{\rho t} - \frac{1}{a} \qquad (5.8.68)$$

$$y = \rho(c_1 \cos \omega t + c_2 \sin \omega t)e^{\rho t} - \omega(c_1 \sin \omega t - c_2 \cos \omega t)e^{\rho t} \qquad (5.8.69)$$

$$z = (\rho^2 - \omega^2)(c_1 \cos \omega t + c_2 \sin \omega t)e^{\rho t}$$
$$-2\omega\rho(c_1 \sin \omega t - c_2 \cos \omega t)e^{\rho t} \qquad (5.8.70)$$

This manifold intersects the plane $x = 0$ along the line D. To determine D, we subtract 2ρ times (5.8.69) from (5.8.70) and obtain

$$z - 2\rho y = -(\rho^2 + \omega^2)(c_1 \cos \omega t + c_2 \sin \omega t)e^{\rho t} \qquad (5.8.71)$$

Setting $x = 0$ in (5.8.68) yields

$$(c_1 \cos \omega t + c_2 \sin \omega t)e^{\rho t} = \frac{1}{a} \qquad (5.8.72)$$

Combining (5.8.71) and (5.8.72) gives the following equation for D:

$$z - 2\rho y = -(\rho^2 + \omega^2)a^{-1} = 2\rho(\rho^2 + \omega^2 - 1)^{-1} \qquad (5.8.73)$$

on account of (5.8.62).

The general solution in the region $x \geq 0$ is given by

$$x = (b_1 \cos \Omega t + b_2 \sin \Omega t)e^{Rt} + b_3 e^{Lt} - \frac{1}{\mu} \qquad (5.8.74)$$

$$y = R(b_1 \cos \Omega t + b_2 \sin \Omega t)e^{Rt}$$
$$-\Omega(b_1 \sin \Omega t - b_2 \cos \Omega t)e^{Rt} + b_3 L e^{Lt} \qquad (5.8.75)$$
$$z = (R^2 - \Omega^2)(b_1 \cos \Omega t + b_2 \sin \Omega t)e^{Rt}$$
$$-2R\Omega(b_1 \sin \Omega t - b_2 \cos \Omega t)e^{Rt} + b_3 L^2 e^{Lt} \qquad (5.8.76)$$

To continue the unstable manifold from the left–half to the right–half plane, we start the trajectory in the right–half plane from the point M at $t = 0$. Hence,

$$b_1 = \frac{(2R - L)La - \lambda\mu(2R - \lambda)}{R^2 + \Omega^2 - 2RL + L^2}$$

$$b_3 = \frac{1}{\mu} - b_1, \quad b_2 = -\frac{R}{\Omega}b_1 - \frac{L}{\Omega}b_3 + \frac{\lambda}{a\Omega} \qquad (5.8.77)$$

For $a = 0.633625$ and $\beta = 0.3375$, $\rho = -0.4$, $\omega = 1.1$, and $\lambda = 0.4625$, and hence $\delta = \rho/\lambda = 0.8649 < 1$. When $R = R_h \approx 0.3982$, there is an orbit homoclinic to the saddle focus A. Although all of the conditions of the Shilnikov theorem are satisfied, Arneodo, Coullet, and Tresser (1982) found that all trajectories quickly wander off to infinity for values of R near R_h. In fact, chaos is first numerically observed at $R \approx 0.1740$, which is very far from the homoclinicity value R_h.

For $a = 0.224635$ and $\beta = 0.3375$, $\rho = -0.27$, $\omega = 1.018$, and $\lambda = 0.2024$. Hence, $\delta = \rho/\lambda = 1.3347 > 1$. Although not all of the conditions in Shilnikov's theorem are fulfilled for chaotic behavior to occur, Arneodo, Coullet, and Tresser (1982) numerically found chaos.

5.9   EXERCISES

5.1. When $\alpha = 1$, use the transformation $x_n = \sin^2(\pi\theta_n)$ to transform the logistic map

$$x_{n+1} = 4\alpha x_n(1 - x_n)$$

into

$$\theta_{n+1} = 2\theta_n \bmod (1)$$

5.2. Consider the one–dimensional map

$$x_{n+1} = b + a\frac{x_n}{1 + x_n^2}$$

Set $a = 11.5$ and show that as b is varied from -5 to 0 there is an incomplete period–doubling cascade (Bier and Bountis, 1984). You will find that there is a finite number of period–doubling bifurcations followed by a finite number of reverse period–doubling bifurcations.

5.3. Consider the Hénon map, which is described by

$$x_{k+1} = 1 + y_k - \alpha x_k^2$$
$$y_{k+1} = 0.3x_k$$

Choose α as a control parameter and gradually decrease it from $\alpha \simeq 1.2263$. Verify that periodic and chaotic attractors occur at $\alpha = 1.2262$ and $\alpha = 1.2260$, respectively. Examine the mechanism leading to the chaotic solution at $\alpha = 1.2260$.

5.4. Consider the following quasiperiodically forced van der Pol oscillator (Kapitaniak, 1991):

$$\dot{x}_1 = x_2$$
$$\dot{x}_2 = -x_1 + \alpha(1 - x_1^2)x_2 + F\cos(\omega_1 t)\cos(\omega_2 t)$$

Set $\alpha = 5.0$, $F = 5.0$, and $\omega_1 = \sqrt{2} + 1.05$ and numerically determine the asymptotic state of this system in each of the following cases: (i) $\omega_2 = 0.002$ and (ii) $\omega_2 = 0.006$. Discuss the characteristics of the two attractors by using Poincaré sections and power spectra and examining sensitivity to initial conditions.

5.5. Consider the circle map (5.5.5):

(a) Noting that the rotation number is zero in the wedge 0/1 of Figure 4.2.1, examine the dynamics of this map.

(b) Set $\Omega = 0.2$ in (5.5.5) and construct the graph of $F(x)$ versus x for each of the following cases: (i) $K = 0.9$, (ii) $K = 1.0$, and (iii) $K = 1.1$. Discuss the nature of these graphs. Also, examine the solutions of the map in each case. Determine the rotation number in the first two cases. (If you try to determine the rotation number in the third case, you will find that the limit (4.2.2) does not exist. You will obtain different values of ρ for different x_0. The rotation number is not unique for $K > 1$.)

(c) In the interval $[0, 1]$ choose an irrational number and verify that there is a continuous curve in the $K - \Omega$ plane that originates from this irrational number on the Ω axis and extends to the line $K = 1$.

5.6. Consider the following two–dimensional map treated by Holmes (1979):

$$x_{k+1} = y_k$$
$$y_{k+1} = -bx_k + dy_k - y_{k+1}^3$$

(a) Set $b = 0.2$, vary d in the range $[2.0, 2.8]$ and display the different attractors obtained in a plot of y_k versus d. Discuss the bifurcation diagram.

(b) Examine the attractor obtained at $d = 2.71$ and discuss its characteristics.

5.7. Consider the Lorenz equations (5.4.5)–(5.4.7) and set $\sigma = 10$ and $\beta = \frac{8}{3}$.

(a) For these parameter values, a subcritical Hopf bifurcation occurs at $\rho_c \simeq 24.74$. Verify that unstable periodic solutions exist for $\rho < \rho_c$.

(b) At $\rho = 28$, numerically ascertain that the asymptotic solution is chaotic by examining the state–space portrait, an associated Poincaré section, and the frequency spectrum of one of the states.

(c) At $\rho = 28$, consider the points (x, y, z) on the chaotic solution that satisfy the condition $xy - \beta z = 0$. These points correspond to the extrema of z because $\dot{z} = 0$. Of these points, choose the maxima, let z_k represent the kth maximum of z, and construct the graph z_{k+1} versus z_k.

(d) Close to the attractor determined at $\rho = 28$, choose two initial points separated by a distance of 10^{-6}. Construct a graph to show how the separation between the evolutions initiated from these two points changes with time. Discuss this graph.

The aperiodic asymptotic solution realized at $\rho = 28$ is called the **Lorenz attractor.** This attractor is also known as the **Lorenz mask** (Abraham and Shaw, 1992, Chapter 9). Maps such as those constructed in part c are called **return maps.** In the present case, you will find, as Lorenz did in 1963, that the return map has a simple structure. The iterates of the map fall on a one–dimensional curve with a maximum and resemble the tent map of Exercise 2.1 for $a < \frac{1}{2}$. This suggests that the features of the dynamics of the Lorenz equations may be captured by the one–dimensional map $z_{k+1} = f(z_k)$. An extensive study of the Lorenz equations was conducted by Sparrow (1982).

5.8. Consider equations (3.4.1)–(3.4.3) and verify that symmetry–breaking and period–doubling bifurcations of periodic solutions take place at $\mu \simeq 0.30$ and $\mu \simeq 0.44$, respectively. Examine the bifurcations that take place as μ is increased beyond 0.45.

5.9. Consider the Rössler equations (5.2.1)–(5.2.3).

(a) Set $b = 2$ and $c = 4$. In the range $0.3 \leq a < 0.4$, verify that a sequence of period–doubling bifurcations occurs, culminating in an aperiodic motion at $a \simeq 0.387$.

(b) Through numerical integration, determine the asymptotic aperiodic solution at $a = 0.2$, $b = 0.2$, and $c = 5.7$. On the chaotic
solution, consider the points that satisfy $y + z = 0$. These points
correspond to the extrema of x because $\dot{x} = 0$. Of these points,
choose the maxima and let x_k represent the kth maximum of x.
Then, plot x_{k+1} versus x_k. Discuss the graph. (You will find that
this graph is remarkably similar to Figure 2.2.2a constructed for
the map [1.1.4].)

Rössler (1976a,b, 1979b) classifies some of the strange attractors of the
Rössler equations as **screw–type** and **spiral–type attractors**. The
attractor obtained in part b is an example of a spiral–type attractor.

5.10. Consider the following externally excited oscillator (Räty,
Isomäki, and von Boehm, 1984; Räty, von Boehm, and Isomäki, 1986);

$$\ddot{x} + 0.4\,\dot{x} + x - \beta x^2 - 4x^3 = 0.115\cos(\Omega t)$$

(a) For $\beta = 0$, as Ω is gradually decreased from 0.54, verify that a
symmetry–breaking bifurcation occurs followed by a sequence of
period–doubling bifurcations culminating in a chaotic motion.

(b) Numerically verify that periodic and chaotic attractors coexist by
using phase portraits and Poincaré sections in the following cases:
(i) $\beta = 0$ and $\Omega = 0.5255$ and (ii) $\beta = 0.1$ and $\Omega = 0.5281$.

5.11. Consider the following parametrically excited single–degree–of–
freedom system (Zavodney and Nayfeh, 1988):

$$\ddot{x} + \omega_0^2 x + \epsilon\left[2\mu\dot{x} + \delta x^2 + gx\cos(\Omega t)\right] + \epsilon^2\alpha x^3 = 0$$

Also,
$$\Omega = \omega_0 + \epsilon^2\sigma$$

Assume that ϵ is a small and positive parameter and μ, δ, α, g, and σ
are all independent of ϵ.

(a) Construct a second–order approximation for periodic oscillations about the origin by using the method of multiple scales.

(b) Compare the perturbation results with the results obtained through direct numerical integrations for $\omega_0 = 1$, $\sigma = 0$, $\alpha = 4$, $\delta = 3$, $\mu = 1$, $g = 11$, and $\epsilon = 0.10$. Discuss the comparison.

(c) Construct the state–space portrait and verify that a periodic attractor and a point attractor coexist for $\omega_0 = 1$, $\sigma = 0$, $\alpha = 4$, $\delta = 5$, $\mu = 1$, $g = 2.5$, and $\epsilon = 0.10$. (You will find that the boundaries of the basins of attraction for the two attractors are smooth.)

(d) Construct the state–space portrait and verify that two periodic attractors and a point attractor coexist for $\omega_0 = 1$, $\sigma = 0$, $\alpha = 4$, $\delta = 5$, $\mu = 1$, $g = 5$, and $\epsilon = 0.10$. (You will find that the boundaries of the basins of attraction for the three attractors are smooth.)

The state–space portraits for $g = 2.5$ and $g = 5.0$ are different because a global bifurcation takes place at the critical value g_c, which lies between 2.5 and 5.0. When the boundaries of basins of attraction are smooth, as in parts c and d, there is a clear demarcation of the domains of attraction of the different attractors. However, this is not so when the boundaries are fractal. Fractal basin boundaries have been found in many studies (e.g., Grebogi, Ott, and Yorke, 1983a,b, 1986, 1987; Moon and Li, 1985; Thompson, Bishop, and Leung, 1987; Ueda and Yoshida, 1987; Zavodney, 1987; Pezeshki and Dowell, 1988; Nayfeh and Sanchez, 1989; Zavodney, Nayfeh, and Sanchez, 1989; Li and Moon, 1990a,b; Nayfeh, Hamdan, and Nayfeh, 1990, 1991; Ueda, 1991; Rega, Salvatori, and Benedettini, 1992).

5.12. Show that
$$I = \int_{-\infty}^{\infty} \text{sech}(t)\tanh(t)\sin(\Omega t)dt = \pi\Omega \, \text{sech}\left(\frac{1}{2}\Omega\pi\right)$$
Hint: Why is the following integral zero?
$$\int_{-\infty}^{\infty} \text{sech}(t)\tanh(t)\cos(\Omega t)dt = 0$$

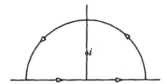

Figure 5.9.1: Contour of integration.

Then, show that

$$I = \frac{1}{i} \int_{-\infty}^{\infty} e^{i\Omega t} \, \mathrm{sech}(t) \tanh(t) dt$$

or

$$I = \frac{2}{i} \int_{-\infty}^{\infty} e^{i\Omega t} \frac{e^t - e^{-t}}{(e^t + e^{-t})^2} dt$$

Let $e^t = u$ and obtain

$$I = \frac{2}{i} \int_{0}^{\infty} u^{i\Omega} \frac{u^2 - 1}{(u^2 + 1)^2} du$$

Show that

$$\int_{-\infty}^{0} u^{i\Omega} \frac{u^2 - 1}{(u^2 + 1)^2} du = e^{-\pi\Omega} \int_{0}^{\infty} v^{i\Omega} \frac{v^2 - 1}{(v^2 + 1)^2} dv$$

and hence

$$I = \frac{2}{i(1 + e^{-\pi\Omega})} \int_{-\infty}^{\infty} u^{i\Omega} \frac{u^2 - 1}{(u^2 + 1)^2} du$$

To evaluate the integral, use the method of residues, choose the path γ in Figure 5.9.1, and show that

$$J = \int_{-\infty}^{\infty} u^{i\Omega} \frac{u^2 - 1}{(u^2 + 1)^2} du = \int_{\gamma} z^{i\Omega} \frac{z^2 - 1}{(z^2 + 1)^2} dz = \int_{\gamma} f(z) dz = 2\pi i R$$

where R is the residue of the integrand at the twofold singularity at $z = i$. Use the formula

$$\text{Residue of } f(z) \text{ at } z = a = \frac{1}{(m - 1)!} \lim_{z \to a} \frac{d^{m-1}}{dz^{m-1}} [(z - a)^m f(z)]$$

where $f(z)$ has an m–fold singularity at $z = a$, to show that

$$J = i\Omega\pi e^{-\frac{1}{2}\Omega\pi}$$

Finally, show that

$$I = \pi\Omega \text{ sech}\left(\frac{1}{2}\Omega\pi\right)$$

5.13. Show that

$$I = \int_{-\infty}^{\infty} \text{sech}^2(t)\cos(\nu t)dt = \frac{\nu\pi}{\sinh\left(\frac{1}{2}\nu\pi\right)}$$

Hint: Show that

$$I = \int_{-\infty}^{\infty} e^{i\nu t} \text{ sech}^2(t)dt$$

Let $e^t = u$ and obtain

$$I = 4\int_0^\infty \frac{u^{i\nu+1}}{(u^2+1)^2}du = \frac{4}{1-e^{-\nu\pi}}\int_{-\infty}^\infty \frac{u^{i\nu+1}}{(u^2+1)^2}du$$

$$= \frac{4}{1-e^{-\nu\pi}}\int_\gamma \frac{z^{i\nu+1}}{(z^2+1)^2}dz$$

$$= \frac{8\pi i}{1-e^{-\nu\pi}}\lim_{z\to i}\frac{d}{dz}\left(\frac{z^{i\nu+1}}{(z+i)^2}\right) = \frac{\nu\pi}{\sinh\left(\frac{1}{2}\nu\pi\right)}$$

5.14. Show that

$$I = \int_{-\infty}^{\infty} \tanh(2t)\text{ sech}^2(2t)\sin(\Omega t)dt = \frac{\Omega^2\pi}{16\sinh\left(\frac{1}{4}\Omega\pi\right)}$$

Hint: Show that

$$I = \frac{1}{i}\int_{-\infty}^{\infty} e^{i\Omega t}\tanh(2t)\text{ sech}^2(2t)dt$$

Then, let $e^{2t} = u$ and show that

$$I = \frac{2}{i} \int_0^\infty u^{iv} \frac{u^3 - u}{(u^2 + 1)^3} du$$

where $v = \frac{1}{2}\Omega$. Then, show that

$$I = \frac{2}{i(1 - e^{-\pi v})} \int_{-\infty}^\infty u^{iv} \frac{u^3 - u}{(u^2 + 1)^3} du$$

Show that

$$J = \int_{-\infty}^\infty u^{iv} \frac{u^3 - u}{(u^2 + 1)^3} du = \int_\gamma z^{iv} \frac{z^3 - z}{(z^2 + 1)^3} dz$$

$$= \frac{2\pi i}{2!} \lim_{z \to i} \frac{d^2}{dz^2} \left[z^{iv} \frac{z^3 - z}{(z + i)^3} \right] = \frac{1}{4} i\pi v^2 e^{-\frac{1}{2}v\pi}$$

Hence, show that

$$I = \frac{\Omega^2 \pi}{16 \sinh\left(\frac{1}{4}\Omega\pi\right)}$$

5.15. Consider an externally excited pendulum modeled by

$$\ddot{x} + \sin x = \epsilon[F \cos(\Omega t) - 2\mu \dot{x}]$$

(a) Show that the unperturbed system has saddle connection orbits $\Gamma^\pm$ between $(\pm\pi, 0)$.

(b) Show that Γ^+ can be expressed as

$$x_1 = x = 4 \tan^{-1}(e^t) - \pi \quad \text{and} \quad x_2 = \dot{x} = 2 \operatorname{sech}(t)$$

where the origin of t is chosen so that $x(0) = 0$.

(c) Show that the Melnikov function is given by

$$M(\theta) = \int_{-\infty}^\infty \left\{ F x_2 \cos[\Omega(t + \theta)] - 2\mu x_2^2 \right\} dt$$

$$= 2F \cos(\Omega\theta) \int_{-\infty}^\infty \operatorname{sech}(t) \cos(\Omega t) dt$$

$$-8\mu \int_{-\infty}^\infty \operatorname{sech}^2(t) dt$$

(d) Use the method of residues and show that

$$\int_{-\infty}^{\infty} \text{sech}(t) \cos(\Omega t)dt = \pi \ \text{sech}\left(\frac{1}{2}\Omega\pi\right)$$

(e) Hence, show that

$$M(\theta) = -16\mu + 2F\pi \cos(\Omega\theta) \ \text{sech}\left(\frac{1}{2}\Omega\pi\right)$$

5.16. Consider a parametrically excited pendulum modeled by

$$\ddot{x} + \sin x = -\epsilon \left[F \sin(x) \cos(\Omega t) + 2\mu\dot{x}\right]$$

(a) Show that the unperturbed system has saddle connection orbits $\Gamma^{\pm}$ between the saddle points $(\pm\pi, 0)$.

(b) Show that Γ^+ can be expressed as

$$x_1 = x = 4\tan^{-1}(e^t) - \pi \ \text{and} \ x_2 = \dot{x} = 2 \ \text{sech}(t)$$

where the origin of t is chosen so that $x(0) = 0$.

(c) Show that the Melnikov function can be expressed as

$$M(\theta) = 4F \sin(\Omega\theta) \int_{\infty}^{\infty} \tanh(t) \ \text{sech}^2(t) \sin(\Omega t)dt$$
$$-8\mu \int_{-\infty}^{\infty} \text{sech}^2(t)dt$$

(d) Use the results of Exercise 5.14 to evaluate the first integral and obtain

$$M(\theta) = 2\pi\Omega^2 F\frac{\sin(\Omega\theta)}{\sinh\left(\frac{1}{2}\pi\Omega\right)} - 16\mu$$

5.17. Consider the damped Sine–Gordon equation

$$\ddot{x} + \sin x = \epsilon(a + F \cos \Omega t - 2\mu\dot{x})$$

Use the expression for the heteroclinic half–orbit of the unperturbed system derived in Exercise 5.16 to show that the Melnikov function is given by

$$M(\theta) = 2a\pi - 16\mu + 2F\pi \cos(\Omega\theta) \operatorname{sech}\left(\frac{1}{2}\Omega\pi\right)$$

5.18. Consider the system

$$\ddot{x} + \frac{1}{2}(x - x^3) = \epsilon[F\cos(\Omega t) - 2\mu\dot{x}]$$

(a) Show that the Melnikov function can be expressed as

$$M(\theta) = 2F \int_{-\infty}^{\infty} \operatorname{sech}^2(2t) \cos[\Omega(t + \theta)]dt - 8\mu \int_{-\infty}^{\infty} \operatorname{sech}^4(2t)dt$$

or

$$M(\theta) = 2F \cos(\Omega\theta) \int_{-\infty}^{\infty} \operatorname{sech}^2(2t) \cos(\Omega t)dt - \frac{16}{3}\mu$$

(b) Use the method of residues and show that

$$\int_{-\infty}^{\infty} \operatorname{sech}^2(2t) \cos(\Omega t)dt = \frac{\Omega\pi}{4 \sinh\left(\frac{1}{4}\Omega\pi\right)}$$

(c) Hence, show that

$$M(\theta) = -\frac{16}{3}\mu + \frac{F\Omega\pi}{2 \sinh\left(\frac{1}{4}\Omega\pi\right)}$$

5.19. Consider the system

$$\ddot{x} - x + x^3 = \epsilon[Fx\cos(\Omega t) - 2\mu\dot{x}]$$

(a) Show that the Melnikov function can be expressed as

$$M(\theta) = -2F \int_{-\infty}^{\infty} \tanh(t) \operatorname{sech}^2(t) \cos[\Omega(t + \theta)]dt$$
$$- 4\mu \int_{-\infty}^{\infty} \operatorname{sech}^2(t) \tanh^2(t)dt$$

(b) Use the results of Exercise 5.14 to evaluate the first integral and
hence show that

$$M(\theta) = -\frac{8}{3}\mu + F\Omega^2\pi\frac{\sin(\Omega\theta)}{\sinh\left(\frac{1}{2}\Omega\pi\right)}$$

5.20. Consider the family of mappings

$$x_{n+1} = 1 - a \mid x_n \mid^N$$

Use a numerical algorithm to determine the universal constant

$$\delta = \lim_{n\to\infty}\frac{a_{n+1} - a_n}{a_{n+2} - a_{n+1}}$$

when $N = 2, 3, 4 \ldots, 8$.

Chapter 6

NUMERICAL METHODS

In this chapter, we present algorithms pertaining to fixed point and periodic solutions of continuous–time systems. In Section 6.1 continuation schemes for tracing branches of fixed point solutions in a given state–control space are discussed. We consider direct schemes for determining static and Hopf bifurcation points in Sections 6.2 and 6.3. In Section 6.4 we present homotopy algorithms for determining the roots of an algebraic system of equations. Numerical construction of periodic solutions of autonomous and nonautonomous systems is discussed in Section 6.5. Finally, in Section 6.6 we consider continuation schemes for tracing branches of periodic solutions in a given state–control space.

6.1   CONTINUATION OF FIXED POINTS

Continuation schemes are used to determine how solutions of a system, such as (2.1.1), vary with a certain parameter. These schemes are based on the **implicit function theorem**. Let us consider the fixed–point solutions determined from (2.1.2). The state vector $\mathbf{x} \in \mathcal{R}^n$, the parameter vector $\mathbf{M} \in \mathcal{R}^m$, and the vector field $\mathbf{F}$ maps $\mathcal{R}^n \times \mathcal{R}^m$ into $\mathcal{R}^n$. Let $\mathbf{F}$ be $\mathcal{C}^r$, where $r \geq 1$, and let $\mathbf{x} = \mathbf{x}_0$ be the solution of (2.1.2) when $\mathbf{M} = \mathbf{M}_0$. Then, according to the **implicit function theorem**, if the Jacobian matrix $D_\mathbf{x}\mathbf{F}(\mathbf{x}_0, \mathbf{M}_0)$ is not singular (i.e., does not have any zero eigenvalues), then there exists a neighborhood around $(\mathbf{x}_0, \mathbf{M}_0)$ in $\mathcal{R}^n \times \mathcal{R}^m$ such that, for each $\mathbf{M}$ in this neighborhood, (2.1.2) has a unique solution $\mathbf{x}$. Further, this solution can be written in the form $\mathbf{x} = \mathbf{G}(\mathbf{M})$, where $\mathbf{x}_0 = \mathbf{G}(\mathbf{M}_0)$. A proof of the implicit function theorem is provided in many textbooks (e.g., Hale, 1969, Chapter 0).

In Section 2.3 where we considered codimension–one bifurcations, we saw that $D_\mathbf{x}\mathbf{F}$ (in this chapter, we use $\mathbf{F}_\mathbf{x}$ to denote $D_\mathbf{x}\mathbf{F}$) is singular at saddle–node, pitchfork, and transcritical bifurcation points. To fix ideas, we revisit Examples 2.12–2.14.

Example 6.1. We find from (2.3.1) that

$$F_x = 0 \quad \text{and} \quad F_\mu = 1$$

at the saddle–node bifurcation point $(0,0)$. Therefore, it follows from the implicit function theorem that the dependence of x on μ is not unique in the neighborhood of the saddle–node bifurcation point. However, there exists a unique function $\mu = \mu(x) = x^2$ such that $\mu(0) = 0$ and $F[x; \mu(x)] = 0$ in a neighborhood of this bifurcation point.

Example 6.2. We find from (2.3.2) that at both supercritical and subcritical pitchfork bifurcation points

$$F_x = 0 \quad \text{and} \quad F_\mu = 0$$

Hence, it follows from the implicit function theorem that neither the dependence of x on μ nor that of μ on x is unique in a neighborhood of subcritical and supercritical pitchfork bifurcation points.

Example 6.3. We find from (2.3.3) that at the transcritical bifurcation point

$$F_x = 0 \quad \text{and} \quad F_\mu = 0$$

Therefore, it follows from the implicit function theorem that neither the dependence of x on μ nor that of μ on x is unique in a neighborhood of the transcritical bifurcation point.

In the literature, the turning points and branch points addressed thus far are referred to as **simple turning points** and **simple branch points**, respectively. We present a method for locating these points in Section 6.2 and Hopf bifurcation points in Section 6.3. Away from turning and branch points, the implicit function theorem implies that a continuation of the solutions is possible. In this section, we describe methods for the continuation of the solutions of (2.1.2) as one of the parameters $\alpha \in \mathbf{M}$ is varied while the rest of the parameters are held constant. Specifically, we consider variation of the solutions of

$$\mathbf{F}(\mathbf{x}; \alpha) = \mathbf{0} \tag{6.1.1}$$

with respect to the scalar parameter $\alpha \in \mathcal{R}^1$. In the $(n+1)$–dimensional $(\mathbf{x}, \alpha)$ space, there may be many branches of fixed points. However, the dependence of $\mathbf{x}$ on α on each branch is unique. A globally convergent homotopy algorithm for calculating all solutions of $\mathbf{F}(\mathbf{x}; \alpha_0) = \mathbf{0}$ at a given α_0 is described in Section 6.4. Here, we assume that at least one solution $\mathbf{x}_0$ has been calculated at α_0 and that the Jacobian matrix $\mathbf{F}_\mathbf{x}$ at $(\mathbf{x}_0, \alpha_0)$ is nonsingular so that the implicit function theorem holds locally. Then, the **continuation** or **path–following method** is an algorithmic procedure for tracing out the branch of fixed points that passes through $(\mathbf{x}_0, \alpha_0)$ in the state–control space.

There are essentially two categories of continuation methods. The first category consists of **predictor–corrector methods**, and the second category consists of **piece–wise–linear** or **simplical methods**. In the predictor–corrector methods, one approximately follows a branch of solutions. On the other hand, in piece–wise–linear methods, one exactly follows a piece–wise–linear curve that approximates a branch of solutions (Allgower and Georg, 1980, 1990). Only predictor–corrector methods are considered here. These methods usually consist of a parameterization strategy, a predictor, a corrector, and a step–length control.

6.1.1 Sequential Continuation

The simplest continuation method is the **sequential scheme** (Kubicek and Marek, 1983). This scheme is also known as **natural parameter continuation** (Doedel, Keller, and Kernevez, 1991a) when α is used as the continuation parameter. The interval of α is divided into closely spaced intervals defined by the grid points $\alpha_0, \alpha_1, \alpha_2, \cdots, \alpha_n$. Then, the solution $\mathbf{x}_j$ at α_j is used as the predicted value or initial guess for the solution $\mathbf{x}_{j+1}$ at α_{j+1}. This predicted value is corrected through a Newton–Raphson iteration scheme. Thus,

$$\mathbf{x}_{j+1}^{k+1} = \mathbf{x}_{j+1}^k + r\Delta\mathbf{x}^k \tag{6.1.2}$$

where the superscript k is the iteration number and

$$\mathbf{x}_{j+1}^1 = \mathbf{x}_j$$

Further, r, which is such that $0 < r \le 1$ or equivalently $r \in (0, 1]$, is called the **relaxation parameter** and $\Delta\mathbf{x}^k$ is the solution of the n linear algebraic equations

$$\mathbf{F}_\mathbf{x}\left(\mathbf{x}_{j+1}^k, \alpha_{j+1}\right) \Delta\mathbf{x}^k = -\mathbf{F}\left(\mathbf{x}_{j+1}^k, \alpha_{j+1}\right) \tag{6.1.3}$$

which are obtained from the Newton–Raphson method. The relaxation parameter r is chosen such that

$$\| \mathbf{F}\left(\mathbf{x}_{j+1}^{k+1}, \alpha_{j+1}\right) \| < \| \mathbf{F}\left(\mathbf{x}_{j+1}^k, \alpha_{j+1}\right) \| \tag{6.1.4}$$

If the relation (6.1.4) is satisfied for $r = 1$, then r is set equal to 1 in (6.1.2); otherwise, r is halved until (6.1.4) is satisfied. If the grid points α_j are sufficiently close, few iterations (one or two) are sufficient for obtaining an accurate solution at each α_j. If the vector function $\mathbf{F}$ is not C^r, where $r \ge 1$, then the components of the matrix $\mathbf{F}_\mathbf{x}$ have to be evaluated numerically using finite–difference schemes.

In Figure 6.1.1a we show a turning point, and in Figure 6.1.1b we show a branch point. Clearly, this sequential scheme will fail at such points where two or more branches meet because the Jacobian $\mathbf{F}_\mathbf{x}$ is singular there, and hence one cannot solve the algebraic set of equations (6.1.3). This problem can be avoided at turning points by choosing a different continuation parameter. Let us suppose that a turning point occurs, say, at $\alpha = \alpha_c$. Then, locally, there are no solutions beyond the turning point at α_c. Consequently, the sequential scheme using α as a continuation parameter would fail for all values of α beyond α_c in the direction of continuation. Because α is a unique or a single–valued

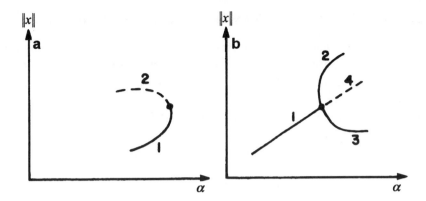

Figure 6.1.1: Branches of fixed points meeting at (a) turning point and (b) branch point.

function of the states x_i and the rank of $[\mathbf{F_x} \mid \mathbf{F_\alpha}]$ is n, the sequential scheme can be modified to carry out the continuation past the turning point. In the modified scheme, one of the variables x_j is used as a continuation parameter instead of α. In general, any independent variable or a parameter can be chosen as a continuation parameter. In Sections 6.1.3 and 6.1.4, the arclength along a branch of solutions is used as a continuation parameter.

In practice, it is sometimes possible to carry out a sequential continuation through a branch point because $\mathbf{F_x}$ is generally not singular at fixed points located in the vicinity of the branch point. A Newton–Raphson method can be used to locate the fixed points surrounding a branch point if the initial estimates are close enough to them. However, this depends on the radius of convergence of the Newton–Raphson method in the neighborhood of the considered branch point. For illustration, let us assume that we are on branch 1 in Figure 6.1.1b and that we take a step to step over the branch point to get onto branch 4. Then, the Newton–Raphson method can converge to a fixed point on branch 4 if the initial estimate is within the radius of convergence of this method.

6.1.2 Davidenko–Newton–Raphson Continuation

In this scheme, the control parameter α is used as the continuation parameter. First, a method devised by Davidenko (1953) is used as a predictor, and then a Newton–Raphson method is used as a corrector. Because the predictor is based on solving a system of ordinary–differential equations, it is also called an **ordinary–differential equation predictor** (e.g., Seydel, 1988).

Differentiating (6.1.1) with respect to α yields

$$\mathbf{F_x}(\mathbf{x}, \alpha)\frac{d\mathbf{x}}{d\alpha} = -\mathbf{F}_\alpha(\mathbf{x}, \alpha) \tag{6.1.5}$$

which constitutes a system of n linear algebraic equations for the unknowns $d\mathbf{x}/d\alpha$. Hence, if the Jacobian $\mathbf{F_x}(\mathbf{x}, \alpha)$ is regular in the interval $[\alpha_0, \alpha_n]$, then

$$\frac{d\mathbf{x}}{d\alpha} = -\mathbf{F_x}^{-1}(\mathbf{x}, \alpha)\,\mathbf{F}_\alpha(\mathbf{x}, \alpha) \tag{6.1.6}$$

Then, given the solution

$$\mathbf{x}(\alpha_0) = \mathbf{x}_0 \tag{6.1.7}$$

one can solve (6.1.6) subject to the initial condition (6.1.7) using any ordinary–differential equation solver, such as a Runge–Kutta method, to determine the dependence of $\mathbf{x}$ on α. The predicted values obtained from the integration are likely to deviate from the true solutions of (6.1.1) due to truncation error. Hence, the predicted values are used as initial guesses for a Newton–Raphson scheme to obtain corrected values. Again, this continuation method will fail at turning and branch points because the Jacobian matrix is singular at such points.

6.1.3 Arclength Continuation

In this scheme, the arclength s along a branch of solutions (see Fig. 6.1.2) is used as the continuation parameter. So, $\mathbf{x}$ and α are considered to be functions of s; that is, $\mathbf{x} = \mathbf{x}(s)$ and $\alpha = \alpha(s)$.

On the path parameterized by the arclength s, we seek $\mathbf{x}$ and α such that

$$\mathbf{F}[\mathbf{x}(s), \alpha(s)] = \mathbf{0} \tag{6.1.8}$$

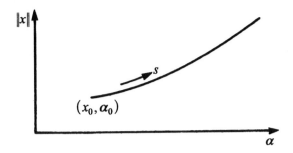

Figure 6.1.2: Illustration for arclength continuation scheme.

Thus, differentiating (6.1.8) with respect to s yields

$$\mathbf{F_x}(\mathbf{x}, \alpha)\mathbf{x}' + \mathbf{F}_\alpha(\mathbf{x}, \alpha)\alpha' = 0 \qquad (6.1.9)$$

where $\mathbf{x}' = d\mathbf{x}/ds$ and $\alpha' = d\alpha/ds$. Equation (6.1.9) may be rewritten as

$$[\mathbf{F_x} \mid \mathbf{F}_\alpha] \left\{ \begin{array}{c} \mathbf{x}' \\ \alpha' \end{array} \right\} = [\mathbf{F_x} \mid \mathbf{F}_\alpha] \, \mathbf{t} = 0$$

where the $(n+1)$ vector $\mathbf{t}$ is the tangent vector at $(\mathbf{x}, \alpha)$ on the path. The system (6.1.9) consists of n linear algebraic equations in the $(n+1)$ unknowns $\mathbf{x}'$ and α'. To specify these unknowns uniquely, we supplement (6.1.9) with a nonhomogeneous equation. A convenient additional equation is specified by the Euclidean arclength normalization

$$\mathbf{x}'^T \mathbf{x}' + \alpha'^2 = x_1'^2 + x_2'^2 + \cdots + x_n'^2 + \alpha'^2 = 1 \qquad (6.1.10)$$

which implies that the tangent vector $\mathbf{t}$ has unit length. Keller (1977) proposed many non–Euclidean normalizations in place of (6.1.10). The initial conditions for (6.1.9) and (6.1.10) are given by

$$\mathbf{x} = \mathbf{x}_0 \text{ and } \alpha = \alpha_0 \text{ at } s = 0 \qquad (6.1.11)$$

If the Jacobian $\mathbf{F_x}$ is nonsingular and $\mathbf{F}_\alpha$ is a zero vector, (6.1.9) and (6.1.10) yield

$$\left[\mathbf{x}'^T \; \alpha'\right] = \pm [0 \; 0 \; \cdots \; 0 \; 1]$$

If the Jacobian $\mathbf{F_x}$ is nonsingular and $\mathbf{F}_\alpha$ is a nonzero vector, one can solve (6.1.9) and (6.1.10) to determine the tangent vector $\mathbf{t}$ as follows. First, one solves the system of n linear algebraic equations

$$\mathbf{F_x}(\mathbf{x}, \alpha)\mathbf{z} = -\mathbf{F}_\alpha(\mathbf{x}, \alpha) \tag{6.1.12}$$

for the vector $\mathbf{z}$. Then, owing to the linearity of (6.1.9) in $\mathbf{x}'$ and α',

$$\mathbf{x}' = \mathbf{z}\alpha' \tag{6.1.13}$$

where α' is still unknown. Substituting (6.1.13) into the arclength condition (6.1.10) yields

$$\alpha' = \pm(1 + \mathbf{z}^T\mathbf{z})^{-1/2} \tag{6.1.14}$$

where the plus and minus signs determine the direction of the continuation. Having determined the tangent vector $\mathbf{t}$, we use it to predict values of $\mathbf{x}$ and α at $s + \Delta s$ by taking an **Euler step**; that is,

$$\mathbf{x} = \mathbf{x}_0 + \mathbf{x}'\Delta s \quad \text{and} \quad \alpha = \alpha_0 + \alpha'\Delta s$$

This predictor is called the **tangent predictor**. The tangent predictor falls under the class of **first-order predictors**, while the predictor used during sequential continuation falls under the class of **zero-order** or **trivial** predictors (e.g., Seydel, 1988, Chapter 4). In order to obtain good predicted values, higher-order predictors have been proposed (e.g., Schwetlick and Cleve, 1987). However, a higher-order predictor may not be desirable in regions of large curvatures on the path.

If a higher-order predictor produces results of sufficient accuracy, then there is no need for a corrector. However, when a tangent predictor is used, a corrector is usually necessary. The predicted values are usually corrected through a Newton–Raphson scheme, and the predictor-corrector scheme is continued until the branch is traced.

The choice of the step size Δs depends on several factors, two of which are the convergence of the corrector and the curvature of the path. The chosen step size should be such that the initial guess or estimate is within the radius of convergence of the corrector. Moreover, if the number of iterations needed by a corrector to achieve the desired accuracy is larger than a specified optimal number, it will be necessary to decrease the step size. To follow the path closely in regions of large curvature, one needs to use small steps. Further, the step size may have to be adaptively varied during the course of the continuation. This has prompted the development of sophisticated algorithms for step–length control (e.g., Den Heijer and Rheinboldt, 1981; Schwetlick and Cleve, 1987).

Like the sequential method, the arclength continuation scheme is bound to break down at turning points and other bifurcation points, where $\mathbf{F_x}$ is singular. Again, continuation past branch points may be achieved by stepping over them, so–to–speak. In this scheme, it is possible to overshoot past turning points and end up at values of α where there are no solutions. To overcome this problem, we describe the so–called **pseudo–arclength continuation scheme** proposed by Keller (1977, 1987) in Section 6.1.4.

In the rest of this section, we describe a modified arclength scheme proposed by Kubicek (1976) to carry out continuation along branches with turning points; this scheme is used in the continuation software DERPAR (Kubicek, 1976; Kubicek and Marek, 1983). The basic idea underlying the scheme of Kubicek is the fact that, although $\mathbf{F_x}$ is singular at a turning point and hence has a rank less than n, the augmented $n \times (n+1)$ matrix $[\mathbf{F_x} \mid \mathbf{F}_\alpha]$ has a rank of n. Therefore, one can find an $n \times n$ nonsingular submatrix by deleting the kth column, which is found by using a Gaussian elimination scheme with pivoting. We treat α as an additional state and let

$$x_{n+1} = \alpha$$

Then, we solve (6.1.9) for $x_1', x_2', \cdots, x_{k-1}', x_{k+1}', \cdots, x_{n+1}'$ in terms of x_k' and obtain

$$x_i' = \beta_i x_k' \text{ for } i = 1, 2, \cdots, k-1, k+1, \cdots, n+1 \qquad (6.1.15)$$

Substituting (6.1.15) into (6.1.10) yields

$$x_k' = \pm \left(1 + \sum_{\substack{i=1 \\ i \neq k}}^{n+1} \beta_i^2 \right)^{-1/2} \qquad (6.1.16)$$

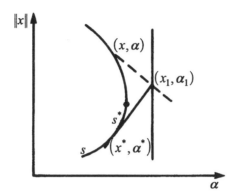

Figure 6.1.3: Illustration for pseudo–arclength continuation scheme.

where, again, the plus and minus signs determine the direction of continuation. After choosing the direction of continuation, we use an ordinary–differential–equation solver to integrate numerically (6.1.15) and (6.1.16) subject to the initial conditions (6.1.11). Kubicek (1976) has suggested the use of the Adams–Bashforth method (e.g., Stoer and Bulirsch, 1980) for carrying out the numerical integration. Again, the predicted values are likely to deviate from the true solutions of (6.1.1) due to truncation error. Hence, the predicted values need to be corrected by using a corrector, such as a Newton–Raphson scheme.

6.1.4   Pseudo–Arclength Continuation

This procedure, which also uses an arclength parameterization, was developed by Keller (1977, 1987) and is used in the continuation software AUTO developed by Doedel (1986).

In Figure 6.1.3, the turning point is marked by a dot. Near the turning point, at $s = s^*$, we obtain $(\mathbf{x}^*, \alpha^*)$ by using (6.1.11), (6.1.13), and (6.1.14). Next, we use the tangent predictor to determine the prediction $(\mathbf{x}_1, \alpha_1)$ at $s^* + \Delta s$; that is,

$$\alpha_1 = \alpha^* + \alpha^{*'} \Delta s \qquad (6.1.17)$$

$$\mathbf{x}_1 = \mathbf{x}^* + \mathbf{x}^{*'} \Delta s \qquad (6.1.18)$$

where the step length Δs can be freely specified.

At this stage, if one were to use a Newton–Raphson method for correction one would seek a solution along the vertical line shown through (x_1, α_1) in Figure 6.1.3. The corrector would break down in the vicinity of the turning point because there are no solutions on the vertical line. To overcome this problem, we seek a correction to (x_1, α_1) along a solution path that is normal to the tangent vector t at (x_1, α_1). This path is shown as a broken line in Figure 6.1.3. On the solution path, (x, α) satisfies

$$F(x, \alpha) = 0 \qquad (6.1.19)$$

and is such that the vector

$$X = \left\{ \begin{array}{c} x - x_1 \\ \alpha - \alpha_1 \end{array} \right\}$$

is normal to the tangent vector t; that is,

$$X^T t = 0 \qquad (6.1.20)$$

Substituting (6.1.17) and (6.1.18) into (6.1.20) and using the definition of t, we obtain

$$(x - x^*)^T x^{*'} + (\alpha - \alpha^*)\alpha^{*'} - [\alpha^{*'2} + (x^{*'})^T x^{*'}]\Delta s = 0$$

or

$$g(x, \alpha) = (x - x^*)^T x^{*'} + (\alpha - \alpha^*)\alpha^{*'} - \Delta s = 0 \qquad (6.1.21)$$

because

$$\alpha^{*'2} + (x^{*'})^T x^{*'} = 1$$

Equations (6.1.19) and (6.1.21) constitute the **pseudo–arclength continuation scheme**. In this scheme, one solves the $n + 1$ nonlinear algebraic equations (6.1.19) and (6.1.21) for the $n + 1$ unknowns (x, α). If we apply a Newton–Raphson scheme to (6.1.19) and (6.1.21), then at each iteration of this scheme the equations are

$$x^{k+1} = x^k + r\Delta x^{k+1} \qquad (6.1.22)$$
$$\alpha^{k+1} = \alpha^k + r\Delta \alpha^{k+1} \qquad (6.1.23)$$

where r is a relaxation parameter and $\Delta \mathbf{x}^{k+1}$ and $\Delta \alpha^{k+1}$ are determined from

$$\mathbf{F_x}(\mathbf{x}^k, \alpha^k)\Delta \mathbf{x}^{k+1} + \mathbf{F}_\alpha(\mathbf{x}^k, \alpha^k)\Delta \alpha^{k+1} = -\mathbf{F}(\mathbf{x}^k, \alpha^k) \quad (6.1.24)$$
$$(\mathbf{x}^{*\prime})^T \Delta \mathbf{x}^{k+1} + \alpha^{*\prime} \Delta \alpha^{k+1} = -g(\mathbf{x}^k, \alpha^k) \quad (6.1.25)$$

If $\mathbf{F_x}$ is nonsingular, one can solve (6.1.24) by using the so–called **bordering algorithm** (Keller, 1987; Doedel, Keller, and Kernevez, 1991a), which is based on the method of superposition. First, one solves the systems

$$\mathbf{F_x}(\mathbf{x}^k, \alpha^k)\mathbf{z}_2 = -\mathbf{F}_\alpha(\mathbf{x}^k, \alpha^k) \quad (6.1.26)$$

and

$$\mathbf{F_x}(\mathbf{x}^k, \alpha^k)\mathbf{z}_1 = -\mathbf{F}(\mathbf{x}^k, \alpha^k) \quad (6.1.27)$$

Then, it follows from (6.1.24) that

$$\Delta \mathbf{x}^{k+1} = \mathbf{z}_1 + \mathbf{z}_2 \Delta \alpha^{k+1} \quad (6.1.28)$$

Substituting (6.1.28) into (6.1.25) yields

$$\Delta \alpha^{k+1} = -\frac{\left[g(\mathbf{x}^k, \alpha^k) + \mathbf{z}_1^T \mathbf{x}^{*\prime}\right]}{[\alpha^{*\prime} + \mathbf{z}_2^T \mathbf{x}^{*\prime}]} \quad (6.1.29)$$

Once $\Delta \alpha^{k+1}$ is known, $\Delta \mathbf{x}^{k+1}$ can be determined from (6.1.28), and the iterations are continued until the required convergence is achieved in the Newton–Raphson procedure. After $(\mathbf{x}, \alpha)$ is determined on the branch, we move on to determine another point on the branch. The step size Δs can be reduced if the convergence is slow and increased if the convergence is rapid.

In the arclength continuation scheme of the software CANDYS/QA, Feudel and Jansen (1992) use a combination of linear and cubic predictors and the corrector described in this section. Furthermore, Feudel and Jansen implement a step–length control based on the curvature of the path being followed.

6.1.5 Comments

The algorithms described in Sections 6.1.3 and 6.1.4 also experience difficulties at a branch point. At such a point $(\mathbf{x}, \alpha)$, $\mathbf{F}_\alpha$ belongs to the range of $\mathbf{F_x}$ and the $n \times (n+1)$ matrix $[\mathbf{F_x} \mid \mathbf{F}_\alpha]$ has a rank of $(n-1)$. Hence, there are at least two independent solutions $\mathbf{v}_1$ and $\mathbf{v}_2$ such that

$$[\mathbf{F_x} \mid \mathbf{F}_\alpha]\mathbf{v}_j = \mathbf{0}$$

Each $n+1$ vector $\mathbf{v}_j$ provides the direction of a branch at the branch point and, thus, choosing a $\mathbf{v}_j$ specifies the direction of continuation. Details on how to determine the $\mathbf{v}_j$ are provided by Kubicek and Marek (1983, Chapter 2), Seydel (1988, Chapter 5), Doedel (1986), and Doedel, Keller, and Kernevez (1991a). The branch–switching algorithms are generally quite involved because they require computation of the second derivatives of $\mathbf{F}(\mathbf{x}; \alpha)$.

A simpler approach to branch switching is to add a small perturbation to $\mathbf{F}(\mathbf{x}; \alpha)$ as the branch point is approached. The perturbation only slightly changes the obtained values, but it breaks pitchfork and transcritical bifurcations because they are structurally unstable, as discussed in Sections 2.3.7 and 2.3.8. (In the perturbed system, there are nonbifurcating branches.) Consequently, the continuation scheme smoothly tracks the solutions from the original branch to the new branch. Once a solution along a new branch has been determined, one can restart the continuation along that branch after removing the perturbation.

Again, for simplical continuation methods, the reader is referred to the works of Allgower and Georg (1980, 1990). The reader can find related material and additional information on continuation methods in the works of Garcia and Zangwill (1981), Keller (1977, 1987), Kubicek and Marek (1983), Seydel (1988), Allgower and Georg (1990), Doedel, Keller, and Kernevez (1991a,b), and Feudel and Jansen (1992), among others.

6.2 SIMPLE TURNING AND BRANCH POINTS

Turning and branch points can be determined either through an **indirect method** or a **direct method**. **Indirect methods** typically yield the bifurcation points as a by–product of the continuation procedure. In these methods, the eigenvalues and/or the rank of the Jacobian matrix $\mathbf{F_x}$ are monitored as the fixed points are determined with respect to the continuation parameter α or s. When one of the eigenvalues is zero, and consequently the rank of the matrix $\mathbf{F_x}$ is $n-1$, the presence of a static bifurcation is detected. Further, the rank of the matrix $[\mathbf{F_x} \mid \mathbf{F_\alpha}]$ is used to determine if the bifurcation point is a turning or a branch point. In comparison to indirect methods, **direct methods** are usually more expensive, but they are more accurate and effective in determining surfaces of bifurcation points in the state–control space.

Here, we describe an algorithm developed by Moore and Spence (1980) for the direct calculation of turning and branch points. The basic idea underlying this method is the fact that $\mathbf{F_x}$ is singular at static bifurcation points so that $\mathbf{F_x u} = \mathbf{0}$ has nontrivial solutions. Thus, these bifurcation points are given by

$$\mathbf{F}(\mathbf{x}, \alpha) = \mathbf{0} \tag{6.2.1}$$

$$\mathbf{F_x}(\mathbf{x}, \alpha)\mathbf{u} = \mathbf{0} \tag{6.2.2}$$

$$\mathbf{u}^T \mathbf{u} = 1 \tag{6.2.3}$$

The normalization condition (6.2.3) ensures that the vector $\mathbf{u}$ is nontrivial. Seydel (1979b) used (6.2.1), (6.2.2), and a non–Euclidean normalization instead of (6.2.3).

Equations (6.2.1)–(6.2.3) are solved by using a Newton–Raphson procedure. Thus, at each iterate k, one obtains the following linear system of equations:

$$\mathbf{F}_{\mathbf{x}}^{k}\Delta\mathbf{x} + \mathbf{F}_{\alpha}^{k}\Delta\alpha = -\mathbf{F}^{k} \tag{6.2.4}$$

$$\left[\mathbf{F}_{\mathbf{xx}}^{k}\mathbf{u}^{k}\right]\Delta\mathbf{x} + \left[\mathbf{F}_{\mathbf{x}\alpha}^{k}\mathbf{u}^{k}\right]\Delta\alpha + \mathbf{F}_{\mathbf{x}}^{k}\Delta\mathbf{u} = -\mathbf{F}_{\mathbf{x}}^{k}\mathbf{u}^{k} \tag{6.2.5}$$

$$2\mathbf{u}^{T}\Delta\mathbf{u} = 1 - (\mathbf{u}^{T}\mathbf{u})^{k} \tag{6.2.6}$$

In (6.2.4)–(6.2.6), the superscript k on the increments has been omitted. In (6.2.5), $\mathbf{F_{xx}}$ represents the matrix of second partial derivatives with respect to $\mathbf{x}$, and $\mathbf{F_{x\alpha}}$ represents the matrix of second partial derivatives with respect to $\mathbf{x}$ and α. As mentioned in Section 2.3, one can check if the bifurcation point in question is a turning or branch point by examining the rank of the matrix $[\mathbf{F_x} \mid \mathbf{F_\alpha}]$.

To solve (6.2.4)–(6.2.6) efficiently, we use the principle of superposition. To this end, first we compute the solutions of

$$\mathbf{F}_\mathbf{x}^k \mathbf{z}_2 = -\mathbf{F}_\alpha^k \tag{6.2.7}$$

and

$$\mathbf{F}_\mathbf{x}^k \mathbf{z}_1 = -\mathbf{F}^k \tag{6.2.8}$$

Then, it follows from (6.2.4) that

$$\Delta\mathbf{x} = \mathbf{z}_1 + \mathbf{z}_2 \Delta\alpha \tag{6.2.9}$$

Substituting (6.2.9) into (6.2.5) yields

$$\left([\mathbf{F}_\mathbf{xx}^k \mathbf{u}^k]\mathbf{z}_2 + [\mathbf{F}_\mathbf{x\alpha}^k \mathbf{u}^k]\right) \Delta\alpha + \mathbf{F}_\mathbf{x}^k \Delta\mathbf{u} = -\mathbf{F}_\mathbf{x}^k \mathbf{u}^k - [\mathbf{F}_\mathbf{xx}^k \mathbf{u}^k]\mathbf{z}_1 \tag{6.2.10}$$

Again, we solve (6.2.10) for $\Delta\mathbf{u}$ in terms of $\Delta\alpha$ by using the principle of superposition. Thus, we calculate the solutions of

$$\mathbf{F}_\mathbf{x}^k \mathbf{z}_4 = -[\mathbf{F}_\mathbf{xx}^k \mathbf{u}^k]\mathbf{z}_2 - [\mathbf{F}_\mathbf{x\alpha}^k \mathbf{u}^k] \tag{6.2.11}$$

and

$$\mathbf{F}_\mathbf{x}^k \mathbf{z}_3 = -[\mathbf{F}_\mathbf{x}^k \mathbf{u}^k] - [\mathbf{F}_\mathbf{xx}^k \mathbf{u}^k]\mathbf{z}_1 \tag{6.2.12}$$

Then, it follows from (6.2.10) that

$$\Delta\mathbf{u} = \mathbf{z}_3 + \mathbf{z}_4 \Delta\alpha \tag{6.2.13}$$

Substituting (6.2.13) into (6.2.6) and solving the resulting equation for $\Delta\alpha$ yields

$$\Delta\alpha = \frac{1 - (\mathbf{u}^T\mathbf{u})^k - 2\mathbf{u}^T\mathbf{z}_3}{2\mathbf{u}^T\mathbf{z}_4} \tag{6.2.14}$$

Having calculated $\Delta\mathbf{x}, \Delta\mathbf{u}$, and $\Delta\alpha$, one updates the values of $\mathbf{x}^k, \mathbf{u}^k$, and α^k as

$$\mathbf{x}^{k+1} = \mathbf{x}^k + r\Delta\mathbf{x} \tag{6.2.15}$$
$$\mathbf{u}^{k+1} = \mathbf{u}^k + r\Delta\mathbf{u} \tag{6.2.16}$$
$$\alpha^{k+1} = \alpha^k + r\Delta\alpha \tag{6.2.17}$$

where $r \in (0, 1]$ is again a relaxation parameter.

There are also other algorithms to determine turning and branch points (e.g., Griewank and Reddien, 1984; Menzel, 1984; Pönisch, 1985; Weber, 1981).

6.3   HOPF BIFURCATION POINTS

One can determine Hopf bifurcation points by using an indirect method as a by–product of a continuation scheme by monitoring the eigenvalues of the Jacobian matrix. Here, we present a direct method proposed by Griewank and Reddien (1983) for numerically calculating Hopf bifurcation points. At a Hopf bifurcation point $(\mathbf{x}^*, \alpha^*)$, the Jacobian matrix $\mathbf{F_x}(\mathbf{x}^*, \alpha^*)$ has a pair of purely imaginary eigenvalues $\pm i\omega$, with all the other eigenvalues having nonzero real parts. To determine $\mathbf{x}^*, \alpha^*$, and ω, we assume that the eigenvector corresponding to the eigenvalue $i\omega$ is $\mathbf{p} + i\mathbf{q}$, where $\mathbf{p}$ and $\mathbf{q}$ are real. Then,

$$\mathbf{F_x}(\mathbf{x}, \alpha)(\mathbf{p} + i\mathbf{q}) = i\omega(\mathbf{p} + i\mathbf{q}) \tag{6.3.1}$$

Separating (6.3.1) into real and imaginary parts and using the fact that $\mathbf{F_x}$ is real, we have

$$\mathbf{F_x}\mathbf{p} + \omega\mathbf{q} = 0 \tag{6.3.2}$$

and

$$\mathbf{F_x}\mathbf{q} - \omega\mathbf{p} = 0 \tag{6.3.3}$$

To ensure that $\mathbf{p}$ and $\mathbf{q}$ are nontrivial and that they are linearly independent, we use the normalization conditions

$$\mathbf{w}^T\mathbf{p} = 0 \tag{6.3.4}$$

$$\mathbf{w}^T \mathbf{q} - 1 = 0 \tag{6.3.5}$$

where $\mathbf{w}$ is a specified vector. Hence, the problem of calculating the Hopf bifurcation points of

$$\mathbf{F}(\mathbf{x}, \alpha) = \mathbf{0} \tag{6.3.6}$$

is transformed into finding the solutions $\mathbf{x}, \mathbf{p}, \mathbf{q}, \alpha$, and ω of (6.3.2)–(6.3.6).

To compute the solutions of (6.3.2)–(6.3.6) efficiently, we use a Newton–Raphson procedure. At the kth iteration, we have

$$\mathbf{x}^{k+1} = \mathbf{x}^k + r\Delta\mathbf{x} \tag{6.3.7}$$
$$\mathbf{p}^{k+1} = \mathbf{p}^k + r\Delta\mathbf{p} \tag{6.3.8}$$
$$\mathbf{q}^{k+1} = \mathbf{q}^k + r\Delta\mathbf{q} \tag{6.3.9}$$
$$\alpha^{k+1} = \alpha^k + r\Delta\alpha \tag{6.3.10}$$
$$\omega^{k+1} = \omega^k + r\Delta\omega \tag{6.3.11}$$

where the superscript k on the increments has been omitted and $r \in (0, 1]$ is a relaxation parameter. Substituting (6.3.7)–(6.3.11) into (6.3.2)–(6.3.6) and linearizing the results in the increments, we have

$$\mathbf{F}_{\mathbf{x}}^k \Delta\mathbf{x} + \mathbf{F}_{\alpha}^k \Delta\alpha = -\mathbf{F}^k \tag{6.3.12}$$

$$[\mathbf{F}_{\mathbf{xx}}\mathbf{p}]^k \Delta\mathbf{x} + [\mathbf{F}_{\mathbf{x}\alpha}\mathbf{p}]^k \Delta\alpha + \mathbf{F}_{\mathbf{x}}^k \Delta\mathbf{p} + \omega^k \Delta\mathbf{q} + \mathbf{q}^k \Delta\omega$$
$$= -[\mathbf{F}_{\mathbf{x}}\mathbf{p} + \omega\mathbf{q}]^k \tag{6.3.13}$$

$$[\mathbf{F}_{\mathbf{xx}}\mathbf{q}]^k \Delta\mathbf{x} + [\mathbf{F}_{\mathbf{x}\alpha}\mathbf{q}]^k \Delta\alpha + \mathbf{F}_{\mathbf{x}}^k \Delta\mathbf{q} - \omega^k \Delta\mathbf{p} - \mathbf{p}^k \Delta\omega$$
$$= -[\mathbf{F}_{\mathbf{x}}\mathbf{q} - \omega\mathbf{p}]^k \tag{6.3.14}$$

$$\mathbf{w}^T \Delta\mathbf{p} = -\mathbf{w}^T \mathbf{p}^k \tag{6.3.15}$$
$$\mathbf{w}^T \Delta\mathbf{q} = 1 - \mathbf{w}^T \mathbf{q}^k \tag{6.3.16}$$

To solve the linear system of equations (6.3.12)–(6.3.16) efficiently, we use the principle of superposition. To this end, we solve the systems of equations

$$\mathbf{F}_{\mathbf{x}}^k \mathbf{z}_1 = -\mathbf{F}^k \tag{6.3.17}$$

and

$$\mathbf{F}_{\mathbf{x}}^k \mathbf{z}_2 = -\mathbf{F}_{\alpha}^k \tag{6.3.18}$$

Then, it follows from (6.3.12) that

$$\Delta x = z_1 + z_2 \Delta \alpha \tag{6.3.19}$$

Substituting (6.3.19) into (6.3.13) and (6.3.14) yields

$$
A \begin{bmatrix} \Delta p \\ \Delta q \end{bmatrix} = - \begin{bmatrix} [F_x p + \omega q]^k + [F_{xx} p]^k z_1 \\ [F_x q - \omega p]^k + [F_{xx} q]^k z_1 \end{bmatrix}
$$
$$
- \begin{bmatrix} [F_{xx} p]^k z_2 + [F_{x\alpha} p]^k \\ [F_{xx} q]^k z_2 + [F_{x\alpha} q]^k \end{bmatrix} \Delta \alpha \tag{6.3.20}
$$
$$
- \begin{bmatrix} q^k \\ -p^k \end{bmatrix} \Delta \omega
$$

where

$$
A = \begin{bmatrix} F_x^k & \omega^k I \\ -\omega^k I & F_x^k \end{bmatrix} \tag{6.3.21}
$$

In (6.3.21), I is the $n \times n$ identity matrix. Multiplying (6.3.20) from the left by A^{-1}, we obtain

$$\Delta p = z_3 + z_4 \Delta \alpha + z_5 \Delta \omega \tag{6.3.22}$$

$$\Delta q = z_6 + z_7 \Delta \alpha + z_8 \Delta \omega \tag{6.3.23}$$

where

$$
A \begin{bmatrix} z_3 \\ z_6 \end{bmatrix} = - \begin{bmatrix} [F_x p + \omega q]^k + [F_{xx} p]^k z_1 \\ [F_x q - \omega q]^k + [F_{xx} q]^k z_1 \end{bmatrix} \tag{6.3.24}
$$

$$
A \begin{bmatrix} z_4 \\ z_7 \end{bmatrix} = - \begin{bmatrix} [F_{xx} p]^k z_2 + [F_{x\alpha} p]^k \\ [F_{xx} q]^k z_2 + [F_{x\alpha} q]^k \end{bmatrix} \tag{6.3.25}
$$

$$
A \begin{bmatrix} z_5 \\ z_8 \end{bmatrix} = - \begin{bmatrix} q^k \\ -p^k \end{bmatrix} \tag{6.3.26}
$$

Substituting (6.3.22) and (6.3.23) into (6.3.15) and (6.3.16), we arrive at

$$w^T z_4 \Delta \alpha + w^T z_5 \Delta \omega = -w^T p^k - w^T z_3 \tag{6.3.27}$$
$$w^T z_7 \Delta \alpha + w^T z_8 \Delta \omega = 1 - w^T q^k - w^T z_6 \tag{6.3.28}$$

which can be solved for $\Delta\alpha$ and $\Delta\omega$. Then, one can calculate $\Delta\mathbf{x}$ from (6.3.19), $\Delta\mathbf{p}$ from (6.3.22), and $\Delta\mathbf{q}$ from (6.3.23). Substituting these increments into (6.3.7)–(6.3.11), one computes $\mathbf{x}, \mathbf{p}, \mathbf{q}, \alpha$, and ω. The procedure is repeated until the specified convergence criterion is satisfied.

Holodniok and Kubicek (1984a,b) and Roose (1985) developed a system of lower dimension than (6.3.2)–(6.3.6) by eliminating either $\mathbf{p}$ or $\mathbf{q}$ from (6.3.2) and (6.3.3). If $\mathbf{q}$ is eliminated, the result is

$$[\mathbf{F_x}]^2\mathbf{p} + \omega^2\mathbf{p} = 0 \tag{6.3.29}$$

Then, the problem of calculating the Hopf bifurcation points of (6.3.6) is reduced to solving (6.3.6) and (6.3.29) for the $2n + 2$ known $x_1, x_2, \ldots,$ $x_n, \alpha, \omega^2, p_1, p_2, \ldots, p_n$. Thus, we can impose two constraints on this system, such as choosing two components of the vector $\mathbf{p}$ arbitrarily. Then, the resulting $2n + 2$ equations can be solved for the $2n + 2$ unknowns by using a Newton–Raphson procedure. The scheme proposed by Holodniok and Kubicek (1984a,b) and Roose (1985) is used in CANDYS/QA (Feudel and Jansen, 1992). Seydel (1979a, 1981) has also proposed a direct scheme for computing Hopf bifurcation points. An indirect scheme for determining Hopf bifurcation points has been discussed by Guckenheimer and Worfolk (1993).

In Sections 6.2 and 6.3, we discussed schemes for determining codimension–one bifurcation points. There are also schemes to determine codimension–two bifurcation points and other more degenerate bifurcation points (e.g., Brindley, Kaas–Petersen, and Spence, 1989; DeDier, Roose, and van Rompay, 1990; Griewank and Reddien, 1984; Pönisch, 1987; Roose and Piessens, 1985; Spence and Werner, 1982; Werner and Janovsky, 1991).

6.4 HOMOTOPY ALGORITHMS

In Section 6.1, we showed that once a solution $\mathbf{x}_0$ of (6.1.1) at $\alpha = \alpha_0$ has been calculated, one can use a continuation scheme to trace the branch passing through $(\mathbf{x}_0, \alpha_0)$ if the Jacobian matrix $\mathbf{F_x}(\mathbf{x}_0, \alpha_0)$ is nonsingular. The question arises as to how one can calculate all possible solutions of (6.1.1) at α_0. A straightforward approach would be to guess values for the solutions and use a Newton–Raphson procedure to improve on these guesses until the procedure converges to within a specified tolerance. This procedure may converge slowly or diverge if the initial guesses are poor. One can randomly generate guesses for x and examine if the procedure converges or diverges for each of these guesses. If a sufficiently large number of guesses are used, then there is a high probability of success in finding some or all of the roots of $F(x; \alpha_0) = 0$. However, for highly nonlinear problems, the number of random initial guesses needed might be very high. Alternatively, one can use **homotopy methods**, which are powerful, robust, accurate, numerically stable, and almost universally applicable, but also often prohibitively expensive (Watson, 1986, 1990; Watson, Billups, and Morgan, 1987; Allgower and Georg, 1990). Homotopy algorithms are applicable when F in (6.1.1) is at least C^2.

The basic idea underlying a **homotopy method** is to deform a simple solvable problem continuously into the given (hard to solve) problem, while solving a continuous sequence of deformed problems. The function that specifies the continuous deformation is called a **homotopy map**. The solutions to the deformed problems are related and tracked as the deformation proceeds.

Let us suppose that the system $G(x, \alpha_0) = 0$ is a simple version of $F(x, \alpha_0) = 0$ and has an easily obtainable unique solution x_0. Then, one possible homotopy map is

$$H(x; \alpha_0, \lambda) = \lambda F(x; \alpha_0) + (1 - \lambda)G(x; \alpha_0) \ , \ \ 0 \le \lambda \le 1 \qquad (6.4.1)$$

By using one of the continuation techniques described in Section 6.1, one tracks the solutions (x, λ) of $H(x; \alpha_0, \lambda) = 0$ starting from $(x, \lambda) = (x_0, 0)$ as λ goes from 0 to 1. If everything works out well, one will obtain a solution $(x, \lambda) = (\bar{x}, 1)$ such that $F(\bar{x}; \alpha_0) = 0$. This approach, which is called the **standard approach**, is likely to fail if any of the following conditions occur: (a) there are no solutions at a particular value of λ, (b) the solutions diverge as $\lambda \to 1$, and (c) the homotopy has branch points. At the values of λ corresponding to the branch points, $H(x; \alpha_0, \lambda) = 0$ is singular and the rank of the matrix $[H_x \mid H_\lambda]$ is less than n.

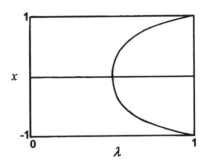

Figure 6.4.1: Variation of the roots of (6.4.3) with λ.

For illustration, we consider the problem of finding the roots of

$$F(x) = x^3 - x = 0 \tag{6.4.2}$$

We choose $G(x) = x$ and construct the homotopy map

$$H = \lambda(x^3 - x) + (1 - \lambda)x \tag{6.4.3}$$

The solutions of $H(x; \lambda) = 0$ are

$$x = 0 \quad \text{and} \quad x = \pm\sqrt{\frac{2\lambda - 1}{\lambda}} \tag{6.4.4}$$

which are plotted in Figure 6.4.1. Clearly, in the $x - \lambda$ space there is a pitchfork bifurcation at $(x, \lambda) = (0, \frac{1}{2})$. At this point, the rank of $[H_x \mid H_\lambda] = [0, 0]$ is zero. Consequently, the problem $H(x, \lambda) = 0$ is singular at $\lambda = \frac{1}{2}$, and the standard approach is expected to fail as $\lambda \to \frac{1}{2}$. To overcome this difficulty, one can use the artificial parameter λ and modify the homotopy map H such that it has smooth nonbifurcating curves for $0 \le \lambda \le 1$. Alternatively, one can construct a homotopy map that involves additional parameters **a** so that bifurcations are not encountered during the continuation. This is essentially what is done in modern homotopy methods.

To describe this procedure, we return to (6.4.3), introduce a scalar parameter a, and obtain

$$H(x; \lambda, a) = \lambda(x^3 - x) + (1 - \lambda)(x - a) \tag{6.4.5}$$

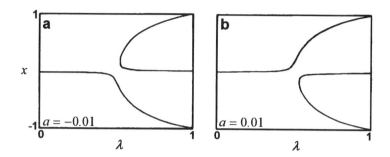

Figure 6.4.2: Zeros of (6.4.5) versus λ: (a) $a = -0.01$ and (b) $a = 0.01$.

When $a \neq 0$ there are no bifurcations in the $x - \lambda$ space for $0 \leq \lambda < 1$. Based on the discussion in Section 2.3.8, this is to be expected because the pitchfork bifurcation in (6.4.3) at $\lambda = \frac{1}{2}$ is structurally unstable to the perturbation $-a(1 - \lambda)$ included in (6.4.5). In Figure 6.4.2, we show the solutions of $H(x; \lambda, a) = 0$ for $a = 0.01$ and $a = -0.01$. This figure illustrates clearly how the inclusion of a eliminates the pitchfork bifurcation seen in Figure 6.4.1. Further, we note that by changing the value of a, one can generate some or all of the solutions of $F(x) = 0$. Moreover, owing to the absence of bifurcations, the continuation scheme will not encounter any singularities, and hence it is guaranteed to converge to one of the solutions of $F(x) = 0$.

Returning to our original problem $\mathbf{F}(\mathbf{x}, \alpha_0) = 0$, we can construct a homotopy map as

$$\mathbf{H}(\mathbf{x}; \alpha_0, \lambda, \mathbf{a}) = \lambda \mathbf{F}(\mathbf{x}; \alpha_0) + (1 - \lambda)(\mathbf{x} - \mathbf{a}) \qquad (6.4.6)$$

Thus, one can randomly pick an $\mathbf{a} \in \mathcal{R}^n$, which uniquely determines $\mathbf{x_0}$ to be $\mathbf{a}$, and then track the solutions $(\mathbf{x}, \lambda)$ of $\mathbf{H}(\mathbf{x}; \alpha_0, \lambda, \mathbf{a}) = 0$ from $(\mathbf{x}, \lambda) = (\mathbf{x_0}, 0)$ to $(\mathbf{x}, \lambda) = (\bar{\mathbf{x}}, 1)$. The **transversality homotopy theorem** from differential topology (e.g., Guillemin and Pollack, 1974; Morgan, 1987) guarantees that the homotopy (6.4.6) will have smooth, nonbifurcating curves for randomly chosen values of $\mathbf{a}$. Hence, modern homotopy methods are called **probability–one homotopy methods**. In addition, because modern homotopy methods converge to a solution for any arbitrarily chosen initial condition, they are said to be **globally convergent**.

Using modern homotopy methods, Watson, Billups, and Morgan (1987) developed a software program called HOMPACK, which includes a suite of codes for Brouwer fixed–point problems, polynomial systems, certain classes of zero–finding and nonlinear programming problems, and discretization of two–point boundary–value problems based on shooting schemes, finite–differences, spline collocation, and finite elements. In particular, the subroutines FIXPDF, FIXPNF, FIXPQF, FIXPDS, FIXPNS, and FIXPQS from HOMPACK can be used to find a solution of $\mathbf{F}(\mathbf{x}; \alpha_0) = \mathbf{0}$. The subroutine POLSYS from HOMPACK can be used to determine all the solutions (including the complex ones) of $\mathbf{F}(\mathbf{x}; \alpha_0) = \mathbf{0}$.

6.5 CONSTRUCTION OF PERIODIC SOLUTIONS

The simplest numerical method of constructing periodic solutions of an autonomous system, such as (3.2.1), or a nonautonomous system, such as (3.2.46), is the so–called **brute–force approach**. In this approach, one chooses an initial condition, integrates the system for a long time, and ultimately converges to an attractor. It is clear that this approach is easy to program and very general (it can locate fixed points, periodic solutions, quasiperiodic solutions, and chaotic solutions). However, there is no guarantee that the integration will converge to the desired attractor. Moreover, the brute–force approach has several disadvantages: (a) the convergence can be very slow for lightly damped systems, (b) only some of the unstable solutions can be realized by reversing the direction of integration, and (c) achievement of steady–state conditions may be difficult to ascertain. To overcome these shortcomings, direct approaches in the frequency domain and time domain have been proposed.

Here, we concentrate on the methods proposed for determining periodic solutions of a system of first–order differential equations. In the **frequency–domain formulation**, the method of harmonic balance (e.g., Mees, 1981; Kundert, Sangiovanni–Vincentelli, and Sugawara, 1987; Kim and Noah, 1990) and generalized versions of this method (Neymeyr and Seelig, 1991) are commonly used. In the **time–domain formulation,** finite–difference schemes (Rinzel and Miller, 1980; Parker and Chua, 1989), shooting techniques (Keller, 1968; Aprille and Trick, 1972; Chua and Lin, 1975; Mees, 1981; Seydel, 1988; Parker and Chua, 1989), and Poincaré map methods (Curry, 1980; Parker and Chua, 1989) are commonly used. As illustrated by Samoilenko and Ronto (1979), collocation methods can also be used to construct periodic solutions. The method of harmonic balance can be used to construct periodic solutions of a weakly nonlinear system.

In what follows, we only describe the time–domain approaches. In these approaches, the initial–value problem is converted into a two–point boundary–value problem. Thus, one seeks an initial condition $\mathbf{x}(0) = \boldsymbol{\eta}$ and a solution $\mathbf{x}(t; \boldsymbol{\eta})$ with a minimal period T such that

$$\mathbf{x}(T, \boldsymbol{\eta}) = \boldsymbol{\eta} \qquad (6.5.1)$$

6.5.1 Finite–Difference Method

In this section, we describe finite–difference schemes for determining periodic solutions of autonomous and nonautonomous systems.

Autonomous Systems

For a periodic solution of an autonomous system, the period T is an unknown quantity, and it is determined along with the states. The two–point boundary–value problem is given by (3.2.1) and (6.5.1). For convenience, one may change the independent variable t in (3.2.1) to a new independent variable τ such that $t = T\tau$, where T is the period of the solution. Then, (3.2.1) and (6.5.1) become

$$\frac{d\mathbf{x}}{d\tau} = T\,\mathbf{F}(\mathbf{x}; \mathbf{M}) \qquad (6.5.2)$$

$$\mathbf{x}(1; \boldsymbol{\eta}, T) = \boldsymbol{\eta} \qquad (6.5.3)$$

where $\mathbf{x}(\tau; \boldsymbol{\eta}, T)$ has the period unity. Thus, the problem of calculating periodic solutions of (3.2.1) is converted into the two–point boundary–value problem defined by (6.5.2) and (6.5.3).

To implement the finite–difference method, first we choose a sufficiently dense set of uniform time steps

$$\tau_0 = 0, \ \tau_1 = h, \ \tau_2 = 2h, \ ..., \ \tau_N = Nh = 1$$

where h is a small number. Then, at the midpoint of an interval, the derivative $dx/d\tau$ is approximated by using a central–difference scheme, and the function $\mathbf{F}$ is approximated by using the trapezoidal rule. The chosen scheme converts (6.5.2) into

$$\mathbf{x}^{i+1} - \mathbf{x}^i = \frac{1}{2}hT\left[\mathbf{F}(\mathbf{x}^{i+1};\mathbf{M}) + \mathbf{F}(\mathbf{x}^i;\mathbf{M})\right] \qquad (6.5.4)$$

where $\mathbf{x}^i$ is the value of $\mathbf{x}$ at the grid point $\tau_i = ih$. The periodicity condition (6.5.3) takes the form

$$\mathbf{x}^N = \mathbf{x}^0 \qquad (6.5.5)$$

Therefore, we have the nonlinear system of nN algebraic equations (6.5.4) to solve for the $(nN + 1)$ unknowns

$$\mathbf{x}^0, \ \mathbf{x}^1, \ \mathbf{x}^2, \ ..., \ \mathbf{x}^{N-1}, \ \text{and} \ T$$

and an additional equation is needed for closure.

We recall that a periodic solution of an autonomous system is invariant to linear shifts in the time origin; that is, if $\mathbf{x}(\tau)$ is a solution, then $\mathbf{x}(\tau+\tau_0)$ is also a solution for any arbitrary τ_0. In other words, the "phase" is arbitrary. To remove this arbitrariness, we use an additional equation to impose a phase condition. To this end, one of the $nN + 1$ variables other than T is fixed, after ensuring that this choice is within the solution range. From a practical point of view, there is really no systematic way for determining the variable to be fixed. Let us suppose that the first component of $\mathbf{x}^0$ is fixed and that $\tilde{\mathbf{x}}^0$ represents the vector of the remaining $n - 1$ components. Then, (6.5.4) is solved for

$$\tilde{\mathbf{x}}^0, \ \mathbf{x}^1, \ \mathbf{x}^2, \ ..., \ \mathbf{x}^{N-1}, \ \text{and} \ T$$

by using a Newton–Raphson scheme. For computational efficiency, one can take advantage of the banded structure of (6.5.4). Of course, this approach will surely fail if the periodic solution does not intersect the hyperplane $x_1 = x_1^0$. On the other hand, specifying x_1^0 may force the algorithm to converge to one attractor instead of another if a priori information about the two attractors is available.

Here, we used a central–difference scheme because it is usually more accurate than a forward– or backward–difference scheme. Further, we have assumed that the chosen time step does not lead to any numerical instabilities. However, in some cases, numerical instabilities may be unavoidable. For example, Rinzel and Miller (1980) found that a central–difference scheme was not appropriate for their problem because of numerical instabilities. Hence, they used a backward–difference scheme.

Nonautonomous Systems

Here, the two–point boundary–value problem is given by (3.2.46) and (6.5.1). Let the explicit time–dependent terms in (3.2.46) have the least common period T_e. Then, the period T of the sought solution is a rational multiple of T_e and, hence, a known quantity. For convenience, again, we change the independent variable t in (3.2.46) to a new independent variable τ such that $t = T\tau$. Then, (3.2.46) and (6.5.1) become

$$\frac{d\mathbf{x}}{d\tau} = T\,\mathbf{F}(\mathbf{x}, T\tau; \mathbf{M}) \tag{6.5.6}$$

$$\mathbf{x}(1; \boldsymbol{\eta}, T) = \boldsymbol{\eta} \tag{6.5.7}$$

where $\mathbf{x}(\tau; \boldsymbol{\eta}, T)$ has the period unity.

Again, to implement the finite–difference method, we first choose a sufficiently dense set of uniform time steps

$$\tau_0 = 0, \ \tau_1 = h, \ \tau_2 = 2h, \ ..., \ \tau_N = Nh = 1$$

where h is a small number. Then, at the midpoint of an interval, the time derivative is approximated by using a central–difference scheme, and the function $\mathbf{F}$ is approximated by using the trapezoidal rule. The chosen scheme converts (6.5.6) into

$$\mathbf{x}^{i+1} - \mathbf{x}^i = \frac{1}{2}hT\left[\mathbf{F}(\mathbf{x}^{i+1}, T\tau_{i+1}; \mathbf{M}) + \mathbf{F}(\mathbf{x}^i, T\tau_i; \mathbf{M})\right] \tag{6.5.8}$$

where x^i is the value of x at the grid point $\tau_i = ih$. The periodicity condition is given by (6.5.5). The system (6.5.8) and (6.5.5) represents nN algebraic equations in the nN unknowns

$$x^0, \, x^1, \, x^2, \, ..., \, x^{N-1}$$

To solve this algebraic system, we use a Newton–Raphson scheme.

In comparison with the shooting method discussed in the next section, in general, a finite–difference scheme is more suitable to determine a highly unstable periodic solution because it is less sensitive to initial conditions.

6.5.2   Shooting Method

In this section, we describe shooting methods for finding periodic solutions of autonomous and nonautonomous systems.

Autonomous Systems

Here, (3.2.1) and (6.5.1) represent the two–point boundary–value problem. In Figure 6.5.1, we graphically illustrate the shooting method when (3.2.1) is two–dimensional. The trajectory that runs from η at $t = 0$ to the same location at $t = T$ represents the desired periodic solution. The other trajectory represents the solution obtained by using the initial guess (T_0, η_0) for (T, η). Because this initial guess is off the mark, it needs correction. The correction is accomplished through a Newton–Raphson scheme as described below.

We seek

$$\delta\eta = \eta - \eta_0 \text{ and } \delta T = T - T_0$$

such that (6.5.1) is satisfied to within a specified tolerance; that is,

$$x(T_0 + \delta T, \eta_0 + \delta\eta) - (\eta_0 + \delta\eta) \simeq 0 \qquad (6.5.9)$$

Expanding (6.5.9) in a Taylor series and keeping only linear terms in $\delta\eta$ and δT, we arrive at

$$\left[\frac{\partial x}{\partial \eta}(T_0, \eta_0) - I \right] \delta\eta + \frac{\partial x}{\partial T}(T_0, \eta_0)\delta T = \eta_0 - x(T_0, \eta_0) \qquad (6.5.10)$$

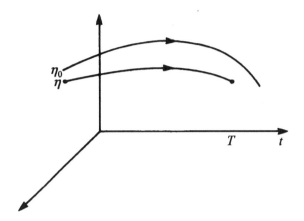

Figure 6.5.1: Depiction of the shooting method for a two–dimensional autonomous system.

where $\partial\mathbf{x}/\partial\boldsymbol{\eta}$ is an $n \times n$ matrix, $\mathbf{I}$ is the $n \times n$ identity matrix, and $\partial\mathbf{x}/\partial T$ is an $n \times 1$ vector. It follows from (3.2.1) that

$$\frac{\partial\mathbf{x}}{\partial T}(T_0, \boldsymbol{\eta}_0) = \mathbf{F}(\mathbf{x}(T_0, \boldsymbol{\eta}_0); \mathbf{M})$$

$$= \mathbf{F}(\mathbf{x}(0, \boldsymbol{\eta}_0); \mathbf{M}) = \mathbf{F}(\boldsymbol{\eta}_0; \mathbf{M}) \qquad (6.5.11)$$

If equations (6.5.2) and (6.5.3) were used instead of (3.2.1) and (6.5.1), then equations (6.5.9)–(6.5.11) would be replaced by

$$\mathbf{x}(1, \boldsymbol{\eta}_0 + \delta\boldsymbol{\eta}, T_0 + \delta T) - (\boldsymbol{\eta}_0 + \delta\boldsymbol{\eta}) \simeq \mathbf{0}$$

$$\left[\frac{\partial\mathbf{x}}{\partial\boldsymbol{\eta}}(1, \boldsymbol{\eta}_0, T_0) - \mathbf{I}\right]\delta\boldsymbol{\eta} + \frac{\partial\mathbf{x}}{\partial T}(1, \boldsymbol{\eta}_0, T_0)\delta T = \boldsymbol{\eta}_0 - \mathbf{x}(1, \boldsymbol{\eta}_0, T_0)$$

$$\frac{\partial\mathbf{x}}{\partial T}(1, \boldsymbol{\eta}_0, T_0) = \mathbf{F}(\boldsymbol{\eta}_0; \mathbf{M})$$

To proceed further, the matrix $\partial\mathbf{x}/\partial\boldsymbol{\eta}$ evaluated at $(T_0, \boldsymbol{\eta}_0)$ needs to be determined. Assuming that the vector function $\mathbf{F}(\mathbf{x}; \mathbf{M})$ is $\mathcal{C}^r$, where $r \geq 1$, we differentiate both sides of (3.2.1) with respect to $\boldsymbol{\eta}$ and obtain

$$\frac{d}{dt}\left(\frac{\partial \mathbf{x}}{\partial \boldsymbol{\eta}}\right) = D_\mathbf{x} \mathbf{F}(\mathbf{x}; \mathbf{M})\frac{\partial \mathbf{x}}{\partial \boldsymbol{\eta}} \qquad (6.5.12)$$

Also, differentiation of the initial condition $\mathbf{x}(0) = \boldsymbol{\eta}$ with respect to $\boldsymbol{\eta}$ leads to

$$\frac{\partial \mathbf{x}}{\partial \boldsymbol{\eta}}(0) = \mathbf{I} \qquad (6.5.13)$$

Thus, after $\mathbf{x}$ is determined, we can solve the linear initial–value problem (6.5.12) and (6.5.13) from zero to T_0 and evaluate $\partial \mathbf{x}/\partial \boldsymbol{\eta}$ at $(\boldsymbol{\eta}_0, T_0)$. Instead of solving (3.2.1) subject to $\mathbf{x}(0) = \boldsymbol{\eta}_0$, saving the solution $\mathbf{x}(t)$ for $0 \le t \le T_0$, and then solving (6.5.12) and (6.5.13), we can solve both problems simultaneously, thereby obtaining $\mathbf{x}(t, \boldsymbol{\eta}_0)$ and $\partial \mathbf{x}/\partial \boldsymbol{\eta}(T_0, \boldsymbol{\eta}_0)$ simultaneously.

If one or more components of $\mathbf{F}(\mathbf{x}; \mathbf{M})$ are nondifferentiable, then the components of the matrix $\partial \mathbf{x}/\partial \boldsymbol{\eta}$ need to be calculated numerically. To this end, we solve (3.2.1) subject to the two initial conditions

$$\mathbf{x}(0) = \boldsymbol{\eta}_0 \qquad (6.5.14)$$
$$\mathbf{x}^k(0) = \boldsymbol{\eta}_0 + \delta \mathbf{e}_k \qquad (6.5.15)$$

where $\mathbf{e}_k$ is the kth column of the identity matrix and δ is a small number. Then,

$$\frac{\partial \mathbf{x}}{\partial \boldsymbol{\eta}_k}(\boldsymbol{\eta}_0, T_0) = \frac{\mathbf{x}^k(T_0, \boldsymbol{\eta}_0 + \delta \mathbf{e}_k) - \mathbf{x}(T_0, \boldsymbol{\eta}_0)}{\delta} \qquad (6.5.16)$$

Once the matrix $\partial \mathbf{x}/\partial \boldsymbol{\eta}$ evaluated at $(\boldsymbol{\eta}_0, T_0)$ is known, then (6.5.10) constitutes a system of n equations in the $n + 1$ unknowns $\delta \boldsymbol{\eta}$ and δT. As mentioned in Section 6.5.1, there is an arbitrariness in the phase associated with a periodic solution of an autonomous system. To remove this arbitrariness, we specify a phase condition in one of the following ways:

(a) By fixing one of the components of the initial vector $\boldsymbol{\eta}_0$, say, η_k (Aluko and Chang, 1984; Holodniok and Kubicek, 1984a).

(b) By setting the kth component $F_k(\mathbf{x}; \mathbf{M})$ of the vector field $\mathbf{F}(\mathbf{x}; \mathbf{M})$ equal to zero (Seydel, 1981, 1988).

(c) By requiring the corrections $\delta\boldsymbol{\eta}$ to be normal to the vector field $\mathbf{F}$ (Mees, 1981); that is, $\mathbf{F}^T \delta\boldsymbol{\eta} = 0$. This condition is called the **orthogonality condition**.

(d) By using an integrated form of the orthogonality condition (Doedel and Heinemann, 1983; Doedel, 1986; Doedel, Keller, and Kernevez, 1991b).

Each of the conditions (a), (b), and (c) essentially defines a Poincaré section $g(x) = 0$ for an orbit of (3.2.1). Hence, the above algorithm will fail to converge if no periodic orbit intersects the chosen section. Therefore, if one of these conditions is chosen, we have the following $(n + 1)$–dimensional system to solve for $(T, \boldsymbol{\eta})$:

$$\mathbf{x}(T, \boldsymbol{\eta}) - \boldsymbol{\eta} = 0$$
$$g(\mathbf{x} = \boldsymbol{\eta}) = 0$$

If we choose $(T_0, \boldsymbol{\eta}_0)$ as the initial guess for a Newton–Raphson scheme, we obtain (6.5.10) and the equation

$$\frac{\partial g}{\partial \boldsymbol{\eta}}(\boldsymbol{\eta}_0)\ \delta\boldsymbol{\eta} = 0$$

to solve for the corrections $(\delta T, \delta\boldsymbol{\eta})$ to $(T_0, \boldsymbol{\eta}_0)$.

When condition (b) is used, we have the following system of $(n+1)$ equations:

$$\left[\begin{array}{cc} \frac{\partial \mathbf{x}}{\partial \boldsymbol{\eta}}(T_0, \boldsymbol{\eta}_0) - \mathbf{I} & \mathbf{F}(\boldsymbol{\eta}_0; \mathbf{M}) \\[2mm] \frac{\partial F_k}{\partial \boldsymbol{\eta}}(\boldsymbol{\eta}_0; \mathbf{M}) & 0 \end{array} \right] \left[\begin{array}{c} \delta\boldsymbol{\eta} \\[2mm] \delta T \end{array} \right] = \left[\begin{array}{c} \boldsymbol{\eta}_0 - \mathbf{x}(T_0, \boldsymbol{\eta}_0) \\[2mm] 0 \end{array} \right]$$

On the other hand, when the orthogonality condition (c) is used, we have the following system:

$$\left[\begin{array}{cc} \frac{\partial \mathbf{x}}{\partial \boldsymbol{\eta}}(T_0, \boldsymbol{\eta}_0) - \mathbf{I} & \mathbf{F}(\boldsymbol{\eta}_0; \mathbf{M}) \\[2mm] \mathbf{F}^T(\boldsymbol{\eta}_0; \mathbf{M}) & 0 \end{array} \right] \left[\begin{array}{c} \delta\boldsymbol{\eta} \\[2mm] \delta T \end{array} \right] = \left[\begin{array}{c} \boldsymbol{\eta}_0 - \mathbf{x}(T_0, \boldsymbol{\eta}_0) \\[2mm] 0 \end{array} \right]$$

After determining the corrections, we check if the convergence criterion $\| \delta\boldsymbol{\eta} \| < \epsilon_1$ and $| \delta T | < \epsilon_2$ is satisfied, where the ϵ_i are specified tolerances. If the criterion is not satisfied, we update the initial guess $(T_0, \boldsymbol{\eta}_0)$ to $(T_0 + \delta T, \boldsymbol{\eta}_0 + \delta \boldsymbol{\eta})$ and repeat the procedure.

We have successfully used the conditions proposed by Seydel (1981, 1988) and Mees (1981). The stable periodic solutions of (3.4.1)–(3.4.3), shown in Figures 3.4.3 and 3.5.1, were obtained by using the condition proposed by Mees. The stable and unstable periodic solutions of (3.4.4)–(3.4.7), shown in Figures 3.4.4 and 3.4.5, were obtained by using the condition proposed by Seydel.

The following remarks regarding the shooting method are worth noting. Because a fixed point of (3.2.1) satisfies (6.5.1), it is possible for the method to converge to a fixed point. Furthermore, even when a periodic solution is located, the computed period T may not always be the minimal period of the determined solution. This is so because we do not impose any condition in the shooting scheme to restrain the period T to be the minimal period. To check whether the computed period is minimal or not, we check whether $\mathbf{x}(T/\text{integer}) = \mathbf{x}(0)$.

In general, the shooting method is more sensitive to initial conditions than the finite–difference method. In particular, the sensitivity of the shooting method is pronounced when the sought periodic solution is highly unstable. Because of roundoff and truncation errors, one is likely to deviate from an unstable periodic solution even if the initial conditions are precisely specified. The shooting method is very effective when the sought solution is not highly unstable and when the different sources of errors are tightly controlled.

As a by–product of the shooting technique, one can obtain the monodromy matrix from

$$\Phi = \frac{\partial \mathbf{x}}{\partial \boldsymbol{\eta}}(T, \boldsymbol{\eta}) \tag{6.5.17}$$

and hence determine the stability of the calculated periodic solution by examining the eigenvalues of Φ. To prove (6.5.17), we proceed along the lines of Seydel (1988, Chapter 7). We let $\boldsymbol{\phi}(t; \boldsymbol{\eta})$ be the solution of (3.2.1) subject to $\mathbf{x}(0) = \boldsymbol{\eta}$ and $\boldsymbol{\phi}(t; \boldsymbol{\eta} + \mathbf{d}_0)$ be the solution of (3.2.1) subject to the perturbed initial condition $\mathbf{x}(0) = \boldsymbol{\eta} + \mathbf{d}_0$. Then, at $t = T$, the separation $\mathbf{d}(T)$ between the two solutions is

$$\mathbf{d}(T) = \boldsymbol{\phi}(T; \boldsymbol{\eta} + \mathbf{d}_0) - \boldsymbol{\phi}(T; \boldsymbol{\eta}) \tag{6.5.18}$$

which for small $\| \mathbf{d}_0 \|$ can be approximated by

$$\mathbf{d}(t) = \frac{\partial \phi}{\partial \eta}(T; \eta)\mathbf{d}_0 = M\mathbf{d}_0 \qquad (6.5.19)$$

Therefore, the matrix M determines whether the initial perturbation $\mathbf{d}_0$ provided to the orbit $\phi(t; \eta)$ grows or decays. Since $\phi(0) = \eta$,

$$\frac{\partial \phi}{\partial \eta}(0) = \mathbf{I} \qquad (6.5.20)$$

Hence, $\partial\phi/\partial\eta$ is the solution of (6.5.12) and (6.5.13). These equations are identical to (3.2.5) and (3.2.8), which define the monodromy matrix Φ.

Nonautonomous Systems

In this case, (3.2.46) and (6.5.1) represent the two–point boundary–value problem. Let the explicit time–dependent terms in (3.2.46) have the least common period T_e. Then, the period T of the sought solution is a rational multiple of T_e.

To implement the shooting method, we proceed along the lines discussed for the autonomous system. However, here, the period of the solution is not an unknown quantity. We provide an initial guess η_0 for η and require the difference $\delta\eta = \eta - \eta_0$ to satisfy a specified tolerance so that

$$\mathbf{x}(T, \eta_0 + \delta\eta) - (\eta_0 + \delta\eta) \simeq \mathbf{0} \qquad (6.5.21)$$

We expand (6.5.21) in a Taylor series, retain only linear terms in $\delta\eta$, and obtain

$$\left[\frac{\partial \mathbf{x}}{\partial \eta}(T, \eta_0) - \mathbf{I}\right] \delta\eta = \eta_0 - \mathbf{x}(T, \eta_0) \qquad (6.5.22)$$

The components of the $n \times n$ matrix $\partial\mathbf{x}/\partial\eta$ evaluated at (T, η_0) can be determined as discussed earlier. Once $\partial\mathbf{x}/\partial\eta$ is known, we have a system of n algebraic equations, namely, (6.5.22), to solve for the n unknowns $\delta\eta$. Then, the initial guess η_0 is updated, and the procedure is repeated until the specified convergence criterion is satisfied. Then, the stability of the computed periodic solution is determined from the eigenvalues of the monodromy matrix $\partial\mathbf{x}/\partial\eta$ evaluated at (T, η_0).

6.5.3   Poincaré Map Method

In the Poincaré map method, the problem of determining a periodic solution of a continuous–time system is reduced to the problem of determining the fixed point of a Poincaré map (Curry, 1980; Parker and Chua, 1989). This map describes the dynamics on an appropriately constructed one–sided Poincaré section.

For nonautonomous systems, the Poincaré map method is identical to the shooting method described above. However, the Poincaré map method differs from the shooting method for autonomous systems. To outline this method, we consider a periodic orbit of (3.2.1) and let $\mathbf{x} = \boldsymbol{\eta}$ represent a point on this orbit. At this point, let the section

$$g(\mathbf{x}) = 0$$

be transverse to the orbit. Restricting this section to be one sided, we let $\mathbf{P}$ represent the corresponding Poincaré map. Then, the fixed point of this map is a solution of the following algebraic system:

$$\mathbf{P}(\boldsymbol{\eta}) - \boldsymbol{\eta} = 0$$

where

$$\mathbf{P}(\boldsymbol{\eta}) = \mathbf{x}(\tau(\boldsymbol{\eta}), \boldsymbol{\eta})$$

and $\tau(\boldsymbol{\eta})$ is the unique return time. Using a Newton–Raphson scheme to solve this system, we obtain

$$\mathbf{x}^{k+1} = \mathbf{x}^k - [D\mathbf{P} - \mathbf{I}]^{-1} \left[\mathbf{P}(\mathbf{x}^k) - \mathbf{x}^k \right]$$

where the matrix $D\mathbf{P}$ is computed as discussed in Section 3.3.2.

Here, one needs to provide an initial guess $\mathbf{x}^0$ that satisfies $g(\mathbf{x}) = 0$. On the other hand, in the shooting method, one needs to provide an initial guess for a point on the orbit and the period of the orbit. Further, Poincaré map methods will not converge to the fixed points of (3.2.1) unless the Poincaré section contains the fixed point (an unlikely event).

6.6   CONTINUATION OF PERIODIC SOLUTIONS

Here, we consider continuation schemes for the periodic solutions of

(3.2.1). These schemes provide algorithmic procedures to generate a continuum of periodic solutions $\mathbf{x}(T; \alpha)$ with respect to a control parameter α. The branch of solutions is typically initiated near a Hopf bifurcation point in the state–control space. By using the analytical methods of Section 3.5, one can determine an approximation for the periodic solutions near the Hopf bifurcation point. This approximation can be used as an initial guess for either the finite–difference scheme or the shooting scheme, described in Section 6.5, to determine the starting point on the branch along which continuation is desired. Next, we discuss three commonly used continuation schemes.

6.6.1 Sequential Continuation

As in the case of fixed points, one can use a sequential scheme to carry out the continuation (e.g., Rinzel and Miller, 1980; Seydel, 1981). In this scheme, the scalar control parameter α is the continuation parameter, and the periodic solution determined in the previous step is used as an initial guess for the periodic solution to be determined in the next step. This algorithm fails to go past a cyclic–fold bifurcation point (a turning point) in the state–control space. To remedy this, we interchange the continuation parameter α and one of the states x_i (e.g., Rinzel and Miller, 1980).

6.6.2 Arclength Continuation

In this section, we discuss the scheme used by Holodniok and Kubicek (1984a) in the software program DERPER. Here, the problem of finding periodic solutions of (3.2.1) is posed as a two–point boundary–value problem. The arclength s is used as the continuation parameter, and a periodic solution $\mathbf{x}[t, \boldsymbol{\eta}(s); \alpha(s)]$ of (3.2.1) with period $T(s)$ is sought such that

$$\mathbf{G}[T(s), \boldsymbol{\eta}(s); \alpha(s)] = \mathbf{x}[T(s), \boldsymbol{\eta}(s); \alpha(s)] - \boldsymbol{\eta}(s) = \mathbf{0} \qquad (6.6.1)$$

with the initial condition

$$\mathbf{x}[T(0), \boldsymbol{\eta}(0); \alpha(0)] = \boldsymbol{\eta}(0) \qquad (6.6.2)$$

Differentiating (6.6.1) with respect to s, we obtain

$$\frac{\partial \mathbf{G}}{\partial T}(T, \boldsymbol{\eta}; \alpha)T' + \frac{\partial \mathbf{G}}{\partial \boldsymbol{\eta}}(T, \boldsymbol{\eta}; \alpha)\boldsymbol{\eta}' + \frac{\partial \mathbf{G}}{\partial \alpha}(T, \boldsymbol{\eta}; \alpha)\alpha' = \mathbf{0} \qquad (6.6.3)$$

where the prime indicates the derivative with respect to s, $\partial \mathbf{G}/\partial \boldsymbol{\eta}$ is an $n \times n$ matrix, $\partial \mathbf{G}/\partial T$ is an $n \times 1$ matrix, and $\partial \mathbf{G}/\partial \alpha$ is an $n \times 1$ matrix. The system (6.6.3) consists of n linear algebraic equations in the $(n + 2)$ unknowns $(\boldsymbol{\eta}'^T, T', \alpha')$. Hence, we need two additional equations for closure.

We recall that in the shooting scheme in Section 6.5.2 we had a system of n equations in the $(n + 1)$ unknowns $(\boldsymbol{\eta}^T, T)$. To close that system, we specified an additional equation in the form of a phase condition. Here, one additional equation specifies a phase condition. This equation fixes one of the states η_k in the vector $\boldsymbol{\eta}$ along the continuation path; that is,

$$\frac{d\eta_k}{ds} = 0 \qquad (6.6.4)$$

An additional equation is specified by the Euclidean arclength normalization

$$\boldsymbol{\eta}'^T \boldsymbol{\eta}' + T'^2 + \alpha'^2 = 1 \qquad (6.6.5)$$

The matrices $\partial \mathbf{G}/\partial \boldsymbol{\eta}$, $\partial \mathbf{G}/\partial T$, and $\partial \mathbf{G}/\partial \alpha$ can be evaluated by using (3.2.1), (6.6.1), and (6.6.2). We note that

$$\frac{\partial \mathbf{G}}{\partial \boldsymbol{\eta}}(T, \boldsymbol{\eta}; \alpha) = \frac{\partial \mathbf{x}}{\partial \boldsymbol{\eta}}(T, \boldsymbol{\eta}; \alpha) - \mathbf{I} \qquad (6.6.6)$$

$$\frac{\partial \mathbf{G}}{\partial \alpha}(T, \boldsymbol{\eta}; \alpha) = \frac{\partial \mathbf{x}}{\partial \alpha}(T, \boldsymbol{\eta}; \alpha) \qquad (6.6.7)$$

$$\frac{\partial \mathbf{G}}{\partial T}(T, \boldsymbol{\eta}; \alpha) = \frac{\partial \mathbf{x}}{\partial T}(T, \boldsymbol{\eta}; \alpha) = \mathbf{F}[\mathbf{x}(T, \boldsymbol{\eta}; \alpha); \alpha] \qquad (6.6.8)$$

To determine $\partial \mathbf{G}/\partial \alpha$, we differentiate (3.2.1) and (6.6.2) with respect to α and obtain

$$\frac{d}{dt}\left(\frac{\partial \mathbf{x}}{\partial \alpha}\right) = D_{\mathbf{x}}\mathbf{F}\frac{\partial \mathbf{x}}{\partial \alpha} + \frac{\partial \mathbf{F}}{\partial \alpha} \qquad (6.6.9)$$

$$\frac{\partial \mathbf{x}}{\partial \alpha} = 0 \text{ at } t = 0 \qquad (6.6.10)$$

Integrating (6.6.9) and (6.6.10) from $t = 0$ to $t = T$, we obtain $\partial \mathbf{x} / \partial \alpha$ and hence $\partial \mathbf{G} / \partial \alpha$.

We let

$$\mathbf{z} = \{\eta_1, \eta_2, \cdots, \eta_{k-1}, \eta_{k+1}, \cdots, \eta_n, T, \alpha\}^T \qquad (6.6.11)$$

and let the $n \times (n - 1)$ matrix obtained from $\partial \mathbf{G} / \partial \boldsymbol{\eta}$ by deleting the kth column be $\partial \hat{\mathbf{G}} / \partial \boldsymbol{\eta}$. Then, using (6.6.4) and (6.6.11), we rewrite (6.6.3) and (6.6.5) as

$$\left[\frac{\partial \hat{\mathbf{G}}}{\partial \boldsymbol{\eta}} \;\middle|\; \frac{\partial \mathbf{G}}{\partial T} \;\middle|\; \frac{\partial \mathbf{G}}{\partial \alpha} \right] \mathbf{z}' = \mathbf{0} \qquad (6.6.12)$$

$$\mathbf{z}'^T \mathbf{z}' = 1 \qquad (6.6.13)$$

Thus, the problem of the continuation of periodic solutions along a branch amounts to determining the continuation of the vector $\mathbf{z}$ from the algebraic system (6.6.12) and (6.6.13). For this purpose, the arclength continuation program DERPAR (see Section 6.1.3) is used. Depending on the problem, the index k in (6.6.4) may have to be adaptively varied so that one stays on a periodic solution.

At a particular point on the branch, the eigenvalues of the matrix $\partial \mathbf{x} / \partial \boldsymbol{\eta}$ provide information about the stability of the corresponding periodic solution. As in Section 6.1.3, the arclength continuation scheme experiences difficulties near branch points. Another point to note is that the branch of periodic solutions being followed might end at a Hopf bifurcation point. The program DERPER is equipped to detect such cases.

6.6.3 Pseudo–Arclength Continuation

This continuation scheme, which is also based on an arclength parameterization, is used by Doedel (1986) in the software program AUTO. A solution $\mathbf{x}[t, \boldsymbol{\eta}(s); \alpha(s)]$ of (3.2.1) with period $T(s)$ is sought such that

$$\mathbf{x}[T(s), \boldsymbol{\eta}(s); \alpha(s)] = \boldsymbol{\eta}(s) \qquad (6.6.14)$$

In this continuation scheme, the system (3.2.1) subject to the boundary condition (6.6.14) constitutes a set of n differential equations in the $(n+2)$ unknowns $(\boldsymbol{\eta}^T, T, \alpha)$. To close this system, we need two additional equations.

The first equation specifies a phase condition in the form of an integral. Let us suppose that the periodic solutions at two consecutive points s_0 and s on the branch are $\mathbf{x}_0 = \mathbf{x}(T_0, \boldsymbol{\eta}_0; \alpha_0)$ and $\hat{\mathbf{x}}$, respectively. Since (3.2.1) is autonomous, if $\hat{\mathbf{x}}(t)$ is a solution, then so is $\hat{\mathbf{x}}(t+\sigma)$ for any σ. Then, the phase condition is obtained by requiring that the distance $\| \hat{\mathbf{x}} - \mathbf{x}_0 \|$ be minimized with respect to the time translation σ. Hence, we desire the solution that minimizes

$$D(\sigma) = \int_0^T \| \hat{\mathbf{x}}(t+\sigma) - \mathbf{x}_0(t) \|^2 \, dt$$

Setting $dD/d\sigma$ to zero, we obtain

$$\int_0^T \frac{d}{d\sigma} [\| \hat{\mathbf{x}}(t+\sigma) - \mathbf{x}_0(t) \|]^2 \, dt = 0$$

which is satisfied, say, at $\sigma = \sigma^*$. Thus, we arrive at

$$\int_0^T [\mathbf{x}(t) - \mathbf{x}_0(t)]^T \dot{\mathbf{x}} dt = \frac{1}{2}\mathbf{x}^T\mathbf{x} \, |_0^T - \int_0^T \mathbf{x}_0^T \dot{\mathbf{x}} dt$$
$$= -\int_0^T \mathbf{x}_0^T \dot{\mathbf{x}} dt = 0 \tag{6.6.15}$$

where $\mathbf{x}(t) \equiv \hat{\mathbf{x}}(t+\sigma^*)$. Integrating (6.6.15) by parts and using (3.2.1), we obtain

$$\int_0^T \mathbf{x}^T \dot{\mathbf{x}}_0 \, dt = \int_0^T \mathbf{x}^T \mathbf{F}[\mathbf{x}(T_0, \boldsymbol{\eta}_0; \alpha_0); \alpha_0] dt = 0 \tag{6.6.16}$$

The second equation is given by the pseudo–arclength constraint

$$\int_0^T (\mathbf{x} - \mathbf{x}_0)^T \mathbf{x}_0' \, dt + (T - T_0)T_0' + (\alpha - \alpha_0)\alpha_0' = \Delta s \tag{6.6.17}$$

where the prime denotes the derivative with respect to the arclength s and Δs represents the step along the continuation path.

In the program AUTO, the derivatives in (6.6.17) are approximated by backward differences. Further, the system (3.2.1), (6.6.14), (6.6.16), and (6.6.17) is discretized by using a collocation scheme. More details on this discretization scheme can be found in the publications of Doedel (1986) and Doedel, Keller, and Kernevez (1991b).

6.6.4   Comments

The software program CANDYS/QA (Feudel and Jansen, 1992) is also
equipped to carry out continuation along a branch of periodic solutions.
All the continuation schemes are likely to experience difficulties at bifurcation points other than cyclic–fold bifurcation or turning points. The
bifurcation points can be determined either indirectly by monitoring
the Floquet multipliers or directly through special schemes (Holodniok
and Kubicek, 1984b, 1987). Indirect schemes are used in DERPER and
AUTO, while direct schemes are used in CANDYS/QA. The program
AUTO has provisions for branch switching at transcritical, pitchfork,
and period–doubling bifurcation points. Furthermore, schemes for numerically computing heteroclinic and homoclinic orbits are also available in AUTO (Doedel, 1986; Doedel, Keller, and Kernevez, 1991b).
Other relevant studies include those by Beyn (1990), Chow and Lin
(1990), and Guckenheimer and Worfolk (1993). The computation of
these orbits is often necessary in global bifurcation studies.

Chapter 7

TOOLS TO ANALYZE MOTIONS

Here, we discuss different tools that can be used to characterize the responses of nonlinear systems. In Chapters 3, 4, and 5 we used time histories, Fourier spectra, state–space plots, and Poincaré sections and maps as tools to study periodic, quasiperiodic, and chaotic motions. There exists a number of other tools to characterize responses; prominent among them are autocorrelation functions, Lyapunov exponents, and dimension calculations. The Melnikov and Shilnikov analyses of Sections 5.7 and 5.8 may be described as analytical tools for predicting parameter values at which chaotic motions are likely to occur. In Section 7.1 we discuss signal types and signal noise and introduce the terminology and basic ideas. Time histories are used to characterize motions in the next section. We illustrate the use of state–space portraits in Section 7.3. In Section 7.4 we discuss the method of delays and the construction of a space of delayed coordinates. In Section 7.5 we discuss how the Fourier transform can be used to analyze motions and point out its limitations. Poincaré sections and details of their practical implementation are addressed in Section 7.6. Autocorrelation functions are treated in the subsequent section. In Section 7.8 we examine Lyapunov exponents and discuss their computation in both analytical and experimental situations. We discuss dimension calculations in Section 7.9 and introduce polyspectra in Section 7.10.

461

7.1   INTRODUCTION

In this section, we introduce some notions, definitions, and topics relevant to the sections that follow. A signal obtained for analysis can be classified as either an **analog** or a **digital** signal. An **analog signal** is continuous in time, whereas a **digital signal** is composed of a discrete number of points, with each point corresponding to a discrete value of time. Digital signals are commonly encountered in digital–computer simulations and/or experimental situations. On the other hand, analog signals are encountered in analytical work, analog–computer simulations, and certain experimental situations. In the context of a system of equations, a signal can be the time history of one of the independent coordinates. In an experimental situation, a signal can be the output from a sensor or a transducer.

The evolutions observed in both digital simulations and experiments are influenced by noise and errors. Finite precision leads to **numerical errors** in digital simulations and **measurement errors** in experiments. For instance, before processing, the continuous analog signal is typically fed into an analog–to–digital (A/D) converter of a computer, which then samples the time series at regular time intervals and converts the signal at predetermined time intervals into a set of binary numbers. The time between samples is referred to as the **sampling time** τ_s. As a consequence, the continuous analog signal is converted into a sequence of numbers that are sampled with a finite precision. When an m bit A/D converter is used over a range of n volts, the number of quantization levels is 2^m and the resolution is $\frac{n}{2^m-1}$ volts. The resolution becomes better as n becomes small or m becomes large. Because A/D converters have finite precision, errors are made in measuring the magnitude of the signal. These errors are called **quantization errors**. Other factors to consider in an experimental situation include **environmental noise** and **instrument noise**. Additional information on A/D converters, quantization errors, and sources of noise can be found in the books of Oppenheim and Schafer (1975) and Horowitz and Hill (1980).

Because the presence of noise limits the amount of information that one can obtain from a signal, methods for noise reduction and signal separation are of considerable interest. Traditionally, for noise reduction, information in the frequency domain is examined, and linear analog or digital filters are used to eliminate the unwanted frequency components from a signal. For instance, a 60 Hz component in the spectrum due to line noise may be eliminated by using a filter. One should note that a linear filter, which is used to reduce noise in a signal that is not highly oversampled, generally distorts the signal.

However, frequency–domain techniques are quite well suited for signal separation in cases where the signal of interest and the noise are confined to different frequency bandwidths. The results of Badii, Broggi, Derighetti, Ravani, Ciliberto, Politi, and Rubio (1988) and Mitschke, Möller, and Lange (1988) indicate that one has to be careful in using filters to separate chaotic signals from noise. For weakly dissipative systems, Badii et al. (1988) examined the dimension values for signals passed through a linear low–pass filter. They found that the dimension value increases as the cut–off frequency is lowered below a certain critical value. Mitschke (1990) discussed how dimension enhancement due to filtering could be avoided through proper scaling of the data. Recently, some novel approaches for noise reduction have been proposed in the physics literature (e.g., Farmer and Sidorowich, 1988, 1991; Kostelich and Yorke, 1990; Hammel, 1990; Grassberger, Schreiber, and Shaffrath, 1991; Abarbanel, Brown, Sidorowich, and Tsimring, 1993; Kostelich and Schreiber, 1993; Schreiber, 1993). In these time–domain approaches, information in the state space of the system is examined, and its dynamics is used to determine and correct errors that result from noise.

The data obtained from a deterministic system can be classified as either **periodic** or **nonperiodic data**. Nonperiodic data may correspond to a quasiperiodic, transient, or chaotic motion. Unlike other motions, transient motions occur only over a finite length of time. The free oscillation of a damped oscillator is an example of a transient motion. When the damping is light, as in space structures, a transient motion can persist over a long interval of time.

A set of equations represents a **random system** if any of the variables and/or parameters has a random character. In a practical situation, one may subject a system to a deterministic excitation and examine the repeatability of a certain outcome under identical conditions. If the data obtained from the experiment are not repeatable within the bounds of the experimental error under identical conditions, then the corresponding system can be called a **random system**. We call the data obtained from a random system **random data**. Deterministic components of random data may be recovered by averaging over a long period of time.

A signal from a deterministic or random system can be classified as being either **stationary** or **nonstationary**. Loosely speaking, for a stationary signal, the relevant properties remain the same for all time. A signal from a deterministic system can be considered to be stationary if the spectral locations of the different peaks and the associated Fourier amplitudes do not change in time. A signal corresponding to a transient motion of a deterministic system is an example of a nonstationary signal.

A signal from a random system is classified as stationary or nonstationary depending on whether its statistical quantities, such as the **mean** and the **autocorrelation function**, associated with this signal are time–independent or time–dependent. Let us suppose that there is an ensemble of time histories obtained from a series of identical experiments. By choosing N different initial conditions, one can obtain N different frames of data. Then, an **ensemble average** is defined as an average over these different records at a given instant in time. The mean value μ_x at any instant t_1 is given by

$$\mu_x(t_1) = \lim_{N \to \infty} \frac{1}{N} \sum_{i=1}^{N} x_i(t_1) \tag{7.1.1}$$

where x_i represents the value of $x(t)$ at the instant $t = t_1$ in the ith frame of data. The autocorrelation function R_{xx} is given by

$$R_{xx}(t_1, \tau) = \lim_{N \to \infty} \frac{1}{N} \sum_{i=1}^{N} x_i(t_1) x_i(t_1 + \tau) \tag{7.1.2}$$

If μ_x and R_{xx} are independent of t_1, then the corresponding data are said to be (weakly) stationary. For stricter definitions of stationarity, we have to consider higher–order averages. Stationary random data are called **ergodic** if the ensemble average is equal to the time average obtained from one of the time histories, and the time average obtained from one frame to the next is the same. It is to be noted that the assumption of ergodicity may be violated in the presence of periodic components. When the random data are stationary and ergodic, a long–time history would suffice to analyze the data in question.

Some nonstationary data can be analyzed by techniques closely related to those used for stationary data. There are specialized techniques available for the analysis of nonstationary data. An example of such techniques is the time–frequency domain technique that uses functions called **wavelets** (e.g., Combes, Grossman, and Tchamitchian, 1990). By and large, the time–domain and frequency–domain techniques presented here are suitable for the analysis of stationary signals.

7.2 TIME HISTORIES

As a first step in characterizing a motion, one can examine the time history of a signal from the system in question. We concentrate on time histories of deterministic motions here. The time series may be examined over a length of time to ascertain if a motion has reached a steady state or not. When the motion settles down to either a constant or periodic state, it should be possible to distinguish steady-state motions from transient motions. However, it will be difficult to do so for nonperiodic motions.

The time series of a periodic motion has the appearance of a uniform trace, and the corresponding spectrum has one basic frequency. Let us suppose that a dissipative system is subjected to a single frequency excitation. If the response spectrum contains a spectral line at the excitation frequency, we refer to the motion as a **linear periodic motion**. On the other hand, if the response spectrum contains spectral lines at a basic frequency and its harmonics and/or subharmonics, we refer to the motion as a **nonlinear periodic motion**. A motion whose associated response spectrum contains a spectral line at a frequency other than the excitation frequency is also called a nonlinear periodic motion. A time history $x_p(t)$ associated with a periodic motion can be expanded in a Fourier series as

$$x_p(t) = \frac{1}{2}a_0 + \sum_{n=1}^{N}(a_n \cos n\omega t + b_n \sin n\omega t) \qquad (7.2.1)$$

where ω is the basic or fundamental radian frequency of the motion. In theory, the series can have an infinite number of terms; that is, $N \to \infty$. However, in practice, N is limited to a finite number.

In some cases, by examining the time series one can infer the presence or absence of harmonics. As discussed in Section 3.4.1, a time series $x(t)$ of a periodic motion with period T is said to possess a reflection or inversion symmetry when

$$x(t) = -x(t + \frac{1}{2}T) \qquad (7.2.2)$$

This condition is satisfied when the signal contains the basic frequency $2\pi/T$ and its odd harmonics.

It is also helpful to examine the envelope of a time series. When the envelope is flat, we have a periodic motion. Otherwise, we have a motion that is periodic, quasiperiodic, or chaotic. An example of a periodic oscillation, whose envelope is not flat, is the mixed–mode oscillation discussed in Section 5.5.2. It is not easy to detect the presence of more than one basic period by examining the time series. So for motions that contain more than one basic period, examining the time series may not be sufficient to characterize the motion. It is difficult to distinguish between three– or higher–period quasiperiodic motions and chaotic motions by examining the time history.

As illustrated in Figure 7.2.1, time histories are used generally along with other tools such as frequency spectra and state–space plots. In Figures 7.2.1 and 7.2.2, we present results obtained by numerically integrating

$$p_1' = -[\mu_1 p_1 + \nu_1 q_1 + \Lambda_1(p_2 q_1 - p_1 q_2)] \qquad (7.2.3)$$
$$q_1' = -[\mu_1 q_1 - \nu_1 p_1 + \Lambda_1(p_1 p_2 + q_1 q_2)] \qquad (7.2.4)$$
$$p_2' = -[\mu_2 p_2 + \nu_2 q_2 - 2\Lambda_2 p_1 q_1] \qquad (7.2.5)$$
$$q_2' = -\left[\mu_2 q_2 - \nu_2 p_2 + \Lambda_2(p_1^2 - q_1^2) - f_2\right] \qquad (7.2.6)$$

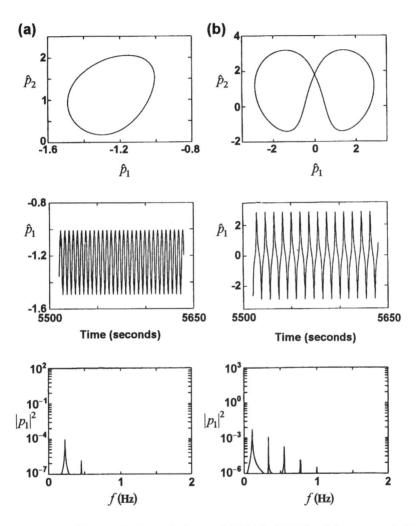

ʒure 7.2.1: Two periodic solutions of (7.2.3)–(7.2.7): (a) $\sigma_2 = -0.4570$
d (b) $\sigma_2 = -0.3800$.

where

$$\nu_1 = \frac{1}{2}(\sigma_1 + \sigma_2) \; ; \; \nu_2 = \sigma_2 \qquad (7.2.7)$$

for $\Lambda_1 = 294.732, \Lambda_2 = 2213.729$, $\mu_1 = 0.09$, $\mu_2 = 0.22$, $\sigma_1 = 1.131$, and $f_2 = 0.003936$. The different parameter values correspond to a structure treated by Nayfeh, Balachandran, Colbert, and Nayfeh (1989) and Balachandran (1990). The results in Figure 7.2.1a correspond to $\sigma_2 = -0.4570$, those in Figure 7.2.1b correspond to $\sigma_2 = -0.3800$, and those in Figure 7.2.2 correspond to $\sigma_2 = -0.2300$. The initial conditions are (0.00059, 0.00119, 0.00300, 0.00350) for the motions shown in Figure 7.2.1 and (0.00010, 0.00000, 0.00599, 0.00574) for the motions shown in Figure 7.2.2. (The integrations were carried out by using the IMSL subroutine DIVPRK from the IMSL MATH/LIBRARY [1989].) In each case, the data obtained during the first 5000 seconds of integration were discarded.

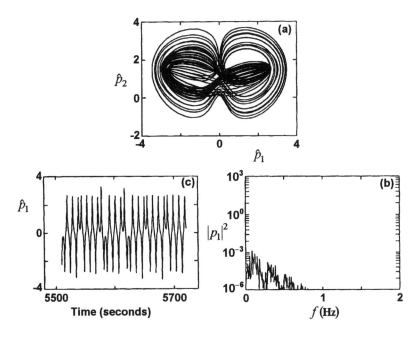

Figure 7.2.2: A chaotic solution of (7.2.3)–(7.2.7) for $\sigma_2 = -0.2300$.

The results in Figures 7.2.1a and 7.2.1b correspond to periodic oscillations, while those in Figure 7.2.2 correspond possibly to nonperiodic
oscillations. In each case, we show the time series for the evolution of
the variable $\hat{p}_1 = \sqrt{\Lambda_1 \Lambda_2} p_1$. Examining the time series in Figure 7.2.1a,
we infer that the envelope is flat and that there is a d.c. (i.e., constant)
offset. Since the form of the signal appears to be sinusoidal, we expect
the harmonics of the basic frequency if any to be small. Due to the
d.c. offset, the signal does not satisfy the inversion symmetry condition. The time series shown in Figure 7.2.1b has a flat envelope and no
d.c. offset. Because the signal is not sinusoidal in form, we expect this
signal to contain harmonics of the basic frequency. The time series in
Figure 7.2.2 is not periodic in the record length shown.

The time history of a two–period quasiperiodic motion is shown in
Figure 7.2.3a. This signal, which has the appearance of a periodically
modulated waveform, can be described by

$$x_{qp}(t) = (a_0 + a_m \cos \omega_m t) \cos(\omega_c t + \gamma_0 + \gamma_m \cos \omega_m t) \qquad (7.2.8)$$

where the two incommensurate frequencies are ω_m and ω_c. The
signal x_{qp} is an example of an **amplitude– and phase–modulated
waveform**. When the frequencies ω_m and ω_c are commensurate, the
corresponding motion is periodic, and the time history of this periodic
motion is similar to that shown in Figure 7.2.3a. The signal

$$x_{qp}(t) = \sum_{i=1}^{n} a_i \cos(\omega_i t + \beta_i) \qquad (7.2.9)$$

where the n frequencies ω_i are incommensurate is an example of a n–
period quasiperiodic signal.

A time trace of a chaotic motion, observed during one of the
experiments of Anderson, Balachandran, and Nayfeh (1992), is shown
in Figure 7.2.3b. This and other chaotic time histories cannot be
described by standard mathematical functions. We note that the
expressions for periodic and quasiperiodic motions consist of discrete
frequency components, which, in turn, correspond to discrete lines in
the corresponding spectra. However, the spectrum associated with a
chaotic motion has a continuous or broadband character. This is also
true of the spectra of transient and random motions.

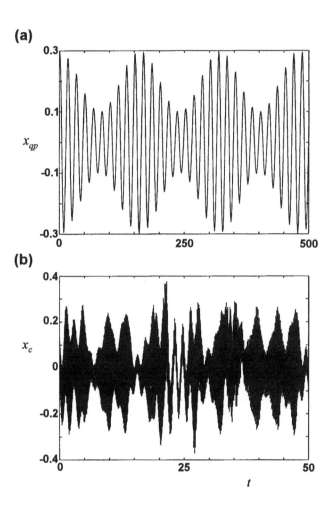

Figure 7.2.3: Time histories of (a) a two–period quasiperiodic motion and (b) a chaotic motion.

The signal

$$x_{ns}(t) = a\cos(\omega_0 t + \frac{1}{2}\alpha t^2 + \beta) \qquad (7.2.10)$$

where ω_0, a, and β are constants, is an example of a **deterministic nonstationary signal**. Here, the frequency is a linear function of time. In the associated spectrum, the location of the peak changes with time. The time histories associated with random motions cannot be described by simple mathematical expressions.

As discussed in Chapter 5, time histories can be very useful in characterization of the types of intermittent transitions to chaos, mixed–mode oscillations and alternating periodic–chaotic sequences, and crises terminating chaotic motions. For example, the time trace shown in Figure 5.4.1a for the Rayleigh–Bénard convection describes a periodic oscillation. The time trace shown in Figure 5.4.1b consists of long laminar stretches of oscillations that appear to be regular and closely resemble the time trace in Figure 5.4.1a, but this regular behavior is intermittently interrupted by chaotic outbreaks at irregular intervals. Consequently, we infer that this intermittent transition to chaos is the result of a cyclic–fold bifurcation, and therefore the intermittency mechanism is of type I.

A time history indicating another type of intermittent transition to chaos in the Rayleigh–Bénard convection is shown in Figure 5.4.5. In this case, the laminar stretches consist of a fundamental harmonic and its subharmonic of order one–half. In each laminar stretch, the amplitude of the fundamental harmonic decreases with each successive oscillation, whereas the amplitude of its subharmonic increases. When the latter exceeds a threshold, a chaotic outbreak occurs. Hence, this intermittent transition to chaos is associated with a period–doubling bifurcation, and consequently the intermittency mechanism is of type III.

A time history indicating a third type of intermittent transition to chaos is shown in Figure 5.4.11. In this case, the laminar stretches consist of quasiperiodic oscillations. Hence, this intermittent transition to chaos is associated with a Hopf bifurcation, and consequently the intermittency mechanism is of type II. A time history indicating an on–off intermittent transition to chaos is shown in Figure 5.4.7. In this case, the laminar stretches correspond to constant motions rather than periodic or quasiperiodic motions.

The time history can also be useful in identifying mixed–mode oscillations and alternating periodic–chaotic responses. Comparing the time trace in part c with those in parts a and b of Figure 5.5.8, one finds a clear disparity between small–amplitude and large–amplitude oscillations, and hence the trace in part c is an example of a mixed–mode oscillation. As discussed in Section 5.5.2, such a mixed–mode oscillation is represented as P_L^S, where L and S are the numbers of large– and small–amplitude oscillations per cycle. Using this notation, one can identify the mixed–mode oscillations in parts b–d in Figure 5.5.11 as P_1^0, P_1^1, and P_1^2. Again, by using the time history, one can identify the chaotic state in Figure 5.5.11e as a motion consisting of a random mixing of P_1^2 and P_1^3 motions.

Further, the time histories can also be useful for characterizing transient motions resulting from crises. The time histories shown in Figures 5.6.3 and 5.6.13 describe transient chaos associated with boundary crises. The time series shown in Figures 5.6.6 and 5.6.9 describe transient chaos resulting from the termination of the period–three and period–five windows of the quadratic map, and the time histories shown in Figure 5.6.18 describe transient chaos resulting from an attractor merging crises.

7.3 STATE SPACE

In this section, we discuss how the state space can be used to characterize different motions either determined as solutions of known systems of equations or observed in experiments. The reader is probably familiar with a state space defined in terms of rectangular coordinates. But in studies of dynamical systems, angular coordinates are also used in defining a state space. When these variables are used, we visualize motions on objects such as cylindrical spaces and tori. Examples of these objects are provided in Figure 1.2.2. Local areas of these objects have features of Euclidean spaces. In general, we study a state–space plot along with its associated time series and frequency spectrum to characterize a motion. State–space plots can be used with ease to visualize motions that evolve in a three–dimensional or lower space. For motions that evolve in higher dimensions, one has to examine their projections onto a two– or three–dimensional space. Often, crossings/twistings are seen in these projections. It is difficult to visualize motions that occur in a k–dimensional space, where k is greater than three. Other tools will be necessary to characterize such motions.

The space defined by the independent coordinates required to describe a motion is called a **state space**, and the independent coordinates are called **state variables**. For a given system of equations, the coordinates of the state space are well defined. The motion of a discrete system, such as a pendulum or a spring–mass system, is governed by a finite–dimensional system of ordinary–differential equations. The associated state space has a finite dimension. The motion of a continuous or distributed–parameter system, such as a beam or a plate or a flow, is governed by a finite system of partial–differential equations. There is an associated finite–dimensional function space. (The elements of a function space are functions.) Because the system of partial–differential equations translates to an infinite–dimensional system of ordinary–differential equations, there is an associated infinite–dimensional state space. In the presence of dissipation, attractors associated with the motions of many distributed–parameter systems exist in low–dimensional spaces. Consequently, to make the problem tractable, in practice, one often models them by a finite number of ordinary–differential equations. This choice, which depends on the problem at hand, is made difficult by nonlinear resonances and interactions (Nayfeh and Balachandran, 1995).

An n–dimensional autonomous system describes a continuous time evolution (trajectory) in the space defined by n state variables. The system of equations (7.2.3)–(7.2.6) is an example of an autonomous system. These equations describe the nonlinear evolution of the variables p_1, q_1, p_2, and q_2 in a four–dimensional space. In this case, it should be apparent that we have four state variables and a four–dimensional state space $\mathcal{R}^4$.

An n–dimensional nonautonomous system describes a continuous time evolution in the space defined by n state variables and time t. In the context of

$$\ddot{v}_1 + \omega_1^2 v_1 = f_1(v_1, \dot{v}_1, v_2, \dot{v}_2) \tag{7.3.1}$$

$$\ddot{v}_2 + \omega_2^2 v_2 = f_2(v_1, \dot{v}_1, v_2, \dot{v}_2) + g \cos \Omega t \qquad (7.3.2)$$

the two–dimensional space constructed with the coordinates v_1 and v_2 is called a **configuration space** because these coordinates define the configuration of the system. We can also rewrite (7.3.1) and (7.3.2) as a system of four first–order ordinary–differential equations using the variables v_1, $\dot{v}_1$, v_2, and $\dot{v}_2$. There is an associated four–dimensional Cartesian space. As discussed in Section 1.2, this four–dimensional nonautonomous system of equations can be made autonomous by introducing the variable $\theta = \Omega t$ mod 2π. Then the coordinates of the associated space are v_1, $\dot{v}_1$, v_2, $\dot{v}_2$, and θ. This space, which is represented by $\mathcal{R}^4 \times S^1$, is an example of a cylindrical state space. Cylindrical spaces are useful for studying motions composed of oscillations at one or more basic frequencies.

Now, we address how different motions can be characterized in the state space. The fixed point (discussed in detail in Chapter 2) of a system of differential equations corresponds to a point in the state space. In the context of an experimental situation, a static equilibrium position of a structure would correspond to a point in the state space. As an example, we consider the static buckling of a perfect rod. Prior to buckling, the rod remains straight and has only one equilibrium position that corresponds to a fixed point in the state space. After buckling, the rod can assume a buckled position about either side of the initially unbuckled position. The buckled states correspond to new fixed points in the state space.

As discussed in Chapter 3, a periodic solution of a given autonomous or nonautonomous system of differential equations corresponds to a closed trajectory in the state space. In Figures 7.2.1 and 7.2.2, we show projections of the four–dimensional motion described by equations (7.2.3)–(7.2.7) onto the two–dimensional space $\hat{p}_2 - \hat{p}_1$, where $\hat{p}_1 = \sqrt{\Lambda_1 \Lambda_2} p_1$ and $\hat{p}_2 = \Lambda_1 p_2$. If the state variables are periodic and contain just one frequency component, then their corresponding cross plot will be a straight line, a circle, or an ellipse. The presence of harmonics leads to distortions in the cross plot, as in Figures 7.2.1a and 7.2.1b. The crossing in Figure 7.2.1b is a consequence of a projection onto a two–dimensional space. Many crossings and loops can be seen in the state space of Figure 7.2.2. The corresponding motion appears to be nonperiodic. The broadband character, which is seen in the associated spectrum, is a signature of chaotic motions.

In Figures 7.3.1a and 7.3.1b, we show projections of a periodic attractor of the four–dimensional system (7.2.3)–(7.2.7) onto three–dimensional and two–dimensional spaces, respectively. The apparent crossings in these figures are a consequence of the projection. When the motion consists of one basic frequency, a single angular coordinate, say, θ_1, can be used to describe the corresponding motion, as is illustrated in Figure 7.3.1c. There is a one–to–one correspondence between the points on the limit cycle and the points in the interval running from 0 to 2π. For example, the point marked by a cross on the limit cycle corresponds to the point labeled θ on the line in Figure 7.3.1c.

Although a point on a periodic orbit can be located by using a single angular coordinate, at least a two–dimensional Euclidean space is required to visualize the geometry of a periodic orbit. The dimension of the Euclidean space should be large enough so that there is no loss

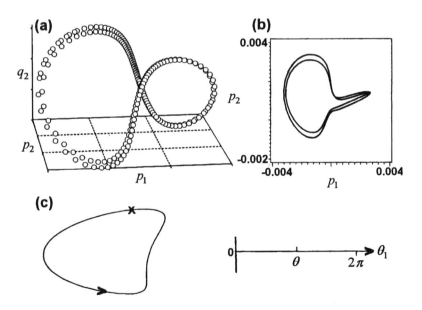

Figure 7.3.1: A periodic attractor of (7.2.3)–(7.2.7) for $\Lambda_1 = 294.732, \Lambda_2 = 2213.729$, $\mu_1 = 0.09$, $\mu_2 = 0.22$, $\sigma_1 = 1.131$, $\sigma_2 = -0.2540$, and $f_2 = 0.003936$.

of information. In Figure 7.3.1b, the crossings in the two–dimensional projection of the periodic orbit indicate that some information is lost. If there are no crossings when a periodic orbit is visualized in a two–dimensional Euclidean space, then this space is large enough to capture the periodic orbit. Otherwise, a higher–dimensional space is required. As discussed in Chapter 4, two angular coordinates can be used to locate a point on a two–period quasiperiodic orbit on a two–torus. To visualize the geometry of a two–torus, a three–dimensional or higher Euclidean space is necessary. The above statements regarding the dimension d of a Euclidean space $\mathcal{R}^d$ required to capture the geometry of the object follow from the **theory of embedding** in differential topology (Whitney, 1936; Guillemin and Pollack, 1974). Considerations for the embedding theory are quite independent of the dynamics of the considered system.

In Figure 7.3.2, some of the experimental observations reported by Nayfeh, Balachandran, Colbert, and Nayfeh (1989) are shown. In this case, two strain–gauge signals labeled as SGV and SGH serve as "independent" coordinates for the construction of a state space. The state space thus constructed is sometimes referred to as a **pseudo–phase plane** because the coordinates of a phase plane are usually displacement and velocity. Since this motion occurs in a space whose dimension is greater than two, the plots in the pseudo–phase plane are projections onto a two–dimensional space. In Figure 7.3.2b, a crossing occurs due to the presence of a one–half subharmonic of the frequency f. The "eight–shaped" pattern is the result of this crossing. During the experiments, the associated pattern in the pseudo–phase plane of a periodic motion was observed to remain steady. The trajectory of a modulated motion is shown in Figure 7.3.2c. In this case, the pattern in the pseudo–phase plane did not remain steady, and an evolving "eight–shaped" pattern was observed in real time. The corresponding Fourier spectrum is indicative of a two–period quasiperiodic motion; this is further discussed in Section 7.5.

The state–space plot in Figure 7.2.2a possibly pertains to a chaotic motion. As discussed in Chapter 5, in dissipative continuous–time systems, such motions can only occur in three or higher dimensions. By studying trajectories in state space, it may be possible to find out if a motion is periodic or otherwise. It is however difficult to

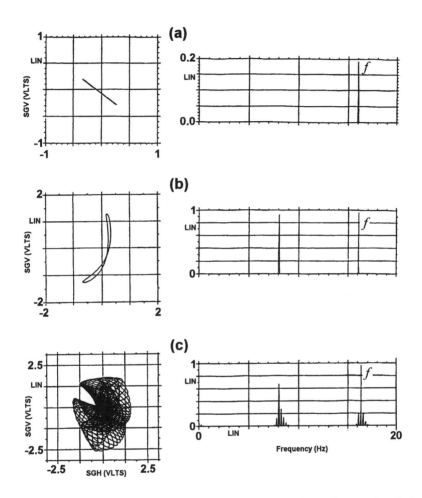

Figure 7.3.2: Cross plots of two strain–gauge signals and associated frequency spectra: (a) linear periodic motion, (b) nonlinear periodic motion, and (c) quasiperiodic motion.

distinguish between "long–period" motions and nonperiodic motions. The trajectories of a nonperiodic attractor do not conform to any simple geometrical form in the state space.

In most experimental situations, the "independent" coordinates required to construct a state space are not well defined, and establishing the "proper" state space is not an easy task. However, as discussed in the next section, certain systematic approaches are available for constructing such spaces.

7.4   PSEUDO–STATE SPACE

In the preceding section, we assumed that one can determine either theoretically or experimentally all of the variables $\mathbf{x}(t)$ that describe the behavior of the dynamical system of interest. However, in experiments, typically one observes only one or at best a few of the dynamical variables that govern the behavior of the desired system. Let us assume that we can only measure one component or, more generally, one scalar function $s_n = s(t) = s(t_0 + n\tau_s)$ of the state vector; that is,

$$s_n = s(t) = S[\mathbf{x}(t)] \qquad (7.4.1)$$

where t_0 is some initial time and τ_s is the sampling time of the instrument used in the experiment. The observable scalar may be a voltage from a strain gauge or an accelerometer attached to a structure, a temperature or pressure in a fluid, a voltage or current in a nonlinear circuit, or a voltage from an optical sensor. Then, the question arises whether one can use this scalar or univariate observation to construct a multivariate state space that describes the attractor, especially in the absence of any a priori knowledge of the dimension of the state space required to describe the motion of the system. In the context of topology, this is an example of an **embedding problem**. The issue is to find a one–to–one mapping between points on the (original) attractor in the full system state space and the attractor in the reconstructed state space. This mapping is called an **embedding**. Here, the mapping should also preserve information about the derivatives of the flow. In 1936, Whitney showed that a smooth $\mathcal{C}^2$ m–dimensional manifold may be embedded in $\mathcal{R}^{2m+1}$ space. This result about the existence of an embedding is quite independent of the dynamics of the system.

Work done over the last decade in the area of dynamical systems theory suggests that a measurement of a scalar signal would suffice to carry out an embedding. From the works of Packard, Crutchfield, Farmer, and Shaw (1980) and Takens (1981), which were motivated by experiments in fluid mechanics, and the many publications that have followed them, it is now known that a state space can be reconstructed from a scalar time signal such that the dynamics in the reconstructed state space is equivalent to the original dynamics. Considering a deterministic three–dimensional autonomous system, Packard et al. (1980) demonstrated the equivalence through numerical simulations. Takens (1981) considered a deterministic finite–dimensional autonomous system and proved the equivalence with mathematical rigor, assuming that an infinite amount of noise–free data was available in the analysis. As a consequence of the equivalence, an attractor in the reconstructed state space has the same invariants, such as Lyapunov exponents and dimension, as the original attractor. Hence, measures such as the Lyapunov exponents and dimension can be obtained for the motion in the reconstructed state space.

When measures, such as Lyapunov exponents, are preserved, the differential structure of the original attractor is also said to be preserved in the reconstruction (Sauer, Yorke, and Casdagli, 1991). For this reason, the associated embedding is called a **differentiable embedding**. When there is only a one–to–one correspondence between the vectors in the reconstructed state space and the vectors in the full state space, the associated embedding is called a **topological embedding**.

Different methods have been proposed for construction of the "independent" coordinates for the state space. According to the method proposed by Packard et al. (1980), the time derivatives of a signal can be used along with it to construct a state space. (An analog differentiator [e.g., Horowitz and Hill, 1980] or a digital computer may be used to obtain the time derivatives.) Thus, using the time series, one can approximate the derivatives of $s(t)$ by using finite differences and hence generate the variables needed to describe the behavior of the system.

For example,

$$\frac{ds}{dt}(t_0 + n\tau_s) \approx \frac{s[t_0 + n\tau_s] - s[t_0 + (n-1)\tau_s]}{\tau_s} \qquad (7.4.2)$$

$$\frac{d^2s}{dt^2}(t_0 + n\tau_s) \approx \frac{s[t_0 + (n+1)\tau_s] - 2s(t_0 + n\tau_s) + s[t_0 + (n-1)\tau_s]}{\tau_s^2}$$
$$(7.4.3)$$

Similarly, one can write down approximations to the higher derivatives. However, with finite τ_s and the presence of noise contamination, (7.4.2) is a crude approximation of the first derivative. Moreover, (7.4.3) represents an even poorer approximation of the second derivative. The quality of the approximation deteriorates even further for the higher derivatives.

Examining the finite–difference formulas for the derivatives, we note that, at each step of the differentiation, we are adding the new information already contained in the measurement at other time steps lagged or advanced by multiples of the sampling time τ_s. This observation led Packard et al. (1980), Ruelle (1989a), and Takens (1981) to conclude that one does not need the derivatives to form a coordinate system that describes the structure of orbits in phase space. Instead, one can use directly the time advanced variables $s(t + n\tau)$, where $n = 1, 2, \ldots, d$ and $\tau = k\tau_s$ is an appropriately chosen time delay, as discussed in Section 7.4.2. and define the so–called **delay–coordinate vectors**

$$\mathbf{y}_n = \{s(t_0 + n\tau_s) \ s(t_0 + n\tau_s + k\tau_s) \ \cdots \ s[t_0 + n\tau_s + k(d-1)\tau_s]\}^T$$

or

$$\mathbf{y}_n = [s_n \ s_{n+k} \ s_{n+2k} \ \cdots \ s_{n+kd-k}]^T \qquad (7.4.4)$$

The space constructed by using the vectors $\mathbf{y}_n$ is called the **reconstructed space**. According to a theory of Takens (1981) and Mané (1981), the geometric structure of the dynamics of the system from which the s_n were measured can be observed in the reconstructed d-dimensional Euclidean space if $d \geq 2d_a + 1$, where d_a is the dimension of the attractor of interest. Recently, Sauer et al. (1991) extended the work of Takens and showed that if d_a is the box–counting dimension, then choosing $d > 2d_a$ suffices for an embedding. The parameter τ is called **time delay**, the integer d is called the **embedding dimension**, the constructed coordinates are called **delayed coordinates**, and this method of constructing coordinates is called the **method of delays**.

Given the equations $\dot{\mathbf{x}} = \mathbf{F}(\mathbf{x})$ describing the (deterministic) system dynamics and given the state $\mathbf{x}(t)$ of the system at $t = t_0$, one can, in principle, integrate these equations forward in time by an amount $k\tau_s$ and obtain $\mathbf{x}(t_0 + k\tau_s)$. In other words, $\mathbf{x}(t_0 + k\tau_s)$ is a unique function of $\mathbf{x}(t_0)$. Hence,

$$\mathbf{x}(t_0 + k\tau_s) = \mathbf{L}[\mathbf{x}(t_0)] \tag{7.4.5}$$

which, upon substitution into (7.4.1), yields

$$s(t_0 + k\tau_s) = S[\mathbf{x}(t_0 + k\tau_s)] = S\{\mathbf{L}[\mathbf{x}(t_0)]\} \tag{7.4.6}$$

Therefore, the reconstructed vector $\mathbf{y}$ is related to the state vector $\mathbf{x}$ by a smooth nonlinear relationship of the form

$$\mathbf{y} = \mathbf{H}(\mathbf{x}) \tag{7.4.7}$$

Consequently, the vector $\mathbf{y}$ can serve as a coordinate basis for the system dynamics because any smooth coordinate transformation can serve the purpose.

Abarbanel, Brown, Sidorowich, and Tsimring (1993) integrated the Lorenz equations (5.8.25)–(5.8.27) for $\rho = 45.92$, $\beta = 4.0$, and $\sigma = 16.0$ using a fourth–order Runge–Kutta scheme with the time step $\tau_s = 0.01$. In Figure 7.4.1a, we show a three–dimensional plot of the obtained chaotic attractor. Using a time delay $\tau = 20\tau_s = 0.2$, Abarbanel et al. (1993) reconstructed the orbit in phase space by using the time series of the variable x and an embedding dimension d of three; that is,

$$\mathbf{y} = [x(t) \ \ x(t + \tau) \ \ x(t + 2\tau)]^T \tag{7.4.8}$$

The reconstructed attractor in the pseudo–state space is shown in Figure 7.4.1b. Comparing Figure 7.4.1b with Figure 7.4.1a, we see that, although distorted as expected because of the nonlinear transformation, the reconstructed geometric object is similar in appearance to the original geometric object obtained by using the state variables.

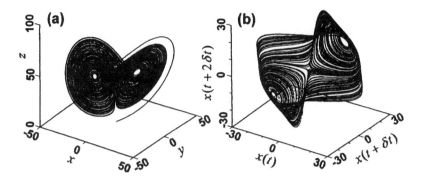

Figure 7.4.1: The Lorenz attractor in three–dimensional state space: (a) constructed from $x(t), y(t)$, and $z(t)$; and (b) constructed from $x(t), x(t+\tau)$, and $x(t+2\tau)$, where $\tau = 0.2$. Reprinted with permission from Abarbanel, Brown, Sidorowich, and Tsimring (1993).

In principle, if d is large enough and if the dimension of the attractor of interest is finite, then one can capture the dynamics by using the delay vectors $\mathbf{y}(t)$. The key idea in the reconstruction is that orbits generated by autonomous systems of equations do not intersect in the full state space. In fact, we have repeatedly noted that the intersections found in some of the plots in Chapters 3–5 are the result of projections onto a subset of the state space. Consequently, the dimension d of the pseudo–state space must be large enough so that the reconstructed orbit does not overlap with itself. When this happens, d is called an **embedding dimension.** In the next section, we discuss methods of choosing the embedding dimension.

In theory, the time delay can be arbitrary if one has an infinite amount of noise–free data. When this is so, it can be said that successive measurements contain new information for any nonzero time interval that separates them. But, in practice, one is limited by factors such as finite data, finite precision, and noise. Therefore, many considerations have to be taken into account in choosing the time delay. In principle, we require a time delay τ that will produce "independent" delayed coordinates. If τ is too small, the trajectories in the pseudo–state space of $\mathbf{y}(t)$ and $\mathbf{y}(t + \tau)$ stack up on the diagonal because the delayed coordinates are highly correlated. On the other hand, if τ is too large, an artificial decorrelation is introduced and the delayed coordinates become uncorrelated. When τ is chosen to be equal to one of the basic periods of the system, one constructs a Poincaré section in the space of the delayed coordinates. So, if τ is close to any of the basic periods of the system, the corresponding periodic component will not be well represented in the space of the delayed coordinates. In Section 7.4.2, we discuss methods of choosing τ.

Assuming that a time delay has been chosen, one can proceed with the construction of the space of delayed coordinates. The issue of state space reconstruction in the presence of noise is addressed in detail by Casdagli, Eubank, Farmer, and Gibson (1991) and Kostelich and Schreiber (1993), among others.

7.4.1 Choosing the Embedding Dimension

The objective of the reconstruction is to find a Euclidean space $\mathcal{R}^d$ that is large enough so that the set of points of dimension d_a, which describe the attractor, can be unfolded without ambiguity. In other words, if two points of the set lie close to each other in some dimension d, they do so because of the property of the attractor rather than because of the small value of d in which the attractor is being examined. This means that the value of d should be large enough so that the asymptotic state of the motion can be captured (embedded) in this d–dimensional space. Based on the general existence theorem for embeddings in Euclidean space, given by Whitney (1936), that a smooth $\mathcal{C}^2$ m–dimensional manifold may be embedded in $\mathcal{R}^{2m+1}$, in theory it is sufficient that $d \geq 2d_a + 1$, where d_a is the dimension associated with the observed motion. Sauer, Yorke, and Casdagli (1991) show that $d > 2d_a$ is sufficient when a delay coordinate embedding is used. For some cases, it has been found that $d \geq d_a$ is sufficient (e.g., Eckmann and Ruelle, 1985; Abarbanel, Brown, and Kadtke, 1990; Buzug, Reimers, and Pfister, 1990). In practice, d_a is not known a priori in most situations. Here, we discuss three approaches for estimating d.

Saturation of System Invariants

The basic idea underlying the saturation of the attractor of system in–

variants is that, if the attractor is unfolded by using a large enough embedding dimension d, then the invariant properties of the attractor, such as the Lyapunov exponents and dimension, calculated from the reconstructed trajectory do not change if one increases d. In other words, there is a dimension d beyond which all invariant properties of the desired attractor saturate.

To carry out this approach, we pick a value for d and then calculate one of the invariants of the attractor. Then, we increase d by one, recalculate the invariant, and compare the result with the preceding value. If the difference is within a specified tolerance, we conclude that d is the appropriate value. Otherwise, we repeat the process until there is relatively little change in the calculated invariant. Methods for calculating Lyapunov exponents and attractor dimensions are discussed in Sections 7.8 and 7.9. In this section we explain this approach by using the moments of the number density.

The number of points on the attractor within a sphere of radius r from the point $\mathbf{x}_m$ in the phase space is given by

$$P_m(r) = \frac{1}{N_0} \sum_{k=1}^{N_0} H(r - |\mathbf{x}_k - \mathbf{x}_m|) \qquad (7.4.9)$$

where N_0 is the total number of sampled points and H is the Heaviside step function defined by

$$H(u) = \begin{cases} 0 & \text{if } u < 0 \\ 1 & \text{if } u \geq 0 \end{cases} \qquad (7.4.10)$$

The average of powers of $P_m(r)$ over all points $\mathbf{x}_m$ yields the correlation function

$$C_q(r) = \frac{1}{N_0} \sum_{m=1}^{N_0} [P_m(r)]^{q-1} \qquad (7.4.11)$$

where q is an integer. Measures, such as (7.4.11), are quite well known in statistics (e.g., Renyi, 1970). Furthermore, since these measures are independent of initial conditions, they can be used to characterize attractors (Abarbanel, Brown, Sidorowich, and Tsimring, 1993).

To determine the embedding dimension d, we calculate (Grassberger and Procaccia, 1983a,b)

$$d_a = \frac{\log[C_q(r)]}{\log(r)} \tag{7.4.12}$$

as a function of d and determine when d_a becomes independent of d. (As discussed in Section 7.9, d_a is defined in the limit as r tends to zero. However, in practice, the finite number of data points and noise place a lower bound on the values of r that can be used to calculate d_a.) Therefore, one often plots $\log[C_q(r)]$ versus $\log(r)$ and estimates d_a from the slope. In Figure 7.4.2, we show variation of $\log[C_2(r)]$ with $\log(r)$ for data generated from the Lorenz equations as a function of d. Clearly, the slope of the plot and hence the value of d_a are independent

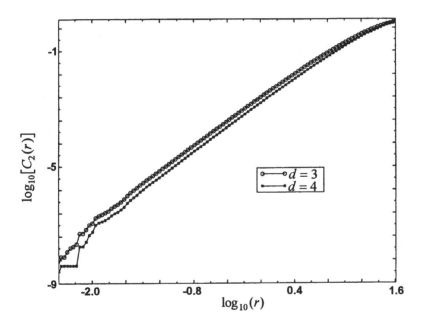

Figure 7.4.2: Variation of $\log[C_2(r)]$ with $\log(r)$ for $z_n = z(t) = z(t_0 + n\tau_s)$ data for the Lorenz attractor using the embedding dimension values $d = 3$ and 4. Reprinted with permission from Abarbanel (1995).

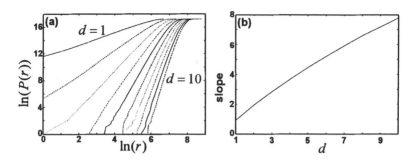

Figure 7.4.3: Illustration of the lack of saturation of the dimension for the dynamics associated with random noise.

of the value of d for $d \geq 3$. Hence, the embedding dimension according to the saturation criterion is 3, in contrast with the sufficient integer dimension of 5. According to the discussion in Section 7.9, the saturated slope is the attractor dimension d_a. In this case, $d_a = 2.06$, and hence the sufficient integer dimension is 5 because $d \geq 2d_a + 1$ according to the embedding theorem.

We note that for random noise, the correlation function continues to increase as the embedding dimension is increased; in other words, the dimension does not saturate. In Figure 7.4.3, we show variation of $\ln[P(r)]$ with $\ln(r)$ for random noise. (The definition for $P(r)$ is provided in Section 7.9.) Clearly, the slope of the plot does not saturate, and in fact it increases with increasing d.

Singular Value Analysis

Broomhead and King (1986) and Broomhead and Jones (1989) developed phase–space reconstructions by using the singular system approach, which is based on the Karhunen–Loeve theorem (Loeve, 1977). This technique was developed to deal with noise and errors that arise from finitely sampling data, and hence, it is ideally suited for experimental data. The key idea underlying this approach is that the mean–square distance between points on the reconstructed attractor should be maximized. The approach used by Landa and Rosenblum (1991), which is based on **Neymark algorithm**, is closely related to the singular system approach.

Using the sampled data s_j, choosing a large enough space dimension n, and assuming noise–contaminated data, we construct the **trajectory matrix** Y as

$$Y = \frac{1}{\sqrt{N}} \left[\mathbf{y}_1^T - \mathbf{y}_{av}^T \quad \mathbf{y}_2^T - \mathbf{y}_{av}^T \cdots \quad \mathbf{y}_N^T - \mathbf{y}_{av}^T \right]^T \qquad (7.4.13)$$

where $N = N_0 - (n-1)$, N_0 is the total number of sampled data points,

$$\mathbf{y}_i = \left[s_i \quad s_{i+1} \quad s_{i+2} \cdots \quad s_{i+n-1} \right]^T \qquad (7.4.14)$$

and

$$\mathbf{y}_{av} = \frac{1}{N} \sum_{i=1}^{N} \mathbf{y}_i \qquad (7.4.15)$$

Next, we calculate the dimension of the subspace of $\mathcal{R}^n$ that contains the reconstructed attractor. To accomplish this, we need to determine the number of linearly independent vectors that can be generated from the $\mathbf{y}_i - \mathbf{y}_{av}$ in the space $\mathcal{R}^n$. We note that

$$\sqrt{N} \mathbf{e}_i^T Y = \mathbf{y}_i^T - \mathbf{y}_{av}^T \qquad (7.4.16)$$

where the $\mathbf{e}_i$ are the standard basis vectors in $\mathcal{R}^N$; they correspond to columns of the $N \times N$ identity matrix. Moreover, we note that any vector $\mathbf{w}$ in $\mathcal{R}^N$ can be expressed in terms of the $\mathbf{e}_i$ as

$$\mathbf{w}^T = \sqrt{N} \sum_{i=1}^{N} w_i \mathbf{e}_i^T \qquad (7.4.17)$$

Hence,

$$\mathbf{w}^T Y = \sum_{i=1}^{N} w_i \left(\mathbf{y}_i^T - \mathbf{y}_{av}^T \right) \qquad (7.4.18)$$

In other words, vectors in $\mathcal{R}^N$ give rise to linear combinations of the $\mathbf{y}_i^T - \mathbf{y}_{av}^T$.

We consider a set of vectors $\mathbf{b}_i \in \mathcal{R}^N$ that give rise to linearly independent vectors $\mathbf{c}_i$ for $i = 1, 2, \ldots, n$ in $\mathcal{R}^n$, which we assume, without loss of generality, to be orthonormalized. Thus,

$$\mathbf{b}_i^T Y = \sigma_i \mathbf{c}_i^T \qquad (7.4.19)$$

where the σ_i are real constants that are chosen to fix the orthonormalization of the c_i. Taking the transpose of (7.4.19) yields

$$Y^T \mathbf{b}_j = \sigma_j \mathbf{c}_j \qquad (7.4.20)$$

Combining (7.4.19) and (7.4.20), we have

$$\mathbf{b}_i^T Y Y^T \mathbf{b}_j = \sigma_i \sigma_j \mathbf{c}_i^T \mathbf{c}_j = \sigma_i \sigma_j \delta_{ij} \qquad (7.4.21)$$

on account of the orthonormality of the c_i. Equation (7.4.21) can be solved by determining the eigenvectors of the $N \times N$ real, symmetric matrix YY^T; that is,

$$YY^T \mathbf{b}_i = \sigma_i^2 \mathbf{b}_i \qquad (7.4.22)$$

The σ_i^2 are the singular values of the so-called **structure matrix** YY^T; they are non-negative definite. It follows from (7.4.19) that, at most, n of the σ_i are nonzero. Therefore, the rank of the structure matrix is less than or equal to n. Next, we show that the associated eigenvalues are also the eigenvalues of the so-called covariance matrix Σ.

Multiplying (7.4.20) from the left with Y, we obtain

$$YY^T \mathbf{b}_j = \sigma_j Y \mathbf{c}_j \qquad (7.4.23)$$

which, upon using (7.4.22), becomes

$$Y \mathbf{c}_j = \sigma_j \mathbf{b}_j \qquad (7.4.24)$$

Multiplying (7.4.24) from the left with Y^T yields

$$Y^T Y \mathbf{c}_j = \sigma_j Y^T \mathbf{b}_j = \sigma_j^2 \mathbf{c}_j \qquad (7.4.25)$$

on account of (7.4.20). The matrix $\Sigma = Y^T Y$ is called the **covariance matrix**. It is an $n \times n$ real symmetric non-negative definite matrix. Hence, its eigenvalues σ_j^2 are non-negative definite. Multiplying (7.4.25) from the left with bfc_j^T and making use of the orthonormality of bfc_j leads to

$$(Y\mathbf{c}_j)^T (Y\mathbf{c}_j) = \sigma_j^2$$

In a perfect world, the number of nonzero eigenvalues σ_j^2 of $Y^T Y$ give the dimensionality of the subspace containing the embedded attractor and hence the embedding dimension d. Moreover, the vectors c_i corresponding to these nonzero eigenvalues span the embedding subspace. (This approach of deriving a set of basis vectors based on the covariance matrix is the essence of the Karhunen–Loeve method used in the areas of signal processing and pattern recoginition.) However, as explained by Broomhead and King (1986), if the data are noisy and the variance of the noise is σ_η^2, then every singular value σ_i will be shifted by σ_η. This shifting means that the noise causes all the singular values of the covariance matrix to be nonzero. Consequently, noise will dominate any eigenvector c_j whose singular value σ_j is comparable to σ_η, and hence such vectors must be discarded. This results in the reduction of the dimension of the embedding subspace from n to $d = n - m$, where m is the number of singular values comparable to σ_η. Broomhead and King (1986) note that m increases as $\tau_w = n\tau_s$ increases, where τ_s is the sampling time. To limit the size of m, they suggest selecting $\tau_w = 2\pi/\omega^*$, where ω^* is the **band–limiting frequency**. The time τ_w corresponds to the first zero crossing of the second derivative of the autocorrelation function

$$C(\tau) =< s(t)s(t + \tau) > \qquad (7.4.26)$$

To illustrate their approach, Broomhead and King (1986) integrated the Lorenz equations (5.8.25)–(5.8.27) for a particular set of parameters by using a Runge–Kutta scheme. The broadband character in the power spectrum of Figure 7.4.4 is indicative of the chaotic nature of $x(t)$.

Using $\tau_s = 0.009$, $\tau_w = 0.063$, and $n = 7$, they calculated the singular values of the covariance matrix $\Sigma = Y^T Y$ and their corresponding eigenvectors. The spectrum of the eigenvalues and the first three singular eigenvectors are shown in Figure 7.4.5. The spectrum of the eigenvalues has two distinct parts: one part, which can be associated with the noise floor, and a second part, which is associated with the deterministic component of the data. The noise floor can be distinguished by its magnitude and flatness; the magnitude corresponds to round–off errors in the computations. Thus, from the data in Figure 7.4.5, we conclude that the dynamics will be confined to a four–dimensional subspace of the embedding seven–dimensional space.

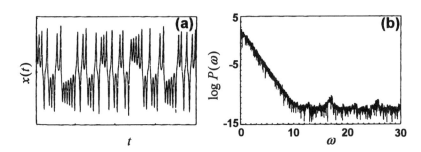

Figure 7.4.4: A time trace and associated power spectrum of the state variable $x(t)$ obtained by integrating the Lorenz equations (5.8.25)–(5.8.27) for $\sigma = 10$, $\beta = \frac{8}{3}$, and $\rho = 28$ by using a fourth–order Runge–Kutta scheme with a time step of 0.009 units. Reprinted with permission from Broomhead and King (1986).

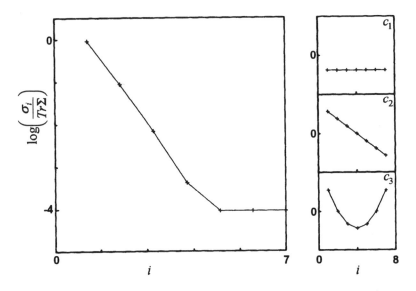

Figure 7.4.5: The first three singular values and their corresponding eigen-vectors calculated for the covariance matrix constructed from the data in Figure 7.4.4. Reprinted with permission from Broomhead and King (1986).

Broomhead and Jones (1989) carried out the above analysis locally by considering the covariance matrix over a neighborhood of a point specified by the reconstructed state vector $\mathbf{y}_n$. We let N_B be the number of nearest neighbors $\mathbf{y}_n^{(r)}$ of $\mathbf{y}_n$. Then, we define the covariance matrix COV_n as

$$\mathrm{COV}_n = \frac{1}{N_B} \sum_{r=1}^{N_B} \left[\mathbf{y}_n^{(r)} - \mathbf{y}_a \right] \left[\mathbf{y}_n^{(r)} - \mathbf{y}_a \right]^T \qquad (7.4.27)$$

where

$$\mathbf{y}_a = \frac{1}{N_B} \sum_{r=1}^{N_B} \mathbf{y}_n^{(r)} \qquad (7.4.28)$$

The matrix COV_n will have d_L eigenvalues arising from the variation of the slightly contaminated real signal about its mean and $n - d_L$ eigenvalues due to the noise. The eigenvectors associated with the d_L eigenvalues can be used to construct a **local pseudo–state space**. The **local embedding dimension** d_L is less than or equal to the **global embedding dimension** d.

False Nearest Neighbors

The basic idea underlying the false nearest neighbors approach is to find an embedding space of dimension d in which all false crossings of the orbit with itself that arise because of the projection onto a low–dimension space are eliminated. When d is not large enough, points that are far apart in the full or original state space are brought close together in the reconstruction space, resulting in **false nearest neighbors**. To determine these neighbors, one needs to examine if two states are neighbors because of the dynamics or because of the projection onto a low–dimension space. Thus, by determining neighbors in increasing embedding dimensions, one can eliminate false neighbors and hence establish the embedding dimension. Kennel, Brown, and Abarbanel (1992) proposed using a kd–tree search routine for finding nearest neighbors among N points. This routine takes $N\log N$ operations.

In dimension d and time advance $k\tau_s$, we assume that the reconstructed vector $\mathbf{y}_n$ given by (7.4.4) has the nearest neighbor specified by the vector

$$\hat{\mathbf{y}}_n = [\hat{s}_n \ \hat{s}_{n+k} \cdots \ \hat{s}_{n+kd-k}]^T \tag{7.4.29}$$

The Euclidean distance $R_n(d)$ between $\mathbf{y}_n$ and $\hat{\mathbf{y}}_n$ is given by

$$R_n^2(d) = \sum_{i=1}^{d} (\hat{s}_{n+ik-k} - s_{n+ik-k})^2 \tag{7.4.30}$$

This distance is assumed to be small. In dimension $d+1$, the distance between these two points becomes

$$R_n^2(d+1) = \sum_{i=1}^{d+1} (\hat{s}_{n+ik-k} - s_{n+ik-k})^2$$

or

$$R_n^2(d+1) = R_n^2(d) + (\hat{s}_{n+kd} - s_{n+kd})^2 \tag{7.4.31}$$

If $R_n(d+1)$ is large compared with $R_n(d)$, we can presume that this is so because $\mathbf{y}_n$ and $\hat{\mathbf{y}}_n$ are false neighbors in dimension d. In the calculations, we select a threshold R_T and decide whether the two neighbors $\mathbf{y}_n$ and $\hat{\mathbf{y}}_n$ are false nearest neighbors or not, depending on whether the following inequality is satisfied or not:

$$\frac{|\hat{s}_{n+kd} - s_{n+kd}|}{R_n(d)} > R_T \tag{7.4.32}$$

Abarbanel, Brown, Sidorowich, and Tsimring (1993) suggest values for R_T in the range $10 \leq R_T \leq 50$.

For the data in Figure 7.4.1, Abarbanel et al. (1993) used the aforementioned procedure to determine variation of the percentage of false nearest neighbors with the embedding dimension. The results are shown in Figure 7.4.6. Clearly, the number of false nearest neighbors drops to zero at $d = 3$, whereas the sufficient dimension from the embedding theorem is 5.

Again, the aforementioned procedure is good in a perfect world where an infinite amount of noise–free data are available. In the presence of noise, Abarbanel et al. (1993) suggest using the criterion

$$\frac{R_n(d+1)}{R_A} \geq 2 \tag{7.4.33}$$

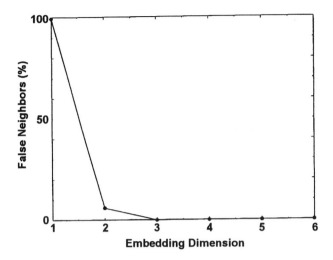

Figure 7.4.6: Variation of the percentage of false nearest neighbors with the embedding dimension calculated for the data in Figure 7.4.1. Reprinted with permission from Abarbanel et al. (1993).

for false nearest neighbors, where

$$R_A^2 = \frac{1}{N} \sum_{n=1}^{N} (s_n - s_{av})^2 \qquad (7.4.34)$$

with s_{av} being the average of the s_n defined by

$$s_{av} = \frac{1}{N} \sum_{n=1}^{N} s_n \qquad (7.4.35)$$

We note that R_A is a measure of the size of the attractor. In Figure 7.4.7, we show, after Abarbanel et al. (1993), the influence of adding uniform random numbers lying in the interval $[-L, L]$ to the x signal from the Lorenz system, shown in Figure 7.4.1. For this system, $R_A \approx 12$ and the different contamination levels L/R_A considered are indicated in the figure. Clearly, for values of L/R_A up to 0.5, a definite indication of a low–dimensional signal is discernible. When the contamination level is low, the residual percentage of false nearest neighbors provides an indication of the noise level. According to

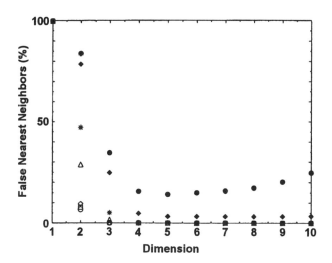

Figure 7.4.7: Influence of noise on variation of the percentage of false nearest neighbors with the embedding dimension calculated by for the data in Figure 7.4.1. The rms values of L/R_A corresponding to the open circles, open squares, open diamonds, open triangles, stars, solid diamonds, and solid circles are 0.0, 0.005, 0.01, 0.05, 0.1, 0.5, and 1.0, respectively. Reprinted with permission from Abarbanel et al. (1993).

Abarbanel et al. (1993), one may choose the embedding dimension to correspond to that at which false nearest neighbors drop to below, say, 1%.

Like the scheme discussed above, the schemes of Aleksic (1991) and Liebert, Pawelzik, and Schuster (1991) are based on topological considerations. Aleksic (1991) determines the minimum embedding dimension to be the dimension at which the dynamics in the reconstructed space is described by a continuous mapping (in a continuous mapping, images of close points are close). Liebert et al. (1991) provide a criterion for optimal embedding in terms of the distances between neighboring points on the reconstructed attractor for different embedding dimensions. Buzug and Pfister (1992) use what is called a **fill–factor method** to carry out a reconstruction. Roux, Simoyi, and Swinney (1983) have suggested that d can be systematically increased until the trajectories in the reconstructed state space no longer appear to cross or intersect. Application of this criterion becomes difficult as d_a becomes large. Typically, we use the first method described in this section, saturation of system invariants. Thus, we calculate the dimension for a particular motion in the space of the delayed coordinates for different values of d and systematically increase d until the dimension values for the different d saturate to a common value.

7.4.2 Choosing the Time Delay

For an infinite amount of noise–free data, one can arbitrarily choose the time delay τ according to the embedding theorem of Mané (1981) and Takens (1981). However, using experimental data, Roux, Simoyi, and Swinney (1983) showed that the quality of the reconstructed portraits depends on the value of τ. For small τ, $s(t)$ and $s(t+\tau)$ are very close to each other in numerical value, and hence they are not independent of each other. On the other hand, for large values of τ, $s(t)$ and $s(t+\tau)$ are completely independent of each other, and any connection between them in the case of chaotic attractors is random because of the butterfly effect. Consequently, we need a criterion for an intermediate choice that is large enough so that $s(t)$ and $s(t+\tau)$ are independent but not so large that $s(t)$ and $s(t+\tau)$ are completely independent in a statistical sense. Moreover, the time delay must be a multiple of the sampling time τ_s because the data are available at these times only and any interpolation may introduce errors, as in the case of estimating the derivatives. For delay coordinate reconstructions with nonuniformly sampled data, we refer the reader to Breedon and Packard (1992). There are many systematic approaches for choosing the time delay. In this section, we discuss three of these approaches.

Autocorrelation Function

The autocorrelation function of the sampled data set $s_i = s(t_0 + i\tau_s)$, where τ_s is the sampling time and $i = 1, 2, \cdots, N_0$, is given by

$$C(\tau) = \frac{\sum_{k=1}^{N_0} [s(t_0 + k\tau_s + \tau) - s_{av}][s(t_0 + k\tau_s) - s_{av}]}{\sum_{k=1}^{N_0} [s(t_0 + k\tau_s) - s_{av}]^2} \qquad (7.4.36)$$

where

$$s_{av} = \frac{1}{N_0} \sum_{k=1}^{N_0} s(t_0 + k\tau_s) \qquad (7.4.37)$$

Then, if the autocorrelation function $C(\tau)$ has a zero crossing at τ, the corresponding value of the time delay is chosen to be τ. Otherwise, the first local minimum of the autocorrelation function is used to specify τ.

We note that the autocorrelation function provides only a linear measure of the independence between the coordinates $s(t_0 + k\tau_s)$ and $s(t_0 + k\tau_s + \tau)$. To illustrate this, we assume that these coordinates are connected by the linear relation

$$s(t_0 + k\tau_s + \tau) - s_{av} = C(\tau)\,[s(t_0 + k\tau_s) - s_{av}] \qquad (7.4.38)$$

Then, we determine $C(\tau)$ by minimizing the average of the square of the error over the observations; that is, by minimizing the mean–square error

$$e = \sum_{k=1}^{N_0} \{[s(t_0 + k\tau_s + \tau) - s_{av} - C(\tau)]\,[s(t_0 + k\tau_s) - s_{av}]\}^2 \qquad (7.4.39)$$

Setting the derivative of e with respect to $C(\tau)$ equal to zero yields (7.4.36). Consequently, choosing τ to be the first zero of $C(\tau)$ would, on the average, make $s(t_0 + k\tau_s + \tau)$ and $s(t_0 + k\tau_s)$ linearly independent.

In Figure 7.4.8, we show variation of the autocorrelation function $C(\tau)$ with the time lag τ for the Lorenz data of Figure 7.4.1. Clearly, the first zero crossing occurs at $\tau \approx 30\tau_s$.

At the current time, the autocorrelation function is quite widely used to determine the time delay. In our limited experience, we have found the delay corresponding to the first zero crossing of the autocorrelation function or integer multiples of it to be adequate for the dimension calculations. We hasten to add that, when the autocorrelation function decays very slowly, there might be problems in choosing a delay (e.g., Gershenfeld, 1992). Moreover, as aforementioned, choosing τ to correspond to the first zero is the optimum linear choice from the point of view of the predictability in a least–squares sense of $s(t_0 + k\tau_s + \tau)$ from a knowledge of $s(t_0 + k\tau_s)$. Consequently, a number of researchers question its adequacy to determine the correlation due to the nonlinear process relating them (e.g., Abarbanel, 1995).

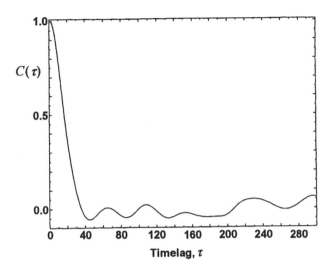

Figure 7.4.8: Variation of the autocorrelation function with the time lag calculated by for the data in Figure 7.4.1. Reprinted with permission from Abarbanel (1995).

Average Mutual Information

Fraser and Swinney (1986) used concepts of information theory and suggested that one should use the time delay corresponding to the first local minimum of the quantity called **mutual information**, which is a function of both linear and nonlinear dependencies between two variables. In the present context, the mutual information is a measure of the information (or predictability) that $s(t)$ can provide about $s(t + \tau)$. The two–dimensional approach of Fraser and Swinney was extended to higher dimensions by Fraser (1989a,b). He introduced a quantity called **redundancy** and suggested choosing a time delay corresponding to minimum redundancy.

Our present discussion follows along the lines of Abarbanel, Brown, Sidorowich, and Tsimring (1993). The idea underlying this approach is to identify how much information we can obtain about a measurement a_i drawn from a set A from a measurement b_j drawn from another set B. We assume that the probability of observing a_i out of the set A is $P_A(a_i)$, that the probability of observing b_j out of the set B is $P_B(b_j)$, and that the joint probability of observing a_i from the set A and b_j from the set B is $P_{AB}(a_i, b_j)$. Then, according to Shannon's idea of mutual information (Gallager, 1968), the amount of information one learns in bits about a measurement of a_i from a measurement of b_j is given by

$$I_{AB}(a_i, b_j) = \log_2 \left[\frac{P_{AB}(a_i, b_j)}{P_A(a_i) P_B(b_j)} \right] \qquad (7.4.40)$$

Then, the **average mutual information** I_{AB} between the sets of measurements A and B is given by

$$I_{AB} = \sum_{i,j} P_{AB}(a_i, b_j) \log_2 \left[\frac{P_{AB}(a_i, b_j)}{P_A(a_i) P_B(b_j)} \right] \qquad (7.4.41)$$

To apply the mutual information theory to the data set $s_k = s(t_0 + k\tau_s)$, we take the set A to be made up of the measurements $s(t_0 + i\tau_s)$ and the set B to be made up of the measurements $s(t_0 + i\tau_s + \tau)$. Then, (7.4.41) becomes

$$
\begin{aligned}
I(\tau) = \sum_i P\left[s(t_0 + i\tau_s), s(t_0 + i\tau_s + \tau) \right] \\
\times \log_2 \left\{ \frac{P\left[s(t_0 + i\tau_s), s(t_0 + i\tau_s + \tau) \right]}{P\left[s(t_0 + i\tau_s) \right] P\left[s(t_0 + i\tau_s + \tau) \right]} \right\}
\end{aligned}
\qquad (7.4.42)
$$

with $I(\tau) \geq 0$. When τ is large, the measurements $s(t_0 + i\tau_s + \tau)$ and $s(t_0 + i\tau_s)$ are completely independent for a chaotic signal and hence

$$P\left[s(t_0 + i\tau_s + \tau), s(t_0 + i\tau_s) \right] = P\left[s(t_0 + i\tau_s + \tau) \right] P\left[s(t_0 + i\tau_s) \right] \qquad (7.4.43)$$

Therefore, $I(\tau) \to 0$ as $\tau \to \infty$.

To evaluate $P[s(t_0 + i\tau_s)]$, we project the time series back onto the s axis. Then, the histogram formed by counting the frequency with which any of the values of s appears, when normalized, yields $P[s(t_0 + i\tau_s)]$. To evaluate $P[s(t_0 + i\tau_s + \tau)]$, we note that, if the time series is long and stationary, then

$$P[s(t_0 + i\tau_s + \tau)] = P[s(t_0 + i\tau_s)] \qquad (7.4.44)$$

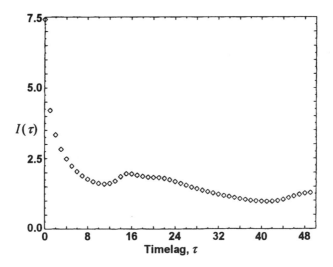

Figure 7.4.9: Variation of the average mutual information with the time lag calculated for the data in Figure 7.4.1. Reprinted with permission from Abarbanel (1995).

To evaluate the joint probability $P[s(t_0 + i\tau_s + \tau), s(t_0 + i\tau_s)]$, we form the two–dimensional histogram and count the number of times a box in the plane $s(t_0 + i\tau_s + \tau)$ versus $s(t_0 + i\tau_s)$ is occupied and then normalize this distribution.

Fraser and Swinney (1986) suggest selecting the value of τ that corresponds to the minimum of $I(\tau)$. In Figure 7.4.9, we show variation of the average mutual information $I(\tau)$ with the time lag τ for the Lorenz data of Figure 7.4.1. The first minimum of the mutual information occurs at $\tau = 10\tau_s$, whereas the first zero of the autocorrelation occurs at $\tau = 30\tau_s$. Fraser and Swinney (1986) found that the visual picture of the reconstructed attractor for low–dimensional systems, such as the Lorenz attractor, changes smoothly for variations around the minimum of $I(\tau)$.

There are situations in which $I(\tau)$ does not have a minimum. These situations include evolutions described by maps and evolutions of continuous–time systems determined with "large" integration steps or sampling times. Abarbanel, Brown, Sidorowich, and Tsimring (1993) suggest using $\tau = 1$ or 2 if the data comes from a map or choosing τ

so that $I(\tau) \approx \frac{1}{5}I(0)$.

Generalized Correlation Integral

Liebert and Schuster (1989) showed that the first minimum of the logarithm of the generalized correlation integral provides an easily calculable criterion for the proper choice of the time delay τ. Further, they examined the relationship between the correlation integral and the average mutual information. Their scheme requires a relatively smaller amount of data and seems to be easier to implement than the average mutual information scheme. As discussed later, the concept of the correlation integral is useful for certain types of dimension calculations.

The probability $P_m(r)$ of finding other d–dimensional states $\mathbf{x}_k$ within a sphere of radius r centered around the state $\mathbf{x}_m$ is given by (7.4.9). This probability is a function of τ because the reconstructed orbit depends on τ. Using (7.4.9), we define the so–called **generalized correlation integral** $C_1^d(r, \tau)$ as

$$C_1^d(r, \tau) = \lim_{q \to 1} \left[\frac{1}{M} \sum_{m=1}^{M} P_m^{q-1}(r, \tau) \right]^{1/(q-1)} \tag{7.4.45}$$

Using data from the Rössler system, the Mackey-Glass equation, and voltage measurements on barium sodium niobate crystals, Liebert and Schuster (1989) concluded that the first local minimum of the correlation integral with respect to τ together with the corresponding minimum of the mean–square deviations around the straight line defined by

$$\log C_1(r) = D_1 \ln(r) \text{ as } r \to 0 \tag{7.4.46}$$

provide a practical and easy way to calculate a criterion for the best choice of the time delay needed to reconstruct an orbit in state space from a scalar time series.

7.4.3   Two or More Measured Signals

In some experimental situations, it may be possible to obtain two or more different signals from the system in question. These signals may be used as coordinates for construction of a state space. Often in structural dynamics studies, signals from transducers mounted at different locations on a structure can be used to construct a state space. This construction may be explained as follows. Let $w(x, t)$ describe the displacement at time t and spatial location s on the structure. Further, let

$$w(x, t) = \sum_{i=1}^{n} \psi_i(x)v_i(t) \tag{7.4.47}$$

where n represents the number of modes used in the approximation, ψ_i represents the mode shape of the ith mode, and v_i represents the temporal function associated with the ith mode. We have implicitly assumed that the spatial and temporal informations are correlated. The n states v_j and their time derivatives represent the $2n$ state variables required to describe the motion of the structure. Let us suppose that a measurement of w taken at location $s = s_j$ yields the signal

$$w_j = \sum_{i=1}^{n} C_i\psi_i(x_j)v_i(t) \tag{7.4.48}$$

where the C_i are calibration constants that depend on the sensor. If the location $x = x_j$ corresponds to the node of a mode shape, there will not be any contribution from the corresponding mode to the signal w_j. The collection of n signals obtained from n different locations essentially represents a set of n coordinates obtained by applying a rotational transformation to the n coordinates v_i.

In structural dynamics, the idea of using different spatial locations to generate "independent" coordinates was first employed in the study of Nayfeh and Zavodney (1988). They used signals obtained from two strain gauges, mounted at different locations on a harmonically forced structure, to construct a state space. Following them, Balachandran (1990), Balachandran and Nayfeh (1991), and Anderson, Balachandran, and Nayfeh (1992) also employed this construction. In all of these studies, projections of motions onto a two–dimensional space were used to characterize the different motions. An example is provided in Figure 7.3.2. In the work of Guckenheimer and Buzyna (1983), simultaneous measurements in a fluid mechanics experiment were used to construct an embedding space and carry out dimension calculations.

Multiple measurements can often be exploited to reduce noise. Let us suppose that there are m measured signals w_m, where the measurement errors associated with the different signals are independent. Then, a linear combination of the m signals can be chosen so that the signal–to–noise ratio is high by using procedures similar to singular value analysis (e.g., Preisendorfer, 1988). Sauer, Yorke, and Casdagli (1991) provide a theoretical basis for state–space reconstruction by using m signals. The integer m should be greater than $2d_a$, where d_a is the box–counting or capacity dimension of the original attractor. Sauer et al. (1991) also discuss how a mixture of independent (measured) coordinates and delayed coordinates can be used to carry out a reconstruction.

7.5   FOURIER SPECTRA

The **Fourier** or **frequency spectra** help in distinguishing among periodic, quasiperiodic, and chaotic motions and are typically used to study stationary signals. The frequency spectrum can be either an **amplitude** or a **power spectrum**. In an amplitude spectrum, the Fourier amplitude is displayed at each frequency. On the other hand, in a power spectrum, the square of the Fourier amplitude per unit time is displayed at each frequency.

The **Fourier transform** of a signal $x(t)$ is defined as

$$X(f) = \int_{-\infty}^{+\infty} x(t)e^{-2i\pi ft}dt \qquad (7.5.1)$$

where f denotes the frequency and $X(f)$ is a complex quantity. In writing (7.5.1), we have assumed that $x(t)$ is integrable; that is,

$$\int_{-\infty}^{+\infty} |x(t)| \, dt < \infty \qquad (7.5.2)$$

In theory, the Fourier transform can be used to determine the frequency content of a signal $x(t)$ if it is known for $-\infty < t < +\infty$ and is integrable. However, a stationary signal that exists for all t is not integrable. Besides, in practice, $x(t)$ is known for only a finite length of time T_c, and hence the so–called **finite Fourier transform** is used. It is given by

$$X(f, T_c) = \int_0^{T_c} x(t) e^{-2i\pi f t} dt \qquad (7.5.3)$$

where, again, $X(f, T_c)$ is a complex quantity.

The finite Fourier transform provides a mechanism for representing a signal as the sum of simple sine and cosine functions. These functions correspond to discrete lines in the frequency spectrum. In the present context, the signal may be a time series obtained from either a physical experiment or a numerical integration of equations. Let us suppose that a time series is collected over a finite time T_c and consists of a discrete number of points obtained at a chosen sampling frequency. We can model these data as a sum of sine and cosine functions of time t with the period being an integer submultiple of T_c. The Fourier transform of these discrete finite–extent data is obtained by using the **discrete Fourier transform (DFT)**. A special case of the DFT is the **fast Fourier transform (FFT)**. It is essentially an efficient computational scheme that takes advantage of certain symmetry properties in the cosine and sine functions at their points of evaluation in order to achieve speed over conventional methods. If the number of data points is N, then the FFT requires $N\log_2 N$ operations, whereas conventional techniques require N^2 operations. The development of FFT is attributed to Cooley and Tukey (1965). There are many commercial software packages available for determining the FFT of a given signal (e.g., IMSL subroutines FFTRF, FFTCF etc; MATLAB, 1989).

A signal may consist of a periodic function whose period T does not exactly equal an integer submultiple of T_c. Further, let $T_c/T \simeq n$, where n is an integer. In such a case, the FFT consists of finite–amplitude peaks at n/T_c and adjacent lines of resolution; that is, there are dominant peaks surrounded by "small" peaks. These peaks at adjacent lines of resolution are called **sidelobes**. This process, where energy at a certain frequency leaks to adjacent frequencies, is referred to as **leakage** and is a consequence of finite–extent data. Of course, if the period of a signal is known, then the time length of the data can be chosen to coincide with an integer multiple of the period of the signal and there would not be any leakage. However, this is usually not the case in practice. To reduce the leakage one uses **windows**, which are weighting functions applied to the data in the time domain. We commonly use flat–top windows for periodic motions and Hanning windows for nonperiodic motions. Besides windows, one needs to take many things into considerations, such as the sampling frequency, noise, filtering, and resolution bias errors, in determining and interpreting the FFT (Oppenheim and Schafer, 1975; Harris, 1978; Bendat and Piersol, 1980; Horowitz and Hill, 1980). Here, we primarily limit ourselves to the interpretation of the FFT in the context of a studied motion.

We next define a power spectrum. **Autospectrum, autospectral density function**, and **power spectral density function** are other names for a power spectrum. Let us consider a stationary signal $x(t)$. The two–sided spectral density function S_{xx} is defined as

$$S_{xx}(f) = \int_{-\infty}^{+\infty} R_{xx}(\tau)e^{-2i\pi f\tau}d\tau \qquad (7.5.4)$$

where R_{xx} is the autocorrelation function. It is defined as

$$R_{xx}(\tau) = \lim_{T\to\infty} \frac{1}{T}\int_0^T x(t)x(t+\tau)dt \qquad (7.5.5)$$

where τ is called the **time delay**. In practice, T is a finite quantity. It follows from definition (7.5.5) that $R_{xx}(-\tau) = R_{xx}(\tau)$, and hence

$$S_{xx}(-f) = S_{xx}(f) \qquad (7.5.6)$$

and

$$S_{xx}(f) = 2\int_0^\infty R_{xx}(\tau)\cos(2\pi f\tau)d\tau \qquad (7.5.7)$$

Equation (7.5.7) implies that S_{xx} is a real–valued function. The inverse transformation yields

$$R_{xx}(\tau) = 2\int_0^\infty S_{xx}(f)\cos(2\pi f\tau)df \qquad (7.5.8)$$

It should be noted that the power spectrum is also defined as

$$S_{xx}(f) = \lim_{T\to\infty} \frac{1}{T} \mid \int_0^T x(t)e^{-2i\pi ft}dt \mid^2$$

The above definition and (7.5.4) are equivalent when R_{xx} decays rapidly in time. The equivalence is established by the Wiener–Khinchin relations (e.g., Bendat and Piersol, 1980, 1986).

It follows from (7.5.8) that

$$R_{xx}(0) = 2 \int_0^\infty S_{xx}(f)df \qquad (7.5.9)$$

which means that the value of the autocorrelation function for zero time delay corresponds to the total power in a spectrum. Also $R_{xx}(0)$ is called the **mean–square value** of a signal $x(t)$, and its square root yields the **root–mean–square (rms)** value.

In reality, one does not have negative frequencies (i.e., $f \geq 0$). So a single–sided spectral density function $G_{xx}(f)$ is defined such that

$$G_{xx}(f) = 2S_{xx}(f) \quad \text{for} \quad f \geq 0$$

$$\qquad (7.5.10)$$

$$G_{xx}(f) = 0 \qquad \text{for} \quad f < 0 \qquad (7.5.11)$$

It follows from (7.5.7) that

$$G_{xx}(f) = 4 \int_0^\infty R_{xx}(\tau) \cos(2\pi f\tau)d\tau \qquad (7.5.12)$$

The spectrum of a periodic motion consists of a single basic frequency. In Figures 7.2.1a and 7.2.1b, we show power spectra of periodic oscillations on a log scale. The power spectrum in Figure 7.2.2 corresponds possibly to a nonperiodic motion. In each case, the ordinate represents the power in units of decibels (the square of the Fourier amplitude, power P, expressed in decibels $= 10 \log_{10}P$) and has a range of 90 decibels. The log scale is useful to discern the presence of harmonics and frequency components with low power. We mention that no windows were used in computing the spectra of Figures 7.2.1 and 7.2.2; this is why the peaks are not sharp in these figures.

When the spectrum has n basic frequencies (i.e., n incommensurate frequencies), the corresponding motion is no longer periodic and is called an n–period quasiperiodic motion. The spectrum of a three–period quasiperiodic motion is shown in Figure 5.5.1. The spectrum of a chaotic motion has a continuous or broadband character. Examples are shown in Figures 5.2.1c, 5.2.3d, and 7.2.2b.

The spectra of random motions such as noise also have a continuous or broadband character, but chaotic motions can be distinguished from noise by using the character of the spectrum and tools, such as dimension calculations and Lyapunov exponents. In embedding–dimension calculations for (purely) random data, the measured dimension always equals the dimension of the embedding space. For the spectrum associated with a chaotic motion, the Fourier amplitudes are frequency dependent in the broadband region. These amplitudes scale as $1/f^\alpha$, where f is the frequency and α is a positive integer. For the spectrum associated with a random motion, the Fourier amplitudes in the broadband region are either frequency independent or frequency dependent and do not follow the $1/f^\alpha$ scaling law. In experimental situations, it is important to keep in mind that the noise generated by elements such as transistors and resistors in electronic circuits also has a $1/f$ spectrum. So, it is advisable to use the Fourier spectrum along with some of the other tools to determine the character of a motion.

The amplitude spectra shown in Figures 7.3.2 and 7.5.1 were obtained by using a signal analyzer during the experiments of Balachandran (1990). In each case, the signal was obtained from a strain gauge mounted on a harmonically excited structure. For each spectrum, a flat top window and 1,280 lines of resolution in a 20 Hz baseband were used. The spectra in Figures 7.3.2a and 7.3.2b correspond to linear and nonlinear periodic motions, respectively. In Figure 7.3.2c, the spectrum of a quasiperiodic motion is displayed. The responses of the structure corresponding to Figures 7.3.2b, 7.3.2c, and 7.5.1 are a consequence of nonlinear interactions.

In Figure 7.5.1a, the spectrum consists of discrete peaks (note that each peak has a width equal to the resolution frequency) at the frequencies f and $2f$. The corresponding signal $x_p(t)$ can be approximated by

$$x_p(t) \simeq \sum_{j=1}^{2} a_j \cos(2j\pi ft + \beta_j) \qquad (7.5.13)$$

where the a_j and β_j are real constants. The approximation sign is

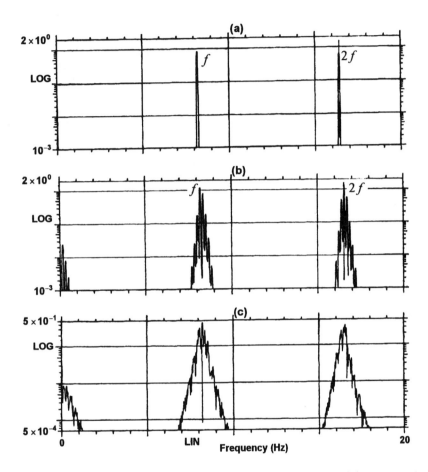

igure 7.5.1: Frequency spectra: (a) periodic motion, (b) two–period uasiperiodic motion, and (c) chaotically modulated motion.

used to reflect the fact that the actual signal contains noise besides the harmonic components.

As discussed in Section 3.4.1, Fourier spectra can also be used to determine if a particular periodic motion possesses an inversion symmetry. A signal does not satisfy this symmetry property in the presence of even harmonics and/or a d.c. (zero frequency) component. The limit cycle shown in Figure 3.4.6a possesses an inversion symmetry, and its power spectrum consists of odd harmonics. On the other hand, the limit cycle shown in Figure 3.4.6b is asymmetric, and its power spectrum consists of odd as well as even harmonics, including a d.c. component. In this regard it is worth noting that, in experimental situations, an improper calibration of a sensor like the strain gauge can also produce a d.c. component in the spectrum.

Although a Fourier spectrum can indicate whether the signal in question is periodic or aperiodic, it may not be able to reveal the source of the different frequency components in the spectrum. To illustrate this point, let us consider a two–degree–of–freedom system that is modeled by the following system of equations:

$$\ddot{v}_1 + \omega_1^2 v_1 + \mu_1 \dot{v}_1 + \delta_1 v_1^2 + \delta_2 v_1 v_2 = 0 \qquad (7.5.14)$$

$$\ddot{v}_2 + \omega_2^2 v_2 + \mu_2 \dot{v}_2 + \alpha_1 v_1^2 + \alpha_2 v_1 v_2 = F \cos(\Omega t) \qquad (7.5.15)$$

The v_i are the modal coordinates (amplitudes), $\omega_2 \simeq 2\omega_1$, and $\Omega \simeq \omega_2$. We assume that the observable w of the system has contributions from both v_1 and v_2 and is given by

$$w = c_1 v_1 + c_2 v_2$$

where the c_i are constants that depend on the measurement location.

We numerically integrated the quadratically coupled differential equations (7.5.14) and (7.5.15) for the following parameter values: $\omega_1 = 3.14$, $\omega_2 = 6.28$, $\mu_1 = 0.1$, $\mu_2 = 0.2$, $\delta_i = \alpha_i = 4.0$, $F = 2.0$, and $\Omega = 6.28$. Further, we chose $w = 0.4v_1 + 0.2v_2$. The initial condition for the integration is $(v_1, \dot{v}_1, v_2, \dot{v}_2) = (1.0, 0.2, 0.0, 0.0)$. The power spectra of v_1, v_2, and w, after the motions had settled down, are shown in Figure 7.5.2. The peaks in the figure are not sharp because we did not use any windows. In all of the three spectra, we observe peaks at $\Omega/4\pi$,

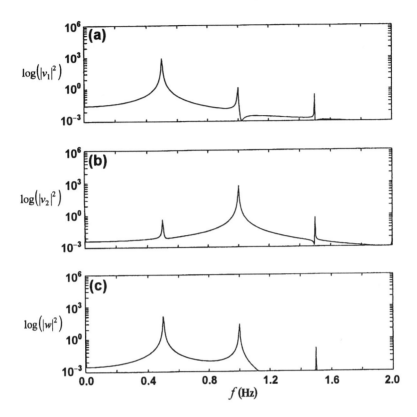

Figure 7.5.2: The power spectra of the response of two nonlinearly coupled oscillators with a two–to–one internal resonance: (a) first mode, (b) second mode, and (c) a linear combination of both modes.

$\Omega/2\pi$, and $3\Omega/4\pi$. In the power spectrum of v_1, the peaks at $\Omega/2\pi$ and $3\Omega/4\pi$ are due to the nonlinearities, and in the power spectrum of v_2, the peaks at $\Omega/4\pi$ and $3\Omega/4\pi$ are due to the nonlinearities. The motions in both modes (i.e., v_1 and v_2) contribute to each of the peaks in the spectrum of w. If one did not have any a priori knowledge of the system, it would be difficult to discern the sources of the different peaks in the w signal. In any case, even with a priori knowledge of the system, some analysis is necessary to determine the contributions of the two modes from the spectrum of w.

In the power spectra of v_1 and v_2, the magnitudes of the peaks due to the nonlinearities depend on the strength of the nonlinearities. If the nonlinearities are weak, in the spectrum of the w signal, the peak at $\Omega/4\pi$ can be assumed to be due to the first mode (v_1) and the peak at $\Omega/2\pi$ can be assumed to be due to the second mode (v_2). For weakly nonlinear motions, it may be assumed that the frequency component close to the natural frequency of a particular mode is due to the participation of that mode in the forced response.

Next, we consider the spectrum of an amplitude– and phase–modulated signal that is given by

$$x(t) = a(t)\cos[\omega t + \beta(t)] \qquad (7.5.16)$$

where

$$a(t) = a_o\left[1 + \alpha\cos(\omega_m t + \theta)\right] \quad \text{and} \quad \beta(t) = \beta_o\cos(\omega_m t) \qquad (7.5.17)$$

The relative phase between the amplitude and phase modulations is expressed by θ. Using properties of the Bessel functions, one can show that

$$
\begin{aligned}
x(t) = a_o &\sum_{n=-\infty}^{\infty} J_n(\beta_o)\cos\left[\frac{1}{2}n\pi + (\omega - n\omega_m)t\right] \\
&+ \frac{1}{2}\alpha a_o\sum_{n=-\infty}^{\infty} J_n(\beta_o)\left(\cos\left\{\frac{1}{2}n\pi + \theta + [\omega - (n-1)\omega_m]t\right\}\right. \\
&\left. + \cos\left\{\frac{1}{2}n\pi - \theta + [\omega - (n+1)\omega_m]t\right\}\right)
\end{aligned}
\qquad (7.5.18)
$$

where J_n is Bessel's function of order n of the first kind. When the signal is either purely amplitude modulated (i.e., $\beta_o = 0$ in the above equation) or purely phase modulated (i.e., $\alpha = 0$ in the above equation), the magnitudes of the peaks at the frequencies $\omega - n\omega_m$ and $\omega + n\omega_m$ would be the same. This would imply a symmetric sideband structure about ω. In the presence of amplitude and phase modulations, the sideband structure about ω is asymmetric.

The spectrum in Figure 7.5.1b contains sidebands, uniformly spaced δf apart, around the frequencies f and $2f$. This spectrum corresponds to a two–period quasiperiodic motion that consists of the incommensurate frequencies f and δf. Further, $\delta f \ll f$, and the ratio of the two frequencies is an irrational number (it needs to be noted that the determination of the frequency ratio is limited by finite precision). The corresponding signal $x_{qp}(t)$ can be approximated by

$$x_{qp}(t) \simeq \sum_{j=0}^{2} a_j \cos(2j\pi f t + \beta_j)$$

where in this case the real quantities a_j and β_j vary periodically at the frequency δf. The motion, whose spectrum is shown in Figure 7.5.1b, is composed of oscillations that occur on two different time scales: One is a fast–time scale and the other is a slow–time scale. Moreover, the motion is composed of two periodically modulated sinusoids. The frequencies f and $2f$ can be called the **carrier frequencies**, while the frequency δf can be called the **modulation frequency**. There may be cases where the ratio $f/\delta f$ is a rational number. In these cases, the corresponding motions are periodic and are called **phase–locked motions**. The spectra of such phase–locked motions resemble the spectrum in Figure 7.5.1b. One should note that "sufficient" frequency resolution is necessary to discern the sideband structure. In addition, one should ensure that the sidebands are well above the sidelobes produced due to leakage before characterizing the signal as quasiperiodic.

There could also be other types of quasiperiodic motions for which all the incommensurate frequencies occur on the fast–time scale. An example of such a spectrum, reported by Nayfeh, Nayfeh, and Mook (1994), is shown in Figure 7.5.3. The corresponding signal was obtained from a strain gauge mounted on a structure. The spectrum consists of the incommensurate frequencies f_1 and f_2. The peaks occur at $| m_1 f_1 + m_2 f_2 |$, where the m_j are integers. If the frequencies f_1 and f_2 are commensurate and their ratio f_1/f_2 is l/k, where l and k are positive integers that do not have a common factor, then the peaks in the spectrum will be harmonics of the frequency $f_b = f_1/l = f_2/k$. In general, a spectrum of a k–period quasiperiodic motion can have peaks at $| m_1 f_1 + m_2 f_2 + \cdots + m_k f_k |$, where the m_j are integers and the f_j are the incommensurate frequencies. An example of the spectrum of a three–period quasiperiodic motion, reported by Gollub and Benson (1980) in a Rayleigh–Bénard convection experiment, is shown in Figure

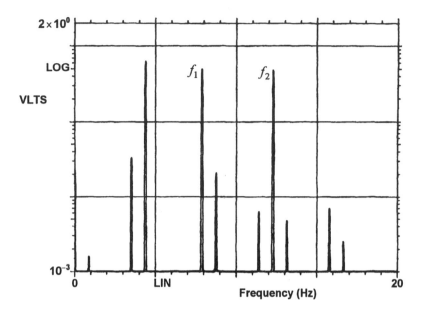

Figure 7.5.3: A power spectrum of a two–period quasiperiodic motion in which the ratio of the two basic frequencies f_1/f_2 is not small.

5.5.1. The spectrum has peaks at $\mid m_1 f_1 + m_2 f_2 + m_3 f_3 \mid$. We note that, for k–period quasiperiodic motions, where k is greater than two, it may not be possible to discern all the k independent frequencies in a physical or numerical experiment.

Examining the spectrum shown in Figure 7.5.1c, we note a broad-band character around each of the frequencies f and $2f$. The broad-band feature (continuous character) is a characteristic of chaotic motions. The corresponding signal has the form

$$x_{qp}(t) \simeq \sum_{j=0}^{2} a_j \cos(2j\pi ft + \beta_j)$$

where the a_j and β_j vary in a chaotic manner. They modulate the sinusoids at the carrier frequencies f and $2f$. The chaotic modulation leads to a continuous character in a narrow bandwidth about each carrier frequency. Further, this motion is a culmination of a series of changes that occur on the slow–time scale. The reader should note that this spectrum is in marked contrast with other "typical" spectra of chaotic motions, like the spectra in Figures 5.2.1c and 7.2.2 (commonly shown in other books and the literature), where one observes a broadband character over the whole frequency bandwidth of interest. In most of these cases, the chaotic motions occur through a series of changes on the fast–time scale. One can argue that the spectrum in Figure 7.5.1c either corresponds to an n–period quasiperiodic motion, where $n \geq 2$, or appears to have a broadband character because of insufficient frequency resolution. The resolution can be increased by examining the spectrum in a zoom span around a carrier frequency. However, to resolve the issue, we have to use tools such as dimension calculations and Lyapunov exponents. It is important to bear in mind that a Fourier spectrum can indicate that a certain motion is aperiodic, but it may not be able to reveal the source of this aperiodicity (i.e., if it is due to noise or to the dynamics of the system in question).

We cite the books of Bendat and Piersol (1980, 1986) as references for discussions on the spectrum. Chapter III of the book by Bergé, Pomeau, and Vidal (1984) contains a detailed discussion of the use of Fourier spectra for characterizing nonlinear motions. The Fourier spectrum serves as an important diagnostic tool for detecting faults in rotating machinery, where one commonly encounters modulated motions of one sort or another (Lyon, 1987).

The following remarks are also worth noting. Fourier analysis is not well suited for signals with transient events that occur over a short period of time because it is not localized in time. This problem can partly be overcome by conducting Fourier analyses in different time windows. The location of the time window adds a time dimension to the overall analysis. For a signal with short–lived transient events, it is desirable to use functions, such as wavelets, that are localized in time and frequency to represent the signal rather than sine and cosine functions that extend over all time. Applications of wavelets in diverse fields can be found in the volume edited by Combes, Grossman, and Tchamitchian (1990).

7.6 POINCARÉ SECTIONS AND MAPS

In Section 3.3, we introduced the concepts of Poincaré sections and maps in the context of stability of periodic solutions. Here, we primarily discuss how Poincaré sections can be used to distinguish among periodic, quasiperiodic, and chaotic motions. By and large, the use of first Poincaré sections is limited to systems whose asymptotic behavior is restricted to an n–dimensional space where typically $n \leq 3$. For $n > 3$, one usually uses two– and three–dimensional projections of the Poincaré section to ascertain whether a motion is chaotic or two–period quasiperiodic. An example of a two–dimensional projection of a three–dimensional Poincaré section is shown in Figure 3.4.15b. It is difficult to distinguish a three– or higher–period quasiperiodic motion from a chaotic motion by using a Poincaré section. However, as discussed in Section 4.1.2, second Poincaré sections can prove useful for characterizing certain aperiodic motions.

7.6.1 Systems of Equations

The construction of a Poincaré section for an orbit of a continuous–time system can be carried out as discussed in Section 3.3. Let us suppose that a known period T is used to construct a Poincaré section. Then, the Poincaré section acts like a stroboscope, freezing the components of the motion commensurate with the period T. If we have a collection of k discrete points on the Poincaré section, the corresponding motion is periodic with the period kT.

For illustration, let us consider a motion that is characterized by the frequencies f_1 and $f_e = \Omega/2\pi$. In addition, let $f_1/f_e = j/k$, where j and k are positive integers that do not have any common factor, $j < k$, and the sampling frequency for the chosen section be f_e. This section will consist of k points, and the order of the locations in which the points fall is determined by the ratio j/k. For $j = k - 1$, it can be verified numerically that the positions are filled up in a sequential manner. When the ratio of the frequencies f_1/f_e is not a rational number, we have a two–period quasiperiodic motion, and the points on

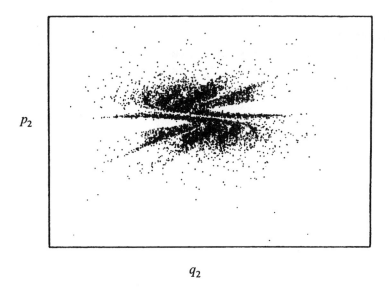

p_2

q_2

Figure 7.6.1: The Poincaré section of the response of a lightly damped system, showing a cloud of points.

a corresponding Poincaré section fill up a closed smooth curve densely. An example is given in Figure 4.1.1.

The pattern on a Poincaré section for either a chaotic motion or a three– or higher–period quasiperiodic motion does not correspond to any simple geometrical form. Therefore, if the Poincaré section does not consist of either a finite number of discrete points or closed curves, the motion may be chaotic. For undamped or lightly damped systems, the Poincaré section of chaotic motions appears as a cloud of unorganized points. Such motions are sometimes called "**stochastic**". In Figure 7.6.1, a Poincaré section of the response of a lightly damped system with a two–to–one internal resonance to a primary excitation is shown (Nayfeh and Nayfeh, 1990). There is some discernible structure in this section. In moderate to heavily damped systems, the pattern of intersections of a chaotic orbit with a Poincaré section is quite well organized. The structure seen in the section is usually scale invariant; that is, the same structure is seen at different levels of magnification. An example of a Poincaré section of a chaotic orbit is shown in Figure

5.2.1.

Because Poincaré maps share the different properties of the associated continuous–time systems, in many situations one can also use Poincaré sections to ascertain if the transients have died out in dissipative systems.

7.6.2 Experiments

In experimental situations, where the coordinates of the state space and its dimension are well defined, a Poincaré section can be constructed along the lines described previously. However, in many experimental situations, the construction is difficult, as one does not have a complete a priori knowledge of the state space associated with a system.

Let us assume that we have constructed a state space with the delayed coordinates $s(t)$, $s(t+\tau)$, ..., $s(t+m\tau)$, where m is an integer. In this space, the surface $s(t+k\tau) = s_{k0}$, where k is an integer and s_{k0} is a constant, can be used to form the Poincaré section. This construction is similar to that described for autonomous systems. Alternatively, if the system in question is being forced by an excitation of period T, one can use this period to construct the Poincaré section

$$\Sigma = \left\{ [s(t),\ s(t+\tau),\ \cdots,\ s(t+m\tau)] \mid t = t_0 + (n-1)T,\ n \in \mathcal{Z}^+ \right\}$$

where t_0 is the initial time and $\mathcal{Z}^+$ is the set of positive integers. Typically, one views the projection of Σ onto a two–dimensional space.

One can also construct a state space with the coordinates $s(t)$, $\dot{s}(t)$, $\ddot{s}(t)$, and so on. In this case, for the Poincaré section, we can collect the points $[s(t), \dot{s}(t), \ddot{s}(t), \cdots]$ for the values of $t = t_0 + (n-1)T$, where n is an integer. The initial time of collection of the points is specified by $t = t_0$. A projection of this section may be examined on the two–dimensional plane with coordinates $s(t)$ and $\dot{s}(t)$. Poincaré sections can also be used in experiments to determine whether transients have died out during a certain time evolution.

Next, we consider the details of construction of a Poincaré section for the motion of a harmonically excited system. If one has access to either a dual–channel analog or a dual–channel digital storage oscilloscope, two signals either corresponding to the displacement $s(t)$ and velocity

$\dot{s}(t)$ or to the delayed coordinates $s(t)$ and $s(t + \tau)$ can be fed into the oscilloscope channels. In the case of a digital oscilloscope, a square wave synchronous with the excitation signal can be used to externally set the sampling (clock) frequency (most function generators or wave synthesizers generate a square wave synchronous with the requested signal). In some digital oscilloscopes, due to the manner in which the points are stored in the scope, the effective sampling frequency used to form the Poincaré section is one–half the clock frequency. In the case of an analog oscilloscope, the synchronous signal can be fed into the input channel for the z–axis. The intensity of the display on the screen is controlled by the signal fed to the z–axis. The display on the screen becomes brighter whenever the synchronous signal externally triggers the oscilloscope. One may also construct Poincaré sections on a plotter by using external triggering.

In the experiments of Balachandran (1990) and Balachandran and Nayfeh (1991), signals were obtained from two strain gauges mounted at different locations on a harmonically forced structure. These strain–gauge signals were used to construct a pseudo–state space, and the excitation frequency f_e was used as the sampling (clock) frequency for constructing a Poincaré section. In Figure 7.6.2, we show two–dimensional projections of the experimentally obtained Poincaré sections. On each section, we have a collection of 512 points.

The points in the sections of Figures 7.6.2a and 7.6.2b lie on a closed curve, indicating that the corresponding motion is two–period quasiperiodic. Further, in Figure 7.6.2b, the points lie on a "figure eight" shaped curve. The apparent intersection or crossing in the two–dimensional space indicates that the corresponding Poincaré map has to be at least three–dimensional and is a consequence of a projection onto a two–dimensional space. In the present case, the motions pertaining to Figures 7.6.2a–c contain two basic frequencies f and δf and $\delta f \ll f$. Further, the ratio of the two frequencies is an irrational number, and the spectrum has the features of Figure 7.5.1b. For the motions corresponding to Figures 7.6.2a and 7.6.2b, the period of the fast–time scale $1/f$ is equal to the period of the clock $1/f_e$, and there is only a loop of points on the Poincaré section. In Figure 7.6.2a, a smooth curve drawn through the collection of points corresponds to the intersection of the associated T^2 torus with the Poincaré section. For the motion

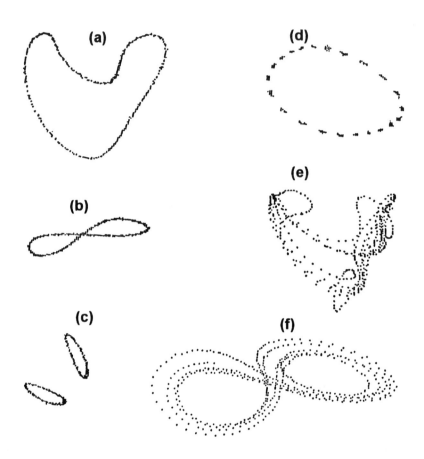

Figure 7.6.2: A collection of experimentally obtained Poincaré sections: (a) quasiperiodic motion, (b) quasiperiodic motion, (c) quasiperiodic motion (d) phase–locked motion, (e) chaotically modulated motion, and (f) chaotically modulated motion.

corresponding to Figure 7.6.2c, the period of the fast–time scale is twice the period of the clock. Hence, Figure 7.6.2c has two loops of points. In Figure 7.6.2d, we see a discrete cluster of points in the Poincaré section, indicating that the corresponding motion is periodic. This motion contains two frequencies f and δf such that $f/\delta f$ is a rational number. The associated spectrum has the features of Figure 7.5.1b.

The pattern of points observed in Figures 7.6.2e and 7.6.2f indicates that the corresponding motion in each case is neither periodic nor two–period quasiperiodic and that it is irregular. When we observed different sets of 512 points on the Poincaré section for the chaotic motions corresponding to Figures 7.6.2e and 7.6.2f, we did not observe any periodicity.

7.6.3   Higher–Order Poincaré Sections

Here, we provide some material to supplement the discussion provided in Section 4.1.2 on second Poincaré sections. Let us consider a system that is being forced at the two incommensurate frequencies f_1 and f_2. In addition, let $f_1/f_2 \simeq j/k$, where j and k are integers and have no common factors. In order to construct a first Poincaré section, we can sample at either of the excitation frequencies. By choosing f_1, we obtain a string of points sampled at time intervals of $1/f_1$. Let the nth point in this string be identified by the positive integer n. In the next step, we choose points from this string for which $n = mj/k$, where m is a positive integer, to obtain the second Poincaré section. A finite number of points in this section indicates a two–period quasiperiodic motion. A dense collection of points on a closed loop indicates a three–period quasiperiodic motion. It should be noted that the construction of higher–order Poincaré sections depends on the availability of a large amount of data and on the closeness of the ratio of the incommensurate frequencies to a rational number.

7.6.4   Comments

As discussed in Chapter 5, Poincaré maps play a key role in global analysis, where typically one is interested in trajectories near homoclinic and/or heteroclinic orbits. A detailed treatment of Poincaré sections is also provided by Bergé, Pomeau, and Vidal (1984, Chapter IV). The books by Moon (1987, 1992) are good references for experimental construction of Poincaré maps.

7.7 AUTOCORRELATION FUNCTIONS

We introduced the autocorrelation function $R_{xx}(\tau)$ or $C(\tau)$ in earlier sections. The autocorrelation function is a measure of correlation between $x(t)$ and $x(t + \tau)$, where τ is a time delay. In the literature, a plot of R_{xx} versus τ is also known as an **autocorrelogram** (e.g., Bendat and Piersol, 1986). For a given signal, the autocorrelation function $R_{xx}(\tau)$ can be determined by using either (7.5.5) or (7.5.8). On the basis of (7.5.8), R_{xx} can be determined as an inverse Fourier transform of the power spectrum. In practice, only finite–length data are available for computing R_{xx}. One uses a discretized version of either (7.5.5) or (7.5.8) for the computation. Typically, in signal analyzers, R_{xx} is determined through an inverse Fourier transform of the power spectrum. Most analyzers use a window, called a **correlation window,** on the input data to get rid of wrap–around effects that arise as a result of the implied periodicity. (In determining the FFT, we assume that the data are periodic with a period equal to the chosen time–window length.) The function R_{xx} may also be determined by using a commercial software package (e.g., MATLAB). The MATLAB function XCORR uses a discretized version of (7.5.5) for computation.

The autocorrelation functions $R_{xx}(\tau)$ for some standard $x(t)$ are as follows:

(1) $x(t) = C$, a constant; $R_{xx} = C^2$

(2) $x(t) = a\sin(\omega t)$; $R_{xx} = \frac{1}{2}a^2\cos(\omega\tau)$

(3) Broadband random noise of spectral density $G_{xx} = G$ in bandwidth B; $R_{xx} = \frac{G}{2\pi\tau}\sin(2\pi B\tau)$

(4) Band limited random noise of spectral density $G_{xx} = G$ in a narrow bandwidth B centered at f_c; $R_{xx} = \frac{G}{\pi\tau}\sin(\pi B\tau)\cos(2\pi f_c\tau)$.

It is instructive to plot R_{xx} versus τ for the above–discussed signals. For any signal, one of the values of τ for which R_{xx} has its maximum value is $\tau = 0$. For random signals, R_{xx} decays to zero as τ increases. The autocorrelation function of an irregular signal decays with τ with the rate of decay giving a measure of the degree of irregularity. However, for a periodic signal or a signal with a periodic component, R_{xx} does not decay to zero and is oscillatory. The envelope of R_{xx} is useful in determining the nature of a motion. The autocorrelation function for a periodic signal may give a less confusing representation of the data than the power spectrum, especially, when there is a high–frequency component in the signal.

For most nonperiodic signals, the autocorrelation function also decays to zero as τ increases. The exceptions include nonperiodic signals with periodic components. As an example, we consider a two–period quasiperiodic signal, which is of the form

$$x(t) = a(t)\cos(\omega t + \beta)$$

where β is a constant and

$$a(t) = a_o\left[1 + \alpha\cos(\omega_m t)\right]$$

We assume that the frequencies ω and ω_m are incommensurate. By using (7.5.5), one can show that

$$R_{xx}(\tau) = \frac{1}{2}a_o^2\left\{\cos(\omega\tau) + \alpha^2\cos\left[(\omega - \omega_m)\tau\right] + \alpha^2\cos\left[(\omega + \omega_m)\tau\right]\right\}$$

From the above expression, it is clear that R_{xx} is oscillatory, is two–period quasiperiodic, and does not decay to zero. If the signal is amplitude– and phase–modulated like (7.5.16), then one would have to use (7.5.16) in (7.5.5) to obtain the corresponding R_{xx}. A phase–modulated signal is a special case of (7.5.16). For both phase–modulated and amplitude– and phase–modulated signals, R_{xx} would be an infinite sum of periodic components. The amplitudes of the different components depend on $J_n(\beta_o)$, where J_n is Bessel's function of order

n of the first kind and β_o is the amplitude of the phase modulation. By induction, it follows that R_{xx} of an n–period quasiperiodic signal is also n–period quasiperiodic.

To fix ideas, we consider the signal

$$x(t) = \sin(\pi t) + \sin(3t)$$

The autocorrelation function for this signal can be analytically computed using (7.5.5) to be

$$R_{xx}(\tau) = 0.5 \cos(\pi \tau) + 0.5 \cos(3\tau)$$

The function R_{xx} is two–period quasiperiodic, does not decay, and has the form of a modulated waveform. Next, we consider a finite-record length of $x(t)$ obtained by collecting 8,192 points at a sampling frequency of 8.0 Hz. The MATLAB function XCORR is used for the computation, and the results are displayed in Figure 7.7.1. The value

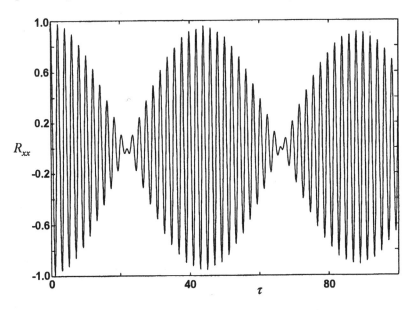

Figure 7.7.1: The autocorrelation function of a quasiperiodic signal.

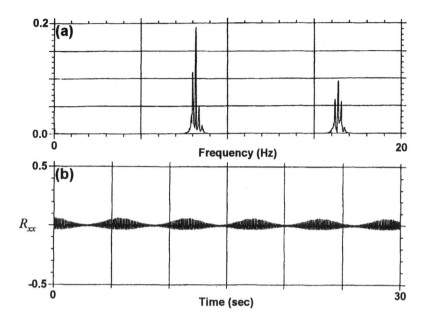

Figure 7.7.2: The spectrum and the autocorrelation function for a periodically modulated motion.

of R_{xx} at $\tau = 0$ is normalized so that it has a value of one. We note that the autocorrelogram has the appearance of a modulated waveform. A careful examination of the envelope indicates a decay in the amplitudes of the peaks. This is a consequence of finite–record length. One should note that R_{xx} of any finite–length data will decay to zero as τ becomes large.

In Figure 7.7.2, we show the spectrum and the associated autocorrelogram for a periodically modulated motion of a structure studied by Balachandran (1990). Both the spectrum and the autocorrelogram were computed by a signal analyzer. The spectrum has 1,280 lines of resolution in the 20 Hz baseband and is shown on a linear scale. It is indicative of a periodically modulated motion, and the associated autocorrelogram has the appearance of a regularly or periodically modulated waveform. Here, R_{xx} serves as a tool to establish the presence of quasiperiodic motions. We expect R_{xx} of a chaotically modulated signal to be an irregularly modulated waveform.

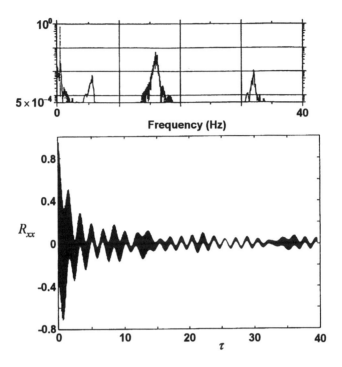

Figure 7.7.3: The spectrum and the autocorrelation function for a chaotically modulated motion.

The spectrum and the autocorrelogram associated with an aperiodic motion of a cantilever beam is shown in Figure 7.7.3. The aperiodic motion was observed in one of the experiments of Anderson, Balachandran, and Nayfeh (1992). The spectrum has 1,280 lines of resolution in the 40 Hz baseband, is shown on a log scale, and was determined by a signal analyzer. The function R_{xx} was determined using the MATLAB function XCORR. We used 40,000 points sampled at 120 Hz for the computation. Examining the spectrum, we note a continuous character in many frequency bandwidths. The envelope of R_{xx} has the appearance of an irregularly modulated waveform. The associated motion may very well be a chaotically modulated motion.

We remark that R_{xx} of a chaotic signal does not always decay to zero. The autocorrelation function R_{xx} of a chaotically modulated signal, which can be obtained from a system of coupled oscillators, does not always decay to zero. Another example is R_{xx} of a signal from a system with two or more equilibrium positions, where the forced oscillations chaotically jump from regular oscillations around one equilibrium position to regular oscillations around another equilibrium position.

7.8   LYAPUNOV EXPONENTS

Let us consider a trajectory described by a certain evolution. The **Lyapunov exponents** (also known as **characteristic exponents**) associated with a trajectory are essentially a measure of the average rates of expansion and contraction of trajectories surrounding it. They are asymptotic quantities, defined locally in state space, and describe the exponential rate at which a perturbation to a trajectory of a system grows or decays with time at a certain location in the state space. Analyses conducted with Lyapunov exponents are called **Lyapunov stability analyses**. Here, we describe how these exponents are determined for solutions of known systems of equations and experimental data. They are useful in characterizing the asymptotic state of an evolution (attractors in dissipative systems). Using Lyapunov exponents, we can distinguish among fixed points, periodic motions, quasiperiodic motions, and chaotic motions.

7.8.1   Concept of Lyapunov Exponents

We begin by defining Lyapunov exponents for a given system of equations. Let $\mathbf{X}(t)$ such that $\mathbf{X}(t = 0) = \mathbf{X}_0$ represent a trajectory of the system governed by the following n–dimensional autonomous system:

$$\dot{\mathbf{x}} = \mathbf{F}(\mathbf{x}; \mathbf{M}) \qquad (7.8.1)$$

where the vector $\mathbf{x}$ is made up of n state variables, the vector function $\mathbf{F}$ describes the nonlinear evolution of the system, and $\mathbf{M}$ represents a vector of control parameters. Denoting the perturbation provided to $\mathbf{X}(t)$ by $\mathbf{y}(t)$ and assuming it to be small, we obtain an equation after linearization in the disturbance terms. The perturbation is governed by

$$\frac{d\mathbf{y}(t)}{dt} = A\mathbf{y}(t) \tag{7.8.2}$$

where, in general, $A = D_{\mathbf{x}}\mathbf{F}[\mathbf{x}(t); \mathbf{M}]$ is an $n \times n$ matrix with time–dependent coefficients. If we consider an initial deviation $\mathbf{y}(0)$, its evolution is described by

$$\mathbf{y}(t) = \Phi(t)\mathbf{y}(0) \tag{7.8.3}$$

where $\Phi(t)$ is the fundamental (transition) matrix solution of (7.8.2) associated with the trajectory $\mathbf{X}(t)$.

The steps carried out thus far are identical to those carried out in a linear stability analyses of fixed points and periodic solutions. In the stability analysis for fixed points, described in Chapter 2, the trajectory $\mathbf{X}(t)$ is the fixed point itself, and the Jacobian matrix A has constant coefficients. The eigenvalues of A provide information about the stability of the associated fixed point. In the stability analysis for limit cycles, described in Chapter 3, the trajectory $\mathbf{X}(t)$ closes on itself, $\mathbf{X}(t) = \mathbf{X}(t + T)$ where T is the period of the limit cycle, and the Jacobian matrix A has periodic coefficients. The information over one period of the limit cycle is used to determine the orbital stability of $\mathbf{X}(t)$. In carrying out the steps, we first formed the fundamental matrix $\Phi(t)$, then chose initial conditions such that $\Phi(t = 0) = \mathbf{I}$, the identity matrix, and determined the matrix $\Phi(T)$, called the **monodromy matrix**. The eigenvalues of $\Phi(T)$ are used to determine the stability of the associated limit cycle.

The eigenvalues of the monodromy matrix provide a measure of the local orbital divergence near the considered limit cycle. For an orbit not constrained to close upon itself, the eigenvalues of the fundamental matrix in (7.8.3) provide a measure of the local divergence near the considered orbit. The procedure used to determine Lyapunov exponents can be considered to be a generalization of linear stability analyses. An interesting and detailed discussion on the relationship between linear stability analyses and Lyapunov stability analyses can be found in the paper of Goldhirsch, Sulem, and Orszag (1987). They argue that the Lyapunov exponents are global quantities associated with an attractor even though they are defined only locally in the state space.

For an appropriately chosen $\mathbf{y}(0)$ in (7.8.3), the rate of exponential expansion or contraction in the direction of $\mathbf{y}(0)$ on the trajectory passing through $\mathbf{X}_0$ is given by

$$\bar{\lambda}_i = \lim_{t \to \infty} \frac{1}{t} \ln \left(\frac{\| \mathbf{y}(t) \|}{\| \mathbf{y}(0) \|} \right) \tag{7.8.4}$$

where the symbol $\| \ \|$ denotes a vector norm and ln stands for the natural logarithm. The asymptotic quantity $\bar{\lambda}_i$ is called the **Lyapunov exponent**.

Since the state space is n–dimensional, we can use a set of n linearly independent vectors $\mathbf{y}_1$, $\mathbf{y}_2$, $\dots$, $\mathbf{y}_n$ to form the basis for this space. Choosing an initial deviation along each of these n vectors, we can determine n Lyapunov exponents $\bar{\lambda}_i(\mathbf{y}_i)$. Following Lyapunov (1947), the basis $\mathbf{y}_1, \mathbf{y}_2, \dots , \mathbf{y}_n$ is called a **normal basis** if

$$\sum_{i=1}^{n} \bar{\lambda}_i(\mathbf{y}_i) \leq \sum_{i=1}^{n} \bar{\lambda}_i(\mathbf{z}_i) \tag{7.8.5}$$

where $\mathbf{z}_1, \mathbf{z}_2, \dots , \mathbf{z}_n$ is any other basis of the n–dimensional state space. There is no unique normal basis. Further, the $\bar{\lambda}_i$ depend on Φ and not on the choice of the normal basis. We can order the $\mathbf{y}_i$ such that

$$\bar{\lambda}_1 \geq \bar{\lambda}_2 \geq \cdots \geq \bar{\lambda}_n \tag{7.8.6}$$

The set of n numbers $\bar{\lambda}_i$ is called the **Lyapunov spectrum**. We have n Lyapunov exponents associated with an n–dimensional autonomous system. It can be shown that if the trajectory $\mathbf{X}(t)$ corresponds to a motion other than a fixed point, then one of the $\bar{\lambda}_i$ is always zero (Haken, 1983).

Following Lyapunov (1947), the fundamental matrix $\Phi(t)$ is called regular if

$$\lim_{t \to \infty} \frac{1}{t} \ln | \det \Phi(t) |$$

exists and is finite and if there exists a normal basis of the n–dimensional state space such that

$$\sum_{i=1}^{n} \bar{\lambda}_i = \lim_{t \to \infty} \frac{1}{t} \ln | \det \Phi(t) | \tag{7.8.7}$$

If $\Phi(t)$ is regular, then, according to a theorem of Oseledec (1968), the asymptotic quantity defined in (7.8.4) exists and is finite for any initial deviation $\mathbf{y}(0)$ belonging to the n–dimensional space. As a consequence of the assumption of ergodicity in Oseledec's theorem, the Lyapunov exponents are independent of the initial condition $\mathbf{X}(t = 0)$ on the attractor. Further, Oseledec's theorem guarantees the existence of higher–order Lyapunov exponents, which are discussed next.

The asymptotic quantity $\bar{\lambda}_i$, given by (7.8.4), is also known as a **one–dimensional exponent**. Extending the notion to p–dimensions, a p–**dimensional Lyapunov exponent** $\bar{\lambda}^p$ is defined as

$$\bar{\lambda}^p = \lim_{t \to \infty} \frac{1}{t} \ln \left[\frac{\| \mathbf{y}_1(t) \wedge \mathbf{y}_2(t) \wedge \ldots \wedge \mathbf{y}_p(t) \|}{\| \mathbf{y}_1(0) \wedge \mathbf{y}_2(0) \wedge \ldots \wedge \mathbf{y}_p(0) \|} \right] \qquad (7.8.8)$$

where $\wedge$ is an exterior or vector cross product. The asymptotic quantity defined by (7.8.8) can be viewed as describing the expansion or contraction rate of a p–dimensional parallelepiped along the trajectory $\mathbf{X}(t)$. Also,

$$\bar{\lambda}^p = \sum_{i=1}^{p} \bar{\lambda}_i \qquad (7.8.9)$$

where the $\bar{\lambda}_i$ have been ordered as in (7.8.6).

Next, we present the relationship between the eigenvalues obtained in a linear stability analysis and the Lyapunov exponents. For a fixed point of an autonomous system, the Lyapunov exponents are defined as

$$\bar{\lambda}_i = \lim_{t \to \infty} \frac{1}{t} \text{Re} \left(\lambda_i t \right) \qquad (7.8.10)$$

where Re denotes the real part and λ_i is an eigenvalue of the Jacobian matrix. When all Lyapunov exponents are negative, the attractor is a stable fixed point.

For a limit cycle of an autonomous system, one of the Lyapunov exponents is always zero. This exponent corresponds to an initial deviation or perturbation $\mathbf{y}_i(0)$ provided along a tangent to the orbit $\mathbf{X}(t)$. Moreover, for a stable periodic orbit, all other Lyapunov exponents are negative. The negative exponents essentially correspond to perturbations provided along directions normal to the orbit $\mathbf{X}(t)$ and imply that, locally, trajectories separated from the limit cycle in these directions converge toward it. For an m–torus, m Lyapunov exponents are zero because there are m tangential directions to the torus along which there is no growth or decay. By using (7.8.7), one can show that the sum of the Lyapunov exponents is negative in the case of dissipative systems. In such systems, chaotic motions are characterized by one or more positive Lyapunov exponents. These positive exponents correspond to perturbations provided along some directions normal to a trajectory and imply that, locally, trajectories separated from the trajectory in question diverge from it along these directions.

7.8.2 Autonomous Systems

It is not straightforward to compute all the n Lyapunov exponents for the motion associated with an n–dimensional autonomous system using (7.8.1), (7.8.2), and (7.8.4). Choosing an initial deviation $y_i(0)$ at random and carrying out a numerical integration of (7.8.2) to determine $y_i(t)$ for large t leads to overflow errors on a digital computer. This is a consequence of exponential expansions along the chosen direction. Even if we choose an initial direction along which we expect only contractions to occur, numerical errors during the integration of (7.8.2) would align the $y_i(t)$ along the exponentially expanding direction. To overcome these difficulties, one needs to carry out the integration in steps and form a new basis of vectors at the end of each step as outlined next (Shimada and Nagashima, 1979; Benettin, Galgani, Giorgilli, and Strelcyn, 1980a,b).

We numerically integrate (7.8.1) and (7.8.2) and subsequently use (7.8.4) to compute the first m Lyapunov exponents associated with an orbit $\mathbf{X}(t)$ initiated at $\mathbf{X}_0$. For the linear system (7.8.2), we choose m orthonormal initial vectors $\mathbf{y}_i$ such that $\mathbf{y}_1 = (1, 0, 0, \dots)$, $\mathbf{y}_2 = (0, 1, 0, \dots)$, and so on. For each of these initial vectors, we integrate (7.8.1) and (7.8.2) for a finite time T_f and obtain a set of vectors $\mathbf{y}_1(T_f), \mathbf{y}_2(T_f), \dots, \mathbf{y}_m(T_f)$. This new set of vectors is orthonormalized using the **Gram–Schmidt procedure** to produce

$$\hat{\mathbf{y}}_1 = \frac{\mathbf{y}_1(T_f)}{\| \mathbf{y}_1(T_f) \|}$$

$$\hat{\mathbf{y}}_2 = \frac{\mathbf{y}_2(T_f) - [\mathbf{y}_2(T_f) \cdot \hat{\mathbf{y}}_1]\hat{\mathbf{y}}_1}{\| \mathbf{y}_2(T_f) - [\mathbf{y}_2(T_f) \cdot \hat{\mathbf{y}}_1]\hat{\mathbf{y}}_1 \|}$$

and

$$\hat{\mathbf{y}}_m = \frac{\mathbf{y}_m(T_f) - \sum_{i=1}^{m-1}[\mathbf{y}_m(T_f) \cdot \hat{\mathbf{y}}_i]\hat{\mathbf{y}}_i}{\| \mathbf{y}_m(T_f) - \sum_{i=1}^{m-1}[\mathbf{y}_m(T_f) \cdot \hat{\mathbf{y}}_i]\hat{\mathbf{y}}_i \|} \tag{7.8.11}$$

where $(\mathbf{x} \cdot \mathbf{y})$ denotes a scalar (dot) product of the vectors $\mathbf{x}$ and $\mathbf{y}$. Subsequently, using $\mathbf{X}(t = T_f)$ as an initial condition for (7.8.1) and using each of the $\hat{\mathbf{y}}_i$ as an initial condition for (7.8.2), we integrate (7.8.1) and (7.8.2) again for a finite time T_f and carry out the Gram–Schmidt procedure to obtain a new set of orthonormal vectors. We denote the norm in the denominator by N_j^k, where the superscript refers to the kth time step and the subscript refers to the jth vector. After repeating the integrations and the Gram–Schmidt orthonormalizations r times, we obtain the Lyapunov exponents from

$$\hat{\lambda}_i = \frac{1}{rT_f} \sum_{k=1}^{r} \ln N_i^k \tag{7.8.12}$$

The number of orthonormalizations r required and the choice of the finite–time length T_f vary from problem to problem. Wolf, Swift, Swinney, and Vastano (1985) use the above described approach in their algorithm to determine the Lyapunov exponents. To determine the largest Lyapunov exponent, Wolf, Swift, Swinney, and Vastano (1985) follow a pair of nearby points to determine N_i^k in (7.8.12).

Implementing this scheme, for a chaotic attractor of the equations (Rössler, 1976a)

$$\dot{x} = -y - z$$
$$\dot{y} = x + 0.15y$$
$$\dot{z} = 0.2 + z(x - 10)$$

Wolf, Swift, Swinney, and Vastano (1985) obtained the following Lyapunov exponents with units of bits/sec. (the logarithm is taken with base two): $\lambda_1 = 0.13, \lambda_2 = 0.00$, and $\lambda_3 = -14.1$. And for a chaotic attractor of the Lorenz system

$$\dot{x} = 16(y - x)$$
$$\dot{y} = x(45.92 - z) - y$$
$$\dot{x} = xy - 4z$$

they obtained $\lambda_1 = 2.16, \lambda_2 = 0.00$, and $\lambda_3 = -32.4$ in bits/sec. And for a chaotic attractor of the equations (Rössler, 1979a)

$$\dot{x} = -y - z$$
$$\dot{y} = x + 0.25y + w$$
$$\dot{z} = 3 + xz$$
$$\dot{w} = 0.05w - 0.5z$$

they obtained $\lambda_1 = 0.16, \lambda_2 = 0.03, \lambda_3 = 0.00$, and $\lambda_4 = -39.0$ in bits/sec.

7.8.3 Maps

Here, we describe the computation of n Lyapunov exponents for an evolution described by an n–dimensional differentiable map. To this end, we consider a differentiable map $\mathbf{F}(\mathbf{x})$ such that

$$\mathbf{x}_{k+1} = \mathbf{F}(\mathbf{x}_k) \tag{7.8.13}$$

where $\mathbf{F}$ maps $\mathcal{R}^n$ into $\mathcal{R}^n$. The sequence of points $\mathbf{x}_0, \mathbf{x}_1, \ldots, \mathbf{x}_k, \ldots$ represents an orbit of the map initiated at the point $\mathbf{x} = \mathbf{x}_0$. We let $D_{\mathbf{x}}\mathbf{F}(\mathbf{x}_k)$ represent the $n \times n$ matrix of first partial derivatives of $\mathbf{F}$ evaluated at $\mathbf{x} = \mathbf{x}_k$.

As in Section 7.8.2, to determine m out of the n Lyapunov exponents, we first choose m orthonormal vectors $\mathbf{y}_i^0$. In the first stage, we compute the set of m vectors $\mathbf{y}_i^1$ using the relation

$$\mathbf{y}_i^1 = D_{\mathbf{x}}\mathbf{F}(\mathbf{x}_0)\mathbf{y}_i^0 \, , \, i = 1, 2, \ldots, m \tag{7.8.14}$$

These m vectors are orthonormalized using the Gram–Schmidt procedure to produce the m orthonormal vectors $\hat{\mathbf{y}}_i^1$. Subsequently, we compute the m vectors $\mathbf{y}_i^2$ from the relation

$$\mathbf{y}_i^2 = D_{\mathbf{x}}\mathbf{F}(\mathbf{x}_1)\hat{\mathbf{y}}_i^1 \, , \, i = 1, 2, \ldots, m \tag{7.8.15}$$

The vectors $\mathbf{y}_i^2$ are orthonormalized using the Gram–Schmidt procedure as described in Section 7.8.2. At the rth stage, we have

$$\mathbf{y}_i^r = D_{\mathbf{x}}\mathbf{F}(\mathbf{x}_{r-1})\hat{\mathbf{y}}_i^{r-1} \, , \, i = 1, 2, \ldots, m \tag{7.8.16}$$

During the orthonormalization process, let the symbol N_j^k represent the norm in the denominator associated with the jth vector in the kth stage. After carrying out the above process r times, we compute the m Lyapunov exponents from

$$\hat{\lambda}_i = \frac{1}{r} \sum_{k=1}^{r} \ln \left(N_i^k \right) \tag{7.8.17}$$

Alternatively, to study the stability of the orbit $\{x_0, x_1, \ldots, x_k, \ldots\}$ of (7.8.13), we add a small perturbation w_k to x_k, linearize in the perturbation, and obtain the variational equation

$$w_{k+1} = DF(x_k)w_k \tag{7.8.18}$$

After L iterations, the evolution of w_k can be written as

$$w_{k+L} = DF\left(x_{k+L-1}\right) \cdot DF\left(x_{k+L-2}\right) \ldots DF(x_k)w_k$$

or

$$w_{k+L} = DF^L(x_k)w_k \tag{7.8.19}$$

where $DF^L(x)$ stands for the composition of L Jacobian matrices $DF(x)$ along the orbit $\{x_0, x_1, \ldots, x_k, \ldots\}$. According to the **multiplicative ergodic theorem** of Oseledec (1968),

$$OS = \lim_{L \to \infty} [OSL(x, L)]^{\frac{1}{2L}} = \lim_{L \to \infty} \left\{ \left[DF^L(x)\right]^T \left[DF^L(x)\right] \right\}^{\frac{1}{2L}}$$

exists and is independent of x for almost all values of x in the basin of attraction of the attractor to which the orbit belongs. Consequently, the eigenvalues $\exp(\lambda_1) \geq \exp(\lambda_2) \geq \ldots \geq \exp(\lambda_n)$ of OS are independent of x for almost all x within the basin of attraction of the attractor. The λ_i are called the **global Lyapunov exponents** (Abarbanel, Brown, Sidorowich, and Tsimring, 1993). Because they are independent of where one starts within the basin of attraction, they are invariant characteristics of the dynamics and not particular to the observed orbit. If one or more of the λ_i are positive, the corresponding orbit is chaotic. For a dissipative system, $\lambda_1 + \lambda_2 + \ldots + \lambda_n < 0$.

Although the above formal definition of the global Lyapunov exponents is straightforward, calculating them using this definition poses some difficulties. The chief difficulty stems from the fact that each Jacobian DF has the eigenvalues $\exp(\lambda_i)$, the composition of L Jacobians has the eigenvalues $\exp(L\lambda_i)$, and, because $\lambda_1 > \lambda_2 > \lambda_3 \ldots > \lambda_n$, the matrix DF^L is terribly ill–conditioned as L becomes large. Because standard QR decomposition routines do not work very well, Eckmann and Ruelle (1985) and Eckmann, Kamphorst, Ruelle, and Ciliberto (1986) developed the following recursive QR decomposition algorithm. Each Jacobian is written in its decomposition form as

$$DF(\mathbf{x}_i) \cdot Q_{i-1} = Q_i \cdot R_i \qquad (7.8.20)$$

where $Q_0 = I$, the identity matrix; Q_i are orthogonal matrices; and R_i are upper triangular elements with positive diagonal elements. Thus,

$$DF(\mathbf{x}_1) = Q_1 \cdot R_1 \qquad (7.8.21)$$

$$DF(\mathbf{x}_2) \cdot Q_1 = Q_2 \cdot R_2 \qquad (7.8.22)$$

It follows from (7.8.21) that

$$Q_1^{-1} \cdot DF(\mathbf{x}_1) = R_1 \qquad (7.8.23)$$

Hence, combining (7.8.22) and (7.8.23) yields

$$DF(\mathbf{x}_2) \cdot DF(\mathbf{x}_1) = Q_2 \cdot R_2 \cdot R_1 \qquad (7.8.24)$$

Continuing the composition, we have

$$DF(\mathbf{x}_3) \cdot DF(\mathbf{x}_2) \cdot DF(\mathbf{x}_1) = Q_3 \cdot R_3 \cdot R_2 \cdot R_1 \qquad (7.8.25)$$

and

$$DF^L(\mathbf{x}_1) = Q_L \cdot \Pi_{k=1}^L R_k \qquad (7.8.26)$$

We note that, at each step of this recursive algorithm, no matrix R_k is much larger than $\exp(\lambda_1)$, and hence the condition number is more or less $\exp(\lambda_1 - \lambda_n)$, a reasonable number for numerical accuracy.

Abarbanel, Brown, and Kennel (1991) defined local rather than global Lyapunov exponents directly from the positive symmetric Oseledec matrix

$$\mathrm{OSL}(\mathbf{x}, L) = \left[DF^L(\mathbf{x})\right] \cdot \left[DF^L(\mathbf{x})\right]^T \qquad (7.8.27)$$

whose eigenvalues behave approximately as $\exp\left[2L\lambda_i(\mathbf{x}, L)\right]$, with the $\lambda_i(\mathbf{x}, L)$ being the **local Lyapunov exponents**. They can be evaluated by using the aforementioned recursive QR decomposition algorithm. According to the multiplicative ergodic theorem of Oseledec,

$$\lim_{L\to\infty} \lambda_i(\mathbf{x}, L) = \lambda_i, \tag{7.8.28}$$

the global Lyapunov exponent. Clearly, the local Lyapunov exponent $\lambda_i(\mathbf{x}, L)$ depends on the time length L and the position $\mathbf{x}$ in phase space and hence the particular orbit being examined. Consequently, one can define average local Lyapunov exponents $\bar{\lambda}_i(L)$ by averaging the $\lambda_i(\mathbf{x}, L)$ over many orbits as

$$\bar{\lambda}_i(L) = \frac{1}{N} \sum_{k=1}^{N} \lambda_i(\mathbf{x}_k, L) \tag{7.8.29}$$

7.8.4   Reconstructed Space

Once the data have been used to reconstruct a pseudo–state space by using the method of delays, we use it to construct a local map at $\mathbf{y}_k$ as follows (Eckmann and Ruelle, 1985; Sano and Sawada, 1985; Eckmann et al., 1986; Abarbanel, 1995). First, we find its N_B nearest neighbors $\mathbf{y}_k^{(r)}, r = 1, 2, \ldots, N_B$. Under the dynamics, each $\mathbf{y}_k^{(r)}$ will evolve into a known state $\mathbf{y}_{k+1}(r)$, which is in the neighborhood of $\mathbf{y}_{k+1}$. We note that $\mathbf{y}_{k+1}(r)$ may not be the rth nearest neighbor to $\mathbf{y}_{k+1}$. Then, we represent $\mathbf{y}_{k+1}(r)$ as a linear combination of a set M of basis functions $\phi_m\left[\mathbf{y}_k^{(r)}\right]$ as

$$\mathbf{y}_{k+1}(r) = \sum_{m=1}^{M} c_{mk}\phi_m\left[\mathbf{y}_k^{(r)}\right] \tag{7.8.30}$$

The basis set ϕ_m may or may not be polynomials (Briggs, 1990; Bryant, Brown, and Abarbanel, 1990; Brown, Bryant, and Abarbanel, 1991; Parlitz, 1992; Abarbanel, Brown, Sidorowich, and Tsimring, 1993). To determine the c_{mk}, we use the method of least squares and minimize the following square of the residuals:

$$\sum_{r=1}^{N_B} \left|\left| \mathbf{y}_{k+1}(r) - \sum_{m=1}^{M} c_{mk}\phi_m\left(\mathbf{y}_k^{(r)}\right) \right|\right|^2$$

This results in the set of algebraic equations

$$\sum_{r=1}^{N_B} \left\{ \phi_n \left[\mathbf{y}_k^{(r)} \right] \right\}^T \mathbf{y}_{k+1}(r)$$

$$- \sum_{r=1}^{N_B} \sum_{m=1}^{M} c_{mk} \left[\phi_n \left(\mathbf{y}_k^{(r)} \right) \right]^T \phi_m \left(\mathbf{y}_k^{(r)} \right) = 0 \tag{7.8.31}$$

for $n = 1, 2, \ldots, M$. Having determined the c_{mk}, we differentiate the local map (7.8.30) and determine its Jacobian as

$$D\mathbf{F}(\mathbf{y}_k) = \sum_{m=1}^{M} c_{mk} D\phi_m(\mathbf{x}) \, |_{\mathbf{x}=\mathbf{y}_k} \tag{7.8.32}$$

If the basis functions are polynomials, then the Jacobian of the local map is the linear term in (7.8.30).

In Figure 7.8.1, we show, after Abarbanel et al. (1993), the variation of the average local Lyapunov exponents $\bar{\lambda}_i(L)$ with L by using the time series x_n obtained for a chaotic attractor of the Lorenz system (5.8.23)–(5.8.27). This time series was used to construct a three–dimensional pseudo–state space, which is used in turn to evaluate the local Jacobians, as indicated above. Then, the local Lyapunov exponents $\lambda_i(\mathbf{y}_k, L)$ were evaluated by using the recursive QR decomposition of the required product of the local Jacobians, which are averaged over the attractor to determine $\bar{\lambda}_i(L)$. Clearly, one of the exponents is zero, corresponding to perturbations along the orbit, one is positive, and the third is negative. When $L = 10$, these exponents are $\lambda_1 = 1.51, \lambda_2 = 0$, and $\lambda_3 = -19.0$. The zero exponent is indicative of the fact that the data comes from a flow instead of a mapping, while the positive exponent indicates that the attractor is chaotic. The Lyapunov dimension calculated in accordance with the definition provided in Section 7.9 is 2.08.

In Figure 7.8.2, we show, after Abarbanel (1995), the variation of $\bar{\lambda}_i(L)$ with L for the Lorenz system obtained by reconstructing a four–dimensional rather than a three–dimensional pseudo–state space from the x_n times series. In this case, we have four rather than three Lyapunov exponents. In Figure 7.8.2, we also show negative of the Lyapunov exponents obtained for time reversed data. The true expon–

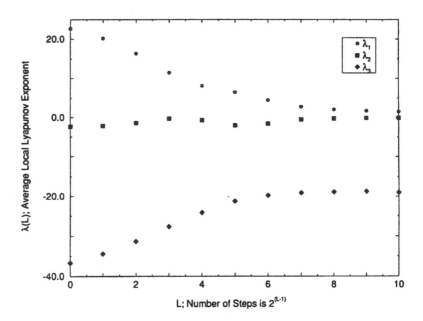

Figure 7.8.1: Average Lyapunov exponents for a chaotic attractor of the Lorenz equations (5.8.23)–(5.8.27). The time history of x used in the calculations consists of 50,000 data points. Local cubic neighborhood to neighborhood maps are made with the linear term giving the required local Jacobian to use in the Oseledec matrix OSL. An embedding dimension $d = 4$ is used in the calculations with a local dimension $d_L = 3$ as determined by local false nearest neighbors. With $d = 3$, the calculated values change little. There are three Lyapunov exponents. One of them is positive, one is zero, and the remaining exponent is negative. Reprinted with permission from Abarbanel (1995).

ents will change sign with time reversal, whereas spurious or false exponents behave otherwise (Parlitz, 1992). Clearly, three of the exponents change sign under time reversal, whereas the one corresponding to λ_3 in forward time and λ_2 in reverse time does not. Consequently, we have three true exponents and one spurious exponent. Several examples of detecting spurious Lyapunov exponents by using the exponents determined for time reversed data are provided by Parlitz (1992).

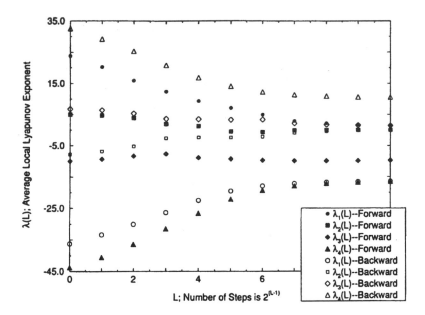

Figure 7.8.2: Average Lyapunov exponents for a chaotic attractor of the Lorenz system (5.8.23)–(5.8.27). The time history of x with 50,000 points is used for the calculations. Local cubic neighborhood to neighborhood maps are made with the linear term giving the required local Jacobian to use in the Oseledec matrix. An embedding dimension $d = 4$ is used in the calculations with a local dimension $d_L = 4$. Four Lyapunov exponents are evaluated for the original data (i.e, forward time), and four Lyapunov exponents are evaluated for the time reversed data. The negative of the exponents determined for the time reversed data are shown. True exponents change sign under this operation, and we see that there are three true exponents. Reprinted with permission from Abarbanel (1995).

7.8.5   Comments

In information theoretic terms, the Lyapunov exponents are a measure of the rates at which information is created or destroyed during an evolution. Because digital information is expressed in bits, in the definition of Lyapunov exponents in (7.8.12) and (7.8.17) some use the base 2 logarithm instead of the natural logarithm. When the base 2 logarithm is used, the units for the exponents are bits/sec or bits/orbit for an autonomous system and bits/iteration for a map. Often, for data obtained in experimental situations, the signal–to–noise levels may not permit one to determine the negative Lyapunov exponents. In the next section, we discuss how the Lyapunov spectrum can be used to determine the dimension of the corresponding motion.

7.9 DIMENSION CALCULATIONS

As discussed in Chapter 5, the geometric structures generated by chaotic systems are extremely complex. Regions in state space are stretched, contracted, folded, and remapped into a compact region of the original space whose volume shrinks to zero for a dissipative system, leaving gaps in the state space. Typically, the Poincaré section of a chaotic attractor consists of an infinite number of infinitely thin layers. Consequently, the orbits tend to fill up less than an integer subspace in state space. Sets with noninteger dimensions are called **fractals** (Mandelbrot, 1983) and attracting sets with noninteger dimensions are called **strange attractors** (e.g., Ruelle and Takens, 1971). In this section, we describe a number of definitions of noninteger or fractal dimension. For comprehensive reviews and critiques of fractal dimensions, we refer the reader to Young (1982, 1983), Farmer, Ott, and Yorke (1983), Badii and Politi (1985), Mayer–Kress (1985), Theiler (1990), Ott (1993), and Abarbanel, Brown, Sidorowich, and Tsimring (1993).

7.9.1 Capacity Dimension

The simplest and most appealing way of assigning a dimension to a set that can yield a fractal dimension to certain kinds of sets is the so–called **capacity** or **box–counting dimension** D_0. We assume that we have a set N_0 of points that lies in a d–dimensional Cartesian space. We cover the set with cubes of edge length r. When $d = 2$, the "cubes" are squares, and when $d = 1$, the "cubes" are intervals. We let $N(r)$ be the minimum number of cubes needed to cover the set.

We repeat the process for successively smaller values of r. Then, the capacity dimension D_0 is given by

$$D_0 = \lim_{r \to 0} \frac{\ln N(r)}{\ln(1/r)} \qquad (7.9.1)$$

if the limit exists; otherwise, D_0 is undefined.

Because an m–dimensional manifold locally resembles $\mathcal{R}^m$, D_0 of such a manifold is equal to m, which is an integer. Thus, the capacity dimension of a limit cycle is 1, that of a two–period quasiperiodic orbit is 2, and that of an m–period quasiperiodic orbit is m. For objects that are not manifolds, D_0 may take noninteger values. Next, we consider two examples with integer and noninteger capacity dimensions.

Example 7.1. We evaluate the capacity dimensions of the three sets lying in the plane (Fig. 7.9.1). The set in Figure 7.9.1a consists of two points, and hence the minimum number of squares needed to cover them is always 2, irrespective of the value of r. Hence,

$$D_0 = \lim_{r \to 0} \frac{\ln 2}{\ln(1/r)} = 0$$

The set in Figure 7.9.1b consists of a curve segment of length l. The minimum number of squares of edge length r needed to cover it is l/r. Hence,

$$D_0 = \lim_{r \to 0} \frac{\ln(l/r)}{\ln(1/r)} = 1$$

The set in Figure 7.9.1c consists of the area A inside a closed curve. The minimum number of squares with edge length r needed to cover this set is A/r^2. Hence,

$$D_0 = \lim_{r \to 0} \frac{\ln(A/r^2)}{\ln(1/r)} = 2$$

Example 7.2. We consider an elementary example of a fractal set, namely, the middle–third Cantor set shown in Figure 7.9.2. The set is

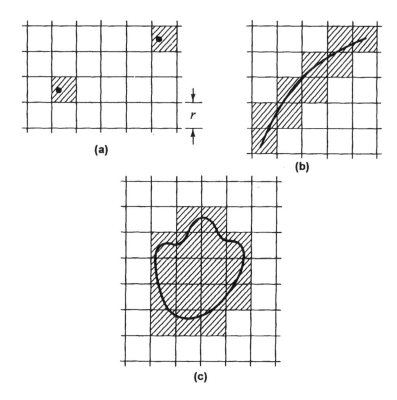

Figure 7.9.1: Illustration of calculation of the capacity dimension for three sets: (a) two points, (b) a curve, and (c) an area enclosed by a closed curve.

constructed iteratively by using a single operation on a straight line of unit length. First, the middle third of the unit interval is removed. Second, the middle third of each of the two remaining intervals is removed. Third, the middle third of each of the remaining four intervals is removed. Fourth, the middle third of each of the remaining eight intervals is removed. At the nth iteration, the number of remaining segments is 2^n, and the length of each remaining segment is $1/3^n$. Hence, as the number of iterations tends to infinity, the number of the remaining segments tends to infinity and the length of each segment

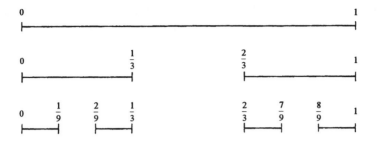

Figure 7.9.2: Construction of the Cantor set.

tends to zero. Thus,

$$D_0 = \lim_{n \to \infty} \frac{\ln(2^n)}{\ln(3^n)} = \frac{\ln 2}{\ln 3} \approx 0.6309$$

7.9.2 Pointwise Dimension

We assume that we have sampled a long–time trajectory in a d–dimensional Cartesian space and obtained a large number N_0 of points $\mathbf{x}_i$. We then place a sphere of radius r at some point $\mathbf{x}_m$ on the trajectory and count the number $N(r, \mathbf{x}_m)$ of points within the sphere; that is,

$$N(r, \mathbf{x}_m) = \sum_{k=1}^{N_0} H(r - |\mathbf{x}_k - \mathbf{x}_m|) \qquad (7.9.2)$$

where H is the Heaviside function defined in (7.4.10). Thus, the probability $P(r, \mathbf{x}_m)$ of finding a point in this sphere is given by

$$P(r, \mathbf{x}_m) = \frac{N(r, \mathbf{x}_m)}{N_0} = \frac{1}{N_0} \sum_{k=1}^{N_0} H(r - |\mathbf{x}_k - \mathbf{x}_m|) \qquad (7.9.3)$$

Then, the **pointwise dimension** is defined by

$$D_p(\mathbf{x}_m) = \lim_{r \to 0} \frac{\ln P(r, \mathbf{x}_m)}{\ln r} \qquad (7.9.4)$$

which, in general, depends on the chosen point $\mathbf{x}_m$ on the attractor. Consequently, an averaged pointwise dimension is usually used. To this end, we randomly select a set of points $M < N_0$ distributed around the attractor and calculate $P(r, \mathbf{x}_m)$ at each of these points. Then, we can calculate an averaged pointwise dimension in two ways. First, we calculate $D_p(\mathbf{x}_m)$ for $m = 1, 2, \ldots, M$, average them over all values of M, and obtain the **averaged pointwise dimension**

$$\hat{D}_p = \frac{1}{M} \sum_{m=1}^{M} D_p(\mathbf{x}_m) \tag{7.9.5}$$

Second, we average $P(r, \mathbf{x}_m)$ over the randomly selected set M and obtain the averaged probability $P(r)$ as

$$P(r) = \frac{1}{M} \sum_{m=1}^{M} P(r, \mathbf{x}_m) = \frac{1}{MN_0} \sum_{m=1}^{M} \sum_{k=1}^{N_0} H(r - |\mathbf{x}_k - \mathbf{x}_m|) \tag{7.9.6}$$

Then, we define the averaged pointwise dimension as

$$D_p = \lim_{r \to 0} \frac{\ln P(r)}{\ln r} \tag{7.9.7}$$

We note that (a) $P(r)$ and hence D_p are functions of the dimension d used to reconstruct the state space; (b) $P(r)$ can be defined for any time series (chaotic, noisy, or mixed); and (c) $P(r)$ is a function of r, which allows us to explore the structure of the attractor at different scales in state space. Moreover, large amounts of data and good signal to noise ratios are necessary to obtain accurate pointwise dimensions. Clearly, the number of data points must scale as some function of the attractor dimension. Furthermore, the number of reference points M must be large enough and reasonably distributed over the attractor in order to produce accurate averaged $P(r)$.

In practice, one cannot estimate D_p from (7.9.7) because, if r is too small, no points will fall within the sphere because the number N_0 of available data points is finite and $P(r)$ is always dominated by noise. Consequently, the standard practice is to plot $\ln P(r)$ versus $\ln r$ for a set of increasing values of d and to identify a scaling region at intermediate values of r where the slope approaches a constant as d increases. In the scaling region, r is smaller than the size of the considered object and larger than the smallest spacing between points. For values of the embedding dimension less than that required to unfold the attractor, the data will fill the entire reconstruction space, and the slope is equal to the embedding dimension. For low–dimensional attractors, the slope will saturate as the embedding dimension increases at a value equal to the attractor dimension.

In Figure 7.9.3, we show a family of plots of $\ln[\hat{P}(r)]$ versus $\ln(r)$, encountered by Balachandran and Nayfeh (1991) when estimating the pointwise dimension of the response of a two–beam two–mass structure

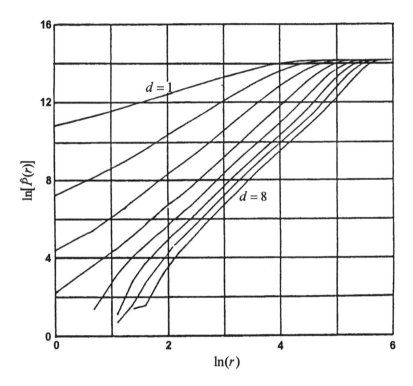

Figure 7.9.3: Variation of $\ln[\hat{P}(r)]$ with $\ln(r)$ for various embedding dimensions for data collected from the response of the structure shown in Figure 7.9.4. Here, $\hat{P}(r) = MN_oP(r)$. Reprinted with permission from Balachandran and Nayfeh (1991).

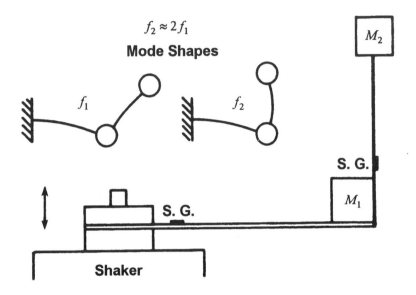

Figure 7.9.4: A two–beam two–mass structure tuned so that $f_2 \approx 2f_1$.

(Fig. 7.9.4) to a primary excitation of the second mode. For this plot, a scalar time series of 60,000 points was collected at a sampling frequency of 120 Hz. A family of d–dimensional pseudo–state spaces were constructed for a delay time $\tau = 3.5$ seconds and for $d = 1, 2, \ldots, 8$. The probability of finding a point in a sphere of radius r was calculated for each of 100 randomly selected points of the 60,000 points. These probabilities were averaged to obtain $\hat{P}(r)$, and $\ln \hat{P}(r)$ is plotted in Figure 7.9.3. For large r (the order of the size of the attractor), all points fall within the sphere, and hence $\hat{P}(r)$ is independent of r and in turn the slope of the curve is constant. For very small values of r, no data points are within the sphere, and again the slope is constant. For intermediate values of r, the slope increases as d increases and saturates at approximately 2.748 when d exceeds 4. Varying the delay time between 1.6 and 4.2 seconds, Balachandran and Nayfeh (1991) found that D_p varied within a range of ± 0.2 around 2.748.

In the algorithm used by Balachandran and Nayfeh (1991), the l_1 norm was used instead of the Euclidean norm. This norm is defined as

$$||\mathbf{x}|| = |x_1| + |x_2| + \ldots + |x_n|$$

This norm was found to improve the computational efficiency of the algorithm. Another norm that is attractive from a computational standpoint is defined as

$$||\mathbf{x}|| = max_{1 \le i \le n}|x_i|$$

The noise level in the data used for Figure 7.9.3 is quite low. In other cases, this may not be so. The presence of a knee in a $\ln - \ln$ plot used for dimension calculations is often due to the presence of noise. If the magnitude of the noise is η, there will be a knee at about η. The slope above the knee will provide the correct dimension, while the slope below the knee will be equal to the considered embedding dimension.

7.9.3   Information Dimension

The capacity dimension is a purely metric concept and does not utilize any information about the dynamical system. It gives the scaling of the number of cubes needed to cover the attractor, irrespective of the number of points in each cube. However, for a strange attractor, the frequency of visitation of the trajectory varies vastly from cube to cube. In fact, as $r \to 0$, typical orbits will spend most of their time in a small number of the cubes needed to cover the attractor. Consequently, the box–counting scheme treats all cubes needed to cover the attractor equally, irrespective of their importance. Moreover, although the box–counting scheme is simple, it is inefficient. As we sweep over the embedding dimension, we need to construct and check the emptiness of the order of $(1/r)^d$ cubes at each resolution r. Clearly, the computation effort increases very rapidly as $r \to 0$ and as d increases. However, it is to be noted that there are fast algorithms to determine the dimension of an attractor by the box–counting scheme (e.g., Leibovitch and Toth, 1989).

To overcome the shortcomings of the box–counting scheme, Grassberger (1983) and Hentschel and Procaccia (1983) introduced other definitions that account for the relative frequency of visitation of a typical trajectory to the different cubes needed to cover the attractor and are also more efficient than the box–counting scheme. In this section we describe the information dimension, in Section 7.9.4 we describe the correlation dimension, and in Section 7.9.5 we describe the generalized correlation dimension.

As in the box–counting scheme, we cover the set N_0 whose dimension is desired by a set $1, 2, \ldots, N$ of spheres of radius r. Then, we count the number $N_i(r)$ of points in each of the spheres i and determine the probability $P_i(r)$ of finding a point in the ith sphere by

$$P_i(r) = \frac{N_i(r)}{N_0} \tag{7.9.8}$$

Clearly,

$$\sum_{i=1}^{N} P_i(r) = 1 \tag{7.9.9}$$

Then, the information entropy is

$$I(r) = -\sum_{i=1}^{N} P_i(r) \ln P_i(r) \tag{7.9.10}$$

The quantity $I(r)$ can be interpreted as the average amount of information required to specify the considered system's state to an accuracy of r. (This state is assumed to be on the attractor of interest.) When the base 2 logarithm is used, $I(r)$ has the units of bits. In terms of $I(r)$, the information dimension D_1 is defined by

$$D_1 = \lim_{r \to 0} \frac{I(r)}{\ln(1/r)} = \lim_{r \to 0} \sum_{i=1}^{N} \frac{P_i(r) \ln P_i(r)}{\ln(r)} \tag{7.9.11}$$

If the points in the set are equally distributed, then

$$P_i(r) = \frac{1}{N(r)}$$

and

$$I(r) = -\sum_{i=1}^{N} \frac{1}{N(r)} \ln \frac{1}{N(r)} = \ln N(r)$$

Hence,

$$D_1 = \lim_{r \to 0} \frac{\ln N(r)}{\ln(1/r)} = D_0$$

according to (7.9.1).

Again, in estimating the information dimension, we cannot use the definition (7.9.11) and take the limit as $r \to 0$ because of the finite amount of data and the fact that for small r the data are dominated by noise. Hence, the standard practice is to plot $-I(r)$ versus $\ln(r)$ and find a scaling region for intermediate values of r such that the slope of the plot approaches a constant as the embedding dimension d increases.

7.9.4 Correlation Dimension

The correlation dimension is another probabilistic measure of dimension. Again, this measure can be calculated more efficiently than the capacity dimension, and it takes into account the relative frequency of visitation of the trajectory to each sphere. Grassberger and Procaccia (1983a,b) give an extensive account of this measure.

To estimate the correlation dimension of a set $x_1, x_2, \ldots, x_N$, where N is large, we first calculate the distances ξ_{ij} between each pair of points x_i and x_j by using, say, the Euclidean measure $\xi_{ij} = ||x_i - x_j||$. Then, we define a correlation function $C(r)$ by

$$C(r) = \lim_{N \to \infty} \frac{1}{N^2} [\text{number of points } (i,j) \text{ with distance } \xi_{ij} < r]$$

$$(7.9.12)$$

Alternatively, the correlation function $C(r)$ can be calculated more efficiently by constructing a sphere of radius r around each point x_i in state space and counting the number of points within the sphere; that is,

$$C(r) = \lim_{N \to \infty} \left[\frac{1}{N^2} \sum_{i=1}^{N} \sum_{j=1}^{N} H(r - |x_i - x_j|) \right] \qquad (7.9.13)$$

where H is the Heaviside function defined in (7.4.10). Then, the correlation dimension D_2 is defined as

$$D_2 = \lim_{r \to 0} \frac{\ln C(r)}{\ln(r)} \qquad (7.9.14)$$

Again, to estimate D_2, we plot $\ln C(r)$ versus $\ln r$ and find a scaling region for intermediate values of r in which the slope of the plot approaches a constant, which is the correlation dimension, as d increases.

Comparing (7.9.6) and (7.9.7) with (7.9.13) and (7.9.14), we conclude that the correlation dimension D_2 differs from the averaged pointwise dimension D_p in that the sum is performed about every point of the set.

7.9.5 Generalized Correlation Dimension

The capacity, information, and correlation dimensions have been consolidated into the following general definition of dimension of order q:

$$D_q = \frac{1}{q-1} \lim_{r \to 0} \frac{\ln \sum_{i=1}^{N} P_i^q(r)}{\ln(r)} \qquad (7.9.15)$$

where $P_i(r)$ is defined in (7.9.8). Thus, for $q > 0$, spheres with larger P have more influence in determining D_q.

When $q = 0$, (7.9.15) reduces to D_0, the capacity dimension defined in (7.9.1). To evaluate (7.9.15) at $q = 1$, we use L'Hospital's rule to determine the limit and obtain

$$D_1 = \lim_{r \to 0} \sum_{i=1}^{N} \frac{P_i(r) \ln P_i(r)}{\ln(r)}$$

in agreement with the information dimension defined in (7.9.11). When $q = 2$, (7.9.15) becomes

$$D_2 = \lim_{r \to 0} \frac{\ln \sum_{i=1}^{N} P_i^2(r)}{\ln(r)}$$

which is equivalent to the correlation dimension defined in (7.9.14) (Parker and Chua, 1989).

We note that D_q decreases as q increases except in the case when the points are distributed fairly uniformly on the attractor. In the latter case, $D_q = D_0$, the capacity dimension.

7.9.6 Lyapunov Dimension

Kaplan and Yorke (1979a,b) and Fredickson, Kaplan, Yorke, and Yorke (1983) proposed an interesting relationship between the attractor dimension and its Lyapunov exponents $\lambda_1 \geq \lambda_2 \geq \lambda_3 \geq \ldots \geq \lambda_n$. For dissipative systems, volumes in state space contract with time, and as $t \to \infty$, the volume of the attractor in the original state space is zero. Consequently, the sum of the Lyapunov exponents is negative.

To define the Lyapunov dimension, we note that if we can find an integer m such that $\sum_{i=1}^{m} \lambda_i = 0$, then the volume in a subspace R^m of R^n is constant and hence there is an m–dimensional attractor in R^n. For a strange attractor, we cannot find such an m. Therefore, we find the largest M such that

$$\sum_{i=1}^{M} \lambda_i > 0 \text{ and } \sum_{i=1}^{M+1} \lambda_i < 0$$

Thus, the volume shrinks in R^{M+1}, but it is undefined in R^M. To make the volume constant in some subspace, Kaplan and Yorke proposed to take into account a fraction of the contracting Lyapunov exponent λ_{M+1} and defined the Lyapunov dimension D_L as

$$D_L = M + \frac{\sum_{i=1}^{M} \lambda_i}{\mid \lambda_{M+1} \mid}$$

Because all other negative Lyapunov exponents have been neglected, D_L is an upper limit for the capacity dimension D_0; that is, $D_L \geq D_0$.

7.9.7 Comments

In the pointwise and correlation dimension calculations discussed above, the number of points in spheres of different chosen radii about any number of arbitrary reference points is counted, and then these counts are used to determine the dimension. In an alternative approach, the dimension of an attractor is estimated as follows (Termonia and Alexandrowicz, 1983). Spheres centered about arbitrary reference points with chosen numbers of points are considered, and the radii corresponding to the different chosen counts are estimated. Then, the dimension is determined from a plot of radius versus count. This latter approach is an example of **constant mass methods**, while the earlier mentioned approaches are examples of **constant volume methods**.

The issue of what the length of a data set should be for a correlation dimension calculation is addressed by Eckmann and Ruelle (1992). They show that the dimension determined for data with N points cannot exceed $2\log_{10}N$. As discussed earlier, noise is also an important factor to reckon with during dimension calculations (e.g., Abarbanel, Brown, Sidorowich, and Tsimring, 1993; Kostelich and Schreiber, 1993). Methods for noise reduction need to be used with caution during dimension calculations (e.g., Badii, Broggi, Derighetti, Ravani, Ciliberto, Politi, and Rubio, 1988; Mitschke, Möller, and Lange, 1988).

7.10   HIGHER–ORDER SPECTRA

As discussed in Section 7.5, one of the most useful tools to characterize nonlinear motions is the power spectrum. We recall that for a real-valued finite–duration stationary signal $x(t)$ with zero mean, the second–order autocorrelation function is defined by

$$R_{xx}(\tau) = E\left[x(t)x(t+\tau)\right] = \lim_{T\to\infty} \frac{1}{T}\int_0^T x(t)x(t+\tau)dt \qquad (7.10.1)$$

where $E[\dots]$ denotes the expected value. The autospectrum S_{xx} is then obtained by taking the Fourier transform of $R_{xx}(\tau)$; that is,

$$S_{xx}(f) = \mathcal{F}\left[R_{xx}(\tau)\right] \qquad (7.10.2)$$

where $\mathcal{F}$ denotes the Fourier transform. Alternatively, the discrete autospectrum, appropriate for discretely sampled data, is given by

$$S_{xx}(f) = \lim_{T\to\infty} \frac{1}{T}E\left[X_T(f)X_T^*(f)\right] \qquad (7.10.3)$$

where $X(f)$ is the complex Fourier transform of $x(t)$, the asterisk represents the complex conjugate, and T is the time–record length.

While the auto–power spectrum gives an estimate of the distribution of power among the frequency components of the signal, it has no phase information and hence it cannot differentiate between modes that are independently excited (i.e., modes whose phases are random relative to each other) from those that are nonlinearly coupled (i.e., modes that interact, resulting in a coupling of their phases). Such a distinction can be made by using **higher–order spectra**, also called **polyspectra**, because they isolate and quantify any nonlinearly induced phase coupling among Fourier modes.

The autospectrum $S_{xx}(f)$ is an example of **second-order spectra**. In the hierarchy of polyspectra, the next higher-order spectrum is the **third-order spectrum** called **auto–bispectrum** (Brillinger, 1965, 1981; Kim and Powers, 1979; Papoulis, 1990; Nikias and Petropulu, 1993). The auto–bispectrum is formally defined as the two–dimensional Fourier transform of the **third–order correlation function**; that is,

$$S_{xxx}(f_i, f_j) = \mathcal{F}\left[R_{xxx}(\tau_i, \tau_j)\right] \qquad (7.10.4)$$

where

$$R_{xxx}(\tau_i, \tau_j) = E\left[x(t)x(t + \tau_i)x(t + \tau_j)\right] \qquad (7.10.5)$$

is the third–order correlation function. Alternatively, the discrete auto–bispectrum, appropriate for discretely sampled data, is

$$S_{xxx}(f_i, f_j) = E\left[X(f_i + f_j)X^*(f_i)X^*(f_j)\right] \qquad (7.10.6)$$

Next, we let

$$X(f_m) = \mid X_m \mid e^{i\theta_m} \qquad (7.10.7)$$

where $\mid X_m \mid$ and θ_m are the magnitude and phase of $X(f_m)$. Then, we rewrite (7.10.6) as

$$S_{xxx}(f_i, f_j) = E\left[\mid X_k \mid\mid X_i \mid\mid X_j \mid e^{i(\theta_k - \theta_i - \theta_j)}\right] \qquad (7.10.8)$$

where $f_k = f_i + f_j$. Clearly, the auto–bispectrum measures the degree of coherence among the modes i, j, and $k = i + j$. If the three modes having the spectral components $X(f_k), X(f_i)$, and $X(f_j)$ are independently excited, then their phases θ_k, θ_i, and θ_j are statistically independent random functions. Hence, the phase difference $\theta_k - \theta_i - \theta_j$ is randomly distributed over $[0, -2\pi)$. Consequently, carrying out the statistical averaging indicated in (7.10.8) yields a zero value for $S_{xxx}(f_i, f_j)$. On the other hand, if the three modes are fully coupled through a quadratic nonlinearity, then the phase difference $\theta_k - \theta_i - \theta_j$ is not random although each of the individual phases θ_k, θ_i, and θ_j is random. Under these conditions, (7.10.8) can be rewritten as

$$S_{xxx}(f_i, f_j) = e^{i(\theta_k - \theta_i - \theta_j)} E\left[\| X_k \| \| X_i \| \| X_j \|\right] \tag{7.10.9}$$

Consequently, carrying out the statistical averaging in (7.10.9) yields a nonzero value for $S_{xxx}(f_i, f_j)$ if $| X_k |, | X_i |$, and $| X_j |$ are not zero. If the three modes are partially coupled, then (7.10.8) can be split into two parts: one part with a random phase difference and the other with a coherent phase difference. Consequently, $S_{xxx}(f_i, f_j) \neq 0$.

Because $X(-f_i) = X^*(f_i)$, the auto–bispectrum possesses the following symmetry relations (Kim and Powers, 1979):

$$S(f_i, f_j) = S(f_j, f_i) = S^*(-f_i, -f_j) = S(-f_i - f_j, f_j) = S(f_i, -f_i - f_j) \tag{7.10.10}$$

Therefore, the auto–bispectrum is uniquely described by its values within a triangle in the $f_1 - f_2$ plane with the vertices $(0,0), (0, f_N)$ and $\left(\frac{1}{2}f_N, \frac{1}{2}f_N\right)$, where $f_N = 1/2\tau_s$ is the Nyquist frequency with τ_s being the sampling time.

The auto–bispectrum is usually normalized with the amplitudes of the individual spectral components to yield the **auto–bicoherence** defined as

$$b^2(f_i, f_j) = \frac{| S_{xxx}(f_i, f_j) |^2}{E[| X(f_i)X(f_j) |^2]E[| X(f_i + f_j) |^2]} \tag{7.10.11}$$

We note that $0 \leq b^2 \leq 1$ by Schwartz inequality. A near zero value for the auto–bicoherence $b^2(f_i, f_j)$ of three modes indicates that they are not quadratically coupled, whereas a value near one for $b^2(f_i, f_j)$ indicates perfect quadratic coupling. Any value for $b^2(f_i, f_j)$ between zero and one indicates partial quadratic coupling.

To illustrate how the bicoherence measures the degree of phase coherence among three modes, we consider the time series

$$s(t) = \cos(2\pi f_1 t + \theta_1) + \cos(2\pi f_2 t + \theta_2) + \cos(2\pi f_3 t + \theta_3) + n(t) \tag{7.10.12}$$

where $n(t)$ is a white Gaussian noise (-20 dB) and the frequencies of the modes are

$$\frac{f_1}{f_N} = \frac{15}{64}, \ \frac{f_2}{f_N} = \frac{20}{64}, \ \text{and} \ \frac{f_3}{f_N} = \frac{35}{64}$$

We generated 64 records of $s(t)$ with each record containing 128 data points. A small–amplitude white Gaussian noise was added to each record. The phases $\theta_1, \theta_2,$ and θ_3 were chosen from a set of random numbers that are uniformly distributed over the interval $[0, 2\pi)$. The resulting time records are expected to have Gaussian statistics. In Figure 7.10.1 we show the power spectrum of one of the records. There are three peaks corresponding to $f_1, f_2,$ and f_3. Because the power spectrum is independent of the phases of the modes, all records will have identical power spectra. On the other hand, it is clear from Figure 7.10.2 that the bicoherence is nearly zero, thereby reflecting the lack of phase coherence among the three modes.

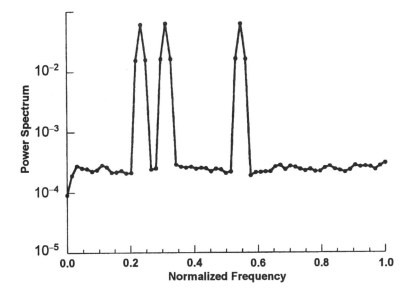

Figure 7.10.1: The power spectrum of the time series defined in (7.10.12): $f_1/f_N = 15/64$, $f_2/f_N = 20/64$, and $f_3/f_N = 35/64$.

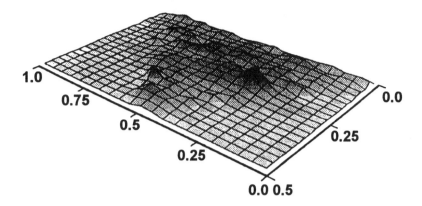

Figure 7.10.2: The bicoherence of the time series defined in (7.10.12) and Figure 7.10.1 when the θ_i are uniformly distributed random numbers over $[0, 2\pi)$.

Next, we generated 64 records with 128 points per record as in the previous case. Again, the phases θ_1 and θ_2 were chosen from a set of random numbers that are uniformly distributed over $[0, 2\pi)$, but θ_3 was chosen to be equal to $\theta_1 + \theta_2$. Because the power spectrum is independent of the phases of the modes, the power spectra in this case are identical to that in Figure 7.10.1. However, because of the phase coherence in this case, the bicoherence is nearly unity, as shown in Figure 7.10.3.

The above definitions of the third–order correlation function and auto–bispectrum can be extended to define a fourth–order correlation function and its Fourier transform, the **auto–trispectrum.** The auto–trispectrum can also be expressed as

$$S_{xxxx}(f_i, f_j, f_k) = \lim_{T \to \infty} \frac{1}{T} E\left[X_T(f_i + f_j + f_k)X_T^*(f_i)X_T^*(f_j)X_T^*(f_k)\right]$$
$$(7.10.13)$$

The auto–trispectrum and its normalized value, the **auto–tricoherence,** can then be used to investigate cubic nonlinearities.

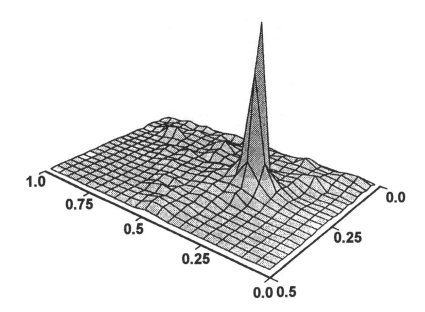

Figure 7.10.3: The bicoherence of the time series defined in (7.10.12) and Figure 7.10.1 when the θ_1 and θ_2 are uniformly distributed random numbers over $[0, 2\pi)$ and $\theta_3 = \theta_1 + \theta_2$.

In many cases, it is required to study the nonlinear relations between two time series, such as the excitation $x(t)$ and the response $y(t)$ of a nonlinear system. In these cases, one can define cross–correlation functions and their Fourier transforms, **cross–bispectra** and **cross-trispectra**. These transforms can be normalized to yield **cross–bicoherence** and **cross–tricoherence**. The cross-bispectrum is defined by

$$S_{yxx}(f_i, f_j) = \lim_{T \to \infty} \frac{1}{T} E\left[Y_T(f_i + f_j) X_T^*(f_j) X_T^*(f_j)\right] \qquad (7.10.14)$$

For illustration, we consider the quadratically coupled oscillators (7.5.14) and (7.5.15) and set $\omega_1 = 1.0$, $\omega_2 = 2.05$, $\mu_1 = \mu_2 = 0.001$, $\delta_1 = 0$, $\delta_2 = 0.002$, $\alpha_1 = 0.001$, $\alpha_2 = 0$, $F = 270$, and $\Omega = 2.03$. We numerically integrated (7.5.14) and (7.5.15) from a chosen initial condition. For computing the bispectra, we collected (steady-state) time histories after $t = 19,000$ units with a sampling frequency of

2 Hz. Thirty two records with 2,048 points each were used for the computations. In Figures 7.10.4a and 7.10.4b the steady-state time history of $u_1(t)$ and associated power spectrum are depicted, respectively. Both the time history and power spectrum indicate that the corresponding oscillations are periodic. In Figure 7.10.5a the following digital cross-bispectrum is depicted:

$$S_{u_2 u_1 u_1} = E\left[U_2(f_1 + f_2)U_1^*(f_1)U_1^*(f_2)\right] \qquad (7.10.15)$$

and in Figure 7.10.5b the cross–bicoherence is depicted. There is

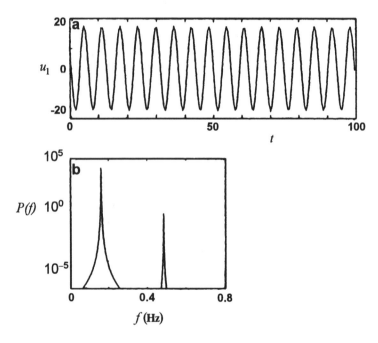

Figure 7.10.4: (a) Time series of u_1 and (b) power spectrum.

a peak at $\left(\frac{\Omega}{4\pi}, \frac{\Omega}{4\pi}\right)$ in Figure 7.10.5a, indicating a quadratic coupling between u_1 and u_2. Although the oscillator governing u_1 is not directly excited by the forcing, u_1 has a non-zero steady-state response due to the quadratic coupling between u_1 and u_2.

There are many studies in which the bispectrum has been used to investigate nonlinear motions. A partial list includes those of

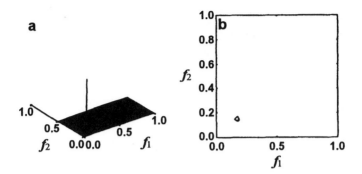

Figure 7.10.5: (a) Cross–bispectrum and (b) cross–bicoherence.

Sato, Sasaki, and Nakamura (1977), Kim and Powers (1979), Kim, Beall, and Powers (1980), Choi, Chang, Stearman, and Powers (1984), Choi, Powers, and Miksad (1985), Miller (1986), Nikias and Raghuveer (1987), Ritz and Powers (1986), Ritz, Powers, Miksad, and Solis (1988), Gifford and Tomlinson (1989), and Hajj, Miksad, and Powers (1992).

7.11   EXERCISES

7.1. Show that if $x(t) = c + a \cos ft + b \sin ft$ then

$$\overline{x(t)x(t+\tau)} = c^2 + \frac{1}{2}(a^2 + b^2) \cos ft$$

and that if $x(t) = c + a_1 \cos f_1 t + a_2 \cos f_2 t$ for $f_2 \neq \pm f_1$ then

$$\overline{x(t)x(t+\tau)} = c^2 + \frac{1}{2}a_1^2 \cos f_1 t + \frac{1}{2}a_2^2 \cos f_2 t$$

7.2. Show that the nonautonomous first–order equation

$$\frac{dx}{dt} = -x + \epsilon \cos t$$

is equivalent to the autonomous second–order system

$$\frac{dx}{dt} = -x + \epsilon \cos \theta, \quad \frac{d\theta}{dt} = 1, \quad \text{with } \theta(t_0) = t_0$$

Represent the orbit $\{x(t)\} \subset S^1 \times \mathcal{R}$ with cylindrical coordinates (r, θ, x) as points $[1, \theta(t), x(t)]$ on the surface of a cylinder with unit radius, so that $(1, \theta, x)$ and $(1, \theta + 2n\pi, x)$ coincide for $n = \pm 1, \pm 2, \dots$. Then construct the Poincaré map $P : \Sigma \to \Sigma$, where $\Sigma = \{(r, \theta, x) : r = 1, \theta = \theta_0\}$ is the line of intersection of the orbits with the half–plane $\theta = \theta_0$, deducing that

$$P(1, \theta_0, x) = \left[1, \; \theta_0, \; x e^{-2\pi} + \frac{1}{2}\epsilon(1 - e^{-2\pi})(\cos \theta_0 + \sin \theta_0) \right]$$

Find the fixed point X of P and show that $X \to 0$ as $\epsilon \to 0$ for all θ_0. For what values of θ_0 and ϵ is X stable?

7.3. Consider the tent map

$$x_{n+1} = \begin{cases} 2a x_n & ; \; 0 \le x_n \le \frac{1}{2} \\ 2a(1 - x_n); & \frac{1}{2} \le x_n \le 1 \end{cases}$$

Use the fact that for a one–dimensional map the Lyapunov exponent is defined by

$$\lambda(x_0) = \lim_{n \to \infty} \frac{1}{n} \ln \left| \frac{d}{dx} f^{(n)}(x) \right|_{x=x_0} |$$

to show that, for $a = 1$, the Lyapunov exponent for the tent map is $\lambda = \ln 2$.

Hint: Use the chain rule of differentiation to show that

$$\lambda(x_0) = \lim_{n \to \infty} \frac{1}{n} \ln \left(\sum_{i=0}^{n-1} \ln | f'(x_i) | \right)$$

7.4. Construct a fractal that is similar to the Cantor set, but instead remove the middle $\frac{1}{2}$ from each previous section. Show that its dimension is $\frac{1}{2}$.

7.5. Construct a fractal that is similar to the Cantor set, but instead remove the middle x fraction from each previous section. Show that its

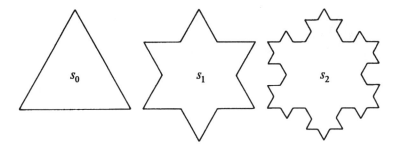

Figure 7.11.1: Construction of the Koch snowflake.

capacity dimension is

$$D_c = \frac{\ln 2}{\ln 2 - \ln(1 - x)}$$

7.6. Show that the capacity dimension of the Koch snowflake, whose construction is indicated in Figure 7.11.1, is

$$D_0 = \frac{2\ln 2}{\ln 3} \approx 1.2618$$

7.7. Let S_0 be the interior of an equilateral triangle with sides of length l. Regard S_0 as the union of the interiors of four equilateral triangles with sides of length $\frac{1}{2}l$, and define S_1 as S_0 less the interior of the middle of four triangles. Similarly, remove the middles of the three triangles of S_1 to construct S_2, and so forth (see Fig. 7.11.2). Define $S = \bigcap_{n=0}^{\infty} S_n$.

Show that the box dimension of S is $D = \ln 3/\ln 2 = 1.58496$.

7.8. Define self–similar sets $S_n \subset \mathcal{R}^{\in}$ as follows for $n = 0, 1, \ldots$. First define $S_0 = \{(x,y): 0 \le x, y \le 1\}$, the unit square; $S_1 = \{(x,y): (x,y) \notin S_0 \text{ and } x \text{ or } y \in \left(\frac{1}{3}, \frac{2}{3}\right)\}$, i.e., S_1 is the union of eight of the nine subsquares of S_0 with sides of length $\frac{1}{3}$, the central subsquare

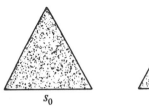  

Figure 7.11.2: A sketch of the sets for Exercise 7.7.

being removed; S_2 as S_1 with the central subsquare of side $\frac{1}{3^2}$ removed from each subsquare of S_1, as shown in Figure 7.11.3; and so forth. Hence, define $S = \bigcap_{n=0}^{\infty} S_n$.

 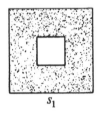 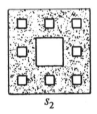

Figure 7.11.3: A sketch of the sets for Exercise 7.8.

Show that the area of S is zero and that the box dimension of S is $D = \ln 8 / \ln 3 = 1.8928$.

7.9. The Cantor set may be used to study some properties of information dimension. Assuming that each line segment is equally probable at each iteration. Show that the information dimension $D_1 = \ln 2 / \ln 3$.

7.10. Repeat the calculation of Exercise 7.8, but do not assume equal probabilities. At every iteration of the set, let the right segment have twice the probability of the left segment. For example, when there are four segments the probabilities are, from left to right, $\frac{1}{9}, \frac{2}{9}, \frac{2}{9}$, and $\frac{4}{9}$. Show that $D_1 = -1 + 2(\ln 2 / \ln 3)$.

7.11. By finding the maximum of the function $I = -\sum_{i=1}^{N} P_i \ln P_i$, show that D_1 is maximized when $D_1 = D_0$.

7.12. Determine the autocorrelation functions of

(a) $x(t) = a \cos(\omega t + \beta)$

(b) $x(t) = a_1 \cos \omega t + a_2 \cos 2\omega t$

(c) $x(t) = a_0 + a_1 \cos \omega t + a_2 \cos 2\omega t$

(d) $x(t) = a_1 \cos \frac{1}{2}\omega t + a_2 \cos \omega t$

(e) $x(t) = a_1 \cos \omega_1 t + a_2 \cos \omega_2 t$

7.13. Show that the Lyapunov exponent of the tent map

$$x_{n+1} = 2rx_n \qquad \text{for } x_n < \frac{1}{2}$$
$$x_{n+1} = 2r(1 - x_n) \qquad \text{for } x_n \geq \frac{1}{2}$$

is $\ln 2r$.

7.14. Show that the Lyapunov exponent of the Bernoulli map

$$x_{n+1} = 2x_n \pmod 1$$

is $\ln 2$.

7.15. Show that the Lyapunov exponent of the logistic map

$$x_{n+1} = 4x_n(1 - x_n)$$

is $\ln 2$.

Chapter 8

CONTROL

8.1 CONTROL OF BIFURCATIONS

As discussed in the preceding chapters, bifurcations can be broadly classified into two categories: (a) continuous and (b) discontinuous or catastrophic bifurcations. During continuous bifurcations, the states of the system vary continuously or in a gradual manner as a system parameter is varied in a quasistationary manner across the bifurcation points. Examples of these bifurcations are supercritical pitchfork, transcritical, and supercritical Hopf bifurcations of fixed points and limit cycles. On the other hand, during discontinuous bifurcations, the state of the system experiences a jump as a system parameter is varied in a quasistationary manner across the bifurcation points. Examples of these bifurcations are saddle–node bifurcations of equilibrium points, cyclic–fold bifurcations of limit cycles, and subcritical pitchfork and Hopf bifurcations of fixed points and limit cycles. In a discontinuous case, the postbifurcation states cannot be determined from local considerations alone. The discontinuous bifurcation may lead to an unbounded motion, an oscillatory behavior, or an intermittent or steady chaotic behavior unless appropriate controls are applied.

In the context of ship motions, the unbounded motions resulting from a discontinuous bifurcation can lead to capsizing (Thompson, Rainey, and Soliman, 1990; Nayfeh and Sanchez, 1990; Sanchez and Nayfeh, 1990; Soliman, 1993). In the context of power systems, discontinuous bifurcations can result in a loss of synchronism (Nayfeh, Hamdan, and Nayfeh, 1990, 1991) or a voltage collapse (Abed, Hamdan, Lee, and Parlos, 1990; Abed, Wang, Alexander, Hamdan, and Lee, 1993; Chiang, Dobson, Thomas, Thorp, and Fekih–Ahmed, 1990; Vu and Liu, 1990; Ajjarapu and Lee, 1992; Chow, Fischl, and Yan, 1990; Dobson and Chiang, 1989; Kwatny, Pasrija, and Bahar, 1986; Rajagopalan, Sauer, and Pai, 1989; Venkatasubramanian, Schättler, and Zaborszky, 1992). Furthermore, discontinuous bifurcations can result in low–performance operations in aircraft systems (Abed and Lee, 1990) and axial flow compressors (Greitzer, 1976; Moore and Greitzer, 1986; Davis and O'Brien, 1987; Abed, Houpt, and Hosny, 1993).

Different methods can be used to control the discontinuous bifurcations to achieve desirable nonlinear dynamics (e.g., Abed and Fu, 1986, 1987; Cibrario and Lévine, 1991; Singer and Bau, 1991; Singer, Wang, and Bau, 1991; Abed, Wang, Alexander, Hamdan, and Lee, 1993; Adomaitis and Abed, 1993; Day, 1993; Wang, Abed, and Hamdan, 1992a,b; Wang and Abed, 1992, 1993, 1994; Abed and Wang, 1994). These methods fall under the category of bifurcation control methods, which are commonly used for one or more of the following purposes: (a) shifting the bifurcation points in the state–control space, (b) suppressing the bifurcations in a sequence, (c) changing the nature of a bifurcation, and (d) changing the character of a bifurcation set. Effectively, the control inputs modify the bifurcation characteristics associated with the system. The control schemes may be implemented either with or without feedback. In the latter case, we have open–loop control. In the study of Tung and Shaw (1988), open–loop control has been suggested for achieving desirable dynamics of impacting print hammers, in which chaotic motions have been observed (Hendriks, 1983). Braiman and Goldhirsch (1991) examined open–loop control of chaotic dynamics of a nonlinear system by applying weakly periodic perturbations. Here, we follow the work of Abed and co–workers to describe feedback–based bifurcation control methods.

8.1.1 Static Feedback Control

In static feedback control, the feedback is used to achieve desirable nonlinear dynamics. For instance, one can suppress discontinuous bifurcations in systems of the form

$$\dot{\mathbf{x}} = \mathbf{F}(\mathbf{x}; \mu) + \mathbf{u}$$

where $\mathbf{x}$ is the n-dimensional state vector, μ is the scalar parameter with respect to which the bifurcations are studied, and $\mathbf{u}$ is the static feedback given by

$$\mathbf{u} = \mathbf{u}(\mathbf{x})$$

Abed and Fu (1986, 1987) illustrated how $\mathbf{u}$ can be chosen to suppress discontinuous bifurcations of fixed points such as subcritical Hopf bifurcations. Here, we consider conversion of a subcritical Hopf bifurcation into a supercritical Hopf bifurcation by using a nonlinear static feedback. In the system (3.5.48), let the matrix J have the pair of purely imaginary eigenvalues $\pm i\omega$. To this system, we add the nonlinear control

$$\epsilon \mathbf{Q}_u(\mathbf{y}, \mathbf{y}) + \epsilon^2 \mathbf{C}_u(\mathbf{y}, \mathbf{y}, \mathbf{y})$$

and obtain

$$\begin{aligned}
\dot{\mathbf{y}} = J\mathbf{y} &+ \epsilon \left[\mathbf{Q}(\mathbf{y}, \mathbf{y}) + \mathbf{Q}_u(\mathbf{y}, \mathbf{y}) \right] \\
&+ \epsilon^2 \left[\mathbf{C}(\mathbf{y}, \mathbf{y}, \mathbf{y}) + \mathbf{C}_u(\mathbf{y}, \mathbf{y}, \mathbf{y}) \right] + \cdots
\end{aligned} \tag{8.1.1}$$

Using the method of multiple scales and carrying out an analysis similar to that conducted in Section 3.5.3, we find that $\mathbf{y}_1$ is still given by (3.5.53) and $\mathbf{y}_2$ is given by

$$\mathbf{y}_2 = 2\left(\mathbf{z}_0 + \mathbf{w}_0\right) A\bar{A} + 2(\mathbf{z}_2 + \mathbf{w}_2)A^2 e^{2i\omega T_0} + cc \tag{8.1.2}$$

where $\mathbf{z}_0$ and $\mathbf{z}_2$ are given by (3.5.59) and (3.5.60), and

$$J\mathbf{w}_0 = -\frac{1}{2}\mathbf{Q}_u(\mathbf{p}, \bar{\mathbf{p}}) \tag{8.1.3}$$

$$(2i\omega - J)\mathbf{w}_2 = \frac{1}{2}\mathbf{Q}_u(\mathbf{p}, \mathbf{p}) \tag{8.1.4}$$

Then, the equation governing A is given by

$$A' = \mu\beta_1 A + 4\hat{\beta}_2 A^2 \bar{A} \tag{8.1.5}$$

where β_1 is given by (3.5.63),

$$\hat{\beta}_2 = \beta_2 + \Delta \tag{8.1.6}$$

and

$$\Delta = 2\mathbf{q}^T \mathbf{Q}_u(\mathbf{p}, \mathbf{z}_0) + \mathbf{q}^T \mathbf{Q}_u(\bar{\mathbf{p}}, \mathbf{z}_2) + 3\mathbf{q}^T \mathbf{C}_u(\mathbf{p}, \mathbf{p}, \bar{\mathbf{p}}) \\ + 2\mathbf{q}^T \mathbf{Q}(\mathbf{p}, \mathbf{w}_0) + \mathbf{q}^T \mathbf{Q}(\bar{\mathbf{p}}, \mathbf{w}_2) \tag{8.1.7}$$

Expressing A in the polar form $\frac{1}{2}a\exp(i\theta)$ and separating real and imaginary parts in (8.1.5), we obtain

$$a' = \mu\beta_{1r}a + \hat{\beta}_{2r}a^3 \tag{8.1.8}$$
$$a\theta' = \mu\beta_{1i}a + \hat{\beta}_{2i}a^3 \tag{8.1.9}$$

Consequently, to transform the subcritical Hopf bifurcation in which $\beta_{2r} > 0$ into a supercritical Hopf bifurcation, we need to choose the feedback control so that Real $(\beta_2 + \Delta) < 0$.

Example 8.1. Following Wang and Abed (1994), we consider the power model of Dobson and Chiang (1989). The system dynamics is governed by the following four–dimensional system:

$$\dot{\delta}_m = \omega \tag{8.1.10}$$
$$\dot{\omega} = \frac{50}{3}V\sin(\delta - \delta_m + 0.0873) - \frac{1}{6}\omega + 1.8807 \tag{8.1.11}$$
$$\dot{\delta} = -\frac{2000}{3}V\cos(\delta - 0.2094) - \frac{500}{3}V\cos(\delta - \delta_m - 0.0873) \\ + 496.8718V^2 - \frac{280}{3}V + \frac{100}{3}Q + \frac{130}{3} \tag{8.1.12}$$
$$\dot{V} = -78.7638V^2 + 26.2172V\cos(\delta - \delta_m - 0.0124) \\ + 104.8689V\cos(\delta - 0.1346) + 14.5229V \\ - 5.2288Q - 7.0327 \tag{8.1.13}$$

In Figure 8.1.1, we show the bifurcation diagram generated by Wang and Abed (1994) when Q is used as the bifurcation parameter. The symbols HB, CFB, PDB, and SNB represent Hopf, cyclic–fold, period–doubling, and saddle–node bifurcations, respectively. The solid and

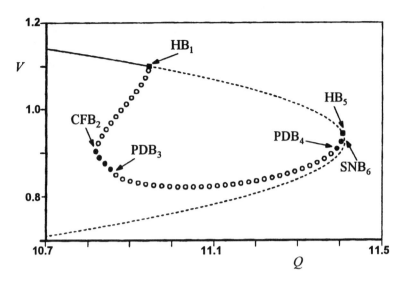

Figure 8.1.1: Bifurcation diagram for the uncontrolled power model. Reprinted with permission from Wang and Abed (1994).

dashed lines are used to depict the loci of stable and unstable equilibrium states, respectively. The solid and open circles are used to denote stable and unstable oscillatory states, respectively. The parameter Q takes the value Q_1 at the subcritical Hopf bifurcation point HB$_1$ and Q_5 at the supercritical Hopf bifurcation point HB$_5$. As Q is varied gradually from 10.70, the system remains in a static state until Q exceeds Q_1, where a subcritical Hopf bifurcation occurs. The postbifurcation state quickly becomes unbounded, resulting in what is called a **voltage collapse.**

To control the subcritical Hopf bifurcation, Wang and Abed (1994) introduced the nonlinear feedback control law

$$\mathbf{u}^T = \begin{bmatrix} 0 & 0 & 0 & -k\omega^3 \end{bmatrix} \qquad (8.1.14)$$

where k is the feedback control gain. Using the gain of 0.7003, Wang and Abed obtained the bifurcation diagram shown in Figure 8.1.2. The subcritical Hopf bifurcation at Q_1 has been transformed into a supercritical Hopf bifurcation, the unstable limit cycles have been eliminated, and the amplitudes of the stable limit cycles born as a

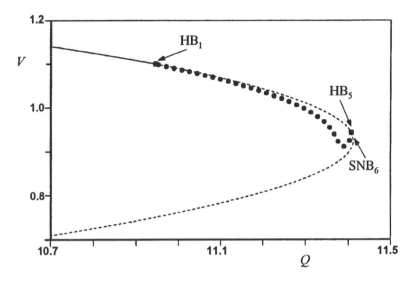

Figure 8.1.2: Bifurcation diagram for the controlled power model. Reprinted with permission from Wang and Abed (1994).

result of the Hopf bifurcation are small. Consequently, voltage collapse has been delayed to the saddle–node bifurcation point.

d'Andréa–Novel and Abichou (1992) used static feedback to control a pitchfork bifurcation in an inverted–pendulum system. For other applications of static feedback control, we mention the studies of Liaw and Abed (1990), Henrich, Mingori, and Monkewitz (1992), Badmus, Chowdhury, Eveker, Nett, and Rivera (1993a,b), and Mohamed and Emad (1993).

8.1.2   Dynamic Feedback Control

In some cases, static feedback is not desirable because the locations of equilibria are affected in the controlled system. When dynamic feedback control is used, it is possible to preserve the equilibrium positions in the controlled system (e.g., Hyötyniemi, 1991; Abed and Wang, 1994; Wang and Abed, 1992, 1994). Abed and co–workers implemented dynamic feedback by using what are called **washout filters**. Let us suppose that we have the following uncontrolled n–

dimensional autonomous system.

$$\dot{\mathbf{x}} = \mathbf{F}(\mathbf{x}; \mu)$$

where μ is the scalar parameter with respect to which we study the bifurcations. When a washout filter is used for dynamic feedback, the controlled system takes the form

$$\dot{\mathbf{x}} = \mathbf{F}(\mathbf{x}; \mu) + \mathbf{u}$$
$$\dot{z}_i = x_i - d_i z_i$$
$$y_i = x_i - d_i z_i$$
$$\mathbf{u} = \mathbf{u}(\mathbf{y})$$

The differential equation describing z_i governs the washout filter used for the state variable x_i. The equation describing y_i is called the **output equation.** It is easy to verify that the equilibrium positions are preserved in the controlled system.

Example 8.2. Again, following Wang and Abed (1994), we consider the convection model (Singer, Wang, and Bau, 1991)

$$\dot{x}_1 = -px_1 + px_2 \qquad\qquad (8.1.15)$$
$$\dot{x}_2 = -x_2 - x_1 x_3 \qquad\qquad (8.1.16)$$
$$\dot{x}_3 = -x_3 + x_1 x_2 - r \qquad\qquad (8.1.17)$$

where p and r are positive constants. In Figure 8.1.3, we show the bifurcation diagram for $p = 4.0$. Again, the solid and dashed curves are used to represent the loci of the stable and unstable equilibria, respectively. The solid and open circles represent the stable and unstable limit cycles, respectively. Clearly, there is a subcritical Hopf bifurcation at $r = 16$. Singer, Wang, and Bau (1991) employed a linear feedback control to delay the occurrence of the Hopf bifurcation and as a result suppressed chaotic motions.

To transform the subcritical Hopf bifurcation into a supercritical Hopf bifurcation, Wang and Abed (1994) used a linear feedback aided with a washout filter for the state x_3 and obtained the closed–loop

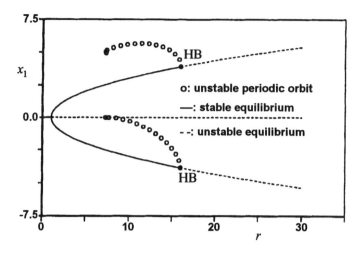

Figure 8.1.3: Bifurcation diagram for the uncontrolled convection model. Reprinted with permission from Wang and Abed (1994).

system

$$\dot{x}_1 = -px_1 + px_2 \tag{8.1.18}$$

$$\dot{x}_2 = -x_2 - x_1 x_3 \tag{8.1.19}$$

$$\dot{x}_3 = -x_3 + x_1 x_2 - r + u \tag{8.1.20}$$

$$\dot{x}_4 = x_3 - dx_4 \tag{8.1.21}$$

where x_4 is the washout filter state and the control law is of the form

$$u = -ky^3 \tag{8.1.22}$$

with y an output variable given by

$$y = x_3 - dx_4 \tag{8.1.23}$$

Here, k is a scalar feedback gain.

For $p = 4.0$, Wang and Abed (1994) generated the bifurcation diagram in Figure 8.1.4 for different values of the feedback gain k. Clearly, using gain values equal to or larger than 0.025 transforms the subcritical Hopf bifurcation into a supercritical Hopf bifurcation. Moreover, increasing the value of k reduces the amplitude of the limit cycles born as a result of the bifurcation.

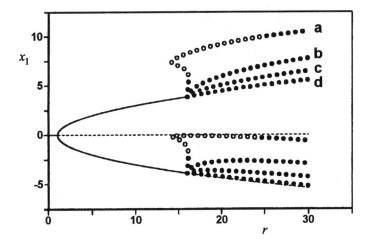

Figure 8.1.4: Bifurcation diagram for the controlled convection model: (a) $k = 0.009$, (b) $k = 0.025$, (c) $k = 0.1$, and (d) $k = 2.5$. Reprinted with permission from Wang and Abed (1994).

8.1.3   Comments

In recent work, Abed, Wang, and Chen (1994) described feedback control schemes for controlling period–doubling bifurcations in discrete systems. By using the Hénon map as an example, they illustrated how a control scheme can be used to suppress a period–doubling cascade to chaos.

8.2   CHAOS CONTROL

In practice, chaos may be desirable or undesirable, depending on the application. In combustion applications, chaos is desirable because it enhances mixing of air and fuel and hence leads to a better performance. On the other hand, in aerodynamic and hydrodynamic applications, chaos (turbulence) is undesirable because it dramatically increases the drag of vehicles and results in increased operational cost. In mechanical and structural systems, chaos may lead to irregular operations and fatigue failure. Moreover, chaos can restrict the operating range of many electronic and mechanical devices. As an example, we mention the study of Hendriks (1983), where the dynamics of impact printers was considered.

In the preceding section, we discussed different bifurcation control methods. These methods can be used to control the bifurcations (e.g., period–doubling, Hopf, cyclic–fold bifurcations) in transitions to chaos and, hence, keep the system dynamics away from chaos. In Section 8.2.1, we describe an approach developed by Ott, Grebogi, and Yorke (1990a,b) for controlling chaos by utilizing the butterfly–effect characteristic of chaos. This approach is usually referred to as the **OGY scheme**. In Section 8.2.2 we describe some experimental implementations of the OGY scheme, and in Sections 8.2.3 and 8.2.4 we discuss the pole placement technique and some of the traditional control schemes.

8.2.1   The OGY Scheme

The OGY approach (Ott, Grebogi, and Yorke, 1990a,b; Shinbrot, Gregobi, Ott, and Yorke, 1993; Ott, 1993) exploits the following characteristics of chaotic attractors: butterfly effect, recurrent behavior, and presence of an infinite number of unstable periodic orbits and the absence of any stable ones. The main goal of this approach is the stabilization of an unstable periodic orbit embedded within the chaotic attractor by means of a small time–dependent change in an accessible system parameter. This can be accomplished because of the extreme sensitivity of the attractor to small perturbations and its recurrent behavior, as discussed next.

The first step in this approach is the identification of the unstable periodic orbits. To accomplish this, we let $x_1, x_2, \cdots, x_n, \cdots, x_N$ be the vectors specifying the intersections of the chaotic trajectory with a chosen one–sided Poincaré section $\sum$. Thus, we define a Poincaré map that maps the nth intersection x_n of the trajectory with this section into the subsequent intersection x_{n+1} by

$$x_{n+1} = P(x_n; M) \tag{8.2.1}$$

with M being a set of control parameters at our disposal. In general, the map $\mathbf{P}$ is not known explicitly. However, local approximations to $\mathbf{P}$ can be determined using the available points $\mathbf{x}_n$, as described below.

We note that a periodic orbit corresponds to a finite number of points on the Poincaré section. Therefore, to determine the locations of the unstable periodic orbits from the above points of intersection of the trajectory with the Poincaré section, we choose a small positive number ϵ. Then, for each vector $\mathbf{x}_i$, we find the smallest index $j > i$ such that

$$\| \mathbf{x}_i - \mathbf{x}_j \| < \epsilon$$

Then, typically there is a period–p orbit, where $p = j - i$, near $\mathbf{x}_i$ and $\mathbf{x}_j$. For example, there is a period–one orbit near $\mathbf{x}_{30}$ if $\| \mathbf{x}_{30} - \mathbf{x}_{31} \| < \epsilon$, and there is a period–two orbit near $\mathbf{x}_{30}$ if $\| \mathbf{x}_{30} - \mathbf{x}_{32} \| < \epsilon$ but $\| \mathbf{x}_{30} - \mathbf{x}_{31} \| > \epsilon$. In this way, we identify approximately the locations of the periodic orbits.

To get a better approximation to the location of a selected orbit of, say, period p, we first find the first pair $\mathbf{x}_i$ and $\mathbf{x}_{i+p}$ where $i + p$ is the smallest index such that $\| \mathbf{x}_i - \mathbf{x}_{i+p} \| < \epsilon$. Then, we search the succeeding data for all pairs $\mathbf{x}_k$ and $\mathbf{x}_{k+p}$ such that $\| \mathbf{x}_k - \mathbf{x}_i \| < \epsilon_2$ and $\| \mathbf{x}_{k+p} - \mathbf{x}_i \| < \epsilon_2$, where ϵ_2 is another chosen small positive number. We label the resulting points by $\boldsymbol{\xi}_1, \boldsymbol{\xi}_2, \cdots, \boldsymbol{\xi}_k$. Then, we fit these points with a linear relationship of the form

$$\boldsymbol{\xi}_{n+p} = A\boldsymbol{\xi}_n + \mathbf{c} \tag{8.2.2}$$

In the presence of noise, a least–squares fit can be used. The fixed point $\boldsymbol{\xi}^*$ of (8.2.2) is given by

$$\boldsymbol{\xi}^* = A\boldsymbol{\xi}^* + \mathbf{c} \tag{8.2.3}$$

or

$$\boldsymbol{\xi}^* = (I - A)^{-1}\mathbf{c} \tag{8.2.4}$$

Moreover, the stable and unstable eigenvectors associated with $\boldsymbol{\xi}^*$ are the eigenvectors of A corresponding to the eigenvalues inside and outside the unit circle, respectively.

We note that the location $\boldsymbol{\xi}^*$ of the periodic orbit on the Poincaré section depends on the control parameters $\mathbf{M}$. To effect the control, we need to vary as many control parameters as there are unstable eigenvectors associated with $\boldsymbol{\xi}^*$. For further description, we assume that

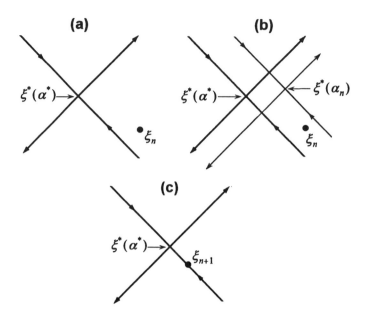

Figure 8.2.1: A schematic of the OGY control method.

the dynamical system is three–dimensional, and hence the Poincaré section is two–dimensional. Consequently, unstable periodic orbits of the dynamical system correspond to saddle points on the section with one stable and one unstable eigenvector. We assume that the accessible control parameter is α and that the fixed point $\xi^*(\alpha^*)$ corresponds to $\alpha = \alpha^*$. In Figure 8.2.1a, we show a schematic of the location of the fixed point and its accompanying stable and unstable directions, which are approximations to the stable and unstable manifolds of the fixed point. Also shown is a state of the system ξ_n approaching the vicinity of $\xi^*(\alpha^*)$ along its stable manifold.

The basic idea is to vary α in the neighborhood of α^* when the uncontrolled trajectory comes close to $\xi^*(\alpha^*)$. The recurrent property of the chaotic attractor guarantees that it will frequently approach the vicinity of the chosen unstable fixed point and then move away again. The approach is always along the stable manifold, and the departure is always along the unstable manifold, as shown in Figure 8.2.1. However, the time taken for the trajectory to come on its own close to $\xi^*(\alpha^*)$

may be long. To shorten this time, one can take advantage of the extreme sensitivity of the chaotic attractor to small perturbations and utilize a so–called targeting scheme to steer the chaotic trajectory to the vicinity of $\xi^*(\alpha^*)$ by using small controls (Shinbrot, Ott, Grebogi, and Yorke, 1990; Shinbrot, Gregobi, Ott, and Yorke, 1992; Kostelich, Grebogi, Ott, and Yorke, 1993). When the state of system comes close to $\xi^*(\alpha^*)$, we want to vary α quickly by a small amount in such a way that the subsequent state of the system will land on the stable manifold of the original fixed point $\xi^*(\alpha^*)$, and consequently the evolution will approach it as $t \to \infty$.

In Figure 8.2.1a, we show the state ξ_n of the system approaching the vicinity of $\xi^*(\alpha^*)$ along the stable manifold. Without control, this state would approach the vicinity of $\xi^*(\alpha^*)$ along the stable manifold and then move away from $\xi^*(\alpha^*)$ along the unstable manifold. Therefore, to prevent this state from leaving the vicinity of $\xi^*(\alpha^*)$ (and in fact forcing it to land on the stable manifold), we quickly perturb the control parameter from α^* to α_n to move the old unstable fixed point and its accompanying manifolds to the new location $\xi^*(\alpha_n)$, as shown in Figure 8.2.1b. Consequently, the state of the system will approach the vicinity of the new fixed point $\xi^*(\alpha_n)$ along its stable manifold and move away from it along its unstable manifold. However, this departure is in the direction of the stable manifold of the old fixed point. By properly choosing α_n, one can make the subsequent state ξ_{n+1} of the system fall precisely on the stable manifold of $\xi^*(\alpha^*)$ and then immediately remove the perturbation of the control parameter, as shown in Figure 8.2.1c. Subsequently, in a perfect world, the state of the system would approach $\xi^*(\alpha^*)$ and stay there forever. However, in the presence of noise or a small error in calculating α_n, the state of the system would not stay at $\xi^*(\alpha^*)$. To overcome these problems, the control is repeated once each period.

To determine α_n, we approximate the map in the neighborhood of $\xi^*(\alpha_n)$ by the linear map

$$\xi_{n+1} - \xi^*(\alpha_n) = D_\xi \mathbf{P} \left[\xi_n - \xi^*(\alpha_n) \right] \qquad (8.2.5)$$

where $D_\xi \mathbf{P}$ is the Jacobian of the map evaluated at α_n. Because α_n is close to α^*, the new fixed point can be related to the old fixed point by

$$\mathbf{g} = \frac{\boldsymbol{\xi}^*(\alpha_n) - \boldsymbol{\xi}^*(\alpha^*)}{\alpha_n - \alpha^*} \tag{8.2.6}$$

or

$$\boldsymbol{\xi}^*(\alpha_n) = \boldsymbol{\xi}^*(\alpha^*) + (\alpha_n - \alpha^*)\mathbf{g} \tag{8.2.7}$$

The vector $\mathbf{g}$ can be calculated ahead of time. The position of the unstable fixed point is determined numerically from a model or is measured experimentally for several values of α just slightly different from α^*. Then, these values are used in (8.2.6) to determine $\mathbf{g}$. Because α_n is close to α^*, $D_{\boldsymbol{\xi}}\mathbf{P}$ can be approximated by its value A at α^*. We denote the stable and unstable unit eigenvectors of A corresponding to the eigenvalues λ_s and λ_u, where $|\lambda_s| < 1$ and $|\lambda_u| > 1$ by $\mathbf{e}_s$ and $\mathbf{e}_u$. Moreover, we denote the corresponding contravariant basis vectors by $\mathbf{f}_s$ and $\mathbf{f}_u$. They are related to $\mathbf{e}_s$ and $\mathbf{e}_u$ by

$$\mathbf{f}_s \cdot \mathbf{e}_s = \mathbf{f}_u \cdot \mathbf{e}_u = 1 \ \text{ and } \ \mathbf{f}_s \cdot \mathbf{e}_u = \mathbf{f}_u \cdot \mathbf{e}_s = 0 \tag{8.2.8}$$

Using the unit vectors and their contravariant basis vectors, we express A as

$$A = \lambda_u \mathbf{e}_u \mathbf{f}_u + \lambda_s \mathbf{e}_s \mathbf{f}_s \tag{8.2.9}$$

Using (8.2.7) and (8.2.9), we rewrite (8.2.5) as

$$\begin{aligned} \boldsymbol{\xi}_{n+1} - \boldsymbol{\xi}^*(\alpha^*) = \delta\alpha_n \mathbf{g} \\ + (\lambda_u \mathbf{e}_u \mathbf{f}_u + \lambda_s \mathbf{e}_s \mathbf{f}_s) \cdot \left[\boldsymbol{\xi}_n - \boldsymbol{\xi}^*(\alpha^*) - \delta\alpha_n \mathbf{g}\right] \end{aligned} \tag{8.2.10}$$

where $\delta\alpha_n = \alpha_n - \alpha^*$.

In order that $\boldsymbol{\xi}_{n+1}$ fall precisely on the stable manifold of $\boldsymbol{\xi}^*(\alpha^*)$, we choose $\delta\alpha_n$ so that

$$\mathbf{f}_u \cdot \left[\boldsymbol{\xi}_{n+1} - \boldsymbol{\xi}^*(\alpha^*)\right] = 0 \tag{8.2.11}$$

Hence, taking the inner product of (8.2.10) with $\mathbf{f}_u$ and using (8.2.11), we obtain

$$\delta\alpha_n(\mathbf{f}_u \cdot \mathbf{g}) + \lambda_u \mathbf{f}_u \cdot [\boldsymbol{\xi}_n - \boldsymbol{\xi}^*(\alpha^*)] - \lambda_u \delta\alpha_n (\mathbf{f}_u \cdot \mathbf{g}) = 0$$

or

$$\delta\alpha_n = \frac{\lambda_u \mathbf{f}_u \cdot \left[\boldsymbol{\xi}_n - \boldsymbol{\xi}^*(\alpha^*)\right]}{(\lambda_u - 1)\mathbf{f}_u \cdot \mathbf{g}} \tag{8.2.12}$$

or

$$\delta\alpha_n = C\mathbf{f}_u \cdot \left[\boldsymbol{\xi}_n - \boldsymbol{\xi}^*(\alpha^*)\right] \tag{8.2.13}$$

where

$$C = \frac{\lambda_u}{(\lambda_u - 1)\mathbf{f}_u \cdot \mathbf{g}} \tag{8.2.14}$$

Equation (8.2.13) yields the perturbation in the control parameter needed to control the system. Due to noise or errors in the calculations, such as the influence of nonlinearities, $\delta\alpha_n$ is updated every period to keep the trajectory on the desired orbit. We note that $\delta\alpha_n$ is proportional to the projection of the distance of the state $\boldsymbol{\xi}_n$ of the system from the unstable fixed point $\boldsymbol{\xi}^*(\alpha^*)$ onto the unstable direction $\mathbf{f}_u$. The gain C can be calculated ahead of time by using $\lambda_u, \mathbf{f}_u$, and $\mathbf{g}$, which can be determined from experimental or numerical data, as described above. Alternatively, if a period of adjustment in an experiment can be tolerated, one might guess a value for C and then adjust it empirically until control is satisfied (e.g., Hunt, 1991; Singer, Wang, and Bau, 1991; Gills, Iwata, Roy, Schwartz, and Triandaf, 1992; Rajarshi, Murphy, Maier, Gills, and Hunt, 1992; Garfinkel, Spano, Ditto, and Weiss, 1992; Petrov, Gaspar, Masere, and Showalter, 1993).

In summary, the characteristics of chaotic attractors permit the use of small feedback perturbations to control trajectories in chaotic systems, thereby effecting a large beneficial change in the long–term system behavior. Moreover, using a similar reasoning, we conclude that one can switch between different orbits by again using small perturbations.

8.2.2   Implementation of the OGY Scheme

Ditto, Rauseo, and Spano (1990) implemented the OGY scheme experimentally with a parametrically excited cantilever ribbon about 100 mm long, 3 mm wide, and 0.025 mm thick, as shown in Figure 8.2.2. This ribbon is made of an amorphous magnetoelastic material whose Young's modulus of elasticity varies nonlinearly with an applied magnetic field.

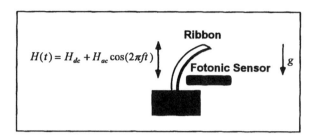

Figure 8.2.2: A schematic of the magnetoelastic ribbon experiment.

Initially, the ribbon buckles under the influence of the gravity field. Ditto et al. (1990) applied a magnetic field H to the ribbon in the vertical direction of the form

$$H = H_{dc} + H_{ac} \cos\left(2\pi ft\right) \qquad (8.2.15)$$

where H_{dc} is a constant field and H_{ac} and f are the amplitude and frequency of the oscillatory field. In this experiment, H_{dc}, H_{ac}, and f are accessible control parameters. Ditto et al. set $H_{ac} = 2.050$ Oe and $f = 0.85$ Hz and used H_{dc} as the accessible control parameter.

The position of the ribbon was measured at a point near its base by means of an optical sensor. The output of the sensor consisted of time–series voltages. The voltages were sampled at the drive period $T = 1/f$ of the oscillatory part of the field (at times $t_n = n/f$) by triggering a voltmeter from the ac signal, resulting in the sampled voltages $V_n = V(t_n)$.

When $H_{dc} = 0.112$, the motion of the ribbon is chaotic. In Figure 8.2.3a, we show variation of the position of the ribbon with respect to time. The data shown for $t_n < 2,350$ correspond to the uncontrolled motion, while the data shown for $t_n > 2,350$ correspond to the controlled motion. Clearly, the uncontrolled motion is chaotic. This is also evident from the associated return map shown in Figure 8.2.3b. Applying the OGY scheme to stabilize the period–one orbit, Ditto et al. obtained the controlled motion illustrated in Figure 8.2.3a. In the associated return map, we have a point attractor. This attractor is located at about (3.4, 3.4) in Figure 8.2.3b.

The empirical version of the OGY scheme has been used to control chaos in a number of physical systems. Hunt (1991) used occasional

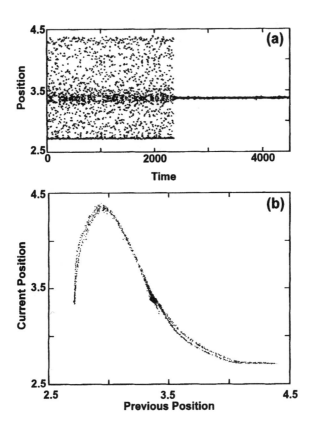

Figure 8.2.3: Experimental observations of the ribbon motion: (a) variation of the ribbon position with time and (b) return map of the ribbon position. Reprinted with permission from Ditto, Rauseo, and Spano (1990).

proportional feedback to stabilize many higher–order periodic orbits in a periodically driven diode resonator. He sampled the peak current I_n, and if it was within a given window, he modulated the drive voltage with a signal proportional to the difference between I_n and the center of the window. Peng, Petrov, and Showalter (1991) and Petrov, Peng, and Showalter (1992) used a similar proportional feedback scheme to stabilize periodic orbits in the chaotic regime of a chemical system, which was modeled by a three–dimensional autonomous system. Petrov, Gaspar, Masere, and Showalter (1993) also used the same approach to control chaos in the Belousov–Zhabotinsky reaction. Roy, Murphy, Maier, and Gills (1992) used a proportional feedback scheme to stabilize complex periodic waveforms in the output intensity of a chaotic multimode laser. As a result, they were able to operate the laser in a stable manner at power levels exceeding by as much as a factor of 15 those that might be attained without control. Singer, Wang, and Bau (1991) used a proportional feedback scheme to suppress chaotic flow in a thermal convection loop. Garfinkel, Spano, Ditto, and Weiss (1992) used an empirical version of the OGY scheme to stabilize cardiac arrhythmias induced by the drug ouabain in rabbit ventricle. By administering electrical stimuli to the heart at irregular times determined by using the OGY scheme, they converted the arrhythmia to periodic beating.

As aforementioned, one can easily use small perturbations in the control and switch from one periodic orbit to another. In Figure 8.2.4, the results of Ditto et al. show switching from chaos to period–four to period–one to period–two to period–one and finally to period–four. There are interludes of chaos between each section of control because Ditto et al. waited until the state of the ribbon came on its own to the vicinity of the desired orbit. These interludes of chaos can be significantly reduced by using the targeting scheme discussed earlier.

8.2.3   Pole Placement Technique

The OGY scheme described in Section 8.2.1 is a special case of the general technique called **pole placement** in control theory. We return to the n–dimensional map (8.2.1) and consider the case in which the state of the dynamical system depends on a single control parameter α. We let $\mathbf{x} = \mathbf{x}(\alpha^*)$ be the unstable fixed point of the map corresponding

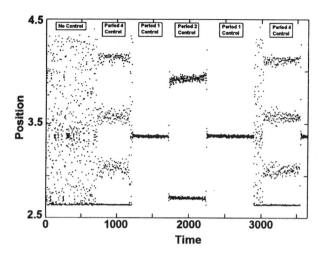

Figure 8.2.4: The ability of the OGY scheme to switch from controlling one periodic orbit to controlling another. Reprinted with permission from Ditto, Rauseo, and Spano (1990).

to the unstable periodic motion we wish to stabilize. Linearizing the map (8.2.1) around $\mathbf{x}^*$ and α^*, we have

$$\delta\mathbf{x}_{n+1} = A\delta\mathbf{x}_n + \mathbf{b}\delta\alpha_n \qquad (8.2.16)$$

where

$$\delta\mathbf{x}_m = \mathbf{x}_m - \mathbf{x}^*(\alpha^*), \quad \delta\alpha_m = \alpha_m - \alpha^* \qquad (8.2.17)$$

Here, A is an $n \times n$ constant matrix and $\mathbf{b}$ is an n–dimensional vector given by

$$A = \frac{\partial \mathbf{P}}{\partial \mathbf{x}}(\mathbf{x}^*, \alpha^*) \text{ and } \mathbf{b} = \frac{\partial \mathbf{P}}{\partial \alpha}(\mathbf{x}^*, \alpha^*) \qquad (8.2.18)$$

We assume that $\mathbf{b}$ satisfies the so–called controllability condition

$$\mathbf{q}^T\mathbf{b} \neq 0 \qquad (8.2.19)$$

where $\mathbf{q}$ stands for any of the left eigenvectors of A. Then, we consider the case of a linear feedback of the form

$$\delta\alpha_n = \mathbf{k}^T\delta\mathbf{x}_n \qquad (8.2.20)$$

where $\mathbf{k}$ is an n–dimensional gain vector. Substituting for $\delta\alpha_n$ from (8.2.20) into (8.2.16) yields the closed loop problem

$$\delta\mathbf{x}_{n+1} = \left[A + \mathbf{b}\mathbf{k}^T\right]\delta\mathbf{x}_n \qquad (8.2.21)$$

To stabilize $\mathbf{x}^*(\alpha^*)$, we need to choose the control gains so that all of the eigenvalues of the $n \times n$ matrix $[A + \mathbf{b}\mathbf{k}^T]$ are inside the unit circle in the complex plane. This is possible because $\mathbf{b}$ satisfies the controllability condition (8.2.19). The choice $\delta\alpha_n$ in Section 8.2.1 corresponds to a special choice of the control gain $\mathbf{k}$, where one of the eigenvalues of $[A + \mathbf{b}\mathbf{k}^T]$ is zero while the other eigenvalue is unaltered. Again, in the implementation of this scheme, one can either wait for the state of the system to come close to the desired unstable state $\mathbf{x}^*$ on its own or use a targeting scheme to steer the state quickly to the vicinity of $\mathbf{x}^*$. Then, one varies the control so that all of the eigenvalues of $[A + \mathbf{b}\mathbf{k}^T]$ are inside the unit circle. Romeiras, Grebogi, Ott, and Dayawansa (1992) implemented the pole placement technique to stabilize fixed points in a chaotic four–dimensional map representing the kicked double rotor.

Dressler and Nitsche (1992) pointed out that, when the attractor is reconstructed from a time series using time delay coordinates, the relevant Poincaré map representing the dynamics of the system depends on both the current value α_n of the control parameter and the previous value α_{n-1}; that is,

$$\mathbf{x}_{n+1} = \mathbf{P}(\mathbf{x}_n, \alpha_n, \alpha_{n-1}) \qquad (8.2.22)$$

Consequently, the proportional feedback control law is modified to

$$\delta\alpha_n = \mathbf{k}^T\delta\mathbf{x}_n + r\delta\alpha_{n-1} \qquad (8.2.23)$$

where $\mathbf{k}$ and r are constant vector and scalar parameters. Rollins, Parmananda, and Sherard (1993) implemented this modified scheme to stabilize periodic orbits in chaotic biological and chemical systems.

8.2.4 Traditional Control Methods

The schemes described in Sections 8.2.1–8.2.3 take advantage of the characteristics of chaotic attractors and utilize small changes in one or more control parameters when the state of the system comes close to the desired orbit on its own or due to targeting. In contrast, one can influence the chaotic dynamics by using traditional control techniques (e.g., Hübler, 1989; Hübler and Lüscher, 1989; Lüscher and Hübler, 1989; Jackson and Hübler, 1990; Jackson, 1991a,b; Jackson and Kodogeorgiou, 1992). To illustrate this approach, we assume that the dynamics of the uncontrolled system is given by

$$\dot{\mathbf{x}} = \mathbf{F}(\mathbf{x}, t; \mathbf{M}) \qquad (8.2.24)$$

and that $\mathbf{x}^*(t)$ represents the goal dynamics; that is, the trajectory that we are interested in stabilizing. Then, we introduce an additive feedback control term $\mathbf{u}(t)$ to (8.2.24) and obtain

$$\dot{\mathbf{x}} = \mathbf{F}(\mathbf{x}, t; \mathbf{M}) + \mathbf{u}(t) \qquad (8.2.25)$$

The objective of the control is to choose $\mathbf{u}(t)$ so that $\mathbf{x}(t) \to \mathbf{x}^*(t)$ as $t \to \infty$. Thus, $\mathbf{u}(t)$ can be expressed as

$$\mathbf{u}(t) = \mathbf{G}(\mathbf{x} - \mathbf{x}^*, t; \mathbf{M}) \qquad (8.2.26)$$

where $\mathbf{G}$ is, in general, nonlinear.

The control objective can be accomplished by the simple choice

$$\mathbf{u}(t) = \dot{\mathbf{x}}^* - \mathbf{F}(\mathbf{x}^*, t; \mathbf{M}) \qquad (8.2.27)$$

With this choice, (8.2.25) becomes

$$\dot{\mathbf{x}} - \dot{\mathbf{x}}^* = \mathbf{F}(\mathbf{x}, t; \mathbf{M}) - \mathbf{F}(\mathbf{x}^*, t; \mathbf{M}) \qquad (8.2.28)$$

Clearly, $\mathbf{x}(t) = \mathbf{x}^*(t)$ is a solution of the controlled problem. However, it is not clear whether $\| \mathbf{x}(t) - \mathbf{x}^*(t) \| \to 0$ as $t \to \infty$. The latter depends on the vector function $\mathbf{F}$ as well as on the initial conditions. Jackson (1991b) addresses the convergence to the goal dynamics in the context of flows, while Jackson and Kodogeorgiou (1992) address convergence to the goal dynamics in the context of two–dimensional maps. The control is initiated only when the system is in the basin of "entrainment" of the desired solution.

8.3   SYNCHRONIZATION

Many physical phenomena can be modeled by dynamical systems driven by external forces and moments. Such systems can be represented by systems of first–order equations having the form

$$\dot{\mathbf{w}} = \mathbf{f}(\mathbf{w}; \mathbf{v}) \qquad\qquad (8.3.1)$$

where the vector $\mathbf{w}$ has k components and the vector $\mathbf{v}$ has m components and stands for the external forces and moments that drive the system. In general, these forces and moments are the output of a second dynamical system, called the **drive system,** which can be represented by a system of first–order equations having the form

$$\dot{\mathbf{u}} = \mathbf{g}(\mathbf{u}, \mathbf{v}) \qquad\qquad (8.3.2)$$

$$\dot{\mathbf{v}} = \mathbf{h}(\mathbf{u}, \mathbf{v}) \qquad\qquad (8.3.3)$$

The vector $\mathbf{u}$ has l components. The system $(8.3.1)$–$(8.3.3)$ is called the **combined system;** it is n–dimensional, where $n = k + m + l$. The system $(8.3.2)$ and $(8.3.3)$ is called the **drive system,** whereas the system $(8.3.1)$ is called the **response system.** The behavior of the response system depends on the behavior of the drive system, but the drive system is not influenced by the response system. We have split the drive system into two parts. The first part represents the variables $\mathbf{u}$ that are not involved in driving the response system, and the second part represents the variables $\mathbf{v}$ that are actually involved in driving the response system.

As discussed in this and the preceding chapters, the behavior of the drive system $(8.3.2)$ and $(8.3.3)$ may be constant, periodic, quasiperiodic, or chaotic. In the accompanying chapters, we have considered systems that are driven by harmonic forces and moments. In such cases, the drive system can be represented by

$$\dot{u} = v \text{ and } \dot{v} = -\Omega^2 u \qquad\qquad (8.3.4)$$

where Ω is the drive or excitation frequency.

In this section, we consider drive systems whose behavior is chaotic so that the response is driven by chaotic forces and moments. Moreover,

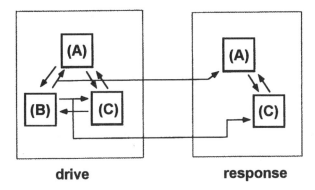

Figure 8.3.1: A schematic diagram of a synchronizing chaotic system. The response system is a duplicate of part of the drive system. The response system is driven by the signal that came from the remaining part of the system.

we investigate whether the behavior of the response system can be synchronized to that of the drive system. Here, by **synchronization** we mean that the two systems follow the same trajectory in phase space, in contrast with the notion we used earlier to describe the phenomenon in which two self–excited oscillators have the same frequency or the response of a self–excited oscillator is at the same frequency as the excitation.

Typically, due to sensitivity to initial conditions, two isolated chaotic systems cannot synchronize. In fact, if we were to construct two chaotic systems that are virtually identical but separate, we would find that they would quickly fall out of step because any minute differences in the initial conditions would cause the two trajectories to diverge exponentially with time. However, Pecora and Carroll (1990, 1991) discovered that a chaotic system could be built in such a way that its parts could act in perfect synchrony. A schematic of such a system is shown in Figure 8.3.1.

The key to synchronization is that the response system is identical to a part of the drive system and that the response system is stable, as described later. As an example, we consider the Lorenz system

$$\dot{x} = \sigma(y - x) \qquad (8.3.5)$$

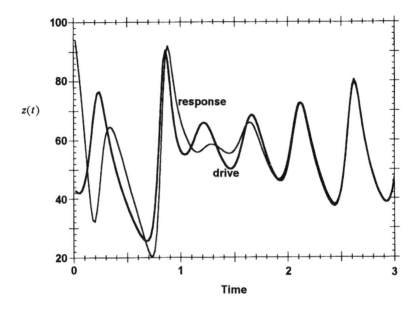

Figure 8.3.2: Comparison of the time series obtained from the response system with that obtained from the drive system. The two converge in a few cycles of the drive. Reprinted with permission from Pecora and Carroll (1991).

$$\dot{y} = -xz + \rho x - y \qquad (8.3.6)$$
$$\dot{z} = xy - \beta z \qquad (8.3.7)$$

where $\sigma = 16, \beta = 4$, and $\rho = 45.92$ as the drive system. For the response system, we choose a subsystem consisting of the x and z equations; that is,

$$\dot{x}' = \sigma(y - x') \qquad (8.3.8)$$
$$\dot{z}' = x'y - \beta z' \qquad (8.3.9)$$

The response system (8.3.8) and (8.3.9) is driven by the variable y obtained from the drive system (8.3.5)–(8.3.7). Pecora and Carroll (1991) found that the z' variable in the response system converges to the z variable in the drive system within a few cycles of the drive, as shown in Figure 8.3.2.

Synchronization could also be achieved if the subsystem consisting of the y and z equations is used as the response system (Pecora and Carroll, 1991); that is,

$$\dot{y}' = -xz' + \rho x - y' \tag{8.3.10}$$

$$\dot{z}' = xy' - \beta z' \tag{8.3.11}$$

In this case, the response system is driven by the x variable from the drive system (8.3.5)–(8.3.7). However, a subsystem consisting of the x and y equations does not synchronize with the drive system because this subsystem is unstable, whereas the subsystems (x, z) and (y, z) are stable.

For the response subsystem to be stable and hence synchronize to the drive system, we demand that, for a fixed set of drive initial conditions, wherever $\mathbf{w}(t)$ is initiated it will always converge to the same trajectory and each point in time will always be at the same predictable position in phase space on that trajectory. In the case of multiple basins of attraction, the $\mathbf{w}$ initiations need to be in the same basin of attraction. Consequently, for a given drive $\mathbf{v}$, if $\mathbf{w}_1$ and $\mathbf{w}_2$ are two trajectories corresponding to two nearby initial conditions, then $\delta\mathbf{w} = \mathbf{w}_2 - \mathbf{w}_1$ must tend to zero as $t \rightarrow \infty$. Letting $\mathbf{w} = \mathbf{w}_1 + \delta\mathbf{w}$ in (8.3.1), expanding for small $\delta\mathbf{w}$, and keeping linear terms, we obtain

$$\delta\dot{\mathbf{w}} = D_{\mathbf{w}}\mathbf{f}(\mathbf{w}_1; \mathbf{v})\delta\mathbf{w} + \ldots \tag{8.3.12}$$

Then, the subsystem (8.3.1) is stable if all of the k Lyapunov exponents calculated from (8.3.1)–(8.3.3) and (8.3.12) are negative. These exponents are called **conditional Lyapunov exponents** because they are neither the same as nor a subset of the Lyapunov exponents calculated from (8.3.1)–(8.3.3) alone. We note that there are n Lyapunov exponents associated with an orbit of (8.3.1)–(8.3.3).

Carroll and Pecora (1993) showed that one can cascade several dynamical systems together and achieve synchronization in all of them by using the same drive. For example, because there are two possible stable subsystems that can be used as response systems in the Lorenz system, one can cascade these systems as shown in Figure 8.3.3. The y signal from the drive system drives the (x', z') subsystem, whose x' signal in turn drives the (y'', z'') subsystem. The signal y'' from

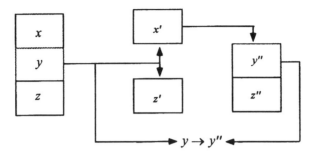

Figure 8.3.3: A block diagram of cascading the synchronized chaotic Lorenz equations.

the (y'', z'') subsystem matches the y signal from the drive system. Consequently, one can think of the cascaded response system as a black box whose input and output are chaotic. If the response systems are synchronized with the drive system, then the chaotic output y'' of the cascaded system matches the input signal y'. However, if a parameter is changed in the drive system, synchronism will be lost and the output and input signals of the black box do not match. Therefore, Carroll and Pecora (1993) propose the use of chaotic signals as carriers for communication.

BIBLIOGRAPHY

Abarbanel, H. D. I. (1995). Tools for analyzing observed chaotic data, in *Stability, Vibration and Control of Structures,* A. Guran and D. J. Inman, eds., World Scientific, Singapore. *485, 496, 497, 499, 534–537*

Abarbanel, H. D. I., R. Brown, and J. B. Kadtke (1990). Prediction in chaotic nonlinear systems: Methods for time series with broadband Fourier spectra, *Phys. Rev. A* 41, 1782–1807. *483*

Abarbanel, H. D. I., R. Brown, and M. B. Kennel (1991). Variation of Lyapunov exponents on a strange attractor, *J. Nonl. Sci.* 1, 175-199. *533*

Abarbanel, H. D. I., R. Brown, J. J. Sidorowich, and L. S. Tsimring (1993). The analysis of observed chaotic data in physical systems, *Rev. Mod. Phys.* 65, 1331–1392. *463, 481, 482, 484, 492–494, 497, 499, 532, 534, 538, 550*

Abarbanel, H. D. I., M. I. Rabinovich, and M. M. Sushchik (1993). *Introduction to Nonlinear Dynamics for Physicists,* World Scientific, Singapore.

Abarbanel, H. D. I. See Brown, Bryant, and Abarbanel; Bryant, Brown, and Abarbanel; Kennel, Brown, and Abarbanel.

Abbott, T. See Tufillaro, Abbott, and Reilly.

Abed, E. H. (1988). A simple proof of stability on the center manifold for Hopf bifurcation, *SIAM Rev.* 30, 487–491.

Abed, E. H. (1994). Bifurcation–theoretic issues in the control of voltage collapse, in *Proceedings of the IMA Workshop on System and Control Theory for Power Systems,* J. H. Chow, P. V. Kokotovic, and R. J. Thomas, eds., Springer–Verlag, New York. *77*

Abed, E. H., and J.-H. Fu (1986). Local feedback stabilization and bifurcation control. I. Hopf bifurcation, *Syst. Control Lett.* 7, 11–17. *219, 564, 565*

Abed, E. H., and J.-H. Fu (1987). Local feedback stabilization and bifurcation control. II. Stationary bifurcation, *Syst. Control Lett.* 8, 467–473. *564, 565*

Abed, E. H., A. M. A. Hamdan, H.-C. Lee, and A. G. Parlos (1990). On bifurcations in power system models and voltage collapse, in *Proceedings of the 29th IEEE Conference on Decision and Control,* Honolulu, Hawaii, 3014–3015. *564*

Abed, E. H., P. K. Houpt, and W. M. Hosny (1993). Bifurcation analysis of surge and rotating stall in axial flow compressors, *ASME J. Turbom.* 115, 817–824. *564*

Abed, E. H., and H.-C. Lee (1990). Nonlinear stabilization of high angle of attack flight dynamics using bifurcation control, in *Proceedings of the 1990 American Control Conference,* San Diego, California, 2235–2238. *564*

Abed, E. H., and P. P. Varaiya (1984). Nonlinear oscillations in power systems, *Elec. Power Energy Syst.* 6, 37–43.

Abed, E. H., and H. O. Wang (1994). Feedback control of bifurcation and chaos in dynamical systems, in *Recent Developments in Stochastic and Nonlinear Dynamics,* a volume commemorating the 60th birthday of Professor S. T. Ariaratnam, N. Sri Namachchivaya and W. Kliemann, eds., CRC Press, Boca Raton, Florida. *564, 568*

Abed, E. H., H. Wang, J. C. Alexander, A. M. A. Hamdan, and H.-C. Lee (1993). Dynamic bifurcations in a power system model exhibiting voltage collapse, *Int. J. Bif. Chaos* 3, 1169–1176. *564*

Abed, E. H., H. O. Wang, and R. C. Chen (1994). Stabilization of period doubling bifurcations and implications for control of chaos, *Physica D* 70, 154–164. *571*

Abed, E. H. See Adomaitis and Abed; Fu and Abed; Lee and Abed; Liaw and Abed; Wang and Abed; Wang, Abed, and Hamdan.

Abhyankar, N. S., E. K. Hall, and S. V. Hanagud (1993). Chaotic vibrations of beams: Numerical solution of partial differential equations, *J. Appl. Mech.* 60, 167–174.

Abichou, A. See d'André–Novel and Abichou.

Abou–Rayan, A. M., A. H. Nayfeh, D. T. Mook, and M. A. Nayfeh (1993). Nonlinear response of a parametrically excited buckled beam,

Nonlinear Dyn. 4, 499–525. *292*

Abraham, N. B., J. P. Gollub, and H. L. Swinney (1984). Meeting report: Testing nonlinear dynamics, *Physica* D11. 252–264. *200*

Abraham, R. H. (1985). Chaostrophes, intermittency, and noise, in *Chaos, Fractals, and Dynamics,* P. Fischer and W. Smith, eds., Marcel Dekker, New York, 3–22. *68, 69, 335, 339*

Abraham, R. H., and K. A. Scott (1985). Chaostrophes of forced van der Pol systems, in *Chaos, Fractals, and Dynamics,* P. Fischer and W. Smith, eds., Marcel Dekker, New York, 123–134. *269*

Abraham, R. H., and C. D. Shaw (1992). *Dynamics: The Geometry of Behavior,* Addison–Wesley, Redwood City, California. *69, 93, 240, 290, 335, 339, 413*

Abraham, R. H., and C. Simó (1986). Bifurcations and chaos in forced van der Pol systems, in *Dynamical Systems and Singularities,* S. Pnevmatikos, ed., North–Holland, Amsterdam, 313–323. *269*

Abraham, R. H., and H. B. Stewart (1986). A chaotic blue sky catastrophe in forced relaxation oscillations, *Physica D* 21, 394–400. *348*

Adomaitis, R. A., and E. H. Abed (1993). Local nonlinear control of stall inception in axial flow compressors, AIAA Paper No. 93–2230. *564*

Afraimovich, V. S., V. V. Bykov, and L. P. Shilnikov (1977). On the origin and structure of the Lorenz attractor, *Sov. Phys. Dokl.* 22, 253–255. *399*

Afraimovich, V. S., and L. P. Shilnikov (1983a). Invariant two–dimensional tori, their destruction and stochasticity, in *Methods of Qualitative Theory of Differential Equations,* Gorki University Press, Gorki, Russia [in Russian]. *323*

Afraimovich, V. S., and L. P. Shilnikov (1983b). On strange attractors and quasi–attractors, in *Nonlinear Dynamics and Turbulence,* G. I. Barenblatt, G. Iooss, and D. D. Joseph, eds., Pitman, Marshfield, Massachusetts.

Ajjarapu, V., and B. Lee (1992). Bifurcation theory and its application to nonlinear dynamical phenomena in an electrical power system, *IEEE Trans. Power Syst.* 7, 424–431. *564*

Aleksic, Z. (1991). Estimating the embedding dimension, *Physica D* 52, 362–368. *494*

Alexander, J. C. See Abed, Wang, Alexander, Hamdan, and Lee.

Alexandrowicz, Z. See Termonia and Alexandrowicz.

Allen, T. (1983). On the arithmetic of phase locking: Coupled neurons as a lattice on R^2, *Physica D* 6, 305–320. *243, 247*

Allgower, E., and K. Georg (1980). Simplical and continuation methods for approximating fixed points and solutions to systems of equations, *SIAM Rev.* 22, 28–85. *425, 435*

Allgower, E., and K. Georg (1990). *Numerical Continuation Methods: An Introduction,* Springer–Verlag, New York. *42, 425, 435, 442*

Alligood, K. T., E. D. Yorke, and J. A. Yorke (1987). Why period-doubling cascades occur: Periodic orbit creation followed by stability shedding, *Physica D* 28, 197–205. *281*

Aluko, M., and H.-C. Chang (1984). PEFLOQ: An algorithm for the bifurcation analysis of periodic solutions of autonomous systems, *Comput. Chem. Eng.* 8, 355–365. *451*

Ames, W. F. See Lee and Ames.

Anderson, P. M., and A. A. Fouad (1977). *Power System Control and Stability,* The Iowa State University Press, Iowa. *56*

Anderson, T. J., B. Balachandran, and A. H. Nayfeh (1992). Observations of nonlinear interactions in a flexible cantilever beam, AIAA Paper No. 92–2332. *469, 501, 524*

Andronov, A. A., and C. E. Chaikin (1949). *Theory of Oscillations,* Princeton University Press, Princeton, New Jersey. *77*

Andronov, A. A., E. A. Leontovich, I. I. Gordon, and A. G. Maier (1971). *Theory of Bifurcations of Dynamic Systems on a Plane,* Israel Program of Scientific Translations, Jerusalem. *391, 396*

Anishchenko, V. S., V. V. Astakhov, T. E. Letchford, and M. A. Safonova (1983a). Structure of a quasihyperbolic stochasticity in an internal self–excited oscillator, *Radiophys. Quant. Electr.* 26, 619–628.

Anishchenko, V. S., V. V. Astakhov, T. E. Letchford, and M. A. Safonova (1983b). Bifurcations in a three–dimensional two–parameter autonomous oscillatory system with a strange attractor, *Radiophys. Quant. Electr.* 26, 135–140.

Anishchenko, V. S., T. E. Letchford, and M. A. Safonova (1985). Effects of synchronization and bifurcations of synchronous and quasisynchronous oscillations in a nonautonomous generator, *Radiophys. Quant. Electr.* 28, 766–776. *323, 333, 334*

Antonsen, T. M. See Romeiras, Bondeson, Ott, Antonsen, and Grebogi.

Aprille, T. J., and T. N. Trick (1972). A computer algorithm to determine the steady–state response of nonlinear oscillators, *IEEE Trans. Circuit Theory* CT–19, 354–360. *446*

Argoul, F., and J.-C. Roux (1985). Quasiperiodicity in chemistry: An experimental path in the neighbourhood of a codimension–two bifurcation, *Phys. Lett. A* 108, 426–430. *323*

Argoul, F., A. Arneodo, P. Richetti, and J.-C. Roux (1987). From quasiperiodicity to chaos in the Belousov–Zhabotinskii reaction. I. Experiment, *J. Chem. Phy.* 86, 3325–3356. *323*

Aris, R. See Aronson, McGehee, Kevrekidis, and Aris; Kevrekidis, Aris, Schmidt, and Pelikan.

Arjona, M. See Pujol, Arjona, and Corbalán.

Armbruster, D. See Rand and Armbruster.

Arneodo, A., P. Coullet, and E. A. Spiegel (1983). Cascade of period doublings of tori, *Phys. Lett. A* 94, 1–6. *334*

Arneodo, A., P. Coullet, and C. Tresser (1980). Occurrence of strange attractors in three–dimensional Volterra equations, *Phys. Lett. A* 79, 259–263.

Arneodo, A., P. Coullet, and C. Tresser (1981). A possible new mechanism for the onset of turbulence, *Phys. Lett. A* 81, 197–201. *399*

Arneodo, A., P. Coullet, and C. Tresser (1982). Oscillators with chaotic behavior: An illustration of a theorem by Shilnikov, *J. Stat. Phys.* 27, 171–182. *404, 407, 410*

Arneodo, A., P. Coullet, C. Tresser, A. Libchaber, J. Maurer, and D. D'Humieres (1983). On the observation of an uncompleted cascade in a Rayleigh–Bénard experiment, *Physica D* 6, 385–392.

Arneodo, A. See Argoul, Arneodo, Richetti, and Roux.

Arnold, V. I. (1973). *Ordinary Differential Equations,* MIT Press, Cambridge, Massachusetts. *10, 12, 17, 48, 158*

Arnold, V. I. (1988). *Geometrical Methods in the Theory of Ordinary Differential Equations,* Springer–Verlag, New York. *42, 43, 77, 97, 116, 121, 128, 175, 187, 239, 242, 254, 255*

Arnold, V. I. (1992). *Ordinary Differential Equations,* Springer–Verlag, New York. *10, 12*

Arnold, V. I., and Yu. S. Il'yashenko (1988). Ordinary differential equations, in *Dynamical Systems I: Ordinary Differential Equations and Smooth Dynamical Systems,* D. V. Anosov and V. I. Arnold, eds., Springer–Verlag, New York.

Aronson, D. G., M. A. Chory, G. R. Hall, and R. P. McGehee (1982). Bifurcations from an invariant circle for two–parameter families of maps of the plane: A computer assisted study, *Comm. Math. Phys.* 83, 303–354. *322*

Aronson, D. G., R. P. McGehee, I. G. Kevrekidis, and R. Aris (1986). Entrainment regions for periodically forced oscillators, *Phys. Rev.* A 33, 2190–2192. *321*

Arrowsmith, D. K., and C. M. Place (1990). *An Introduction to Dynamical Systems,* Cambridge University Press, Cambridge, England. *243, 367*

Asfar, K. R. See Nayfeh and Asfar.

Asrar, W. See Nayfeh, Asrar, and Nayfeh.

Astakhov, V. V. See Anishchenko, Astakhov, Letchford, and Safonova.

Aubry, N., P. Holmes, J. L. Lumley, and E. Stone (1988). The dynamics of coherent structures in the wall region of a turbulent boundary layer, *J. Fluid Mech.* 192, 115-173. *299*

Auerbach, D., C. Grebogi, E. Ott, and J. A. Yorke (1992). Controlling chaos in high dimensional systems, *Phy. Rev. Lett.* 69, 3479–3482.

Avriel, M. (1976). *Nonlinear Programming: Analysis and Methods,* Prentice Hall, Englewood Cliffs, New Jersey, Chapter 9.

Awrejcewicz, J. (1989). Two kinds of evolution of strange attractors for the example of a particular non–linear oscillator, *J. Appl. Math. Phys. (ZAMP)* 40, 375–386.

Awrejcewicz, J. (1990). Bifurcation portrait of the human vocal cord oscillations, *J. Sound Vib.* 136, 151–156.

Awrejcewicz, J. (1991). Numerical analysis of the oscillations of human vocal cords, *Nonlinear Dyn.* 2, 35–52.

Awrejcewicz, J., and W. D. Rheinhardt (1990). Quasiperiodicity, strange non–chaotic and chaotic attractors in a forced two–degree–of–freedom system, *J. Appl. Math. Phys. (ZAMP)* 41, 713–727. *295*

Bachelart, S. See Pomeau, Roux, Rossi, Bachelart, and Vidal; Roux, Rossi, Bachelart, and Vidal.

Badii, R., and A. Politi (1985). Statistical description of chaotic attractors: The dimension function, *J. Stat. Phys.* 40, 725–750. *538*

Badii, R., G. Broggi, B. Derighetti, M. Ravani, S. Ciliberto, A. Politi, and S. A. Rubio (1988). Dimension increase in filtered chaotic signals, *Phys. Rev. Lett.* 60, 979–982. *463, 550*

Badmus, O. O., S. Chowdhury, K. M. Eveker, C. N. Nett, and C. J. Rivera (1993a). A simplified approach for control of rotating stall, part 1: Theoretical development, AIAA Paper No. 93–2229. *568*

Badmus, O. O., S. Chowdhury, K. M. Eveker, C. N. Nett, and C. J. Rivera (1993b). A simplified approach for control of rotating stall, part 2: Experimental results, AIAA Paper No. 93–2234. *568*

Bahar, L. Y. See Kwatny, Pasrija, and Bahar.

Baier, G., K. Wegmann, and J. L. Hudson (1989). An intermittent type of chaos in the Belousov–Zhabotinsky reaction, *Phys. Lett. A* 141, 340–345. *299*

Baier, G. See Klein, Baier, and Rössler.

Bajaj, A. K. (1991). Examples of boundary crisis phenomenon in structural dynamics, *Int. Ser. Num. Math.* 97, 27–36. *336*

Bajaj, A. K., and J. M. Johnson (1990). Asymptotic techniques and complex dynamics in weakly non–linear forced mechanical systems, *Int. J. Non–Linear Mech.* 25, 211–226. *334, 336*

Bajaj, A. K., and J. M. Johnson (1992). On the amplitude dynamics and crisis in resonant motion of stretched strings, *Philos. Trans. Roy. Soc. Lond. A* 338, 1–41. *334, 336*

Bajaj, A. K., and S. Tousi (1990). Torus doublings and chaotic amplitude modulations in a two degree–of–freedom resonantly forced mechanical system, *Int. J. Non–Linear Mech.* 25, 625–642. *334*

Bajaj, A. K. See Johnson and Bajaj; Restuccio, Krousgrill, and Bajaj; Sethna and Bajaj; Streit, Bajaj, and Krousgrill; Tousi and Bajaj.

Bak, P. (1986). The devil's staircase, *Phys. Today* 39, 38–45.

Bak, P. See Bohr, Bak, and Jensen; Jensen, Bak, and Bohr.

Baker, G. L., and J. P. Gollub (1990). *Chaotic Dynamics,* Cambridge University Press, Cambridge, England.

Balachandran, B. (1990). *A Theoretical and Experimental Investigation of Modal Interactions in Resonantly Forced Structures,* Ph.D. Dissertation, Virginia Polytechnic Institute and State University, Blacksburg, Virginia. *468, 501, 506, 517, 523*

Balachandran, B., and A. H. Nayfeh (1991). Observations of modal interactions in resonantly forced beam–mass structures, *Nonlinear Dyn.* 2, 77–117. *333, 334, 501, 512, 543, 544*

Balachandran, B. See Anderson, Balachandran, and Nayfeh; Bikdash, Balachandran, and Nayfeh; Nayfeh and Balachandran; Nayfeh, Balachandran, Colbert, and Nayfeh.

Barnsley, M. (1988). *Fractals Everywhere,* Academic Press, New York.

Bassett, M. R., and J. L. Hudson (1989). Experimental evidence of period doubling of tori during an electrochemical reaction, *Physica D* 35, 289–298. *323, 333*

Bau, H. H. See Singer and Bau; Singer, Wang, and Bau.

Beall, J. M. See Kim, Beall, and Powers.

Beaman, J. J., and J. K. Hedrick (1980). Freight car harmonic response: A simplified nonlinear method, in *Nonlinear System Analysis and Synthesis,* Vol. 2, R. V. Ramnath, J. K. Hedrick, and H. M. Paynter, eds., ASME, New York, 177-195. *221*

Beasley, M. R. See D'Humieres, Beasley, Huberman, and Libchaber.

Bélair, J., and L. Glass (1985). Universality and self–similarity in the bifurcations of circle maps, *Physica D* 16, 143–154. *323*

Belogorstev, A. B. (1992). Bifurcations of tori and chaos in the quasiperiodically forced Duffing oscillator, *Nonlinearity* 5, 889–897.

Bendat, J. S., and A. G. Piersol (1980). *Engineering Applications of Correlation and Spectral Analysis,* Wiley, New York. *504, 505, 513*

Bendat, J. S., and A. G. Piersol (1986). *Random Data: Analysis and Measurement Procedures,* Wiley, New York. *505, 513*

Bender, C. M., and S. A. Orszag (1978). *Advanced Mathematical Methods for Scientists and Engineers,* McGraw–Hill, New York. *144*

Benedettini, F. See Rega, Benedettini, and Salvatori; Rega, Salvatori, and Benedettini.

Benettin, G., L. Galgani, A. Giorgilli, and J. Strelcyn (1980a). Lyapunov characteristic exponents for smooth dynamical systems and for Hamiltonian systems; a method for computing all of them. Part 1: Theory, *Meccanica* 15, 9–20. *529*

Benettin, G., L. Galgani, A. Giorgilli, and J. Strelcyn (1980b). Lyapunov characteristic exponents for smooth dynamical systems and for Hamiltonian systems; a method for computing all of them. Part 2: Numerical application, *Meccanica* 15, 21–30. *529*

Ben–Jacob, E., I. Goldhirsch, Y. Imry, and S. Fishman (1982). Intermittent chaos in Josephson junctions, *Phys. Rev. Lett.* 49, 1599–1602. *299*

Ben–Mizrachi, A., and I. Procaccia (1984). Characterization of experimental (noisy) strange attractors, *Phys. Rev. A* 29, 975–977.

Benson, S. V. See Gollub and Benson.

Bergé, P. (1982). Study of the phase space diagrams through experimental Poincaré sections in prechaotic and chaotic regimes, *Physica Scripta* T1, 71–72.

Bergé, P., M. Dubois, P. Manneville, and Y. Pomeau (1980). Intermittency in Rayleigh–Bénard covection, *J. Phys. Lett.* 41, L341–L345. *297, 298*

Bergé, P., Y. Pomeau, and C. Vidal (1984). *Order Within Chaos: Towards a Deterministic Approach to Turbulence*, Wiley, New York. *197, 203, 281, 298, 301, 322, 323, 513, 520*

Bergé, P. See Dubois, Rubio and Bergé.

Berger, B. S., and M. Rokni (1987). Lyapunov exponents and the evolution of normals, *Int. J. Eng. Sci.* 25, 1393–1396.

Berger, B. S., M. Rokni, and I. Minis (1992). The nonlinear dynamics of metal cutting, *Int. J. Eng. Sci.* 30, 1433–1440.

Berger, B. S. See Rokni and Berger.

Beyn, W. J. (1990). The numerical computation of connecting orbits in dynamical systems, *IMA J. Numer. Anal.* 9, 169–181. *460*

Bier, M., and C. Bountis (1984). Remerging Feigenbaum trees in dynamical systems, *Phys. Lett. A* 104, 239–244. *411*

Bikdash, M. U., B. Balachandran, and A. H. Nayfeh (1994). Melnikov analysis for a ship with a general damping model, *Nonlinear Dyn.* 6, 101–124. *52, 380*

Billups, S. C. See Watson, Billups, and Morgan.

Birkhoff, G. D. (1927). *Dynamical Systems*, American Mathematical Society, Providence, Rhode Island.

Birkhoff, G. D. (1950). *Collected Mathematical Papers, Vols. 1–3*, American Mathematical Society, Providence, Rhode Island.

Bishop, S. R. See Thompson, Bishop and Leung.

Bitmead, R. R. See Mareels and Bitmead.

Bloch, A. M., and J. E. Marsden (1989). Controlling homoclinic orbits, *Theor. Comput. Fluid Dyn.* 1, 179–190.

Bogoliubov, N. N., and Y. A. Mitropolsky (1961). *Asymptotic Methods in the Theory of Nonlinear Oscillations,* Gordon and Breach, New York.

Bohr, T., P. Bak, and M. H. Jensen (1984). Transition to chaos by interaction of resonances in dissipative systems. II. Josephson junctions, charge–density waves, and standard maps, *Phys. Rev. A* 30, 1970–1981. *248, 321*

Bohr, T. See Jensen, Bak, and Bohr.

Bondeson, A. See Romeiras, Bondeson, Ott, Antonsen, and Grebogi.

Botez, R. M. See Paidoussis and Botez.

Bouguerra, H. See Nayfeh and Bouguerra.

Bountis, C. See Bier and Bountis.

Brackx, F., and D. Constales (1991). *Computer Algebra with LISP and REDUCE: An Introduction to Computer-Aided Pure Mathematics,* Kluwer, Dordrecht, Netherlands.

Braiman, Y., and I. Goldhirsch (1991). Taming chaotic dynamics with weak periodic perturbations, *Phys. Rev. Lett.* 66, 2545–2548. *564*

Breedon, J. L., and N. H. Packard (1992). Nonlinear analysis of data sampled nonuniformly in time, *Physica D* 58, 273–283. *495*

Breitenberger, E., and R. D. Mueller (1981). The elastic pendulum: A nonlinear paradigm, *J. Math. Phys.* 22, 1196–1210.

Briggs, K. (1990). An improved method for estimating Liapunov exponents of chaotic time series, *Phys. Lett. A* 151, 27–32. *534*

Brillinger, D. R. (1965). An introduction to polyspectra, *Ann. Math. Stat.* 36, 1351–1374. *551*

Brillinger, D. R. (1981). *Time Series,* Holden-Day, San Francisco, California. *551*

Brindley, J., C. Kaas–Petersen, and A. Spence (1989). Path–Following methods in bifurcation problems, *Physica D* 34, 456–461. *441*

Brindley, J., and T. Kapitaniak (1991a). Existence and characterization of strange nonchaotic attractors in nonlinear systems, *Chaos Sol. Fract.* 1, 323–337. *295*

Brindley, J., and T. Kapitaniak (1991b). Analytic predictors for strange nonchaotic attractors, *Phys. Lett. A* 155, 361–364. *295*

Brindley, J., T. Kapitaniak, and M. S. El Naschie (1991). Analytical conditions for strange chaotic and nonchaotic attractors of quasiperiodically forced van der Pol's equation, *Physica D* 51, 28–38. *295*

Broggi, G. See Badii, Broggi, Derighetti, Ravani, Ciliberto, Politi, and Rubio.

Broomhead, D. S., and R. Jones (1989). Time–series analysis, *Proc. R. Soc. Lond. A* 423, 103–121. *486, 491*

Broomhead, D. S., and G. P. King (1986). Extracting qualitative dynamics from experimental data, *Physica D* 20, 217–236. *486, 489, 490*

Broomhead, D. S. See Healey, Broomhead, Cliffe, Jones, and Mullin.

Brorson, S. D., D. Dewey, and P. S. Linsay (1983). Self–replicating attractor of a driven semiconductor oscillator, *Phys. Rev. A* 28, 1201–1203. *336*

Brown, R. (1993). Orthogonal polynomials as prediction functions in arbitrary phase space dimensions, *Phys. Rev. E* 47, 3962–2969.

Brown, R., P. Bryant, and H. D. I. Abarbanel (1990). Computing the Lyapunov spectrum of a dynamical system from an observed time series, *Phys. Rev. A* 43, 2787–2806.

Brown, R. See Abarbanel, Brown, and Kadtke; Abarbanel, Brown, and Kennel; Abarbanel, Brown, Sidorowich, and Tsimring; Bryant, Brown, and Abarbanel; Kennel, Brown, and Abarbanel.

Brunsden, V., and P. Holmes (1987). Power spectra of strange attractors near homoclinic orbits, *Phys. Rev. Lett.* 58, 1699–1702.

Bryant, P., R. Brown, and H. D. I. Abarbanel (1990). Lyapunov exponents from observed time series, *Phys. Rev. Lett.* 65, 1523–1526. *534*

Bryant, P. See Brown, Bryant, and Abarbanel.

Bulirsch, R. See Stoer and Bulirsch.

Buzug, Th., and G. Pfister (1992). Comparison of algorithms calculating optimal embedding parameters for delay time coordinates, *Physica D* 58, 127–137. *494*

Buzug, Th., T. Reimers, and G. Pfister (1990). Optimal reconstruction of strange attractors from purely geometrical arguments, *Euro. Phys. Lett.* 13, 605–610. *483*

Buzyna, G. See Guckenheimer and Buzyna.

Bykov, V. V. See Afraimovich, Bykov, and Shilnikov.

Carr, J. (1981). *Applications of Centre Manifold Theory,* Springer–Verlag, New York. *51 97, 98, 99, 102, 128*

Carroll, T. L., and L. M. Pecora (1993). Communicating with chaos, *Naval Res. Rev.* 55, 4–11. *587*

Carroll, T. L., L. M. Pecora, and F. J. Rachford (1987). Chaotic transients and multiple attractors in spin–wave experiments, *Phys. Rev. Lett.* 59, 2891–2894. *336, 588*

Carroll, T. L. See Pecora and Carroll.

Cartwright, M. L., and J. E. Littlewood (1945). On nonlinear differential equations of the second order, *J. Lond. Math. Soc.* 20, 180–189. *269*

Casdagli, M., S. Eubank, J. D. Farmer, and J. Gibson (1991). State space reconstruction in the presence of noise, *Physica D* 51, 52–98. *483*

Casdagli, M. See Sauer, Yorke, and Casdagli.

Cenys, A., and K. Pyragas (1988). Estimation of the number of degrees of freedom from chaotic time series, *Phys. Lett. A* 129, 227–230.

Chaikin, C. E. See Andronov and Chaikin.

Chang, H.-C. See Aluko and Chang.

Chang, J.-H. See Choi, Chang, Stearman, and Powers.

Char, B. W., K. O. Geddes, G. H. Gonnet, B. L. Leong, M. B. Monagan, and S. W. Watt (1991). *Maple V Language Reference,* Springer–Verlag, New York.

Chen, R. C. See Abed, Wang, and Chen.

Chencier, A., and G. Iooss (1979). Bifurcations de tores invariants, *Arch. Rational Mech. Anal.* 69, 108–198.

Chiang, H.-D., M. W. Hirsch, and F. F. Wu (1988). Stability regions of nonlinear autonomous dynamical systems, *IEEE Trans. Auto. Control* 33, 16–27. *46, 55*

Chiang, H.-D., I. Dobson, R. J. Thomas, J. S. Thorp, and L. Fekih-Ahmed (1990). On voltage collapse in electric power systems, *IEEE Trans. Power Syst.* 5, 601–611. *564*

Chiang, H.-D. See Dobson and Chiang.

Chin, C., A. H. Nayfeh, and D. T. Mook (1993). Parametrically excited nonlinear two–degree–of–freedom systems with repeated natural frequencies, in *Nonlinear Vibrations,* ASME DE–Vol. 54, 105–125. *404*

Chin, C. See Nayfeh and Chin.

Choi, D.-W., J.-H. Chang, R. O. Stearman, and E. J. Powers (1984). Bispectral identification of nonlinear mode interactions, in *Proceedings of the 2nd IMAC Conference,* Orlando, Florida, 602–609. *557*

Choi, D.-W., R. W. Miksad, E. J. Powers, and P. J. Fischer (1985). Application of digital cross-bispectral analysis techniques to model the non-linear response of a moored vessel system in random seas, *J. Sound Vib.* 99, 309–326. *557*

Choi, S.-K., and S. T. Noah (1992a). Mode–locking and chaos in a modified Jeffcott rotor with a bearing clearance, in *Nonlinear Vibrations,* ASME DE–Vol. 50, 21–28. *248, 254, 255, 323*

Choi, S.-K., and S. T. Noah (1992b). Response and stability analysis of piecewise–linear oscillators under multi–forcing frequencies, *Nonlinear Dyn.* 3, 105–121.

Chory, M. A. See Aronson, Chory, Hall, and McGehee.

Chow, J. C., R. Fischl, and H. Yan (1990). On the evaluation of voltage collapse criteria, *IEEE Trans. Power Syst.* 5, 612–620. *564*

Chow, S.-N., J. K. Hale, and J. Mallet–Paret, An example of bifurcation to homoclinic orbits, *J. Diff. Eqns.* 37, 351–373.

Chow, S.-N., and X. B. Lin (1990). Bifurcation of a homoclinic orbit with a saddle–node equilibrium, *Diff. Integral Eq.* 3, 435–466. *460*

Chowdhury, S. See Badmus, Chowdhury, Eveker, Nett, and Rivera.

Chua, L. O., and P.-M. Lin (1975). *Computer–Aided Analysis of Electronic Circuits: Algorithms & Computational Techniques,* Prentice-Hall, Englewood Cliffs, New Jersey. *446*

Chua, L. O., and A. Ushida (1981). Algorithms for computing almost periodic steady–state response of nonlinear systems to multiple input frequencies, *IEEE Trans. Circuit Syst.* CAS–28, 953–971. *252*

Chua, L. O. See Matsumoto, Chua, and Tokunaga; Parker and Chua; Ushida and Chua.

Cibrario, M., and J. Lévine (1991). Saddle–node bifurcation control with application to thermal runaway of continuous stirred tank reactors, in *Proceedings of the 30th IEEE Conference on Decision and Control,* Brighton, England, 1551–1552. *564*

Ciliberto, S. See Badii, Broggi, Derighetti, Ravani, Ciliberto, Politi, and Rubio; Eckmann, Kamphorst, Ruelle, and Ciliberto; Nobili, Ciliberto, Cocciaro, Faetti, and Fronzoni.

Clauss, W. See Richter, Peinke, Clauss, Rau, and Parisi.

Cleve, J. See Schwetlick and Cleve.

Cliffe, K. A. See Healey, Broomhead, Cliffe, Jones, and Mullin.

Cocciaro, B. See Nobili, Ciliberto, Cocciaro, Faetti, and Fronzoni.

Coddington, E. A., and N. Levinson (1955). *Theory of Ordinary Differential Equations,* McGraw–Hill, New York. *10, 44, 155, 291*

Colbert, M. A. See Nayfeh, Balachandran, Colbert, and Nayfeh.

Cole, J. D. See Kevorkian and Cole.

Collet, P., and J.–P. Eckmann (1980). *Iterated Maps on the Interval as Dynamical Systems,* Birkhaüser, Boston, Massachusetts. *283, 295*

Combes, J. M., A. Grossman, and P. Tchamitchian (1990). *Wavelets: Time–Frequency Methods and Phase Space,* Springer–Verlag, New York.

Constales, D. See Brackx and Constales.

Cooley, T. W., and J. W. Tukey (1965). An algorithm for the machine calculation of complex Fourier series, *Math. Comp.* 19, 297–301. *503*

Cooperrider, N. K. (1980). Nonlinear behavior in rail vehicle dynamics, in *New Approaches to Nonlinear Problems in Dynamics,* P. J. Holmes, ed., SIAM, Philadelphia, Pennsylvania, 173–194. *221*

Corbalán, R. See Pujol, Arjona, and Corbalán.

Coullet, P. See Arneodo, Coullet, and Spiegel; Arneodo, Coullet, and Tresser; Arneodo, Coullet, Tresser, Libchaber, Maurer, and D'Humieres.

Crutchfield, J. P., D. Farmer, N. Packard, R. Shaw, G. Jones, and R. J. Donnelly (1980). Power spectral analyses of a dynamical system, *Phys. Lett. A* 76, 1–4. *281*

Crutchfield, J. P., and B. A. Huberman (1980). Fluctuations and the onset of chaos, *Phys. Lett. A* 77, 407–410. *296*

Crutchfield, J. P., M. Nauenberg, and J. Rudnick (1981). Scaling for external noise at the onset of chaos, *Phys. Rev. Lett.* 46, 933–935. *296*

Crutchfield, J. P. See Froehling, Crutchfield, Farmer, Packard, and Shaw; Huberman and Crutchfield; Packard, Crutchfield, Farmer, and Shaw.

Cumming, A., and P. S. Linsay (1988). Quasiperiodicity and chaos in a system with three competing frequencies, *Phys. Rev. Lett.* 60, 2719–2722. *317*

Curry, J. H. (1980). An algorithm for finding closed orbits, in *Global Theory of Dynamical Systems,* Z. Nitecki and C. Robinson, eds., Springer–Verlag, New York, 111–120. *183, 446, 455*

Curry, J., and J. A. Yorke (1977). A transition from Hopf bifurcation to chaos: Computer experiments with maps in $\mathcal{R}^2$, in *The Structure of Attractors in Dynamical Systems,* Vol. 668, N. G. Markley, J. C. Martin, and W. Perrizo, eds., Springer Notes in Mathematics, Springer–Verlag, New York, 48–66. *322*

Cvitanović, P. (1989). *Universality in Chaos,* Adam Hilger, New York.

d'André–Novel, B., and A. Abichou (1992). Center manifold theory for stabilizing a flexible mechanical system with a pitchfork bifurcation, in *Proceedings of the 2nd IFAC Workshop on System Structure and Control,* V. Strejc., ed., Prague, Czechoslovokia, 232–235. *568*

Dangelmayr, G., and J. Guckenheimer (1987). On a four parameter family of planar vector fields, *Arch. Rational Mech. Anal.* 97, 321–352. *396*

Dangoisse, D., P. Glorieux, and D. Hennequin (1986). Laser chaotic attractors in crisis, *Phys. Rev. Lett.* 57, 2657–2660. *336*

Darbyshire, A. G. See Mullin and Darbyshire.

Davidenko, D. F. (1953). On a new method of numerical solution of systems of nonlinear equations, *Dokl. Akad. Nauk. SSSR* 88, 601–602; *Math. Rev.* 14, 906. *428*

Davis, M. W., Jr., and W. F. O'Brien (1987). A stage–by–stage post–stall compression system modeling technique, AIAA Paper No. 87–2088. *564*

Day, I. J. (1993). Active suppression of rotating stall and surge in axial compressors, *ASME J. Turbom.* 115, 40–47. *564*

Dayawansa, W. P. See Romeiras, Grebogi, Ott, and Dayawansa.

DeDier, B., D. Roose, and P. van Rompay (1990). Interaction between fold and Hopf curves leads to new bifurcation phenomena, in *Continuation Techniques and Bifurcation Problems,* H. D. Mittleman and D. Roose, eds., Birkhaüser Verlag, Boston, Massachusetts, 171–186. *441* deGrassie, S. See Ikezi, deGrassie, and Jensen.

DeKepper, P. See Richetti, Roux, and Swinney.

Deneef, P., and H. Lashinsky (1973). Van der Pol model for unstable waves on a beam–plasma system, *Phys. Rev. Lett.* 31, 1039–1041. *221*

Den Heijer, C., and W. C. Rheinboldt (1981). On steplength algorithms for a class of continuation methods, *SIAM J. Num. Anal.* 18, 925–948. *431*

Derighetti, B. See Badii, Broggi, Derighetti, Ravani, Ciliberto, Politi, and Rubio.

Devaney, R. L. (1977). Blue sky catastrophes in reversible and Hamiltonian systems, *Indiana Univ. Math. J.* 26, 247–263.

Devaney, R. L. (1989). *An Introduction to Chaotic Dynamical Systems*, Addison–Wesley, New York.

Devaney, R. L., and Z. Nitecki (1979). Shift automorphisms in the Hénon mapping, *Commun. Math. Phys.* 67, 137–148.

Dewey, D. See Brorson, Dewey, and Linsay.

D'Humieres, D., M. R. Beasley, B. A. Huberman, and A. Libchaber (1982). Chaotic states and routes to chaos in the forced pendulum, *Phys. Rev. A* 26, 3483–3496. *281*

D'Humieres, D. See Arneodo, Coullet, Tresser, Libchaber, Maurer, and D'Humieres.

Diener, M. (1984). The canard unchained, or how fast/slow dynamical systems bifurcate, *Math. Intell.* 6, 38–49. *396*

Ding, M., C. Grebogi, and E. Ott (1989). Dimensions of strange nonchaotic attractors, *Phys. Rev. A* 39, 2593–2598. *295*

Ding, W. X., H. Q. She, W. Huang, and C. X. Yu (1994). Controlling chaos in a discharge plasma, *Phys. Rev. Lett.* 72, 96–99.

Ditto, W. L., S. N. Rauseo, and M. L. Spano (1990). Experimental control of chaos, *Phys. Rev. Lett.* 65, 3211–3214. *577–581*

Ditto, W. L. See Garfinkel, Spano, Ditto, and Weiss; Sommerer, Ditto, Grebogi, Ott, and Spano; Spano and Ditto.

Dobson, I., and H.–D. Chiang (1989). Towards a theory of voltage collapse in electric power systems, *Syst. Control Lett.* 13, 253–262. *564, 566*

Dobson, I. See Chiang, Dobson, Thomas, Thorp, and Fekih–Ahmed.

Doedel, E. J. (1986). AUTO– Software for continuation and bifurcation problems in ordinary differential equations. California

Institute of Technology, Pasadena, California. *432, 435, 452, 458, 459, 466*

Doedel, E. J., and R. F. Heinemann (1983). Numerical computation of periodic solution branches and oscillatory dynamics of the stirred reactor with **A** → **B** → **C** reactions, *Chem. Eng. Sci.* 38, 1493–1499. *452*

Doedel, E. J., H. B. Keller, and J. P. Kernevez (1991a). Numerical analysis and control of bifurcation problems (I) Bifurcation in finite dimensions, *Int. J. Bif. Chaos* 1, 493–520. *425, 434, 435*

Doedel, E. J., H. B. Keller, and J. P. Kernevez (1991b). Numerical analysis and control of bifurcation problems (II) Bifurcation in infinite dimensions, *Int. J. Bif. Chaos* 1, 745–772. *435, 452, 459, 460*

Donnelly, R. J. See Crutchfield, Farmer, Packard, Shaw, Jones, and Donnelly.

Dowell, E. H. (1975). *Aeroelasticity of Plates and Shells,* Noordhoff, Leyden. *221*

Dowell, E. H., and C. Pezeshki (1988). On necessary and sufficient condition for chaos to occur in Duffing's equation: An heuristic approach, *J. Sound Vib.* 121, 195–200. *292*

Dowell, E. H. See Pezeshki and Dowell.

Drazin, P. G. (1992). *Nonlinear Systems,* Cambridge University Press, Cambridge, England. *131, 224, 225*

Dressler, U., and G. Nitsche (1992). Controlling chaos using time delay coordinates, *Phys. Rev. Lett.* 68, 1–4. *582*

Dubois, M., M. A. Rubio, and P. Bergé (1983). Experimental evidence of intermittencies associated with a subharmonic bifurcation, *Phys. Rev. Lett.* 51, 1446–1449. *299, 305, 306*

Dubois, M. See Bergé, Dubois, Manneville, and Pomeau.

Dugundji, J. (1966). *Topology,* Allyn and Bacon, Boston, Massachusetts. *2*

Dugundji, J. See Tseng and Dugundji.

Dumortier, F. (1993). Techniques in the theory of local bifurcations: Blow–up, normal forms, nilpotent bifurcations, singular perturbations, in *Bifurcations and Periodic Orbits of Vector Fields*, D. Schlomiuk, ed., Kluwer, Boston, Massachusetts, 19–74.

Eckmann, J.-P. (1981). Roads to turbulence in dissipative dynamical systems, *Rev. Mod. Phys.* 53, 643–654. *29, 296, 317*

Eckmann, J.-P., and D. Ruelle (1985). Ergodic theory of chaos and strange attractors, *Rev. Mod. Phys.* 57, 617–656. *283, 483, 533, 534*

Eckmann, J.-P., and D. Ruelle (1992). Fundamental limitations for estimating dimensions and Lyapunov exponents in dynamical systems, *Physica D* 56, 185–187. *550*

Eckmann, J.-P., S. O. Kamphorst, D. Ruelle, and S. Ciliberto (1986). Liapunov exponents from time series, *Phys. Rev. A* 34, 4971–4979. *533, 534*

Eckmann, J.-P. See Collet and Eckmann.

El Naschie, M. S. (1990). *Stress, Stability and Chaos,* McGraw–Hill, New York.

El Naschie, M. S. See Brindley, Kapitaniak, and El Naschie.

Elzebda, J. M., A. H. Nayfeh, and D. T. Mook (1989). Development of an analytical model of wing rock for slender delta wings, *J. Aircraft* 26, 737–743. *139*

Emad, F. P. See Mohamed and Emad.

Eubank, S. See Casdagli, Eubank, Farmer, and Gibson.

Evan–Iwanowski, R. M. (1976). *Resonance Oscillations in Mechanical Systems,* Elsevier, New York.

Evans, J. W., N. Feinchel, and J. A. Feroe (1982). Double impulse solutions in nerve axon equations, *SIAM J. Appl. Math.* 42, 219–234.

Eveker, K. M. See Badmus, Chowdhury, Eveker, Nett, and Rivera.

Faetti, S. See Nobili, Ciliberto, Cocciaro, Faetti, and Fronzoni.

Fallside, F., and M. R. Patel (1965). Step–response behaviour of a speed–control system with a back–e.m.f. nonlinearity, *Proc. IEE* 112, 1979–1984. *134*

Falzarano, J. M. (1990). *Predicting Complicated Dynamics Leading to Vessel Capsizing,* Ph.D. Dissertation, University of Michigan, Ann Arbor, Michigan. *52, 380, 383*

Falzarano, J. M., S. W. Shaw, and A. W. Troesh (1992). Application of global methods for analyzing dynamical systems to ships rolling motion and capsizing, *Int. J. Bif. Chaos* 2, 101–115. *380*

Farmer, J. D., E. Ott, and J. A. Yorke (1983). The dimensions of chaotic attractors, *Physica D* 7, 153–180. *538*

Farmer, J. D., and J. J. Sidorowich (1988). Exploiting chaos to predict the future and reduce noise, in *Evolution, Learning, and Cognition,* Y. C. Lee, ed., World Scientific, Singapore. *463*

Farmer, J. D., and J. J. Sidorowich (1991). Optimal shadowing and noise reduction, *Physica D*, 47, 373–392. *463*

Farmer, J. D. See Casdagli, Eubank, Farmer, and Gibson; Crutchfield, Farmer, Packard, Shaw, Jones, and Donnelly; Froehling, Crutchfield, Farmer, Packard, and Shaw; Packard, Crutchfield, Farmer, and Shaw.

Fauve, S. See Libchaber, Fauve, and Laroche.

Feigenbaum, M. J. (1978). Quantitative universality for a class of nonlinear transformations, *J. Stat. Phys.* 19, 25–52. *203, 280, 295, 296*

Feigenbaum, M. J., L. P. Kadanoff, and S. J. Shenker (1982). Quasiperiodicity in dissipative systems: A renormalization group analysis, *Physica D* 5, 370–386. *322*

Feinchel, N. See Evans, Feinchel, and Feroe.

Fekih–Ahmed, L. See Chiang, Dobson, Thomas, Thorp, and Fekih–Ahmed.

Feldberg, R. See Knudsen, Feldberg, and Jaschinski.

Fenstermacher, P. R., H. L. Swinney, and J. P. Gollub (1979). Dynamical instabilities and transition to chaotic Taylor vortex flow, *J. Fluid Mech.*, 94, 103–128. *323*

Feroe, J. A. See Evans, Feinchel, and Feroe.

Feudel, U., and W. Jansen (1992). CANDYS/QA– A software for qualitative analysis of nonlinear dynamical systems, *Int. J. Bif. Chaos* 2, 773–794. *434, 435, 441, 460*

Field, R. J., and L. Györgyi (1993). *Chaos in Chemistry and Biochemistry*, World Scientific, Singapore. *329, 331*

Fischer, P. J. See Choi, Miksad, Powers, and Fischer.

Fischl, R. See Chow, Fischl, and Yan.

Fishman, S. See Ben–Jacob, Goldhirsch, Imry, and Fishman.

Fletcher, W. H. W. See Keen and Fletcher.

Floquet, G. (1883). Sur les équations différentielles linéaires á coefficients périodiques, *Annal Sci. école Norm. Sup.* 12, 47–89. *159*

Fouad, A. A. See Anderson and Fouad.

Fowler, A. C. (1986). Analytic methods for predicting chaos, in *Nonlinear Phenomena and Chaos*, S. Sarkar, ed., Adam Hilger, Boston, Massachusetts.

Fowler, A. C. (1990a). Homoclinic bifurcations in n dimensions, *Stud. Appl. Math.* 83, 193–209. *404*

Fowler, A. C. (1990b). Homoclinic bifurcations for partial differential equations in unbounded domains, *Stud. Appl. Math.* 83, 329–353. *404*

Fowler, A. C., and C. T. Sparrow (1991). Bifocal homoclinic orbits in four dimensions, *Nonlinearity* 4, 1159–1182.

Franceschini, V. (1983). Bifurcations of tori and phase locking in a dissipative system of differential equations, *Physica D* 6, 285–304. *248, 334*

Fraser, A. M. (1989a). Information and entropy in strange attractors, *IEEE Trans. Inf. Theory* 35, 245–262. *497*

Fraser, A. M. (1989b). Reconstructing attractors from scalar time series: A comparison of singular system and redundancy criteria, *Physica D* 34, 391–404. *497*

Fraser, A. M., and H. L. Swinney (1986). Independent coordinates for strange attractors from mutual information, *Phys. Rev. A* 33, 1134–1140. *497, 499*

Fraser, S. See Schell, Fraser, and Kapral.

Frederickson, P., J. L. Kaplan, E. D. Yorke, and J. A. Yorke (1983). The Liapunov dimension of strange attractors, *J. Diff. Eqns.* 49, 185–207. *549*

Frelich, R. G. See Novak and Frelich.

Froehling, H., J. P. Crutchfield, D. Farmer, N. H. Packard, and R. Shaw (1981). On determining the dimension of chaotic flows, *Physica D* 3, 605–617.

Fronzoni, L. See Nobili, Ciliberto, Cocciaro, Faetti, and Fronzoni.

Froude, W. (1863). Remarks on Mr. Scott Russell's paper on rolling, *Trans. Inst. Naval Arch.* 4, 232–275.

Froyland, J. (1992). *Introduction to Chaos and Coherence,* Institute of Physics Publishing, Bristol, England.

Fu, J.-H., and E. H. Abed (1993a). Families of Lyapunov functions for nonlinear systems in critical cases, *IEEE Trans. Auto. Control* 38, 3–16.

Fu, J.-H., and E. H. Abed (1993b). Linear feedback stabilization of nonlinear systems with an uncontrollable mode, *Automatica* 4, 999–1010.

Fu, J.-H. See Abed and Fu.

Fujii, T. See Tsuda, Tamura, Sueoka, and Fujii.

Fujisaka, H. See Ishii, Fujisaka and Inoue.

Fung, Y. C. (1955). *An Introduction to the Theory of Aeroelasticity,* Wiley, New York. *221*

Galgani, L. See Benettin, Galgani, Giorgilli, and Strelcyn.

Gallager, R. G. (1968). *Information Theory and Reliable Communication,* Wiley, New York. *498*

Gang, H., and Zhilin, Q. (1994). Controlling spatiotemporal chaos in coupled map lattice systems, *Phys. Rev. Lett.* 72, 68–71.

Garcia, C. B., and W. I. Zangwill (1981). *Pathways to Solutions, Fixed Points, and Equilibria,* Prentice–Hall, Englewood Cliffs, New Jersey. *435*

Garfinkel, A., M. L. Spano, W. L. Ditto, and J. N. Weiss (1992). Controlling cardiac chaos, *Science* 257, 1230–1235. *577, 580*

Gaspar, V. See Petrov, Gaspar, Masere, and Showalter.

Gaspard, P. (1983). Generation of a countable set of homoclinic flows through bifurcation, *Phys. Lett. A* 97, 1–4. *403*

Gaspard, P., R. Kapral, and G. Nicolis (1984). Bifurcation phenomena near homoclinic systems: A two–parameter analysis, *J. Stat. Phys.* 35, 697–727. *404*

Gaspard, P., and G. Nicolis (1983). What can we learn from homoclinic orbits in chaotic dynamics? *J. Stat. Phys.* 31, 499–518. *336*

Gear, C. W. (1971). *Numerical Initial Value Problems in Ordinary Differential Equations,* Prentice–Hall, Englewood Cliffs, New Jersey.

Geddes, K. O. See Char, Geddes, Gonnet, Leong, Monagan, and Watt.

Geest, T. See Larter, Olsen, Steinmetz, and Geest.

Genesio, R., and A. Tesi (1992). Control techniques for chaotic dynamical systems, in *Proceedings of the 2nd IFAC Workshop on System Structure and Control,* V. Strejc, ed., Prague, Czechoslovokia, 260–263.

Genesio, R., and A. Tesi (1993). Distortion control of chaotic systems: The Chua's circuit, *J. Circuit Syst. Comp.* 3, 151–171.

Georg, K. See Allgower and Georg.

Gershenfeld, N. A. (1992). Dimension measurement on high-dimensional systems, *Physica D* 55, 135–154. *496*

Gibson, J. See Casdagli, Eubank, Farmer, and Gibson.

Gifford, S. J., and G. R. Tomlinson (1989). Recent advances in the application of functional series to non-linear structures, *J. Sound Vib.* 135, 289–317. *557*

Giglio, M., S. Musazzi, and V. Perine (1981). Transition to chaotic behavior via a reproducible sequence of period–doubling bifurcations, *Phys. Rev. Lett.* 47, 243–246. *296*

Gills, Z., C. Iwata, R. Roy, I. B. Schwartz, and I. Triandaf (1992). Tracking unstable steady states: Extending the stability regime of a multimode laser system, *Phys. Rev. Lett.* 69, 3169–3172. *577*

Gills, Z. See Rajarshi, Murphy, Maier, Gills, and Hunt; Roy, Murphy, Maier, and Gills.

Gilsinn, D. E. (1993). Constructing invariant tori for two weakly coupled van der Pol oscillators, *Nonlinear Dyn.* 4, 289–308. *251*

Giorgilli, A. See Benettin, Galgani, Giorgilli, and Strelcyn.

Glass, L., M. R. Guevara, A. Shrier, and R. Perez (1983). Bifurcation and chaos in a periodically stimulated cardiac oscillator, *Physica D* 7, 89–101. *323*

Glass, L. See Bélair and Glass; Guevara, Glass and Shrier; Perez and Glass.

Glazier, J. A., and A. Libchaber (1988). Quasi–periodicity and dynamical systems: An experimentalist's view, *IEEE Trans. Circuit Syst.* CAS–35, 790–809. *248*

Glendinning, P. (1984). Bifurcations near homoclinic orbits with symmetry, *Phys. Lett. A* 103, 163–166. *403*

Glendinning, P., and C. Sparrow (1984). Local and global behavior near homoclinic orbits, *J. Stat. Phys.* 35, 645–696. *403, 407*

Glendinning, P., and C. Sparrow (1986). T–points: A codimension two heteroclinic bifurcation, *J. Stat. Phys.* 43, 479–488. *407*

Glorieux, P. See Dangoisse, Glorieux, and Hennequin.

Goldhirsch, I., P. Sulem, and S. A. Orszag (1987). Stability and Lyapunov stability of dynamical systems: A differential approach and a numerical method, *Physica D* 27, 311–337. *526*

Goldhirsch, I. See Ben–Jacob, Goldhirsch, Imry, and Fishman; Braiman and Goldhirsch.

Gollub, J. P., and H. L. Swinney (1975). Onset of turbulence in a rotating fluid, *Phys. Rev. Lett.* 35, 927-930. *316*

Gollub, J. P., and S. V. Benson (1980). Many routes to turbulent convection, *J. Fluid Mech.* 100, 449–470. *316, 511*

Gollub, J. P. See Baker and Gollub; Fenstermacher, Swinney, and Gollub; Swinney and Gollub; Abraham, Gollub, and Swinney.

Golubitsky, M., and D. G. Schaeffer (1985). *Singularities and Groups in Bifurcation Theory, Volume I*, Springer–Verlag, New York. *76*

Gonnet, G. H. See Char, Geddes, Gonnet, Leong, Monagan, and Watt.

Gordon, I. I. See Andronov, Leontovich, Gordon and Maier.

Gorman, M., L. A. Reith, and H. L. Swinney (1980). Modulation patterns, multiple frequencies, and other phenomena in circular Couette flow, in *Nonlinear Dynamics*, R. H. G. Hellemann, ed., New York Academy of Sciences, New York. *317*

Gorman, M., P. J. Widmann, and K. A. Robbins (1984). Chaotic flow regimes in a convection loop, *Phys. Rev. Lett.* 52, 2241–2244.

Gouesbet, C. See Ringuet, Roźe, and Gouesbet.

Grassberger, P. (1983). Generalized dimensions of strange attractors, *Phys. Lett. A* 97, 227–230. *545*

Grassberger, P., and I. Procaccia (1983a). Characterization of strange attractors, *Phys. Rev. Lett.* 50, 346–349. *485, 547*

Grassberger, P., and I. Procaccia (1983b). Measuring the strangeness of strange attractors, *Physica D* 9, 189–208. *485, 547*

Grassberger, P., and I. Procaccia (1984). Dimensions and entropies of strange attractors from a fluctuating dynamics approach, *Physica D* 13, 34–54.

Grassberger, P., T. Schreiber, and C. Shaffrath (1991). Nonlinear time sequence analysis, *Int. J. Bif. Chaos* 1, 521–547. *463*

Grebogi, C., E. Ott, S. Pelikan, and J. A. Yorke (1984). Strange attractors that are not chaotic, *Physica D* 13, 261–268.

Grebogi, C., E. Ott, F. Romeiras, and J. A. Yorke (1987). Critical exponents for crisis–induced intermittency, *Phys. Rev. A* 36, 5365–5380. *335, 339, 340*

Grebogi, C., E. Ott, and J. A. Yorke (1982). Chaotic attractors in crisis, *Phys. Rev. Lett.* 48, 1507–1510.

Grebogi, C., E. Ott, and J. A. Yorke (1983a). Crises, sudden changes in chaotic attractors and transient chaos, *Physica D* 7, 181–200. *334–336, 339, 342, 415*

Grebogi, C., E. Ott, and J. A. Yorke (1983b). Fractal basin boundaries, long lived chaotic transients and unstable–unstable pair bifurcation, *Phys. Rev. Lett.* 50, 935–938. *415*

Grebogi, C., E. Ott, and J. A. Yorke (1983c). Are three–frequency quasiperiodic orbits to be expected in typical dynamical systems? *Phys. Rev. Lett.* 51, 339–342. *317*

Grebogi, C., E. Ott, and J. A. Yorke (1985). Attractors on an n–torus: Quasiperiodicity versus chaos, *Physica D* 15, 354–373. *255, 295, 317*

Grebogi, C., E. Ott, and J. A. Yorke (1986). Metamorphoses of basin boundaries in nonlinear dynamical systems, *Phys. Rev. Lett.* 56, 1011–1014. *362, 415*

Grebogi, C., E. Ott, and J. A. Yorke (1987). Chaos, strange attractors, and fractal basin boundaries in nonlinear dynamics, *Science* 238, 632–638. *339, 340, 345, 415*

Grebogi, C. See Auerbach, Grebogi, Ott, and Yorke; Ding, Grebogi, and Ott; Kostelich, Grebogi, Ott, and Yorke; Ott, Grebogi, and Yorke; Romeiras, Bondeson, Ott, Antonsen, and Grcbogi; Romeiras, Grebogi, Ott, and Dayawansa; Shinbrot, Ott, Grebogi, and Yorke; Shinbrot, Grebogi, Ott, and Yorke; Sommerer, Ditto, Grebogi, Ott, and Spano.

Green, J. M., and J. S. Kim (1987). The calculation of Lyapunov sepctra, *Physica D* 24, 213–225.

Greenspan, B. D., and P. J. Holmes (1983). Homoclinic orbits, sub-harmonics and global bifurcations in forced oscillations, in *Nonlinear Dynamics and Turbulence*, G. Barenblatt, G. Iooss, and D. D. Joseph, eds., Pitman, Marshfield, Massachusetts, 172–214.

Greitzer, E. M. (1976). Surge and rotating stall in axial flow compressor, Part I: Theoretical compression system model, *ASME J. Eng. Power* 190–198. *564*

Greitzer, E. M. See Moore and Greitzer.

Griewank, A., and G. W. Reddien (1983). The calculation of Hopf bifurcation points by a direct method, *IMA J. Num. Anal.* 3, 295–303. *438*

Griewank, A., and G. W. Reddien (1984). Characterization and computation of generalized turning points, *SIAM J. Num. Anal.* 21, 176–185. *438, 441*

Grossman, A. See Combes, Grossman, and Tchamitchian.

Gu, X. M., and P. R. Sethna (1987). Resonant surface waves and chaotic phenomena, *J. Fluid Mech.* 183, 543–565. *334, 336*

Guckenheimer, J., and G. Buzyna (1983). Dimension measurements for geostrophic turbulence, *Phys. Rev. Lett.* 51, 1438–1441. *501*

Guckenheimer, J., and P. Holmes (1983). *Nonlinear Oscillations, Dynamical Systems, and Bifurcations of Vector Fields,* Springer–Verlag, New York. *43, 51, 121, 128, 269, 288, 359, 367, 390*

Guckenheimer, J., and P. Worfolk (1992). Instant chaos, *Nonlinearity* 5, 1211–1222. *460*

Guckenheimer, J., and P. Worfolk (1993). Dynamical systems: Some computational problems, in *Bifurcations and Periodic Orbits of Vector Fields*, D. Schlomiuk, ed., Kluwer, Boston, Massachusetts, 241–278. *61, 441, 460*

Guckenheimer, J. See Dangelmayr and Guckenheimer.

Guevara, M. R., L. Glass, and A. Shrier (1981). Phase locking, period–doubling bifurcations, and irregular dynamics in periodically stimulated cardiac cells, *Science* 214, 1350–1353. *323*

Guevara, M. R. See Glass, Guevara, Shrier, and Perez.

Guillemin, V., and V. Pollack (1974). *Differential Topology,* Prentice–Hall, Englewood Cliffs, New Jersey. *9, 53, 444, 476*

Guyader, J. L. See Mevel and Guyader.

Gwinn, E. G., and R. M. Westervelt (1985). Intermittent chaos and low–frequency noise in the driven damped pendulum, *Phys. Rev. Lett.* 54, 1613–1616.

Gwinn, E. G., and R. M. Westervelt (1986). Frequency locking, quasiperiodicity, and chaos in extrinsic Ge, *Phys. Rev. Lett.* 57, 1060–1063.

Gwinn, E. G., and R. M. Westervelt (1987). Scaling structure of attractors at the transition from quasiperiodicity to chaos in electronic transport in Ge, *Phys. Rev. Lett.* 59, 157–160. *336*

Györgyi, L. See Field and Györgyi.

Hagedorn, P. (1988). *Non–Linear Oscillations,* Clarendon Press, Oxford, England. *31*

Hajj, M. R., R. W. Miksad, and E. J. Powers (1992). Subharmonic growth by parametric resonance, *J. Fluid Mech.* 236, 385–413. *557*

Haken, H. (1983). At least one Lyapunov exponent vanishes if the trajectory of an attractor does not contain a fixed point, *Phys. Lett. A* 94, 71–72. *527*

Haken, H. (1987). *Advanced Synergetics,* Springer–Verlag, Berlin, Germany.

Haken, H. See Mayer–Kress and Haken.

Hale, J. K. (1963). *Oscillations in Nonlinear Systems,* McGraw–Hill, New York. *157, 232, 249, 250, 254*

Hale, J. K. (1969). *Ordinary Differential Equations,* Wiley, New York. *155, 158, 291, 424*

Hale, J. K., and H. Kocak (1991). *Dynamics and Bifurcations,* Springer–Verlag, New York. *145, 155*

Hale, J. K. See Chow, Hale, and Mallet–Paret.

Hall, E. K. See Abhyankar, Hall, and Hanagud.

Hall, G. R. See Aronson, Chory, Hall, and McGehee.

Hamdan, A. M. A. See Abed, Wang, Alexander, Hamdan, and Lee; Abed, Hamdan, Lee, and Parlos; Nayfeh, Hamdan and Nayfeh; Wang, Abed, and Hamdan.

Hammel, S. M. (1990). A noise reduction method for chaotic systems. *Phys. Lett. A* 148, 421–428. *463*

Hanagud, S. V. See Abhyankar, Hall, and Hanagud.

Harris, F. J. (1978). On the use of windows for harmonic analysis with the discrete fourier transform, *Proc. IEEE* 66, 51–83. *504*

Harrison, J. A. See Mingori and Harrison.

Hassard, B. D. (1980). Computation of invariant manifolds, in *New Approaches to Nonlinear Problems in Dynamics,* P. J. Holmes, ed., SIAM, Philadelphia, Pennsylvania, 27–42.

Hassard, B. D., N. D. Kazarinoff, and Y.-H. Wan (1981). *Theory and Applications of Hopf Bifurcation,* Cambridge University Press, Cambridge, England. *61, 79, 217*

Hassler, M. (1987). Electrical circuits with chaotic behavior, *Proc. IEEE* 75, 1009–1021.

Hastings, S. (1982). Single and multiple pulse waves for the Fitzhugh–Nagumo equations, *SIAM J. Appl. Math.* 42, 247–260.

Healey, J. J., D. S. Broomhead, K. A. Cliffe, R. Jones, and T. Mullin (1991). The origins of chaos in a modified van der Pol oscillator, *Physica D* 48, 322–339.

Hedrick, J. K. See Beaman and Hedrick.

Heinemann, R. F. See Doedel and Heinemann.

Henderson, D. See Miles and Henderson.

Hendriks, F. (1983). Bounce and chaotic motion in impact print hammers, *IBM J. Res. Dev.* 27, 273–280. *564, 572*

Hennequin, D. See Dangoisse, Glorieux, and Hennequin.

Hénon, M. (1976). A two–dimensional mapping with a strange attractor, *Comm. Math. Phys.* 50, 69–77. *4*

Hénon, M. (1982). On the numerical computation of Poincaré maps, *Physica D* 5, 412–414. *184*

Henrich, E., D. Mingori, and P. Monkewitz (1992). Control of pitchfork and Hopf bifurcations, in *Proceedings of the American Control Conference*, Chicago, Illinois, 2217–2221. *568*

Hentschel, H. G. E., and I. Procaccia (1983). The infinite number of generalized dimensions of fractals and strange attractors, *Physica D* 8, 435–444. *545*

Herzel, H., P. Plath, and P. Svensson (1991). Experimental evidence of homoclinic chaos and type–II intermittency during the oxidation of methanol, *Physica D* 48, 340–352. *299, 311, 313*

Heslot, F. See Stavans, Heslot, and Libchaber.

Hill, W. See Horowitz and Hill.

Hirsch, M. W., and S. Smale (1974). *Differential Equations, Dynamical Systems, and Linear Algebra,* Academic Press, New York. *182*

Hirsch, M. W. See Chiang, Hirsch, and Wu.

Holden, A. V. (1986). *Chaos,* Princeton University Press, Princeton, New Jersey.

Holmes, P. J. (1977). Bifurcations to divergence and flutter in flow–induced oscillations: A finite–dimensional analysis, *J. Sound Vib.* 53, 471–503. *221*

Holmes, P, J. (1979). A nonlinear oscillator with a strange attractor, *Philos. Trans. R. Soc. Lond. A* 292, 419–448. *132, 292, 412*

Holmes, P. J. (1980). Averaging and chaotic motions in forced oscillations, *SIAM J. Appl. Math.* 38, 65–80. *390*

Holmes, P. J. (1981). Center manifolds, normal forms, and bifurcations of vector fields, *Physica D* 2, 449–481.

Holmes, P. J. (1986). Chaotic motions in a weakly nonlinear model for surface waves, *J. Fluid Mech.* 162, 365–388.

Holmes, P. J. (1990a). Nonlinear dynamics, chaos, and mechanics, *Appl. Mech. Rev.* 43, 23–39.

Holmes, P. J. (1990b). Poincaré celestial mechanics, dynamical systems–theory and "chaos," *Phys. Rept.* 193, 138–163.

Holmes, P. J., and J. E. Marsden (1981). A partial differential equation with infinitely many periodic orbits: Chaotic oscillations of forced beam, *Arch. Rational Mech. Anal.* 76, 135–166. *390*

Holmes, P. J., and J. E. Marsden (1982). Horseshoes in perturbations of Hamiltonian systems with two degrees of freedom, *Comm. Math. Phys.* 82, 523–544. *390*

Holmes, P. J., and F. C. Moon (1983). Strange attractors and chaos in nonlinear mechanics, *J. Appl. Mech.* 50, 1021–1032. *292*

Holmes, P. J., and D. A. Rand (1978). Bifurcations of the forced van der Pol oscillator, *Q. Appl. Math.* 35, 495–509. *269*

Holmes, P. J. See Aubry, Holmes, Lumley, and Stone; Brunsden and Holmes; Greenspan and Holmes; Guckenheimer and Holmes; Rand and Holmes; Shaw and Holmes; Wiggins and Holmes.

Holodniok, M., and M. Kubicek (1984a). DERPER: An algorithm for the continuation of periodic solutions in ordinary differential equations, *J. Comp. Phys.* 55, 254–267. *196, 441, 451, 456*

Holodniok, M., and M. Kubicek (1984b). Computation of period doubling bifurcation points in ordinary differential equations, *Institut für Mathematik Report TUM-8406*, Technical University, München, Germany. *441, 460*

Holodniok, M., and M. Kubicek (1987). Determination of invariant tori bifurcation points, *Math. Comp. Sim.* 29, 33–39. *460*

Holodniok, M. See Kubicek and Holodniok.

Holzfuss, J., and G. Mayer–Kress (1985). An approach to error-estimation in the application of dimension algorithms, in *Dimensions and Entropies in Chaotic Systems*, G. Mayer–Kress, ed., Springer-Verlag, New York, 114–122.

Hopf, E. (1948). A mathematical example displaying the features of turbulence, *Comm. Pure Appl. Math.* 1, 303–322. *315*

Horowitz, P., and W. Hill (1980). *The Art of Electronics,* Cambridge University Press, Cambridge, England. *462, 479, 504*

Horsthemke, W. See Tam, Vastano, Swinney, and Horsthemke.

Hosny, W. M. See Abed, Houpt, and Hosny.

Hough, M. E. (1993). Analytic continuation of quasiperiodic orbits from non–linear periodic modes, *Cel. Mec. Dyn. Astr.* 57, 587–617.

Houpt, P. K. See Abed, Houpt, and Hosny.

Howard, L. N. (1979). Nonlinear oscillations, in *Nonlinear Oscillations in Biology,* F. C. Hoppensteadt, ed., *American Mathematical Society,* Providence, Rhode Island, pp. 1–68. *219*

Hsieh, J.-C. See Szemplińska–Stupnicka, Plaut, and Hsieh.

Hsu, C. S. (1980). A theory of cell–to–cell mapping for nonlinear dynamical systems, *J. Appl. Mech.* 47, 931–939. *363*

Hsu, C. S. (1981). A generalized theory of cell–to–cell mapping for nonlinear dynamical systems, *J. Appl. Mech.* 48, 634–642. *363*

Hsu, C. S. (1987). *Cell–to–Cell Mapping: A Method of Global Analysis for Nonlinear Systems,* Springer–Verlag, New York. *363*

Hsu, C. S. (1992). Global analysis by cell mapping, *Int. J. Bif. Chaos* 2, 727–771. *363*

Hsu, C. S. See Troger and Hsu.

Hu, B., and J. Rudnick (1986). Differential–equation approach to functional equations: Exact solutions for intermittency, *Phys. Rev. A* 34, 2453–2457. *302*

Hu, Q. See Iansiti, Hu, Westervelt, and Tinkham.

Huang, J.-Y., and J.-J. Kim (1987). Type–II intermittency in a coupled nonlinear oscillator: Experimental observation, *Phys. Rev. A* 36, 1495–1497. *299*

Huang, W. See Ding, She, Huang, and Yu.

Huberman, B. A., and J. P. Crutchfield (1979). Chaotic states of anharmonic systems in periodic fields, *Phys. Rev. Lett.* 43, 1743–1747. *281, 335*

Huberman, B. A. See Crutchfield and Huberman; D'Humieres, Beasley, Huberman, and Libchaber.

Hübler, A. (1989). Adaptive control of chaotic systems, *Helv. Phys. Acta* 62, 343–346. *583*

Hübler, A., and E. Lüscher (1989). Resonant stimulation and control of nonlinear oscillators, *Naturwissenschaften* 76, 67–69. *583*

Hübler, A. See Jackson and Hübler; Lüscher and Hübler.

Hudson, J. L. See Baier, Wegmann, and Hudson; Bassett and Hudson.

Hunt, E. R. (1991). Stabilizing high–period orbits in a chaotic system: The diode resonator, *Phys. Rev. Lett.* 67, 1953–1955. *577, 578*

Hunt, E. R., and R. W. Rollins (1984). Exactly solvable model of a physical system exhibiting multidimensional chaotic behavior, *Phys. Rev. A* 29, 1000–1002.

Hunt, E. R. See Rajarshi, Murphy, Maier, Gills, and Hunt; Rollins and Hunt.

Huntley, I. (1972). Observations on a spatial resonance phenomenon, *J. Fluid Mech.* 53, 209–216.

Hüpper, G. See Rein, Hüpper, and Schöll.

Hyötyniemi, H. (1991). Postponing chaos using a robust stabilizer, in *Proceedings of the 1st IFAC Symposium on Design Methods of Control Systems*, Zurich, Switzerland, 568–572. *568*

Iansiti, M., Q. Hu, R. M. Westervelt, and M. Tinkham (1985). Noise and chaos in a fractal basin boundary regime of a Josephson junction, *Phys. Rev. Lett.* 55, 746–749. *336*

Ikeda, T. See Ishida, Ikeda, and Yamamoto.

Ikezi, H., S. deGrassie, and T. H. Jensen (1983). Observation of multiple–valued attractors and crises in a driven nonlinear circuit, *Phys. Rev. A* 28, 1207–1209. *336*

Il'yashenko, Y. S. (1993). Local dynamics and nonlocal bifurcations, in *Bifurcations and periodic orbits of vector fields*, D. Scholmiuk, ed., Kluwer, Boston, Massachusetts, 279–320.

Il'yashenko, Y. S. See Arnold and Il'yashenko.

Imry, Y. See Ben–Jacob, Goldhirsch, Imry, and Fishman.

IMSL MATH/LIBRARY (1989). *IMSL Problem–Solving Software Systems*, IMSL, Houston, Texas. *96, 115, 468, 503*

Inoue, M. See Ishii, Fujisaka and Inoue.

Iooss, G. (1979). *Bifurcation of Maps and Applications*, North–Holland, Amsterdam, Netherlands. *128*

Iooss, G., and D. D. Joseph (1980). *Elementary Stability and Bifurcation Theory*, Springer–Verlag, New York. *254*

Iooss, G., and W. F. Langford (1980). Conjectures on the routes to turbulence via bifurcations, in *Nonlinear Dynamics*, H. G. Helleman, ed., New York Academy of Sciences, New York.

Iooss, G. See Chencier and Iooss.

Ishida, Y., T. Ikeda, and T. Yamamoto (1990). Nonlinear forced oscillations caused by quartic nonlinearity in a rotating shaft system, *J. Vib. Acoust.* 112, 288–297.

Ishii, H., H. Fujisaka, and M. Inoue (1986). Breakdown of chaos symmetry and intermittency in the double–well potential system, *Phys. Lett. A* 116, 257–263. *336, 353, 355*

Isomäki, H. M. See Räty, Isomäki, and von Boehm; Räty, von Boehm, and Isomäki.

Iwata, C. See Gills, Iwata, Roy, Schwartz, and Triandaf.

Jackson, E. A. (1989). *Perspectives of Nonlinear Dynamics, Vol. 1,* Cambridge University Press, Cambridge, England. *281*

Jackson, E. A. (1990). *Perspectives of Nonlinear Dynamics, Vol. 2,* Cambridge University Press, Cambridge, England.

Jackson, E. A. (1991a). On the control of complex dynamical systems, *Physica D* 50, 341–346. *583*

Jackson, E. A. (1991b). Control of dynamic flows with attractors, *Phys. Rev. A* 44, 4839–4853. *583*

Jackson, E. A., and A. Hübler (1990). Periodic entrainment of chaotic logistic map dynamics, *Physica D* 44, 404–420. *583*

Jackson, E. A., and A. Kodogeorgiou (1992). Entrainment and migration controls of two–dimensional maps, *Physica D* 54, 253–265. *583*

Jain, P. C., and V. Srinivasan (1975). A review of self–excited vibrations in oil film journal bearings, *Wear* 31, 219–225. *221*

Janovsky, V. See Werner and Janovsky.

Jansen, W. See Feudel and Jansen.

Jaschinski, A. See Knudsen, Feldberg, and Jaschinski.

Jebril, A. E. S. See Nayfeh and Jebril.

Jeffries, C., and J. Perez (1982). Observation of a Pomeau–Manneville intermittent route to chaos in a nonlinear oscillator, *Phys. Rev. A* 26, 2117–2122. *299*

Jeffries, C., and J. Perez (1983). Direct observation of crises of the chaotic attractor in a nonlinear oscillator, *Phys. Rev. A* 27, 601–603.

335, 336

Jeffries, C., and K. Wiesenfeld (1985). Observations of noisy precursors of dynamical instabilities, *Phys. Rev. A* 31, 1077–1084.

Jeffries, C. See Testa, Perez, and Jeffries; Van Buskirk and Jeffries.

Jensen, M. H., P. Bak, and T. Bohr (1983). Complete devil's staircase, fractal dimension, and universality of mode–locking structure in the circle map, *Phys. Rev. Lett.* 50, 1637–1639. *245*

Jensen, M. H., P. Bak, and T. Bohr (1984). Transition to chaos by interaction of resonances in dissipative systems. I. Circle maps, *Phys. Rev. A* 30, 1960–1969. *245, 246, 319, 320*

Jensen, M. H., L. P. Kadanoff, A. Libchaber, I. Procaccia, and J. Stavans (1985). Global universality at the onset of chaos: Results of a forced Rayleigh–Bénard experiment, *Phys. Rev. Lett.* 55, 2798–2801. *248*

Jensen, M. H. See Bohr, Bak, and Jensen.

Jensen, T. H. See Ikezi, deGrassie, and Jensen.

Jezequel, L. See Malasoma, Lamarque, and Jezequel.

Johnson, J. M., and A. K. Bajaj (1989). Amplitude modulated and chaotic dynamics in resonant motion of strings, *J. Sound Vib.* 128, 87–107. *334*

Johnson, J. M. See Bajaj and Johnson.

Jones, G. See Crutchfield, Farmer, Packard, Shaw, Jones, and Donnelly.

Jones, R. See Broomhead and Jones; Healey, Broomhead, Cliffe, Jones, and Mullin.

Jorgensen, D. V., and A. Rutherford (1983). On the dynamics of a stirred tank with consecutive reactions, *Chem. Eng. Sci.* 38, 45–53. *296*

Joseph, D. D. See Iooss and Joseph.

Kaas–Petersen, C. (1985a). Computation of quasi–periodic solutions of forced dissipative systems, *J. Comp. Phys.* 58, 395–403. *253*

Kaas–Petersen, C. (1985b). Computation of quasi–periodic solutions of forced dissipative systems II, *J. Comp. Phys.* 64, 433–442. *253*

Kaas–Petersen, C. (1987). Computation, continuation, and bifurcation of torus solutions for dissipative maps and ordinary differential equations, *Physica D* 25, 288–306. *239–241, 253*

Kaas–Petersen, C. See Brindley, Kaas–Petersen, and Spence.

Kadanoff, L. P. See Feigenbaum, Kadanoff, and Shenker; Jensen, Kadanoff, Libchaber, Procaccia, and Stavans.

Kadtke, J. B. See Abarbanel, Brown, and Kadtke.

Kambe, T. See Umeki and Kambe.

Kamphorst, S. O. See Eckmann, Kamphorst, Ruelle, and Ciliberto.

Kaneko, K. (1984). Oscillation and doubling of a torus, *Prog. Theor. Phys.* 72, 202–215. *334*

Kaneko, K. (1986). *Collapse of Tori and Genesis of Chaos in Dissipative Systems,* World Scientific, Singapore.

Kao, Y. H. See Yeh and Kao.

Kapitaniak, T. (1991). On strange nonchaotic attractors and their dimensions, *Chaos Sol. Fract.* 1, 67–77. *295, 411*

Kapitaniak, T., and W.–H. Steeb (1991). Transition to hyperchaos in coupled generalized van der Pol equations, *Phys. Lett.* 152A, 33–36. *290*

Kapitaniak, T. See Brindley and Kapitaniak; Brindley, Kapitaniak, and El Naschie.

Kaplan, H. (1993). Type–I intermittency for the Hénon–map family, *Phys. Rev. E* 48, 1655–1669.

Kaplan, J. L., and J. A. Yorke (1979a). Chaotic behavior of multi-dimensional difference equations, in *Functional Differential Equations and Approximation of Fixed Points,* H.–O. Peitgen and H.–O. Walter, eds., Lecture Notes in Mathematics, Vol. 730, Springer–Verlag, New York, 228–237. *549*

Kaplan, J. L., and J. A. Yorke (1979b). The onset of chaos in a fluid flow model of Lorenz, *Annal. N.Y. Acad. Sci.* 316, 400–407. *549*

Kaplan, J. L., and J. A. Yorke (1979c). Preturbulence, a regime observed in a fluid flow model of Lorenz, *Comm. Math. Phys.* 67, 93–108. *399, 401*

Kaplan, J. L. See Frederickson, Kaplan, Yorke, and Yorke.

Kapral, R. See Gaspard, Kapral, and Nicolis; Schell, Fraser, and Kapral.

Karamcheti, K. (1976). *Principles of Ideal–Fluid Aerodynamics,* Wiley, New York. *15, 16*

Kazarinoff, N. D. See Hassard, Kazarinoff, and Wan.

Keen, B. E., and W. H. W. Fletcher (1970). Suppression of a plasma instability by the method of "asynchronous quenching," *Phys. Rev. Lett.* 24, 130–134. *221*

Keller, H. B. (1968). *Numerical Methods for Two–Point Boundary–Value Problems,* Blaisdell, Waltham, Massachusetts. *446*

Keller, H. B. (1977). Numerical solutions of bifurcation and nonlinear eigenvalue problems, in *Applications of Bifurcation Theory,* P. H. Rabinowitz, ed., Academic Press, New York. *429, 431, 432, 435*

Keller, H. B. (1987). *Lectures on Numerical Methods in Bifurcation Problems,* Springer–Verlag, New York. *85, 431, 432, 434, 435*

Keller, H. B. See Doedel, Keller, and Kernevez.

Kelley, A. (1967). The stable, center–stable, center, center–unstable, and unstable manifolds, *J. Diff. Eqns.* 3, 546–570. *51*

Kennel, M. B., R. Brown, and H. D. I. Abarbanel (1992). Determining minimum embedding dimension using a geometrical construction, *Phys. Rev. A* 45, 3403–3411. *491*

Kennel, M. B. See Abarbanel, Brown, and Kennel.

Kernevez, J. P. See Doedel, Keller, and Kernevez.

Ketema, Y. (1992). A physical interpretation of Melnikov's method, *Int. J. Bif. Chaos* 2, 1–9.

Kevorkian, J., and J. D. Cole (1981). *Perturbation Methods in Applied Mechanics,* Springer–Verlag, New York.

Kevrekidis, I. G., R. Aris, L. D. Schmidt, and S. Pelikan (1985). Numerical computation of invariant circles of maps, *Physica D* 16, 243–251. *253*

Kevrekidis, I. G. See Aronson, McGehee, Kevrekidis, and Aris.

Khadra, L. See Kim, Khadra, and Powers.

Khdeir, A. A. See Nayfeh and Khdeir.

Kim, J. H., and J. Stringer (1992). *Applied Chaos,* Wiley, New York.

Kim, J.-J. See Huang and Kim.

Kim, J. S. See Green and Kim.

Kim, Y. B., and S. T. Noah (1990). Bifurcation analysis for a modified Jeffcott rotor with bearing clearances, *Nonlinear Dyn.* 1, 221–241. *445*

Kim, Y. B., and S. T. Noah (1991). Stability and bifurcation analysis of oscillators with piecewise linear characteristics: A general approach, *J. Appl. Mech.* 58, 545–553. *281*

Kim, Y. C., J. M. Beall, and E. J. Powers (1980). Bispectrum and nonlinear wave coupling, *Phys. Fluids* 23, 258–263. *557*

Kim, Y. C., L. Khadra, and E. J. Powers (1980). Wave modulation in a nonlinear dispersive medium, *Phys. Fluids* 23, 2250–2257.

Kim, Y. C., and E. J. Powers (1979). Digital bispectral analysis and its applications to nonlinear wave interactions, *IEEE Trans. Plasma Sci.* PS-7, 120–131. *551, 552, 557*

Kimura, K. See Yagasaki, Sakata, and Kimura.

King, G. P. See Broomhead and King.

Klein, M., G. Baier, and O. E. Rössler (1991). From n-tori to hyperchaos, *Chaos Sol. Fract.* 1, 105–118. *134*

Knobloch, E., and N. O. Weiss (1981). Bifurcations in a model of double–diffusive convection, *Phys. Lett. A* 85, 127–130.

Knobloch, E., and N. O. Weiss (1983). Bifurcations in a model of magnetoconvection, *Physica D* 9, 379–407.

Knobloch, H. W. (1990). Construction of center manifolds, *ZAMM* 70, 215–233. *99*

Knudsen, C., R. Feldberg, and A. Jaschinski (1991). Non–linear dynamic phenomena in the behaviour of a railway wheelset model, *Nonlinear Dyn.* 2, 389–404.

Kocak, H. See Hale and Kocak.

Koch, B. P. See Levin, Pompe, Wilke, and Koch.

Kodogeorgiou, A. See Jackson and Kodogeorgiou.

Kolodner, P. See Walden, Kolodner, Passner, and Surko.

Kostelich, E. J., C. Grebogi, E. Ott, and J. A. Yorke (1993). Higher–Dimensional targeting, *Phys. Rev. E* 47, 305–310. *575*

Kostelich, E. J., and T. Schreiber (1993). Noise reduction in chaotic time–series data: A survey of common methods, *Phys. Rev. E* 48, 1752–1763. *463, 483, 550*

Kostelich, E. J., and J. A. Yorke (1990). Noise reduction: Finding the simplest dynamical system consistent with the data, *Physica D* 41, 183–196. *463*

Kostelich, E. J. See Vastano and Kostelich.

Kovacic, G. (1993). Homoclinic and heteroclinic orbits in resonant systems, in *Recent Development in Stability, Vibrations, and Control of Strutural Systems*, AMD–Vol. 167, 237–253.

Krasovskii, N. N. (1963). *Stability of Motion,* Stanford University Press, Stanford, California. *27, 28, 31*

Kreider, M. A. (1992). *A Numerical Investigation of Global Stability of Ship Roll: Invariant Manifolds, Melnikov's Method, and Transient Basins,* M.S. Thesis, Virginia Polytechnic Institute and State University, Blacksburg, Virginia. *52, 380, 383*

Kreisberg, N., W. D. McCormick, and H. L. Swinney (1991). Experimental demonstration of subtleties in subharmonic intermittency, *Physica D* 50, 463–477. *299*

Krousgrill, C. M. See Restuccio, Krousgrill, and Bajaj; Streit, Bajaj, and Krousgrill.

Kubicek, M. (1976). Algorithm 502: Dependence of solution of nonlinear systems on a parameter, *ACM Trans. Math. Software* 2, 98–107. *431, 432*

Kubicek, M., and M. Holodniok (1984). Evaluation of Hopf bifurcation points in parabolic equations describing heat and mass transfer in chemical reactors, *Chem. Eng. Sci.* 39, 593–599.

Kubicek, M., and M. Marek (1983). *Computational Methods in Bifurcation Theory and Dissipative Structures,* Springer–Verlag, New York. *85, 95, 425, 431, 435*

Kubicek, M. See Holodniok and Kubicek.

Kundert, K. S., A. Sangiovanni–Vincentelli, and T. Sugawara (1987). *Techniques for Finding the Periodic Steady–State Response of Circuits,* Marcel Dekker, New York. *445*

Kundert, K. S., G. B. Sorkin, and A. Sangiovanni–Vincentelli (1983). Applying harmonic balance to almost–periodic circuits, *IEEE Trans. Microw. Theory Tech.* 36, 366–378. *253*

Kwatny, H. G., A. K. Pasrija, and L. Y. Bahar (1986). Static bifurcation in electric power networks: Loss of steady–state stability and voltage collapse, *IEEE Trans. Circuit Syst.* CAS–33, 981–991. *564*

Lamarque, C.-H. See Malasoma, Lamarque, and Jezequel.

Lamb, W. E. (1964). Theory of an optical maser, *Phys. Rev. A* 34, 1429–1450. *221*

Landa, P. S., and M. G. Rosenblum (1991). Time series analysis for system identification and diagnostics, *Physica D* 48, 232–254.

Landau, L. D. (1944). On the problem of turbulence, *Dokl. Akad. Nauk SSSR* 44, 339–342. *315*

Landau, L. D., and E. Lifschitz (1959). *Fluid Mechanics*, Pergamon, Oxford, England. *315*

Lange, W. See Mitschke, Möller, and Lange.

Langford, W. F. (1979). Periodic and steady–state mode interactions lead to tori, *SIAM J. Appl. Math.* 37, 22–48. *233*

Langford, W. F. (1983). A review of interactions of Hopf and steady–state bifurcations, in *Nonlinear Dynamics and Turbulence*, G. I. Barenblatt, G. Iooss, and D. D. Joseph, eds., Pitman, Boston, Massachusetts, 215–237. *233*

Langford, W. F. (1985). Unfolding of degenerate bifurcations, in *Chaos, Fractals, and Dynamics*, P. Fisher and W. Smith, eds., Marcel Dekker, New York, 87–103. *269*

Langford, W. F. See Iooss and Langford.

Lansbury, A. N. See Stewart and Lansbury.

Laroche, C. See Libchaber, Fauve, and Laroche.

Larter, R., L. F. Olsen, C. G. Steinmetz, and T. Geest (1993). Chaos in biochemical systems: The peroxidase reaction as a case study, in *Chaos in Chemistry and Biochemistry*, R. J. Field and L. Györgyi, eds., World Scientific, Singapore. *323, 325, 329, 331*

Larter, R., and C. G. Steinmetz (1991). Chaos via mixed–mode oscillations, *Philos. Trans. R. Soc. Lond.* A 337, 291–298. *327, 328*

Larter, R. See Steinmetz and Larter.

Lashinsky, H. (1969). Periodic pulling and the transition to turbulence in a system with discrete modes, in *Turbulence of Fluids and Plasmas*, J. Fox, ed., Wiley, New York, 29–46. *221*

Lashinsky, H. See Deneef and Lashinsky.

Leber, H. See Martin, Leber, and Martienssen.

Lee, B. See Ajjarapu and Lee.

Lee, H.-C., and E. H. Abed (1991). Washout filters in the bifurcation control of high alpha flight dynamics, in *Proceedings of the American Control Conference*, Boston, Massachusetts, 206–211.

Lee, H.-C. See Abed, Wang, Alexander, Hamdan, and Lee; Abed, Hamdan, Lee, and Parlos; Abed and Lee.

Lee, S. Y., and W. F. Ames (1973). A class of general solutions to the nonlinear dynamic equations of elastic strings, *J. Appl. Mech.* 40,

1035–1039.

Lefschetz, S. (1956). Linear and nonlinear oscillations, in *Modern Mathematics for the Engineer,* E. F. Beckenback, ed., McGraw–Hill, New York, 7–30.

Leibovitch, L. S., and T. Toth (1989). A fast algorithm to determine fractal dimensions by box counting, *Phys. Lett. A* 141, 386–390. *545*

Leong, B. L. See Char, Geddes, Gonnet, Leong, Monagan, and Watt.

Leontovich, E. A. See Andronov, Leontovich, Gordon, and Maier.

Letchford, T. E. See Anishchenko, Astakhov, Letchford, and Safonova; Anishchenko, Letchford, and Safonova.

Leung, L. M. See Thompson, Bishop and Leung.

Levi, M. (1981). Qualitative analysis of the periodically forced relaxation oscillations, *Mem. Am. Math. Soc.* 214, 1–147. *269*

Levin, R. W., B. Pompe, C. Wilke, and B. P. Koch (1985). Experiments on periodic and chaotic motions of a parametrically forced pendulum, *Physica D* 16, 371–384. *296*

Lévine, J. See Cibrario and Lévine.

Levinson, N. (1949). A second order differential equation with singular solutions, *Annal Math.* 50, 127–153. *269*

Levinson, N. See Coddington and Levinson.

Li, G.-X., and F. C. Moon (1990a). Criteria for chaos of a three–well potential oscillator with homoclinic and heteroclinic orbits, *J. Sound Vib.* 136, 17–34. *364, 380, 415*

Li, G.-X., and F. C. Moon (1990b). Fractal basin boundaries in a two–degree–of–freedom nonlinear system, *Nonlinear Dyn.* 1, 209–219. *380, 381, 415*

Li, G.-X. See Moon and Li.

Li, M. Y. See Tang, Li, and Weiss.

Li, Y., and J. S. Muldowney (1993). On Bendixson's criterion, *J. Diff. Eqs.* 106, 27–39. *156*

Liaw, D.-C., and E. H. Abed (1990). Stabilization of tethered satellites during station keeping, *IEEE Trans. Auto. Control* AC–35, 1186–1196. *568*

Libchaber, A., S. Fauve, and C. Laroche (1982). Period doubling cascade in mercury: A quantitative measurement, *J. Phys. Lett.* 43, L211–L216. *296*

Libchaber, A. See Arneodo, Coullet, Tresser, Libchaber, Maurer, and D'Humieres; D'Humieres, Beasley, Huberman, and Libchaber; Glazier and Libchaber; Jensen, Kadanoff, Libchaber, Procaccia, and Stavans; Maurer and Libchaber; Stavans, Heslot, and Libchaber.

Lichtenberg, A. J., and M. A. Lieberman (1983). *Regular and Stochastic Motion,* Springer–Verlag, New York.

Lichtenberg, A. J., and M. A. Lieberman (1992). *Regular and Chaotic Dynamics,* Springer–Verlag, New York. *17, 242*

Lieberman, M. A. See Lichtenberg and Lieberman.

Liebert, W., K. Pawelzik, and H. G. Schuster (1991). Optimal embeddings of chaotic attractors from topological considerations, *Europhys. Lett* 14, 521–526. *494*

Liebert, W., and H. G. Schuster (1989). Proper choice of the time delay for the analysis of chaotic time series, *Phys. Lett. A* 142, 107–111. *500*

Lifschitz, E. See Landau and Lifschitz.

Lin, P.–M. See Chua and Lin.

Lin, X. B. See Chow and Lin.

Ling, F. H. (1991). Quasi-periodic solutions calculated with the simple shooting technique, *J. Sound Vib.* 144, 291–304. *254*

Linkens, D. A. (1974). Analytical solution of large numbers of mutually coupled nearly sinusoidal oscillators, *IEEE Trans. Circuit Syst.* CAS–21, 294–300. *221, 230, 255, 270*

Linkens, D. A. (1976). Stability of entrainment conditions for a particular form of mutually coupled van der Pol oscillators, *IEEE Trans. Circuit Syst.* CAS–23, 113–121. *221, 255*

Linsay, P. S. (1981). Period doubling and chaotic behavior in a driven anharmonic oscillator, *Phys. Rev. Lett.* 47, 1349–1352. *281, 296*

Linsay, P. S. See Brorson, Dewey, and Linsay; Cumming and Linsay.

Littlewood, J. E. See Cartwright and Littlewood.

Liu, C. C. See Vu and Liu.

Loeve, M. (1977). *Probability Theory,* Springer–Verlag, New York. *486*

Lorenz, E. N. (1963). Deterministic nonperiodic flow, *J. Atmos. Sci.* 20, 130–141. *31, 141*

Lorenz, E. N. (1984). The local structure of a chaotic attractor in four dimensions, *Physica D* 13, 90-104. *240*

Lumley, J. L. See Aubry, Holmes, Lumley, and Stone.

Lüscher, E., and A. Hübler (1989). Resonant stimulation of complex systems, *Helv. Phys. Acta* 62, 544–551. *583*

Lüscher, E. See Hübler and Lüscher.

Lyapunov, A. M. (1947). *The General Problem of the Stability of Motion,* Princeton University Press, Princeton, New Jersey. *27, 527*

Lyon, R. H. (1987). *Machinery Noise and Diagnostics,* Butterworths, Boston, Massachusetts. *513*

Mackay, R. S., and C. Tresser (1984). Transition to chaos for two–frequency systems, *J. Phys. Lett.* 45, L741–L746. *323*

MACSYMA User's Guide (1988), Symbolics Inc., Cambridge, Massachusetts. *115*

Maier, A. G. See Andronov, Leontovich, Gordon, and Maier.

Maier, T. D. See Rajarshi, Murphy, Maier, Gills, and Hunt; Roy, Murphy, Maier, and Gills.

Malasoma, J.-M., C.-H. Lamarque, and L. Jezequel (1994). Chaotic behavior of a parametrically excited nonlinear mechanical system, *Nonlinear Dyn.* 5, 1994, 153–160. *299, 300, 308–313*

Mallet–Peret, J. See Chow, Hale, and Mallet–Peret.

Mandelbrot, B. B. (1977). *Fractals, Form, Chance, and Dimension,* W. H. Freeman, San Francisco, California. *285*

Mandelbrot, B. B. (1983). *The Fractal Geometry of Nature,* W. H. Freeman, New York. *285, 538*

Mané, R. (1981). On the dimension of compact invariant sets of certain nonlinear maps, in *Dynamical Systems and Turbulence, Lecture Notes in Mathematics,* Vol. 898, D. A. Rand and L. S. Young, eds., Springer–Verlag, New York, 230–242. *480, 495*

Manneville, P., and Y. Pomeau (1979). Intermittency and the Lorenz model, *Phys. Lett. A* 75, 1–2. *197, 298, 302*

Manneville, P., and Y. Pomeau (1980). Different ways to turbulence in dissipative dynamical systems, *Physica D* 1, 219–226. *197, 201, 206*

Manneville, P. See Bergé, Dubois, Manneville, and Pomeau; Pomeau and Manneville.

Mareels, I. M. Y., and R. R. Bitmead (1988). Bifurcation effects in robust adaptive control, *IEEE Trans. Circuit Syst.* CAS–35, 835–841.

Marek, M., and I. Schreiber (1991). *Chaotic Behavior of Deterministic Dissipative Systems*, Cambridge University Press, Cambridge, England. *329, 331*

Marek, M. See Kubicek and Marek.

Marsden, J. E., and M. McCracken (1976) *The Hopf Bifurcation and Its Applications,* Springer–Verlag, New York. *76*

Marsden, J. E. See Bloch and Marsden; Holmes and Marsden.

Marshfield, W. B. See Wright and Marshfield.

Martienssen, W. See Martin, Leber, and Martienssen.

Martin, P. C. See Shraiman, Wayne, and Martin.

Martin, S., H. Leber, and W. Martienssen (1984). Oscillatory and chaotic states of the electrical conduction in barium sodium niobate crystals, *Phys. Rev. Lett.* 53, 303–306. *317*

Maselko, J., and H. Swinney (1986). Complex periodic oscillations and Farey arithmetic in the Belousov–Zhabotinskii reaction, *J. Chem. Phys.* 85, 6430–6441. *331*

Maselko, J., and H. Swinney (1987). A Farey triangle in the Belousov–Zhabotinskii reaction, *Phys. Lett. A* 119, 403–406. *331*

Masere, J. See Petrov, Gaspar, Masere, and Showalter.

Mathieu, E. (1868). Mémoire sur le mouvement vibratoire d'une membrane de forme elliptique, *J. Math.* 13, 137–203.

MATLAB (1989). *User's Guide*, The Math Works, Inc., South Natick, Massachusetts. *115, 503, 520*

Matsumoto, T. (1987). Chaos in electronic circuits, *Proc. IEEE* 75, 1033–1057.

Matsumoto, T., L. O. Chua, and R. Tokunaga (1987). Chaos via torus breakdown, *IEEE Trans. Circuit Syst.* CAS–34, 240–253. *296, 323*

Maurer, J., and A. Libchaber (1979). Rayleigh–Bénard experiment in liquid helium: Frequency locking and the onset of turbulence, *J. Phys. Lett.* 40, L419–L423. *296*

Maurer, J. See Arneodo, Coullet, Tresser, Libchaber, Maurer, and D'Humieres.

May, R. M. (1976). Simple mathematical models with very complicated dynamics, *Nature* 261, 459–467. *4, 281*

Mayer–Kress, G. (1985). *Dimensions and Entropies in Chaotic Systems: Quantification of Complex behavior*, Springer–Verlag, New

York. *538*

Mayer–Kress, G., and H. Haken (1981). The influence of noise on the Logistic model, *J. Stat. Phys.* 26, 149–171. *296*

Mayer–Kress, G. See Holzfuss and Mayer–Kress.

McCormick, W. D. See Kreisberg, McCormick, and Swinney; Turner, Roux, McCormick, and Swinney.

McCracken, M. See Marsden and McCracken.

McGehee, R. P. See Aronson, Chory, Hall, and McGehee; Aronson, McGehee, Kevrekidis, and Aris.

McNamara, B. See Wiesenfeld and McNamara.

Medvéd, M. (1992). *Fundamentals of Dynamical Systems and Bifurcation Theory*, Adam Hilger, New York. *172, 176*

Mees, A. (1981). *Dynamics of Feedback Systems*, Wiley, Chichester, England. *445, 446, 452, 453*

Mees, A., and C. Sparrow (1987). Some tools for analyzing chaos, *Proc. IEEE* 75, 1058–1070.

Meirovitch, L. (1970). *Methods of Analytical Dynamics*, McGraw–Hill, New York. *80*

Melnikov, V. K. (1963). On the stability of the center for time periodic perturbations, *Trans. Moscow Math. Soc.* 12, 1–57. *366*

Menzel, R. (1984). Numerical determination of multiple bifurcation points, in *Numerical Methods for Bifurcation Problems*, T. Kuepper, H. D. Mittleman, and H. Weber, eds., Birkhaüser Verlag, Boston, Massachusetts, 310–318. *438*

Metropolis, N., M. L. Stein, and P. R. Stein (1973). On finite limit sets for transformations on the unit interval, *J. Comb. Theory* 15, 25–44. *281*

Mevel, B., and J. L. Guyader (1993). Routes to chaos in ball bearings, *J. Sound Vib.* 162, 471–487.

Michel, A. N., R. K. Miller, and B. H. Nam (1982). Stability analysis of interconnected systems using computer generated Lyapunov functions, *IEEE Trans. Circuit Syst.* CAS–29, 431–440. *27, 46*

Michel, A. N., B. H. Nam, and V. Vittal (1984). Computer generated Lyapunov functions for interconnected systems: Improved results with applications to power systems, *IEEE Trans. Circuit Syst.* CAS–31, 189–198. *27, 46*

Miksad, R. W. See Choi, Miksad, Powers, and Fischer; Hajj, Miksad, and Powers; Ritz, Powers, Miksad, and Solis.

Miller, M. (1986). Bispectral analysis of Sine-Gordon chain, *Phys. Rev. B* 34, 6326–6333. *557*

Miller, R. K. See Michel, Miller, and Nam.

Miller, R. N. See Rinzel and Miller.

Miles, J. W. (1984). Resonant motion of a spherical pendulum, *Physica D* 11, 309–323. *334*

Miles, J., and D. Henderson (1990). Parametrically forced surface waves, *Annal Rev. Fluid Mech.* 22, 143–165. *334*

Mingori, D. L., and J. A. Harrison (1974). Circularly constrained particle motion in spinning and coning bodies, *AIAA J.* 12, 1553–1558. *140*

Mingori, D. See Henrich, Mingori, and Monkewitz.

Minis, I. See Berger, Rokni, and Minis.

Minorsky, N. (1947). *Introduction to Non–Linear Mechanics,* J. W. Edwards, Ann Arbor, Michigan.

Mitropolsky, Y. A. (1967). Averaging method in non–linear mechanics, *Int. J. Non–Linear Mech.* 2, 69–96.

Mitropolsky, Y. A. See Bogoliubov and Mitropolsky.

Mitschke, F. (1990). Acausal filters for chaotic signals, *Phys. Rev. A,* 41, 1169–1171. *463*

Mitschke, F., M. Möller, and W. Lange (1988). On systematic errors in characterizing chaos, *J. Phys. Colloq.* 49, C2397–C2400. *463, 550*

Mohamed, A. M., and F. P. Emad (1993). Nonlinear oscillations in magnetic bearing systems, *IEEE Trans. Auto. Control* 38, 1242–1245. *568*

Möller, M. See Mitschke, Möller, and Lange.

Molteno, T. C. A. (1993). *Chaos and Crises in Strings,* Ph.D. Thesis, University of Otago, Australia. *334*

Molteno, T. C. A., and N. B. Tufillaro (1990). Torus doubling and chaotic string vibrations: Experimental results, *J. Sound Vib.* 137, 327–330. *334*

Monagan, M. B. See Char, Geddes, Gonnet, Leong, Monagan, and Watt.

Monkewitz, P. See Henrich, Mingori, and Monkewitz.

Mook, D. T. See Abou-Rayan, Nayfeh, Mook, and Nayfeh; Chin, Nayfeh, and Mook; Elzebda, Nayfeh, and Mook; Nayfeh and Mook; Nayfeh, Nayfeh, and Mook.

Moon, F. C. (1987). *Chaotic Vibrations: An Introduction for Applied Scientists and Engineers,* Wiley, New York. *281, 520*

Moon, F. C. (1992). *Chaotic and Fractal Dynamics: An Introduction for Applied Scientists and Engineers,* Wiley, New York. *520*

Moon, F. C., and G.-X. Li (1985). Fractal basin boundaries and homoclinic orbits for periodic motion in a two well potential, *Phys. Rev. Lett.* 55, 1439–1442. *292, 415*

Moon, F. C., and R. H. Rand (1985). Parametric stiffness control of flexible structures, in *Proceedings of the Workshop on Identification and Control of Flexible Space Structures,* Vol. II, California Institute of Technology, Pasadena, California, 329–342. *190*

Moon, F. C. See Holmes and Moon; Li and Moon.

Moore, F. K., and E. M. Greitzer (1986). A theory of post–stall transients in axial compression systems: Part II–Application, *ASME J. Eng. Gas Turb. Power* 108, 231–239. *564*

Moore, G., and A. Spence (1980). The calculation of turning points of nonlinear equations, *SIAM J. Num. Anal.* 17, 567–576. *436*

Morgan, A. P. (1987). *Solving Polynomial Systems using Continuation for Engineering and Scientific Problems,* Prentice–Hall, Englewood Cliffs, New Jersey. *444*

Morgan, S. P. See Watson, Billups, and Morgan.

Moroz, I. (1986). Amplitude expansions and normal forms in a model for thermohaline convection, *Studies Appl. Math.* 74, 155–170. *115*

Mueller, R. D. See Breitenberger and Mueller.

Muldowney, J. S. See Li and Muldowney.

Mullin, T. (1993). *The Nature of Chaos,* Oxford University, Oxford, England.

Mullin, T., and A. G. Darbyshire (1989). Intermittency in a rotating annular flow, *Europhys. Lett.* 9, 669–673. *299*

Mullin, T. See Healey, Broomhead, Cliffe, Jones, and Mullin; Price and Mullin.

Murdock, J. A. (1991). *Perturbations: Theory and Methods,* Wiley, New York.

Murphy, T. W. See Rajarshi, Murphy, Maier, Gills, and Hunt; Roy, Murphy, Maier, and Gills.

Musazzi, S. See Giglio, Musazzi, and Perine.

Nagashima, T. See Shimada and Nagashima.

Nakamura, Y. (1971). Suppression and excitation of electron oscillation in a beam–plasma system, *J. Phys. Soc. Jpn.* 31, 273–279. *221*

Nakamura, Y. See Sato, Sasaki, and Nakamura.

Nam, B. H. See Michel, Miller, and Nam; Michel, Nam, and Vittal.

Namachchivaya, N. S. (1991). Co–dimension two bifurcations in the presence of noise, *J. Appl. Mech.* 58, 259–265.

Nauenberg, M. See Crutchfield, Nauenberg, and Rudnick.

Nayfeh, A. H. (1968). Forced oscillations of van der Pol oscillator with delayed amplitude limiting, *IEEE Trans.* CT–15, 192–200. *138, 271*

Nayfeh, A. H. (1970). Nonlinear stability of a liquid jet, *Phys. Fluids* 13, 841–847. *115*

Nayfeh, A. H. (1973). *Perturbation Methods,* Wiley, New York. *83, 166, 169, 171, 215*

Nayfeh, A. H. (1981). *Introduction to Perturbation Techniques,* Wiley, New York. *25, 83, 110, 166, 169, 171, 215*

Nayfeh, A. H. (1987a). Parametric excitation of two internally resonant oscillators, *J. Sound Vib.* 119, 95–109. *334*

Nayfeh, A. H. (1987b). Surface waves in closed basins under parametric and internal resonances, *Phys. Fluids* 30, 2976–2983. *334*

Nayfeh, A. H. (1988). On the undesirable roll characteristics of ships in regular seas, *J. Ship Res.* 32, 92–100. *334*

Nayfeh, A. H. (1993). *Method of Normal Forms,* Wiley, New York. *43, 213*

Nayfeh, A. H., and K. R. Asfar (1988). Non–stationary parametric oscillations, *J. Sound Vib.* 124, 529–537. *90*

Nayfeh, A. H., and B. Balachandran (1990a). Motion near a Hopf bifurcation of a three–dimensional system, *Mech. Res. Comm.* 17, 191–198. *190, 203, 209, 213*

Nayfeh, A. H., and B. Balachandran (1990b). Experimental investigation of resonantly forced oscillations of a two–degree–of–freedom structure, *Int. J. Non–Linear Mech.* 25, 199–209.

Nayfeh, A. H., and B. Balachandran (1995). *Nonlinear Interactions,* Wiley, New York, in press. *473*

Nayfeh, A. H., B. Balachandran, M. A. Colbert, and M. A. Nayfeh (1989). An experimental investigation of complicated responses of a two–degree–of–freedom structure, *J. Appl. Mech.* 56, 960–967. *468, 476*

Nayfeh, A. H., and H. Bouguerra (1990). Non–linear response of a fluid valve, *Int. J. Non–Linear Mech.* 25, 433–439.

Nayfeh, A. H., and C. Chin (1994). A parametrically excited system with widely spaced frequencies and cubic nonlinearities, *Nonlinear Dyn.*, in press. *336, 405*

Nayfeh, A. H., and A. E. S. Jebril (1987). The response of two–degree–of–freedom systems with quadratic and cubic nonlinearities to multifrequency excitations, *J. Sound Vib.* 115, 83–101.

Nayfeh, A. H., and A. A. Khdeir (1986a). Nonlinear rolling of biased ships in regular beam seas, *Int. J. Shipbldg. Prog.* 33, 84–93. *52, 351*

Nayfeh, A. H., and A. A. Khdeir (1986b). Nonlinear rolling of ships in regular beam seas, *Int. J. Shipbldg. Prog.* 33, 40–49. *52, 351*

Nayfeh, A. H., and D. T. Mook (1979). *Nonlinear Oscillations,* Wiley, New York. 130–136. *83, 88, 90, 220, 221, 224, 257, 269*

Nayfeh, A. H., and J. F. Nayfeh (1990). Surface waves in closed basins under principal and autoparametric resonances, *Phys. Fluids A.* 2, 1635–1648. *192–194, 198, 334, 515*

Nayfeh, A. H., and R. A. Raouf (1987). Nonlinear forced response of infinitely long circular cylindrical shells, *J. Appl. Mech.* 54, 571–577. *334*

Nayfeh, A. H., R. A. Raouf, and J. F. Nayfeh (1991). Nonlinear response of infinitely long circular cylindrical shells to subharmonic radial loads, *J. Appl. Mech.* 58, 1033–1041. *334*

Nayfeh, A. H., and N. E. Sanchez (1989). Bifurcations in a forced softening Duffing oscillator, *Int. J. Non–Linear Mech.* 24, 483–497. *194, 199, 203, 281, 362, 363, 415, 503*

Nayfeh, A. H., and N. E. Sanchez (1990). Stability and complicated rolling responses of ships in regular beam seas, *Int. J. Shipbldg. Prog.* 37, 331–352. *52, 563*

Nayfeh, A. H., and L. D. Zavodney (1986). The response of two–degree–of–freedom systems with quadratic non–linearities to a combination parametric resonance, *J. Sound Vib.* 107, 329–350. *334*

Nayfeh, A. H., and L. D. Zavodney (1988). Experimental observation of amplitude– and phase–modulated responses of two internally coupled oscillators to a harmonic excitation, *J. Appl. Mech.* 55, 706–710. *501*

Nayfeh, A. H. See Abou-Rayan, Nayfeh, Mook, and Nayfeh; Anderson, Balachandran, and Nayfeh; Balachandran and Nayfeh; Bikdash, Balachandran, and Nayfeh; Chin, Nayfeh, and Mook; Elzebda, Nayfeh, and Mook; Nayfeh, Asrar, and Nayfeh; Nayfeh, Hamdan and Nayfeh; Nayfeh, Nayfeh, and Mook; Pai and Nayfeh; Raouf and Nayfeh; Sanchez and Nayfeh; Zavodney and Nayfeh; Zavodney, Nayfeh, and Sanchez.

Nayfeh, J. F. See Nayfeh and Nayfeh; Nayfeh, Raouf, and Nayfeh.

Nayfeh, M. A., A. M. A. Hamdan, and A. H. Nayfeh (1990). Chaos and instability in a power system– primary resonant case, *Nonlinear Dyn.* 1, 313–339. *56, 57, 203, 281, 352, 353, 415, 564*

Nayfeh, M. A., A. M. A. Hamdan, and A. H. Nayfeh (1991). Chaos and instability in a power system– subharmonic–resonant case, *Nonlinear Dyn.* 2, 53–72. *56, 415, 564*

Nayfeh, M. A. See Abou-Rayan, Nayfeh, Mook, and Nayfeh; Nayfeh, Balachandran, Colbert, and Nayfeh.

Nayfeh, T. A. (1991). *Dynamics of Three–Degree–of–Freedom Systems with Quadratic Nonlinearities,* M.S. Thesis, Virginia Polytechnic Institute and State University, Blacksburg, Virginia.

Nayfeh, T. A., W. Asrar, and A. H. Nayfeh (1992). Three–mode interactions in harmonically excited systems with quadratic nonlinearities, *Nonlinear Dyn.* 3, 385–410. *206–208*

Nayfeh, T. A., A. H. Nayfeh, and D. T. Mook (1994). A theoretical–experimental investigation of a three–degree–of–freedom structure, *Nonlinear Dyn.* 6, 353–374. *511*

Nemytskii, V. V., and V. V. Stepanov (1960). *Qualitative Theory of Differential Equations,* Princeton University Press, Princeton, New Jersey. *223*

Nett, C. N. See Badmus, Chowdhury, Eveker, Nett, and Rivera.

Neu, J. C. (1979). Coupled chemical oscillators, *SIAM J. Appl. Math.* 37, 307–315. *251*

Newhouse, S. E., D. Ruelle, and F. Takens (1978). Occurrence of strange axiom A attractors near quasiperiodic flows on T^m, $m \geq 3$, *Comm. Math. Phys.* 64, 35–40. *255, 315, 316*

Neymeyr, K., and F. F. Seelig (1991). Determination of unstable limit cycles in chaotic systems by the method of unrestricted harmonic balance, *Z. Naturforsch.* 46, 499–502. *446*

Nicolis, G., and I. Prigogine (1989). *Exploring Complexity, an Introduction,* W. H. Freeman, San Francisco, California.

Nicolis, G. See Gaspard, Kapral, and Nicolis; Gaspard and Nicolis.

Nikias, C. L., and A. P. Petropulu (1993). *Higher-Order Spectra Analysis*, Prentice-Hall, Englewood Cliffs, New Jersey. *551*

Nikias, C. L., and M. R. Raghuveer (1987). Bispectrum estimation: A digital signal processing framework, *Proc. IEEE* 75, 869–891. *557*

Nitecki, Z. See Devaney and Nitecki.

Nitsche, G. See Dressler and Nitsche.

Noah, S. T. See Choi and Noah; Kim and Noah.

Nobili, M., S. Ciliberto, B. Cocciaro, S. Faetti, and L. Fronzoni (1988). Time–dependent surface waves in a horizontally oscillating container, *Europhys. Lett.* 7, 587–592.

Novak, S., and R. G. Frelich (1982). Transition to chaos in the Duffing oscillator, *Phys. Rev. A* 26, 3660–3663. *203, 281*

O'Brien, W. F. See Davis and O'Brien.

Olsen, L. F. See Larter, Olsen, Steinmetz, and Geest.

Oppenheim, A. V., and R. W. Schafer (1975). *Digital Signal Processing,* Prentice–Hall, Englewood Cliffs, New Jersey. *462, 504*

Orszag, S. A. See Bender and Orszag; Goldhirsch, Sulem, and Orszag.

Oseledec, V. I. (1968). A multiplicative ergodic theorem, Lyapunov characteristic numbers for dynamical systems, *Trans. Moscow Math. Soc.* 19, 197–231. *528, 532*

Ostlund, S., D. Rand, J. Sethna, and E. Siggia (1983). Universal properties of the transition from quasiperiodicity to chaos in dissipative systems, *Physica D* 8, 303–342. *322*

Ostlund, S. See Rand, Ostlund, Sethna, and Siggia.

Ott, E. (1981). Strange attractors and chaotic motions of dynamical systems, *Rev. Mod. Phys.* 53, 655–671. *281, 283*

Ott, E. (1993). *Chaos in Dynamical Systems,* Cambridge University Press, Cambridge, England. *281, 283, 538, 572*

Ott, E., C. Grebogi, and J. A. Yorke (1990a). Controlling chaotic dynamical systems, in *CHAOS: Soviet–American Perspectives on Nonlinear Science,* D. K. Campbell, ed., 153–172. *572*

Ott, E., C. Grebogi, and J. A. Yorke (1990b). Controlling chaos, *Phys. Rev. Lett.* 64, 1196–1199. *572*

Ott, E., T. Sauer, and J. A. Yorke (1994). *Coping with Chaos,* Wiley, New York.

Ott, E. See Auerbach, Grebogi, Ott, and Yorke; Ding, Grebogi, and Ott; Farmer, Ott, and Yorke; Grebogi, Ott, Pelikan, and Yorke; Grebogi, Ott, Romeiras, and Yorke; Grebogi, Ott, and Yorke; Kostelich, Grebogi, Ott, and Yorke; Romeiras, Bondeson, Ott, Antonsen, and Grebogi; Romeiras, Grebogi, Ott, and Dayawansa; Romeiras and Ott; Shinbrot, Ott, Grebogi, and Yorke; Shinbrot, Grebogi, Ott, and Yorke; Sommerer, Ditto, Grebogi, Ott, and Spano.

Packard, N. H., J. P. Crutchfield, J. D. Farmer, and R. S. Shaw (1980). Geometry from a time series, *Phys. Rev. Lett.* 45, 712–716. *479, 480*

Packard, N. H. See Breedon and Packard; Crutchfield, Farmer, Packard, Shaw, Jones, and Donnelly.

Pai, M. A. (1989). *Energy Function Analysis for Power System Stability,* Kluwer, Boston, Massachusetts. *27*

Pai, M. A. See Rajagopalan, Sauer, and Pai.

Pai, P. F., and A. H. Nayfeh (1990). Nonlinear nonplanar oscillations of a cantilever beam under lateral base excitations, *Int. J. Nonl. Mech.* 25, 455–474. *334*

Paidoussis, M. P., and R. M. Botez (1995). Three routes to chaos for a three–degree–of–freedom articulated cylinder system subjected to annular axial flow and impacting on the outer pipe, *Nonlinear Dyn.,* in press. *299*

Palis, J., and F. Takens (1993). *Hyperbolicity & Sensitive Chaotic Dynamics at Homoclinic Bifurcations,* Cambridge University Press, Cambridge, England. *404*

Papoulis, A. (1990). Bispectra and system identification, in *Recent Advances in Fourier Analysis and Its Applications,* J. S. Byrnes and J. F. Byrnes, eds., Kluwer, Dordrecht, Netherlands. *551*

Parisi, J. See Richter, Peinke, Clauss, Rau, and Parisi.

Parker, T. S., and L. O. Chua (1987). Chaos: A tutorial for engineers, *Proc. IEEE* 75, 982–1008. *339*

Parker, T. S., and L. O. Chua (1989). *Practical Numerical Algorithms for Chaotic Systems,* Springer–Verlag, New York. *61, 64, 95, 184, 240, 241, 281, 339, 364, 365, 446, 455, 548, 534, 536*

Parlitz, U. (1992). Identification of true and spurious Lyapunov exponents from time series, *Int. J. Bif. Chaos* 2, 155–165.

Parlos, A. G. See Abed, Hamdan, Lee, and Parlos.

Parmananda, P. See Rollins, Parmananda, and Sherard.

Pasrija, A. K. See Kwatny, Pasrija, and Bahar.

Passner, A. See Walden, Kolodner, Passner, and Surko.

Patel, M. R. See Fallside and Patel.

Pawelzik, K. See Liebert, Pawelzik, and Schuster.

Pecora, L. M., and T. L. Carroll (1990). Synchronization in chaotic systems, *Phys. Rev. Lett.* 64, 821–824. *585*

Pecora, L. M., and T. L. Carroll (1991). Driving systems with chaotic signals, *Phys. Rev. A* 44, 2374–2383. *585, 586*

Pecora, L. M. See Carroll and Pecora; Carroll, Pecora, and Rachford.

Peinke, J. See Richter, Peinke, Clauss, Rau, and Parisi.

Pelikan, S. See Grebogi, Ott, Pelikan, and Yorke; Kevrekidis, Aris, Schmidt, and Pelikan.

Peitgen, H.–O., and P. H. Richter (1986). *Beauty of Fractals,* Springer–Verlag, New York.

Peng, B., V. Petrov, and K. Showalter (1991). Controlling chemical chaos, *J. Phys. Chem.* 95, 4957–4959. *580, 581*

Peng, B. See Petrov, Peng, and Showalter.

Perez, J. See Jeffries and Perez; Testa, Perez, and Jeffries.

Perez, R., and L. Glass (1982). Bistability, period doubling bifurcations and chaos in a periodically forced oscillator, *Phys. Lett. A* 90, 441–443. *323*

Perez, R. See Glass, Guevara, Shrier, and Perez.

Perine, V. See Giglio, Musazzi, and Perine.

Perkins, N. C. (1992). Modal interactions in the non–linear response of elastic cables under parametric/external excitation, *Int. J. Non-Linear Mech.* 27, 233–250.

Petropulu, A. P. See Nikias and Petropulu.

Petrov, V., V. Gaspar, J. Masere, and K. Showalter (1993). Controlling chaos in the Belousov–Zhabotinsky reaction, *Nature* 361, 240–243. *577, 580*

Petrov, V., B. Peng, and K. Showalter (1992). A map–based algorithm for controlling low–dimensional chaos, *J. Chem. Phys.* 96, 7506–7513. *580*

Petrov, V., S. K. Scott, and K. Showalter (1992). Mixed–mode oscillations in chemical systems, *J. Chem. Phys.* 97, 6191–6198. *331*

Petrov, V. See Peng, Petrov, and Showalter.

Pezeshki, C., and E. H. Dowell (1988). On chaos and fractal behavior in a generalized Duffing's system, *Physica D* 32, 194–209. *415*

Pezeshki, C. See Dowell and Pezeshki.

Piersol, A. G. See Bendat and Piersol.

Pfister, G. See Buzug and Pfister; Buzug, Reimers, and Pfister.

Piessens, R. See Roose and Piessens.

Place, C. M. See Arrowsmith and Place.

Plath, P. See Herzel, Plath, and Svensson.

Platt, N., E. A. Spiegel, and C. Tresser (1993). On–off intermittency: A mechanism for bursting, *Phys. Rev. Let.* 70, 279–282. *299*

Plaut, R. H. See Szemplińska–Stupnicka, Plaut, and Hsieh.

Poincaré, H. (1899). *Les Méthods Nouvelles de la Mécanique Céleste, Vols. I–III,* Gauthier–Villars, Paris, France, (reprinted by Dover, New York, 1957). English translation, in *New Methods of Celestial Mechanics, Parts 1, 2, and 3,* D. L. Goroff, ed., 1993. *77*

Politi, A. See Badii and Politi; Badii, Broggi, Derighetti, Ravani, Ciliberto, Politi, and Rubio.

Pollack, V. See Guillemin and Pollack.

Pomeau, Y., and P. Manneville (1980). Intermittent transition to turbulence in dissipative dynamical systems, *Comm. Math. Phys.* 74, 189–197. *197, 201, 206, 298*

Pomeau, Y., J.–C. Roux, A. Rossi, S. Bachelart, and C. Vidal (1981). Intermittent behaviour in the Belousov–Zhabotinsky reaction, *J. Phys. Lett.* 42, L271–L273. *298*

Pomeau, Y. See Bergé, Dubois, Manneville, and Pomeau; Bergé, Pomeau, and Vidal; Manneville and Pomeau.

Pompe, B. See Levin, Pompe, Wilke, and Koch.

Pönisch, G. (1985). Computing simple bifurcation points using a minimally extended system of nonlinear equations, *Computing* 35, 277–294. *438*

Pönisch, G. (1987). Computing hysteresis points of nonlinear equations depending on two parameters, *Computing* 39, 1–17. *441*

Pontryagin, L. S. (1962). *Ordinary Differential Equations,* Addison-Wesley, Reading, Massachusetts. *78*

Powers, E. J. See Choi, Chang, Stearman, and Powers; Choi, Miksad, Powers, and Fischer; Hajj, Miksad, and Powers; Kim, Beall, and Powers; Kim, Khadra, and Powers; Kim and Powers; Ritz and Powers; Ritz, Powers, Miksad, and Solis.

Preisendorfer, R. W. (1988). *Principal Component Analysis in Meteorology and Oceanography*, Elsevier, Amsterdam, Holland. *502*

Price, T. J., and T. Mullin (1991). An experimental observation of a new type of intermittency, *Physica D* 48, 29–52. *299*

Prigogine, I. See Nicolis and Prigogine.

Procaccia, I. See Ben–Mizrachi and Procaccia; Grassberger and Procaccia; Hentschel and Procaccia; Jensen, Kadanoff, Libchaber, Procaccia, and Stavans.

Pujol, J., M. Arjona, and R. Corbalán (1993). Type–III intermittency in a four–level coherently pumped laser, *Phys. Rev. A* 48, 2251–2255. *299*

Pujol, J. See Tang, Pujol, and Weiss.

Pyragas, K. (1992). Continuous control of chaos by self–controlling feedback, *Phys. Lett. A* 170, 421–428.

Pyragas, K. See Cenys and Pyragas.

Rabinovich, M. I. See Abarbanel, Rabinovich, and Sushchik.

Rachford, F. J. See Carroll, Pecora, and Rachford.

Rainey, R. C. T. See Thompson, Rainey, and Soliman.

Raghuveer, M. R. See Nikias and Raghuveer.

Rajagopalan, C., P. W. Sauer, and M. A. Pai (1989). Analysis of voltage control systems exhibiting Hopf bifurcation, in *Proceedings of the 28th IEEE Confrence on Decision Control,* Tampa, Florida, 332–335. *564*

Rajarshi, R., T. W. Murphy, Jr., T. D. Maier, Z. Gills, and E. R. Hunt (1992). Dynamical control of a chaotic laser: Experimental stabilization of a globally coupled system, *Phys. Rev. Lett.* 68, 1259–1262. *577*

Ramaswamy, R. See Sinha, Ramaswamy, and Rao.

Rand, D. A., S. Ostlund, J. Sethna, and E. Siggia (1982). A universal transition from quasiperiodicity to chaos in dissipative systems, *Phys. Rev. Lett.* 49, 132–135. *322*

Rand, D. A. See Holmes and Rand; Ostlund, Rand, Sethna, and Siggia.

Rand, R. H. (1989). Analytical approximation for period–doubling following a Hopf bifurcation, *Mech. Res. Comm.* 16, 117–123. *190, 211*

Rand, R. H. (1994). *Topics in Nonlinear Dynamics with Computer Algebra*, Gordon and Breach Science Publishers, Langhorne, Pennsylvania.

Rand, R. H., and D. Armbruster (1987). *Perturbation Methods, Bifurcation Theory, and Computer Algebra,* Springer–Verlag, New York.

Rand, R. H., and P. J. Holmes (1980). Bifurcation of periodic motions in two weakly coupled van der Pol oscillators, *Int. J. Non-Linear Mech.* 15, 387–399. *251*

Rand, R. H. See Moon and Rand; Shaw and Rand; Storti and Rand.

Rao, J. S. See Sinha, Ramaswamy, and Rao.

Raouf, R. A., and A. H. Nayfeh (1990a). Nonlinear axisymmetric response of closed spherical shells to a radial harmonic excitation, *Int. J. Nonl. Mech* 25, 475–492. *334*

Raouf, R. A., and A. H. Nayfeh (1990b). One-to-one autoparametric resonances in infinitely long cylindrical shells, *Comp. Struc.* 35, 163–173. *334*

Raouf, R. A. See Nayfeh and Raouf; Nayfeh, Raouf, and Nayfeh.

Rasband, S. N. (1990). *Chaotic Dynamics of Nonlinear Systems,* Wiley, New York.

Räty, R., H. M. Isomäki, and J. von Boehm (1984). Chaotic motion of a classical anharmonic oscillator, in *Acta Polytechnica Scandinavia, Mechanical Engineering Series No. 85,* Vol. 30, Helsinki, Finland. *203, 281, 414*

Räty, R., J. von Boehm, and H. M. Isomäki (1986). Chaotic motion of a periodically driven particle in an asymmetric potential well, *Phys.*

Rev. A 34, 4310–4315. *414*

Rau, U. See Richter, Peinke, Clauss, Rau, and Parisi.

Rauseo, S. N. See Ditto, Rauseo, and Spano.

Ravani, M. See Badii, Broggi, Derighetti, Ravani, Ciliberto, Politi, and Rubio.

Reddien, G. W. See Griewank and Reddien.

Rega, G., F. Benedettini, and A. Salvatori (1991). Periodic and chaotic motions of an unsymmetrical oscillator in nonlinear structural dynamics, *Chaos Sol. Fract.* 1, 39–54.

Rega, G., A. Salvatori, and F. Benedettini (1992). Basin bifurcation and chaotic attractor in an elastic oscillator with quadratic and cubic nonlinearities, in *Nonlinear Vibrations,* ASME DE–Vol. 50, 7–13. *415*

Reilly, J. See Tufillaro, Abbott, and Reilly.

Reimers, T. See Buzug, Reimers, and Pfister.

Rein, A., G. Hüpper, and E. Schöll (1993). Fluctuations and critical scaling in type–I intermittency, *Europhys. Lett.* 21, 7–12.

Reith, L. A. See Gorman, Reith, and Swinney.

Renyi, A. (1970). *Probability Theory.* North–Holland, Amsterdam. *484*

Restuccio, J. M., C. M. Krousgrill, and A. K. Bajaj (1991). Nonlinear nonplanar dynamics of a parametrically excited inextensional elastic beam, *Nonlinear Dyn.* 2, 263–289. *334*

Rheinboldt, W. C. See Den Heijer and Rheinboldt.

Rheinhardt, W. D. See Awrejcewicz and Rheinhardt.

Richetti, P., P. DeKepper, J.–C. Roux, and H. L. Swinney (1987). A crisis in Belousov–Zhabotinskii reaction: Experiment and Simulation, *J. Stat. Phys.* 48, 977–990.

Richetti, P. See Argoul, Arneodo, Richetti, and Roux.

Richter, P. H. See Peitgen and Richter.

Richter, R., J. Peinke, W. Clauss, U. Rau, and J. Parisi (1991). Evidence of type–III intermittency in the electric breakdown of p–type germanium, *Europhys. Lett.* 14, 1–6. *299*

Ringuet, E., C. Roźc, and C. Gouesbet (1993). Experimental observation of type–II intermittency in a hydrodynamic system, *Phys. Rev E.* 47, 1405–1410. *299*

Rinzel, J., and R. N. Miller (1980). Numerical calculation of stable and unstable periodic solutions to the Hodgkin–Huxley equations,

Math. Biosci. 49, 27–59. *446, 448, 456*

Ritz, C. P., and E. J. Powers (1986). Estimation of nonlinear transfer functions for fully developed turbulence, *Physica D* 20, 320–334. *557*

Ritz, C. P., E. J. Powers, R. W. Miksad, and R. S. Solis (1988). Nonlinear spectral dynamics of a transitioning flow, *Phys. Fluids* 31, 3577–3588. *557*

Rivera, C. J. See Badmus, Chowdhury, Eveker, Nett, and Rivera.

Robbins, K. A. See Gorman, Widmann, and Robbins.

Rokni, M., and B. S. Berger (1987). Lyapunov exponents and subspace evolution, *Q. Appl. Math.* XLV, 789–793.

Rokni, M. See Berger and Rokni; Berger, Rokni, and Minis,

Rollins, R. W., and E. R. Hunt (1982). Exactly solvable model of a physical system exhibiting universal chaotic behavior, *Phys. Rev. Lett.* 49, 1295–1298.

Rollins, R. W., and E. R. Hunt (1984). Intermittent transient chaos at interior crises in the diode resonator, *Phys. Rev. A* 29, 3327–3334. *336*

Rollins, R. W., P. Parmananda, and P. Sherard (1993). Controlling chaos in highly dissipative systems: A simple recursive algorithm, *Phys. Rev. E* 47, R780–R783. *582*

Rollins, R. W. See Hunt and Rollins.

Romeiras, F. J., A. Bondeson, E. Ott, T. M. Antonsen, and C. Grebogi (1987). Quasiperiodically forced dynamical systems with strange nonchaotic attractors, *Physica D* 26, 277–294. *295*

Romeiras, F. J., C. Grebogi, E. Ott, and W. P. Dayawansa (1992). Controlling chaotic dynamical systems, *Physica D* 58, 165–192. *582*

Romeiras, F. J., and E. Ott (1987). Strange nonchaotic attractors of the damped pendulum with quasiperiodic forcing, *Phys. Rev. A* 35, 4404–4413. *295*

Romeiras, F. See Grebogi, Ott, Romeiras, and Yorke.

Ronto, N. I. See Samoilenko and Ronto.

Roose, D. (1985). An algorithm for the computation of Hopf bifurcation points in comparison with other methods, *J. Comp. Appl. Math.* 12 & 13, 517–529. *441*

Roose, D., and R. Piessens (1985). Numerical computation of nonsimple turning points and cusps, *Num. Math.* 46, 189–211. *441*

Roose, D. See DeDier, Roose, and van Rompay.

Rosenblum, M. G. See Landa and Rosenblum.

Rossi, A. See Pomeau, Roux, Rossi, Bachelart, and Vidal; Roux, Rossi, Bachelart, and Vidal.

Rössler, O. E. (1976a). An equation for continuous chaos, *Phys. Lett. A* 57, 397–398. *32, 141, 289, 290, 414*

Rössler, O. E. (1976b). Different types of chaos in two simple differential equations, *Z. Naturforsch.* 31a, 1664–1670. *335, 414*

Rössler, O. E. (1979a). An equation for hyperchaos, *Phys. Lett.* 71A, 155–157. *283, 290*

Rössler, O. E. (1979b). Continuous chaos–four prototype equations, *Annal N.Y. Acad. Sci.* 316, 376–392. *414*

Rössler, O. E. See Klein, Baier, and Rössler.

Roux, J.-C. (1983). Experimental studies of bifurcations leading to chaos in the Belousof–Zhabotinsky reaction, *Physica D* 7, 57–68. *323*

Roux, J.-C., A. Rossi, S. Bachelart, and C. Vidal (1981). Experimental observations of complex dynamical behavior during a chemical reaction, *Physica D* 2, 395–403. *299*

Roux, J.-C., R. H. Simoyi, and H. L. Swinney (1983). Observation of a strange attractor, *Physica D* 8, 257–266. *494, 495*

Roux, J.-C. See Argoul and Roux; Argoul, Arneodo, Richetti, and Roux; Pomeau, Roux, Rossi, Bachelart, and Vidal; Richetti, Roux, and Swinney; Turner, Roux, McCormick, and Swinney.

Roy, R., T. W. Murphy, T. D. Maier, and Z. Gills (1992). Dynamical control of a chaotic laser: Experimental stabilization of a globally coupled system, *Phys. Rev. Lett.* 68, 1259–1262. *558, 580*

Roy, R. See Gills, Iwata, Roy, Schwartz, and Triandaf.

Roże, C. See Ringuet, Roże, and Gouesbet.

Rubio, M. A. See Dubois, Rubio and Bergé.

Rubio, S. A. See Badii, Broggi, Derighetti, Ravani, Ciliberto, Politi, and Rubio.

Rudnick, J. See Crutchfield, Nauenberg, and Rudnick; Hu and Rudnick.

Ruelle, D. (1989a). *Chaotic Evolution and Strange Attractors,* Cambridge University Press, Cambridge, England. *283, 315, 480*

Ruelle, D. (1989b). *Elements of Differentiable Dynamics and Bifurcation Theory,* Academic Press, New York.

Ruelle, D., and F. Takens (1971). On the nature of turbulence, *Comm. Math. Phys.* 20, 167–192. *255, 315, 316, 538*

Ruelle, D. See Eckmann and Ruelle; Eckmann, Kamphorst, Ruelle, and Ciliberto; Newhouse, Ruelle, and Takens.

Rutherford, A. See Jorgensen and Rutherford.

Safonova, M. A. See Anishchenko, Astakhov, Letchford, and Safonova; Anishchenko, Letchford, and Safonova.

Sakata, M. See Yagasaki, Sakata, and Kimura.

Salam, F. A. (1988). The applicability of Melnikov's method to (highly) dissipative systems, in *Dynamical Systems Approaches to Nonlinear Problems in Systems and Circuits,* F. M. A. Salam and M. L. Levi, eds., SIAM, Philadelphia, Pennsylvania, 15–28. *390*

Salvatori, A. See Rega, Benedettini, and Salvatori; Rega, Salvatori, and Benedettini.

Samoilenko, A. M., and N. I. Ronto (1979). *Numerical–Analytical Methods of Investigating Periodic Solutions,* Mir Publishers, Moscow, Russia. *446*

Sanchez, N. E., and A. H. Nayfeh (1990). Nonlinear rolling motions of ships in longitudinal waves, *Int. Shipbldg. Prog.* 37, 247–272. *563*

Sanchez, N. E. See Nayfeh and Sanchez; Zavodney, Nayfeh, and Sanchez.

Sanders, J. A. (1982). Melnikov's method and averaging, *Celestial Mech.* 28, 171–181. *390, 563*

Sanders, J. A., and F. Verhulst (1985). *Averaging Methods in Nonlinear Dynamical Systems,* Springer–Verlag, New York. *44*

Sangiovanni–Vincentelli, A. See Kundert, Sangiovanni–Vincentelli, and Sugawara; Kundert, Sorkin, and Sangiovanni–Vincentelli.

Sano, M., and Y. Sawada (1983). Transition from quasiperiodicity to chaos in a system of coupled nonlinear oscillator, *Phys. Lett. A* 97, 73–76. *322*

Sano, M., and Y. Sawada (1985). Measurement of the Lyapunov spectrum from a chaotic time series, *Phys. Rev. Lett.* 55, 1082–1085. *534*

Sasaki, K. See Sato, Sasaki, and Nakamura.

Sato, T., K. Sasaki, and Y. Nakamura (1977). Real-time bispectral analysis of gear noise and its application to contactless diagnosis, *J. Acous. Soc. Amer.* 62, 382–387. *557*

Sauer, P. W. See Rajagopalan, Sauer, and Pai.

Sauer, T., J. A. Yorke, and M. Casdagli (1991). Embedology, *J. Stat. Phys.* 65, 579–616. *479, 480, 483, 502*

Sauer, T. See Ott, Sauer, and Yorke.

Sawada, Y. See Sano and Sawada.

Schaeffer, D. G. See Golubitsky and Schaeffer.

Schafer, R. W. See Oppenheim and Schafer.

Schättler, H. See Venkatasubramanian, Schättler, and Zaborszky.

Schell, M., S. Fraser, and R. Kapral (1983). Subharmonic bifurcation in the sine map: An infinite hierarchy of cusp bistabilities, *Phys. Rev. A* 28, 373–378. *323*

Schell, M. See Xu and Schell.

Schmidt, G., and A. Tondl (1986). *Non–Linear Vibrations,* Akademie-Verlag, Berlin, Germany.

Schmidt, J. See Steintuch and Schmidt.

Schmidt, L. D. See Kevrekidis, Aris, Schmidt, and Pelikan.

Schöll, E. See Rein, Hüpper, and Schöll.

Schreiber, I. See Marek and Schreiber.

Schreiber, T. (1993). Determination of the noise level of chaotic time series, *Phys. Rev. E* 48, R13–R16. *463*

Schreiber, T. See Grassberger, Schreiber, and Shaffrath; Kostelich and Schreiber.

Schroeder, M. (1986). *Number Theory in Science and Communication,* Springer–Verlag, New York. *247*

Schroeder, M. (1991). *Fractal, Chaos and Power Laws: Minutes from an Infinite·Paradise,* W. H. Freeman, San Francisco, California.

Schuster, H. G. (1988). *Deterministic Chaos,* VHC, Weinheim, Germany.

Schuster, H. G. See Liebert, Pawelzik, and Schuster; Liebert and Schuster.

Schwartz, I. B. See Gills, Iwata, Roy, Schwartz, and Triandaf.

Schwetlick, H., and J. Cleve (1987). Higher order predictors and adaptive steplength control in path following algorithms, *SIAM J. Num. Anal.* 24, 1382–1393. *430, 431*

Scott, K. A. See Abraham and Scott.

Scott, S. K. See Petrov, Scott, and Showalter.

Seelig, F. F. See Neymeyr and Seelig.

Seifert, H. (1983). Intermittent chaos in Josephson junctions represented by stroboscopic maps, *Phys. Lett. A* 98, 213–216. *299*

Sethna, J. See Ostlund, Rand, Sethna, and Siggia; Rand, Ostlund, Sethna, and Siggia.

Sethna, P. R. (1963). Coupling in certain classes of weakly nonlinear vibrating systems, in *Nonlinear Differential Equations and Nonlinear Mechanics,* J. P. Lasalle and S. Lefschetz, eds., Academic Press, New York, 58–70.

Sethna, P. R., and A. K. Bajaj (1978). Bifurcations in dynamical systems with internal resonance, *J. Appl. Mech.* 45, 895–902.

Sethna, P. R., and S. W. Shaw (1987). On codimension–three bifurcations in the motion of articulated tubes conveying a fluid, *Physica D* 24, 305–327.

Sethna, P. R. See Gu and Sethna; Yang and Sethna.

Seydel, R. (1979a). Numerical computation of branch points in ordinary differential equations, *Num. Math.* 32, 51–68. *95, 441*

Seydel, R. (1979b). Numerical computation of branch points in nonlinear equations, *Num. Math.* 33, 339–352. *436*

Seydel, R. (1981). Numerical computation of periodic orbits that bifurcate from stationary solutions of ordinary differential equations, *Appl. Math. Comp.* 9, 257–271. *441, 451, 453, 456*

Seydel, R. (1988). *From Equilibrium to Chaos: Practical Bifurcation and Stability Analysis,* Elsevier, New York. *31, 95, 281, 428, 430, 435, 446, 451, 453*

Shaffrath, C. See Grassberger, Schreiber, and Shaffrath.

Shaw, C. D. See Abraham and Shaw

Shaw, J., and S. W. Shaw (1990). The effects of unbalance on oil whirl, *Nonlinear Dyn.* 1, 293–311.

Shaw, R. S. (1981). Strange attractors, chaotic behavior, and information flow, *Z. Naturforsch. A* 36, 80–112. *269*

Shaw, R. S. See Crutchfield, Farmer, Packard, Shaw, Jones, and Donnelly; Froehling, Crutchfield, Farmer, Packard, and Shaw; Packard, Crutchfield, Farmer, and Shaw.

Shaw, S. W. (1985a). The dynamics of a harmonically excited system having rigid amplitude constraints, Part I: Subharmonic motions and local bifurcations, *J. Appl. Mech.* 52, 453–458.

Shaw, S. W. (1985b). The dynamics of a harmonically excited system having rigid amplitude constraints, Part II: Chaotic motions and global bifurcations, *J. Appl. Mech.* 52, 459–464.

Shaw, S. W., and P. J. Holmes (1983). A periodically forced piecewise linear oscillator, *J. Sound Vib.* 90, 129–155.

Shaw, S. W., and R. H. Rand (1989). The transition to chaos in a simple mechanical system, *Int. J. Non–Linear Mech.* 24, 41–56. *138, 226*

Shaw, S. W., and S. Wiggins (1988). Chaotic dynamics of a whirling pendulum, *Physica D* 31, 190–211.

Shaw, S. W. See Falzarano, Shaw, and Troesh; Sethna and Shaw; Shaw and Shaw; Tung and Shaw; Wiggins and Shaw.

She, H. Q. See Ding, She, Huang, and Yu.

Shenker, S. J. (1982). Scaling behavior in a map of a circle onto itself: Empirical results, *Physica D*, 5, 405–411. *322*

Shenker, S. J. See Feigenbaum, Kadanoff, and Shenker.

Sherard, P. See Rollins, Parmananda, and Sherard.

Shilnikov, L. P. (1965). A case of the existence of a denumerable set of periodic motions, *Soviet Math. Dokl.* 6, 163–166. *402, 403*

Shilnikov, L. P. (1967). Existence of a countable set of periodic motions in a four–dimensional space in an extended neighborhood of a saddle–focus, *Soviet Math. Dokl.* 8, 54–58. *407*

Shilnikov, L. P. (1968). On the generation of a periodic motion from trajectories doubly asymptotic to an equilibrium state of saddle–type, *Math. USSR Sb.* 6, 427–438. *402, 403*

Shilnikov, L. P. (1970). A contribution to the problem of the structure of an extended neighborhood of a rough equilibrium state of saddle–focus type, *Math. USSR Sb.* 10, 91–102. *402, 403, 407*

Shilnikov, L. P. (1976). Theory of the bifurcation of dynamical systems and dangerous boundaries, *Soviet Phys. Dokl.* 20, 674–676. *68*

Shilnikov, L. P. See Afraimovich, Bykov, and Shilnikov; Afraimovich and Shilnikov.

Shimada, I. and T. Nagashima (1979). A numerical approach to ergodic problem of dissipative dynamic systems, *Prog. Theor. Phys.* 61, 1605–1616. *529*

Shinbrot, T., E. Ott, C. Grebogi, and J. A. Yorke (1990). Using chaos to direct trajectories to targets, *Phys. Rev. Lett.* 65, 3215–3218. *575*

Shinbrot, T., C. Grebogi, E. Ott, and J. A. Yorke (1992). Using chaos to target stationary states of flows, *Phys. Lett. A* 169, 349–354. *575*

Shinbrot, T., C. Grebogi, E. Ott, and J. A. Yorke (1993). Using small perturbations to control chaos, *Nature* 363, 411–417. *572*

Shoshitaishvili, A. N. (1991). Structural control of nonlinear systems, *Auto. Remote Control* 8, 1089–1096.

Showalter, K. See Peng, Petrov, and Showalter; Petrov, Gaspar, Masere, and Showalter; Petrov, Peng, and Showalter; Petrov, Scott, and Showalter.

Shraiman, B., C. E. Wayne, and P. C. Martin (1981). Scaling theory for noisy period–doubling transitions to chaos, *Phys. Rev. Lett.* 46, 935–939. *296*

Shrier, A. See Glass, Guevara, Shrier, and Perez; Guevara, Glass and Shrier.

Sidorowich, J. J. See Abarbanel, Brown, Sidorowich, and Tsimring; Farmer and Sidorowich.

Siggia, E. See Ostlund, Rand, Sethna, and Siggia; Rand, Ostlund, Sethna, and Siggia.

Simó, C. (1979). On the Hénon–Pomeau attractor, *J. Stat. Phys.* 21, 465–494. *335*

Simó, C. See Abraham and Simó.

Simoyi, R. H. See Roux, Simoyi, and Swinney.

Singer, J., and H. H. Bau (1991). Active control of convection, *Phys. Fluids* 3, 2859–2865. *564*

Singer, J., Y.–Z. Wang, and H. H. Bau (1991). Controlling a chaotic system, *Phys. Rev. Lett.* 66, 1123–1125. *564, 569, 577, 580*

Singleton, R. C. (1969). An algorithm for computing the mixed radix fast Fourier transform, *IEEE Trans. Audio Elect.* AU–17, 93–103.

Sinha, S., R. Ramaswamy, and J. S. Rao (1990). Adaptive control in nonlinear dynamics, *Physica D* 43, 118–128.

Sipcic, R. S. (1990). The chaotic response of a fluttering panel: The influence of maneuvering, *Nonlinear Dyn.* 1, 243–264.

Smale, S. (1967). Differentiable dynamical systems, *Bull. Am. Math. Soc.* 73, 748–817. *286, 288*

Smale, S. (1980). *The Mathematics of Time: Essays on Dynamical Systems, Economic Processes, and Related Topics,* Springer–Verlag, New York. *286, 288*

Smale, S. See Hirsch and Smale.

Soliman, M. S. (1993). Jumps to resonance: Long chaotic transients, unpredictable outsome, and the probability of restabilization, *J. Appl. Mech.* 60, 669–676. *563*

Soliman, M. S., and J. M. T. Thompson (1989). Integrity measures quantifying the erosion of smooth and fractal basins of attraction, *J. Sound Vib.* 135, 453–475. *362*

Soliman, M. S. See Thompson, Rainey, and Soliman; Thompson and Soliman.

Solis, R. S. See Ritz, Powers, Miksad, and Solis.

Sommerer, J. C., W. L. Ditto, C. Grebogi, E. Ott, and M. L. Spano (1991). Experimental confirmation of the theory for critical exponents of crises, *Phys. Lett. A* 153, 105–109.

Sorkin, G. B. See Kundert, Sorkin, and Sangiovanni–Vincentelli.

Sotomayor, J. (1973). Generic bifurcations of dynamical systems, in *Proceedings of the Salvador Symposium on Dynamical Systems,* M. M. Piexoto, ed., Academic Press, New York.

Spano, M. L. (1990). Putting chaos theory to practice in R40, *On Surface* 13, 1–5.

Spano, M. L., and W. L. Ditto (1993). Controlling chaos, *Naval Res. Rev.* 55, 12–21.

Spano, M. L. See Ditto, Rauseo, and Spano; Garfinkel, Spano, Ditto, and Weiss; Sommerer, Ditto, Grebogi, Ott, and Spano.

Sparrow, C. T. (1982). *The Lorenz Equations: Bifurcations, Chaos, and Strange Attractors,* Springer–Verlag, New York. *399, 413*

Sparrow, C. T. See Fowler and Sparrow; Glendinning and Sparrow; Mees and Sparrow.

Spek, J. A. W. van der (1994). *Cell Mapping Methods: Modifications and Extensions,* Ph.D. Thesis, The Eindhoven University of Technology, Eindhoven, Netherlands.

Spence, A., and B. Werner (1982). Non–simple turning points and cusps, *IMA J. Num. Anal.* 2, 413–427. *441*

Spence, A. See Brindley, Kaas–Petersen, and Spence; Moore and Spence.

Spiegel, E. A. See Arneodo, Coullet, and Spiegel; Platt, Spiegel, and Tresser.

Spirig, F. (1983). Sequence of bifurcations in a three–dimensional system near a critical point, *J. Appl. Math. Phys. (ZAMP)* 34, 259–276. *233*

Srinivasan, V. See Jain and Srinivasan.

Stavans, J., F. Heslot, and A. Libchaber (1985). Fixed winding number and the quasiperiodic route to chaos in a convective fluid, *Phys. Rev. Lett.* 55, 596–599. *248, 323*

Stavans, J. See Jensen, Kadanoff, Libchaber, Procaccia, and Stavans.

Stearman, R. O. See Choi, Chang, Stearman, and Powers.

Steeb, W.–H. See Kapitaniak and Steeb.

Stein, M. L. See Metropolis, Stein, and Stein.

Stein, P. R. See Metropolis, Stein, and Stein.

Steindl, A., and H. Troger (1991). Chaotic oscillations in the motion of a simple robot, in *Engineering Applications of Dynamics of Chaos*, W. Szemplińska–Stupnicka and H. Troger, eds., Springer–Verlag, New York, 196–204. *334*

Steindl, A. See Troger and Steindl.

Steinmetz, C. G., and R. Larter (1991). The quasiperiodic route to chaos in a model of the peroxidase–oxidase reaction, *J. Chem. Phys.* 94, 1388–1396. *323, 324, 326, 327, 330, 331*

Steinmetz, C. G. See Larter and Steinmetz; Larter, Olsen, Steinmetz, and Geest.

Steintuch, M., and J. Schmidt (1988). Bifurcations to periodic and aperiodic solutions during ammonia oxidation on a Pt wire, *J. Phys. Chem.* 92, 3404–3411. *299*

Stepanov, V. V. See Nemytskii and Stepanov.

Stewart, H. B., and A. N. Lansbury (1992). Forecasting catastrophe by exploring chaotic dynamics, in *Applied Chaos*, J. H. Kim and J. Stringer, eds., Wiley, New York, 393–410. *338*

Stewart, H. B., and Y. Ueda (1991). Catastrophes with intermediate outcome, *Proc. R. Soc. Lond. A* 432, 113–123.

Stewart, H. B. See Abraham and Stewart; Thompson and Stewart; Thompson, Stewart, and Ueda; Ueda, Yoshida, Stewart, and Thompson.

Stoer, J., and R. Bulirsch (1980). *Introduction to Numerical Analysis,* Springer–Verlag, New York. *96, 432*

Stone, E. See Aubry, Holmes, Lumley, and Stone.

Storti, D. W., and R. H. Rand (1982). Dynamics of two strongly coupled van der Pol oscillators, *Int. J. Non–Linear Mech.* 17, 143–152. *251, 255*

Streit, D. A., A. K. Bajaj, and C. M. Krousgrill (1988). Combination parametric resonance leading to periodic and chaotic response in two–degree–of–freedom systems with quadratic non–linearities, *J. Sound Vib.* 124, 297–314. *334, 336*

Strelcyn, J. See Benettin, Galgani, Giorgilli, and Strelcyn.

Stringer, J. See Kim and Stringer.

Sueoka, A. See Tsuda, Tamura, Sueoka, and Fujii.

Sugawara, T. See Kundert, Sangiovanni–Vincentelli, and Sugawara.

Sulem, P. See Goldhirsch, Sulem, and Orszag.

Surko, C. M. See See Walden, Kolodner, Passner, and Surko.

Sushchik, M. M. See Abarbanel, Rabinovich, and Sushchik.

Svensson, P. See Herzel, Plath, and Svensson.

Swift, J. B. See Wolf, Swift, Swinney, and Vastano.

Swift, J. W., and K. Wiesenfeld (1984). Suppression of period doubling in symmetric systems, *Phys. Rev. Lett.* 52, 705–708. *194*

Swinney, H. L. (1983). Observations of order and chaos in nonlinear systems, *Physica D* 7, 3–15. *197, 203, 281, 331*

Swinney, H. L., and J. P. Gollub (1978). The transition to turbulence, *Phys. Today* 31, 41–49. *316*

Swinney, H. L. See Abraham, Gollub, and Swinney; Fenstermacher, Swinney, and Gollub; Fraser and Swinney; Gollub and Swinney; Gorman, Reith, and Swinney; Kreisberg, McCormick, and Swinney; Maselko and Swinney; Richetti, Roux, and Swinney; Roux, Simoyi, and Swinney; Tam, Vastano, Swinney, and Horsthemke; Turner, Roux, McCormick, and Swinney; Wolf, Swift, Swinney, and Vastano; Abrahim, Gollub, and Swinney.

Szemplińska–Stupnicka, W. (1990). *The Behavior of Nonlinear Vibrating Systems, Vols. I and II,* Kluwer, Dordrecht, Netherlands.

Szemplińska–Stupnicka, W. (1992). Cross–well chaos and escape phenomena in driven oscillators, *Nonlinear Dyn.* 3, 225–243.

Szemplińska–Stupnicka, W., R. H. Plaut, and J.-C. Hsieh (1989). Period–doubling and chaos in unsymmetric structures under parametric excitation, *J. Appl. Mech.* 56, 947–952. *292*

Tabor, M. (1989). *Chaos and Integrability in Nonlinear Dynamics,* Wiley, New York.

Takens, F. (1981). Detecting strange attractors in turbulence, in *Dynamical Systems and Turbulence, Lecture Notes in Mathematics, Vol. 898,* D. A. Rand and L. S. Young, eds., Springer–Verlag, New York, 366–381. *479, 480, 495*

Takens, F. See Newhouse, Ruelle, and Takens; Palis and Takens; Ruelle and Takens.

Tam, W. Y., J. A. Vastano, H. L. Swinney, and W. Horsthemke (1988). Regular and chaotic chemical spatiotemporal problems, *Phys. Rev. Lett.* 61, 2163–2166.

Tamura, H. See Tsuda, Tamura, Sueoka, and Fujii.

Tamura, T., and Y. Yorino (1987). Possibility of auto– and hetero-parametric resonances in power systems and their relationship with long–term dynamics, *IEEE Trans. Power Syst.* PWRS-2, 890–897. *56*

Tang, D. Y., J. Pujol, and C. O. Weiss (1991). Type–III intermittency of a laser, *Phys. Rev. A* 44, R35–R38.

Tang, D. Y., M. Y. Li, and C. O. Weiss (1992). Laser dynamics of type–I intermittency, *Phys. Rev. A* 46, 676–678. *299*

Tavakol, R. K., and A. S. Tworkowski (1984). On the occurrence of quasiperiodic motion on three tori, *Phys. Lett. A* 100, 65–67. *317*

Tchamitchian, P. See Comber, Grossman, and Tchamitchian.

Termonia, Y., and Z. Alexandrowicz (1983). Fractal dimension of strange attractors from radius versus size of arbitrary clusters, *Phys. Rev. Lett.* 51, 1265–1268. *549*

Tesi, A. See Genesio and Tesi.

Testa, J., J. Perez, and C. Jeffries (1982). Evidence of universal chaotic behavior of a driven nonlinear oscillator, *Phys. Rev. Lett.* 48, 714–717. *281, 296*

Theiler, J. (1990). Estimating fractal dimensions, *J. Opt. Soc. Am. A* 7, 1055–1073. *538*

Thom, R. (1975). *Structural Stability and Morphogenesis,* Benjamin, Reading, Massachussetts. *132*

Thomas, R. J. See Chiang, Dobson, Thomas, Thorp, and Fekih–Ahmed.

Thompson, J. M. T., S. R. Bishop, and L. M. Leung (1987). Fractal basins and chaotic bifurcations prior to escape from a potential well, *Phys. Lett. A* 121, 116–120. *415*

Thompson, J. M. T., R. C. T. Rainey, and M. S. Soliman (1990). Ship stability criteria based on chaotic transients from incursive fractals, *Philos. Trans. R. Soc. Lond.* 332, 149–167. *563*

Thompson, J. M. T., and M. S. Soliman (1990). Fractal control boundaries of driven oscillators and their relevance to safe engineering design, *Proc. R. Soc. Lond. A* 428, 4–13. *362, 389*

Thompson, J. M. T., and H. B. Stewart (1986). *Nonlinear Dynamics and Chaos,* Wiley, Chichester, England. *68, 69, 269, 281, 335, 349, 352, 354, 398*

Thompson, J. M. T., H. B. Stewart, and Y. Ueda (1994). Safe, explosive, and dangerous bifurcations in dissipative dynamical systems, *Phys. Rev. E* 49, 1019–1027. *68, 93, 335, 352*

Thompson, J. M. T. See Soliman and Thompson; Ueda, Yoshida, Stewart, and Thompson.

Thorp, J. S. See Chiang, Dobson, Thomas, Thorp, and Fekih–Ahmed.

Tinkham, M. See Iansiti, Hu, Westervelt, and Tinkham.

Tokunaga, R. See Matsumoto, Chua, and Tokunaga.

Tomlinson, G. R. See Gifford and Tomlinson.

Tondl, A. (1965). *Some Problems of Rotor Dynamics,* Chapman and Hall, London, England.

Tondl, A. See Schmidt and Tondl.

Tongue, B. H. (1987). On obtaining global nonlinear system characteristics through interpolated cell mapping, *Physica D* 28, 401–408.

Toth, T. See Leibovitch and Toth.

Tousi, S., and A. K. Bajaj (1985). Period–doubling bifurcations and modulated motions in forced mechanical systems, *J. Appl. Mech.* 52, 446–452. *334*

Tousi, S. See Bajaj and Tousi.

Trenogin, V. A. See Vainberg and Trenogin.

Tresser, C. (1984a). About some theorems by L. P. Shilnikov, *Annal I.H.P.* 40, 441–461. *404*

Tresser, C. (1984b). Homoclinic orbits for flows in R^3, *J. Physique* 45, 837–841. *404*

Tresser, C. See Arneodo, Coullet, and Tresser; Arneodo, Coullet, Tresser, Libchaber, Maurer, and D'Humieres; Mackay and Tresser; Platt, Spiegel, and Tresser.

Triandaf, I. See Gills, Iwata, Roy, Schwartz, and Triandaf.

Trick, T. N. See Aprille and Trick.

Troesh, A. W. See Falzarano, Shaw, and Troesh.

Troger, H., and C. S. Hsu (1977). Response of a nonlinear system under combined parametric and forcing excitation, *J. Appl. Mech.* 44, 179–181.

Troger, H., and A. Steindl (1991). *Nonlinear Stability and Bifurcation Theory: An Introduction for Engineers and Applied Scientists,* Springer–Verlag, New York.

Troger, H. See Steindl and Troger.

Tseng, W. Y., and J. Dugundji (1971). Nonlinear vibrations of a buckled beam under harmonic excitation, *J. Appl. Mech.* 38, 467–476.

Tsimring, L. S. See Abarbanel, Brown, Sidorowich, and Tsimring.

Tsuda, Y., H. Tamura, A. Sueoka, and T. Fujii (1992). Chaotic behaviour of a nonlinear vibrating system with a retarded argument, *JSME Int. J., Series III* 35, 259–267.

Tufillaro, N. B. (1989). Nonlinear and chaotic string vibrations, *Am. J. Phys.* 57, 408–414.

Tufillaro, N. B., T. Abbott, and J. Reilly (1992). *An Experimental Approach to Nonlinear Dynamics and Chaos,* Addison–Wesley, Redwood City, California.

Tufillaro, N. B. See Molteno and Tufillaro.

Tukey, J. W. See Cooley and Tukey.

Tung, P.–C. (1992). Dynamics of a nonharmonically forced impact oscillator, *JSME Int. J., Series III* 35, 378–385.

Tung, P.–C., and S. W. Shaw (1988). The dynamics of an impact print hammer, *J. Vib. Acoust. Stress Reliab. Design* 110, 193–200. *564*

Turner, J. S., J.-C. Roux, W. D. McCormick, and H. L. Swinney (1981). Alternating periodic and chaotic regimes in a chemical reaction–experiment and theory, *Phys. Lett. A* 85, 9–12. *331, 332*

Tworkowski, A. S. See Tavakol and Tworkowski.

Ueda, Y. (1980a). Explosions of strange attractors exhibited by Duffing's equation, in *Nonlinear Dynamics,* R. H. G. Helleman, ed., New York Academy of Sciences, New York, 422–434. *335, 342*

Ueda, Y. (1980b). Steady motions exhibited by Duffing's equation: A picture book of regular and chaotic motions, in *New Approaches to Nonlinear Problems in Dynamics,* P. J. Holmes, ed., SIAM, Philadelphia, Pennsylvania, 311–322. *342*

Ueda, Y. (1991). Survey of regular and chaotic phenomena in the forced Duffing oscillator, *Chaos Sol. Fract.* 1, 199–231. *342, 415*

Ueda, Y. (1992). *The Road to Chaos,* Aerial Press, Santa Cruz, California.

Ueda, Y., and S. Yoshida (1987). Attractor–Basin phase portraits of the forced Duffing's oscillator, in *Proceedings of the European Conference on Circuit Theory Design,* Vol. 1, Paris, France, 281–286. *415*

Ueda, Y., S. Yoshida, H. B. Stewart, and J. M. T. Thompson (1990). Basin explosions and escape phenomena in the twin–well Duffing oscillator: Compound global bifurcations organizing behavior, *Philos. Trans. R. Soc. Lond. A* 332, 169–186.

Ueda, Y. See Stewart and Ueda; Thompson, Stewart, and Ueda.

Umeki, M., and T. Kambe (1989). Nonlinear dynamics and chaos in parametrically excited surface waves, *J. Phys. Soc.* 58, 140–154. *334*

Urabe, M. (1967). *Nonlinear Autonomous Oscillations,* Academic Press, New York.

Urabe, M. (1974). Quasiperiodic solutions of ordinary differential equations, *Nonlinear Vib. Prob. (Zagadnienia Drgań Nieliniowych)* 18, 85–93. *232*

Ushida, A., and L. O. Chua (1984). Frequency–domain analysis of nonlinear circuits driven by multi–tone signals, *IEEE Trans. Circuit Syst.* CAS–31, 766–779. *252*

Ushida, A. See Chua and Ushida.

Ushiki, S. (1980). Analytic expressions of unstable manifolds, *Proc. Jpn. Acad. A* 56, 239–244. *64*

Vainberg, M. M., and V. A. Trenogin (1962). The method of Lyapunov and Schmidt in the theory of nonlinear equations and their further development, *Russian Math. Surv.* 17, 1–60. *108*

Vainberg, M. M., and V. A. Trenogin (1974). *Theory of Branching of Solutions of Non-Linear Equations,* Noordhoff, Leyden, Netherlands. *108*

van Boehm, J. See Räty, Isomäki, and von Boehm; Räty, von Boehm, and Isomäki.

Van Buskirk, R., and C. Jeffries (1985). Observation of chaotic dynamics of coupled nonlinear oscillators, *Phys. Rev. A* 31, 3332–3357. *296*

van der Mark, J. See van der Pol and van der Mark.

van der Pol, B., and J. van der Mark (1927). Frequency demultiplication, *Nature* 120, 363–364.

van Rompay, P. See DeDier, Roose, and van Rompay.

Varaiya, P. P. See Abed and Varaiya.

Vastano, J. A., and E. J. Kostelich (1985). Comparison of algorithms for determining Lyapunov exponents from experimental data, in *Dimensions and Entropies in Chaotic Systems,* G. Mayer–Kress, ed., Springer–Verlag, New York, 100–107.

Vastano, J. A. See Tam, Vastano, Swinney, and Horsthemke; Wolf, Swift, Swinney, and Vastano.

Venkatasubramanian, V., H. Schättler, and J. Zaborszky (1992). Voltge dynamics: Study of a generator with voltage control, transmission and matched MW load, *IEEE Trans. Auto. Control* 37, 1717–1733. *564*

Verhulst, F. See Sanders and Verhulst.

Vidal, C. See Bergé, Pomeau and Vidal; Pomeau, Roux, Rossi, Bachelart, and Vidal; Roux, Rossi, Bachelart, and Vidal.

Vincent, T. L., and J. Yu (1991). Control of a chaotic system, *Dyn. Control* 1, 35–52.

Vittal, V. See Michel, Nam, and Vittal.

Vu, K. T., and C. C. Liu (1990). Dynamic mechanisms of voltage collapse, *Syst. Control Lett.* 15, 329–338. *564*

Walden, R. W., P. Kolodner, A. Passner, and C. M. Surko (1984). Nonchaotic Rayleigh–Bénard convection with four and five incommensurate frequencies, *Phys. Rev. Lett.* 53, 242–245. *312, 317*

Wan, Y.-H. See Hassard, Kazarinoff, and Wan.

Wang, H. O., and E. H. Abed (1992). Bifurcation control of dynamic systems, in *Proceedings of the 2nd IFAC Nonlinear Systems Design Symposium,* Bordeaux, France, 57–62. *564, 568*

Wang, H. O., and E. H. Abed (1993). Control of nonlinear phenomena at the inception of voltage collapse, in *Proceedings of the 1993 American Control Conference,* San Francisco, California, 2071–2075. *564, 568*

Wang, H. O., and E. H. Abed (1994). Bifurcation control of a chaotic system, Tech. Rept., Inst. Syst. Res., University of Maryland, College Park, Maryland. *564, 566–571*

Wang, H. O., E. H. Abed, and A. M. A. Hamdan (1992a). Bifurcations, chaos and crises in power system voltage collapse, *Systems Research Center Report,* University of Maryland, College Park, Maryland. *564*

Wang, H., E. H. Abed, and A. M. A. Hamdan (1992b). Is voltage collapse triggered by the boundary crisis of a strange attractor, in *Proceedings of the 1992 American Control Conference,* Chicago, Illinois, 2084–2088. *564*

Wang, H. O., E. H. Abed, and A. M. A. Hamdan (1994). Bifurcations, chaos and crises in voltage collapse of a model power system, *IEEE Trans. Circuit Syst.,* in press. *336*

Wang H. O. See Abed, Wang, Alexander, Hamdan, and Lee; Abed and Wang; Abed, Wang, and Chen.

Wang, Y.-Z. See Singer, Wang, and Bau.

Watson, L. T. (1986). Numerical linear algebra aspects of globally convergent homotopy methods, *SIAM Rev.* 28, 529–545. *442*

Watson, L. T. (1990). Globally convergent homotopy algorithms for nonlinear systems of equations, *Nonlinear Dyn.* 1, 143–191. *96, 442*

Watson, L. T., S. C. Billups, and S. P. Morgan (1987). HOMPACK: A suite of codes for globally convergent homotopy algorithms, *ACM Trans. Math. Software* 13, 281–310. *96, 442, 445*

Watt, S. W. See Char, Geddes, Gonnet, Leong, Monagan, and Watt.

Wayne, C. E. See Shraiman, Wayne, and Martin.

Weber, H. (1981). On the numerical approximation of secondary bifurcation problems, in *Numerical Solution of Nonlinear Equations,*

E. L. Allgower, K. Glashoff, and H.-O. Peitgen, eds., Springer–Verlag, New York, 407–425. *438*

Wegmann, K. See Baier, Wegmann, and Hudson.

Wiesenfeld, K. See Swift and Wiesenfeld.

Weiss, C. O. See Tang, Pujol, and Weiss; Tang, Li, and Weiss.

Weiss, J. N. See Garfinkel, Spano, Ditto, and Weiss.

Weiss, N. O. See Knobloch and Weiss.

Werner, B., and V. Janovsky (1991). Computation of Hopf branches bifurcating from Takens–Bogdanov points for problems with symmetries, in *Bifurcation and Chaos,* R. Seydel, F. W. Schneider, T. Küpper, and H. Troger, eds., Birkhaüser Verlag, Boston, Massachusetts, 377–388. *441*

Werner, B. See Spence and Werner.

Westervelt, R. M. See Gwinn and Westervelt; Iansiti, Hu, Westervelt, and Tinkham.

Whitney, H. (1936). Differentiable manifolds, *Annal Math.* 37, 645–680. *476, 478, 483*

Widmann, P. J. See Gorman, Widmann, and Robbins.

Wiesenfeld, K. (1985). Noisy precursors of nonlinear instabilities, *J. Stat. Phys.* 38, 1071–1097.

Wiesenfeld, K., and B. McNamara (1986). Small–Signal amplification in bifurcating dynamical systems, *Phys. Rev. A* 33, 629–642.

Wiesenfeld, K. See Jeffries and Wiesenfeld; Swift and Wiesenfeld.

Wiggins, S. (1988). *Global Bifurcations and Chaos,* Springer–Verlag, New York. *176, 359, 367, 390, 398, 399, 404*

Wiggins, S. (1990). *Introduction to Applied Nonlinear Dynamical Systems and Chaos,* Springer–Verlag, New York. *42, 77, 128, 176, 288, 359, 367, 390*

Wiggins, S., and P. Holmes (1987). Homoclinic orbits in slowly varying oscillators, *SIAM J. Math. Anal.* 18, 612–629; Errata, 19, 1254–1255.

Wiggins, S., and S. W. Shaw (1988). Chaos and three–dimensional horseshoes in slowly varying oscillators, *J. Appl. Mech.* 55, 959–968.

Wiggins, S. See Shaw and Wiggins.

Wilke, C. See Levin, Pompe, Wilke, and Koch.

Wolf, A., J. B. Swift, H. L. Swinney, and J. A. Vastano (1985). Determining Lyapunov exponents from a time series, *Physica D* 16,

285–317. *530*

Wolfram, S. (1991). *Mathematica: A System For Doing Mathematics by Computer,* Addison–Wesley, Redwood City, California.

Worfolk, P. See Gucknheimer and Worfolk.

Wright, J. (1984). Method for calculating a Lyapunov exponent, *Phys. Rev. A* 29, 2924–2927.

Wright, J. H. G., and W. B. Marshfield (1979). Ship roll response and capsize behaviour in beam seas, *Trans. R. Inst. Naval Arch.* 122, 129–148. *52*

Wu, F. F. See Chiang, Hirsch, and Wu.

Xu, Y., and M. Schell (1990). Bistability and oscillations in the electrocatalyzed oxidation of formaldehyde, *J. Phys. Chem.* 94, 7137–7143. *323*

Yagasaki, K. (1991). Chaos in a weakly nonlinear oscillator with parametric and external resonances, *J. Appl. Mech.* 58, 244–250.

Yagasaki, K., M. Sakata, and K. Kimura (1990). Dynamics of a weakly nonlinear system subjected to combined parametric and external excitation, *J. Appl. Mech.* 57, 209–217.

Yamamoto, T. See Ishida, Ikeda, and Yamamoto.

Yan, H. See Chow, Fischl, and Yan.

Yang, X. L., and P. R. Sethna (1991). Local and global bifurcations in parametrically excited vibrations of nearly square plates, *Int. J. Non–Linear Mech.* 26, 199–220. *299*

Yeh, W. J., and Y. H. Kao (1983). Intermittency in Josephson junctions, *Appl. Phys. Lett.* 42, 299–301. *299*

Yorino, Y. See Tamura and Yorino.

Yorke, E. D. See Alligood, Yorke, and Yorke; Frederickson, Kaplan, Yorke, and Yorke; Yorke and Yorke.

Yorke, J. A., and E. D. Yorke (1979). Mestable chaos: The transition to sustained chaotic behavior in the Lorenz model, *J. Stat. Phys.* 21, 263–278. *339, 401*

Yorke, J. A. See Alligood, Yorke, and Yorke; Auerbach, Grebogi, Ott, and Yorke; Curry and Yorke; Farmer, Ott, and Yorke; Frederickson, Kaplan, Yorke, and Yorke; Grebogi, Ott, Pelikan, and Yorke; Grebogi, Ott, Romeiras, and Yorke; Grebogi, Ott, and Yorke; Kaplan and Yorke; Kostelich, Grebogi, Ott, and Yorke; Kostelich and Yorke; Ott, Grebogi, and Yorke; Ott, Sauer, and Yorke; Sauer, Yorke, and

Casdagli; Shinbrot, Ott, Grebogi, and Yorke; Shinbrot, Grebogi, Ott, and Yorke.

Yoshida, S. See Ueda and Yoshida; Ueda, Yoshida, Stewart, and Thompson.

Young, L.-S. (1982). Dimension, entropy, and Lyapunov exponents, *Erg. Theory Dyn. Syst.* 2, 109–124. *538*

Young, L.-S. (1983). Entropy, Lyapunov exponents, and Hausdorff dimension in differentiable dynamic systems, *IEEE Trans. Circuit Syst.* 8, 599–607. *538*

Yu, C. X. See Ding, She, Huang, and Yu.

Yu, J. See Vincent and Yu.

Zaborszky, J. See Venkatasubramanian, Schättler, and Zaborszky.

Zangwill, W. I. See Garcia and Zangwill.

Zavodney, L. D. (1987). *A Theoretical and Experimental Investigation of Parametrically Excited Nonlinear Mechanical Systems*, Ph.D. Dissertation, Virginia Polytechnic University and State University, Blacksburg, Virginia. *415*

Zavodney, L. D., and A. H. Nayfeh (1988). The response of a single-degree-of-freedom system with quadratic and cubic non-linearities to a fundamental parametric resonance, *J. Sound Vib.* 120, 63–93. *281, 292, 414*

Zavodney, L. D., A. H. Nayfeh, and N. E. Sanchez (1989). The response of a single-degree-of-freedom system with quadratic and cubic non-linearities to a principal parametric resonance, *J. Sound Vib.* 129, 417–442. *281, 291, 415*

Zavodney, L. D. See Nayfeh and Zavodney.

Zeeman, E. C. (1982). Bifurcation and catastrophe theory, in *Papers in Algebra, Analysis, and Statistics*, R. Lidl, ed., American Mathematical Society, Providence, Rhode Island, 207–272. *68*

Zhilin, Q. See Gang and Zhilin.

SUBJECT INDEX

section, 178
stability, 37
Locking, 243–248, 37
  frequency, 315, 318
  phase, 318
Logistic map, 4, 64–66, 126, 128,
    129, 410
  chaos in, 278–284
  intermittency in, 304, 305
  Lyapunov exponent of, 561
  period–doubling in, 279, 280
Lorenz attractor, 482
Lorenz equations, 31, 141,
    399–402, 412, 585–587
  data for, 482, 490
  intermittency in, 302, 303
Lorenz mask, 413
Lorenz sections, 240
Loss of synchronism, 588
Lunar orbital dynamics, 77
Lyapunov,
  dimension, 535, 549
  function, 27–29, 45
  second method, 31
  stability, 20–23
  stability through, function,
  27–29
  theorem, 27
Lyapunov exponents, 281, 295, 525
  for autonomous systems, 529
  for Bernoulli map, 561
  calculation of, 529–538
  conditional, 587
  for dissipative systems, 532
  fixed points, 528
  global, 532
  for Hénon map, 285

for limit cycle, 528
local, 534
for logistic map, 282, 561
for Lorenz equations, 531
for m–torus, 529
numerical errors, 529
for one–dimensional maps, 528
p–dimensional, 528
for Rössler equations, 530, 531
spectrum, 527
spurious or false, 536
for tent map, 558, 561
true, 535
Lyapunov–Schmidt reduction, 108
Lyapunov spectrum, 527
Lyapunov stability analyses, 525

Mackey–Glass equation, 500
Magnetic field, 577
Magnetoelastic, 577
Manifold,
  analytical construction of,
  58–61
  center, 50, 97
  definition of, 8
  global, 51
  intersections, 363–390
  invariant, 47–61, 98
  local, 51, 98
  of periodic solution, 172
  stable, 50
  unstable, 50
Map, 2–6
  bifurcation of solutions of,
  121–128